Lecture Notes in Engineering

Edited by C. A. Brebbia and S. A. Orszag

2

Bernard Amadei

Rock Anisotropy and the Theory of Stress Measurements

Springer-Verlag
Berlin Heidelberg New York Tokyo 1983

Author
Bernard Amadei
Professor
University of Colorado, Boulder
Department of Civil, Environmental, and
Architectural Engineering
Boulder, Colorado 80309

ISBN 3-540-12388-1 Springer-Verlag Berlin Heidelberg New York
ISBN 0-387-12388-1 Springer-Verlag New York Heidelberg Berlin

Library of Congress Cataloging in Publication Data
Amadei, Bernard, 1954-
Rock anisotropy and the theory of stress measurements.
(Lecture notes in engineering ; 2)
Bibliography: p.
1. Rock mechanics. I. Title. II. Series.
TA706.A546 1983 624.1'5132 83-673

Printed in Germany

Printing and binding: Beltz Offsetdruck, Hemsbach/Bergstr.
2061/3020-543210

FOREWORD BY RICHARD GOODMAN

This important work by Bernard Amadei provides the solution to a problem central to interpretation of stress measurements in anisotropic rock. Although numerical methods have the capability of describing the stress and strain distributions in anisotropic bodies subjected to very general systems of loading, it is not feasible to perform a full numerical analysis for each new stress measurement configuration. In the overcoring methods of stress measurement, in which an inclusion is emplaced in a pilot borehole and then overcored with a concentric, larger borehole, the boundary conditions are sufficiently well defined and regular that a "closed-form" solution could be obtained. Previous authors have produced useful analyses of subsets of the overcoring problem in which an inclusion is emplaced in rock having either isotropic elastic properties or anisotropic properties with symmetry aligned to the boreholes. Dr. Amadei's results, on the other hand, allow the concentric boreholes to have any orientation with respect to both the initial stress field and the axes of elastic symmetry of the rock. The thickness and properties of the elastic inclusion and the system used to measure strains or displacements in the overcored inclusion, are also described with general parameters. Thus the results have real value and are a significant advance for practical rock mechanics.

To the uninitiated, the measurement of the absolute state of stress in rock might seem to be an esoteric exercise, albeit an interesting one. There is no doubt that it is interesting, and challenging, but in-situ stress measurement is far from esoteric. It figures prominently in engineering and mining for: removal of valuable minerals; retorting of oil-shale; gasification of coal; disposal of nuclear wastes; conversion into work of energy stored as underground air pressure or heat; and many other real problems. In-situ stress affects rock properties and safety of underground ventures, controls caving mechanisms above old mines, and affects the performance of civil engineering excavations below ground. Stresses in-situ describe the environment in which earthquakes are born and then propagate, in which folds are tightened, and volcanoes are enlarged.

The new analytical solutions developed by Bernard Amadei provide a sound basis for reducing the field data from the standard overcoring techniques applied to new situations, for example in orientations cutting diagonally across the rock structure. They also offer the basis for interpreting the results of the most modern methods, for example using instrumented hollow-cylinder inclusions for complete determination of the stress state from one experiment. Because the analysis is general, it also plants seeds for new stress measurement schemes yet to be devised.

This work was written in partial fulfillment for a doctoral degree at the University of California, Berkeley. It includes a complete, and logical development of the important aspects of mechanics and rock mechanics relating to anisotropy and to the measurement of stresses by inclusion techniques. The elaboration of the analytical methods employed provides a clear exposition of and original contributions to the theory of anisotropic elasticity. As such it deserves a place among important developments in applied mechanics.

ACKNOWLEDGEMENTS

The author is indebted to numerous people for their assistance throughout the course of this work.

I wish to express my sincere thanks to Professor R. E. Goodman for his guidance, encouragement and supervision with all aspects of this work. I also wish to acknowledge Professors M. M. Carroll and J. M. Duncan for serving on my thesis committee and reviewing this manuscript.

The author is grateful to Drs. F. E. Heuze, R. M. Nolting, Gen-Hua Shi, Professors T. L. Brekke, D. R. Salcedo, N. Sitar for their helpful suggestions and advice.

Special acknowledgement is made to Professor R. Houpert and P. M. Naghdi for initially stimulating the author's interest in continuum mechanics.

The friendship and the constructive discussions of a number of my fellow students were supplementary in the crystallization of the ideas of this thesis. Special thanks go to: Jess Albino, Douglas Blankenship, Bill Boyle, Ian Brown, Lap-Yan Chan, Sandra Green, Joe Ratigan, Dave Rogers, Kari Saari, Bjorn Stefanson, Neil Thomsen, Richard Thorpe, Brenda Myers-Bohlke.

My sincere thanks go to Marcia Golner for her diligence and patience in typing this thesis and to Glenn Boyce for his work on the illustrations.

This research was financed by the National Science Foundation by means of grant CME-7824439, together with grants from the Geological Engineering Foundation and the Jane Lewis Fellowship fund.

TABLE OF CONTENTS

ACKNOWLEDGEMENTS i

TABLE OF CONTENTS ii

LIST OF ILLUSTRATIONS vii

LIST OF TABLES xi

LIST OF SYMBOLS xii

Chapter 1: Introduction 1

Chapter 2: Deformability of Anisotropic Rocks 6

2.1 Introduction 6

2.2 Constitutive Relations 6

2.2.1 Generalized Hooke's Law 6

2.2.2 Stratified Media 16

2.2.3 Regularly Jointed Rocks 18

2.2.4 Effective Stress Laws 20

2.2.5 Transformation of Elastic Constants 21

2.3 Testing of Anisotropic Rocks 26

2.3.1 Laboratory Testing 26

2.3.2 In Situ Testing 41

Chapter 3: Strength of Anisotropic Rocks 48

3.1 Introduction 48

3.2 Experimental Observations 48

3.2.1 Laboratory Testing 48

3.2.2 In Situ Testing 57

3.3 Analytical Models 58

3.3.1 Discontinuous Models 58

3.3.2 Continuous Models 62

Chapter 4: Elastic Equilibrium of an Anisotropic Homogeneous Body Bounded Internally by a Cylindrical Surface of Arbitrary Cross Section 81

4.1 Introduction 81

4.2 Geometry and Definition of the Problem 81

4.3 Formulation of the Problem 83

4.3.1 Basic Equations 83

4.3.2 Sign Conventions 88

4.3.3 Beltrami Michell Equations of Compatibility 89

4.3.4 Boundary Conditions 92

4.3.5 The General Expressions for the Stress Functions 93

4.3.6 The General Expressions for the Components of Stress and Displacement 95

4.4 Special Case of Anisotropy: A Plane of Elastic Symmetry Perpendicular to the Hole Axis 98

4.5 Plane Strain and Plane Stress Formulations 101

4.5.1 Plane Problems of the Theory of Elasticity 101

4.5.2 Plane Strain Formulations 103

4.5.3 Plane Stress Formulations 107

4.5.4 Remarks 119

4.6 Particular Solution for an Infinite Cylinder with a Circular Cross Section 121

4.6.1 Fourier Series Boundary Conditions 121

4.6.2 General Expressions for the Analytic Functions $\phi_k(z_k)$ $(k=1,2,3)$. 123

4.6.3 Special Cases of Fourier Distributions 124

4.6.4 Closed Form Solutions 126

4.6.5 Remarks 133

4.6.6 Numerical Examples 135

Chapter 5: Elastic Equilibrium of an Anisotropic Homogeneous Body Bounded Internally by an Isotropic Inclusion of Circular Cross Section 143

5.1 Introduction 143

5.2 Geometry and Definition of the Problem 144

5.3 Formulation of Problem (A) 148

5.3.1 Basic Equations 148

5.3.2 Beltrami Michell Equations of Compatibility 150

5.3.3 Fourier Series Boundary Conditions 151

5.3.4 Formulation of the Plane Problem 152

5.3.5 Formulation of the Antiplane Problem 156

5.4 Formulation of Problem (B) 156

5.5 Condition of Continuity 158

5.6 Closed Form Solutions 163

5.7 Remarks 167

5.7.1 Rigid Body Translations 167

5.7.2 Variations in Length of Oblique Distances 168

5.7.3 Induced Stress Field within the Anisotropic Body 171

5.7.4 Limiting Cases 171

5.8 Numerical Examples 172

5.8.1 Isotropic Solution 173

5.8.2 Anisotropic Solution 178

Chapter 6: Influence of Rock Anisotropy on Stress Measurements by Overcoring Techniques 189

6.1 Introduction 189

6.2 In Situ Determination of Stress by Relief Techniques 190

6.3 Information Obtained from Measuring Techniques 194

6.4 General Formulas for Overcoring and Undercoring Techniques 198

6.4.1 Overcoring Techniques 199

6.4.2 Undercoring Techniques 205

6.4.3 Absolute Stresses - Changes in Stress 207

6.5 General Results for Overcoring in Anisotropic Media 208

6.5.1 Introduction 208

6.5.2 Isotropic Solution 209

6.5.3 Anisotropic Solution; Literature Review 213

6.5.4 Anisotropic Solution; Present Analysis 216

a) Number of Independent Measurements in a Single Borehole 221

b) Number of Boreholes 222

c) Influence of Rock Anisotropy on the Determination of the In Situ Stress Field 233

Chapter 7: Summary and Conclusions 242

References 253

Appendix 2.1 267

Appendix 4.1 287

Appendix 4.2	293
Appendix 4.3	303
Appendix 4.4	307
Appendix 4.5	310
Appendix 4.6	338
Appendix 4.7	345
Appendix 4.8	351
Appendix 5.1	356
Appendix 5.2	368
Appendix 5.3	373
Appendix 5.4	380
Appendix 6.1	383
Appendix 6.2: Program Berni 1	387
Appendix 6.3: Program Berni 2	398
Appendix 6.4: Program Berni 3	403
Appendix 6.5: Program listings	411

LIST OF ILLUSTRATIONS

Figure	Title	Page
2.1	Bounds for elastic constants of a transversely isotropic material (Pickering, 1970).	15
2.2	Geometry of the problem. Definition of the coordinate systems.	23
2.3	Example of strain gage distributions.	30
2.4	Variation of the apparent Young's modulus E_y and apparent Poisson's ratios ν_{yx} and ν_{yz} with the angle θ. Comparison between experimental data and theoretical prediction (Pinto, 1970).	32
2.5	Uniaxial compressive tests for three different inclinations of rock anisotropy.	34
2.6	Sample orientations for the determination of the elastic constants of an orthotropic rock.	36
2.7	Schematic model of jointed rock mass. Three orthogonal joint sets KK1, KK2, KK3 with orientation defined by angles α and β. Multiaxial stress field $\sigma_{xx}, \sigma_{yy}, P_z$ (Reik and Zacas, 1978).	40
2.8	Variation of the modulus of deformation (tangent modulus at 50% of max. substainable axial stress) with the angle β ($\alpha = 0°$) (Reik and Zacas, 1978).	40
2.9	Determination of normal and shear stiffnesses of a joint set striking parallel to a borehole.	44
3.1	Differential stress at failure vs. inclination of cleavages ψ. Martinsburg slate (Donath, 1964).	51
3.2	Differential stress at failure vs. confining pressure. Martinsburg slate (Donath, 1964).	53
3.3	Effect of cleavage on the angle of the failure plane. Martinsburg slate (Donath, 1964).	53
3.4	Variation of tensile strength $T_{o\psi}$ measured by direct tension ($(T_o)_{\psi=0°}$ = 7.3 MPa) for a gneiss (Barla, 1974).	56
3.5	Bray's diagram for a joint set with $\phi_j = 40°$ in rock with $\phi = 50°$ (Goodman, 1976).	60

Figure | Title

Figure	Title	
3.6	Anisotropy of a rock with one joint set: $\phi_j = 40^\circ$, $\phi = 50^\circ$ (Goodman, 1976).	60
3.7	Jaeger's theory.	63
3.8	(a) Differential stress at failure vs. confining pressure. (b) Variation of c_ψ and $\tan\phi_\psi$ with respect to ψ (Green River Shale-2; McLamore and Gray, 1967)	66
3.9	Geometry of the problem.	71
3.10	Differential stress at failure vs. inclination of anisotropy (theoretical example).	76
3.11	(a) Variation of c_ψ with inclination of anisotropy ψ. (b) Variation of $\tan\phi_\psi$ with inclination of anisotropy ψ.	77
3.12	Bray's diagram for a joint set with $\phi_j = 30^\circ$ in a transversely isotropic rock.	80
4.1	Geometry of the problem.	82
4.2	Orientation of the coordinate system attached to the contour of the hole.	87
4.3	Sign convention for stresses. Orientation of the positive shear stresses.	87
4.4	Plate theory. Geometry of the problem.	112
4.5	(a) Distribution of total tangential stresses.	137
	(b) Distribution of induced radial displacements.	138
	(c) Distribution of induced tangential displacements	138
4.6	(a) Distribution of induced tangential stresses.	139
	(b) Distribution of induced radial displacements.	140
5.1	Geometry of the problem.	145
5.2	Decomposition of the problem.	147
5.3	Variation of $E(h_n + k_n)$ and $E(s_n + t_n)$ with n.	155
5.4	Variation in length of the oblique distance P_1P_2. Geometry of the problem.	169
5.5	Distribution of the principal stress σ_1 along the contour $r/a = 1$. (a) $E_{inc} = 2\ 10^4$ MPa (b) $E_{inc} = 2\ 10^3$ MPa. Isotropic solution.	175

Figure	Title	
5.6	Orientation and magnitude of the principal stresses within a solid inclusion for $E_{inc} = 2\ 10^4$ and $E_{inc} = 2\ 10^3$ MPa. Lower hemisphere stereographic projection. Isotropic solution.	176
5.7	Distribution of the radial stress σ_r along the contour $r/a = 1$. (a) $E_{inc} = 2\ 10^4$ MPa (b) $E_{inc} = 2\ 10^3$ MPa. Isotropic solution.	177
5.8	(a) Radial stress distribution induced by an inclusion within an isotropic body ($\theta = 0°$).	179
	(b) Distribution of tangential stress $\tau_{r\theta}$ induced by an inclusion within an isotropic body ($\theta = 0°$).	180
	(c) Distribution of tangential stress τ_{rz} induced by an inclusion within an isotropic body ($\theta = 0°$).	180
5.9	Distribution of the radial displacement u_r/a along the contour $r/a = 1$. Hollow inclusion with $a/b = 2$. Isotropic solution.	181
5.10	Orientation and magnitude of the principal stress field within a solid inclusion for $E_{inc} = 2\ 10^4$ and $E_{inc} = 2\ 10^3$ MPa. Lower hemisphere stereographic projection. Anisotropic solution.	183
5.11	Radial stress distribution induced by a solid inclusion within an anisotropic body ($\theta = 0°$).	184
5.12	Radial stress distribution induced by a hollow inclusion ($a/b = 2.$)within an anisotropic body ($\theta = 0°$).	185
5.13	(a) Distribution of tangential stress $\tau_{r\theta}$ induced by a solid inclusion within an anisotropic body ($\theta = 0°$).	187
	(b) Distribution of tangential stress τ_{rz} induced by a solid inclusion within an anisotropic body ($\theta = 0°$).	188
6.1	Steps commonly involved in any overcoring technique (see explanations of the procedure in the text).	192
6.2	Set up for the undercoring technique.	195
6.3	Terminology (Bielenstein and Barron, 1971).	197
6.4	Hollow epoxy probe. Rosette configuration and definition of coordinate systems.	211
6.5	(a) Diametral measurements. (b) Strain measurements.	218
	(c) Combination of measurements of changes in diameter and changes in length of oblique distances within an instrumented hollow inclusion.	219

Figure	Title	
6.6	Illustrative examples for Table 6.1.	225
6.7	Orientation and magnitude of the principal in situ stress field components for different cases of transverse isotropy using two or three boreholes.	231
6.8	Magnitude of the principal in situ stress field components for different orientations of a plane of transverse isotropy.	235
6.9	Orientation of the principal in situ stress field components for different strike angles β of a plane of transverse isotropy ($\psi = 30°$) and for a horizontal transverse isotropy. Lower hemisphere stereographic projection.	237
6.10	Orientation of the principal in situ stress field components for different strike angles β of a plane of transverse isotropy ($\psi = 90°$). Lower hemisphere stereographic projection.	238
6.11	Determination of the principal in situ stress field for different cases of transverse isotropy ($\beta = 0°$, $\psi = 30°$).	241

Appendix 4.1:

1. Orientation of the coordinate system x',y',z' with respect to the global one X,Y,Z. 288

2. Orientation of the coordinate system x,y,z with respect to the global one X,Y,Z. 288

Appendix 4.4:

1. Definition of the coordinate system r, θ, z. 308

Appendix 4.5:

1. Schematic view of a long circular hole to be excavated in a triaxial stress field. 311

2. Definition of the radial and tangential displacement components. 322

Appendix 5.4:

1. Definition of point P_1 and displacement components u_r, v_θ, w. 381

2. Variation in length of the oblique distance P_1P_2 . Geometry of the problem. 381

LIST OF TABLES

Table	Title	Page
4.1	Classical plane strain formulation . Generalized plane stress formulation .	116
4.2	Types of plane strain formulation.	131
6.1	Number of boreholes required for the determination of the complete state of stress from diametral measurements only.	224
1.	Generalized plane strain formulation. Present formulation (conditions (14) and (20) being satisfied). Appendix 4.2.	301

LIST OF SYMBOLS

Principal Symbols and Notation

$(\ )$	Matrix brackets
$(\ )^t$	Transpose of a matrix
$(\ )^{-1}$	Inverse of a matrix
xyz, x'y'z', XYZ	Cartesian coordinate systems
r, θ, z	Cylindrical coordinate system
a	radius of a circular hole or outer radius of a hollow inclusion
b	Inner radius of a hollow inclusion
X_n, Y_n, Z_n	Components in the x,y,z directions of surface forces per unit area.
X, Y, Z	Components in the x,y,z directions of body forces per unit volume (Chapter 4)
u, v, w	Components of displacement in the x,y,z directions
u_r, v_θ, w	Components of displacement in the r, θ, z directions
$\underset{\sim}{n}, \underset{\sim}{t}$	Normal and tangential unit vectors along a contour
s	Arc length along a contour
Δd	Change in diameter of a circular hole
$s_\theta, \Delta s_\theta$	Initial length and change in length between two points located in two different cross sections of an inclusion with a distance 2 L apart
q	Uniform internal pressure in a circular hole
τ_{ij}	Stress tensor
ε_{kl}	Strain tensor

δ_{ij}	Kronecker Delta
$(\sigma)_{ijk}$, $(\varepsilon)_{ijk}$	Matrix representation for the stress and strain tensors in the ijk coordinate system (xyz, $x'y'z'$, XYZ, $r\theta z$)
C_{ijkl}	Tensor of elastic constants
A_{ijkl}	Tensor of compliances
(A)	Matrix representation of the tensor of compliances in the xyz coordinate system
a_{ij} $(i,j = 1, 6)$	Components of matrix (A)
σ_1 σ_2 σ_3	Major, intermediate and minor principal stresses
(σ_o)	3D stress field applied at infinity with components in the xyz coordinate system: $\sigma_{x,o}$; $\sigma_{y,o}$; $\sigma_{z,o}$; $\tau_{yz,o}$; $\tau_{xz,o}$; $\tau_{xy,o}$.
ε_{z_o}	Constant longitudinal strain component
$\overline{\sigma_{x,o}}$ $\overline{\sigma_{y,o}}$ $\overline{\tau_{xy,o}}$	Average components of a 2D stress field acting parallel to a plate in the x0y coordinate system
β, ψ	Orientation angles of a plane with respect to a fixed arbitrary global coordinate system
β_h, δ_h	Orientation angles of a hole with respect to a fixed arbitrary global coordinate system
F, Ψ, ϕ	Stress functions
L_2, L_3, L_4	Differential operators of the second, third and fourth orders
∇^2	Laplace operator
∇^4 ; $\nabla^2\nabla^2$	Biharmonic operator
μ_k, λ_k, z_k	Complex numbers (k = 1,2,3)
p_k, q_k, r_k	Functions of μ_k and λ_k (k = 1,2,3)

$\ell_2(\mu_k), \ell_3(\mu_k), \ell_4(\mu_k)$	Polynomials of the second, third and fourth order in μ_k (k = 1,2,3)
$\phi_k(z_k)$	Analytic functions of the complex variable z_k (k = 1,2,3)
$\phi'_k(z_k)$	Derivative of $\phi_k(z_k)$ with respect to z_k (k = 1,2,3)
$\xi_k(z_k)$	Analytic functions of z_k and μ_k (k = 1,2,3)
Re	Real part of a complex function

Chapter 1: Introduction

Any undisturbed rock mass is subject to natural stresses including gravitational stresses due to the mass of the overburden and possibly tectonic stresses due to the straining of the earth's crust and remanent stresses due to past tectonism. Knowledge of the in situ stress field must be integrated into any rock engineering design along with general rock mass characteristics such as deformability, strength, permeability and time dependent behavior. For example, the choice of optimum orientation and shape of deep underground caverns or complex underground works will be controlled by the orientation and the magnitude of the in situ stress field if it is necessary to minimize stress concentration problems. Long term variation of the in situ stress field may also help to evaluate the potential hazard of earthquake occurences.

The magnitude and orientation of the stress field at a point within a rock mass can be measured but there is no known method by which the state of stress at a point can be accurately determined by instruments located remotely. In general, measurements are made inside boreholes, on outcrops or on the internal surfaces of underground cavities. Most of the measuring techniques intentionally disturb the state of stress in the rock and then measure consequent strains and displacements. Measured strains or displacements are then related to the stresses through assumptions of material behavior. A common procedure is to assume that the rock mass is <u>linearly elastic</u>, <u>isotropic</u>, <u>continuous</u> and <u>homogeneous</u>. Beside the measurement inaccuracies, that can be minimized by proper care, another source of inaccuracies is associated with the imperfection of the analytical formulation used to convert the full data into rock stresses. The reliability of the in situ stress determination can be increased by increasing the fidelity of the mechanical model involved for the analytical formulation. Any rock is to a certain extent anisotropic and/or heterogeneous and/or discontinuous and somewhat nonlinearly elastic. It is important to understand first the meaning of each part of this statement in the light of the in situ determination of stresses in rock.

A medium is anisotropic if its properties vary with direction, it is heterogeneous if they vary from point to point and it is discontinuous if there are separations or gaps in the stress field. These three definitions are in general scale dependent. They depend upon the relative size of the smallest structural feature of the problem of interest with respect to the largest structural feature of the medium.

A medium possesses the property of elasticity if the deformations associated with its loading are fully recovered during unloading. In terms of a load deformation or stress strain curve, there is a one to one correspondence between load and deformation or stress and strain. If stress and strain are linearly related the material is said to be "linear elastic". A particularly attractive property of the linear elastic theory is the principle of superposition of effects. Furthermore, the applicability of the theory of elasticity is a function of the duration of loading. Elasticity assumes an immediate response upon the application or removal of loads.

Anisotropy is the usual characteristic of schists, slates, gneisses, phyllites and other metamorphic rocks with fabrics having parallel arrangments of flat or long minerals. Bedded rocks like shale/sandstone sequences that include interlayered mixtures of different components also display anisotropy. Even isotropic rocks like some granites, sandstones or limestones will behave anisotropically if cut by regular joint sets. Some rocks may display simultaneously several types of anisotropy such as planes of discontinuities parallel to foliation planes, or foliation and stratification. Cleavage and bedding can also be associated with other features such as lineations. A general classification of rock anisotropy proposed by Barla (1974) consists of two classes, A and B. Class A rocks exhibit anisotropic properties despite apparent isotropy. Some granitic rocks belong to this group. In comparison, Class B rocks display clear evidence of anisotropy and show apparent directions of symmetry.

The purpose of this research is to study the influence of rock anisotropy on in situ stress measurements. The rock is assumed to belong to Class B and to present a planar or rectilinear anisotropy. The present study is limited to stress measurements by overcoring techniques for which strains and displacements are recorded on the walls of a borehole or within instrumented hollow or solid inclusions perfectly bonded to the surface of that borehole. The rock is also described as a continuous and homogeneous medium, the size of the smallest structural feature of the problem of interest being the diameter of the boreholes where measurements take place. Furthermore, the disturbance associated with the overcoring techniques is assumed to be instantaneous and the rock response is considered essentially linear elastic.

In practice, several questions often arise concerning the measurements of strains and displacements.

(i) How many independent measurements can we make in a single borehole and therefore how many boreholes do we need to determine the in situ stress field?

(ii) Are these numbers influenced by the anisotropic character of the rock?

(iii) For a given set of measurements, how do anisotropy type and orientation influence the determination of the in situ stress field?

(iv) How large an error is involved by neglecting rock anisotropy?

In order to answer these questions we need a closed form solution connecting the in situ stress field components and the measurements of strains and displacements. Since the rock material is assumed to be linear elastic and anisotropic, its constitutive relation can be included in that closed form solution.

Chapter 2 reviews the constitutive relations available for anisotropic media and analyzes their applicability to rocks. Laboratory and in situ techniques used to determine the elastic constants entering those relations are also discussed. A prerequisite for the applicability of stress measurement techniques is that measurements must take place on a rock whose strength has not been totally mobilized. This is particularly important when the stresses to be measured are expected to be high. Intuitively both rock deformability and strength must be affected by rock anisotropy. Chapter 3 reviews the failure criteria applicable to anisotropic media and analyzes their applicability to rocks. Following the line of Chapter 2, the laboratory and in situ techniques used to evaluate the directional character of the strength of anisotropic rocks are discussed.

Closed form solutions between strain and displacement components at any point along the walls of a circular hole or within a hollow inclusion perfectly bonded to that hole and a 3D stress field applied at infinity are proposed respectively in Chapters 4 and 5. The hole is located within an infinite linear elastic, anisotropic, continuous and homogeneous medium. No restrictions are made on the type or orientation of the anisotropy or on the orientation of the applied stress field with respect to the hole.

The influence of rock anisotropy on stress measurements has been investigated by several authors. A literature review is presented in Chapter 6. In that chapter, the closed form solutions mentioned previously are applied to calculate 3D in situ stress field components from measurements of strains and displacements using several existing overcoring techniques. New techniques are also proposed involving hollow or solid inclusions. Questions (i) to (iv) are answered with special emphasis to rocks that can be described as transversely isotropic or orthotropic materials.

Chapter 2: Deformability of Anisotropic Rocks

2.1 Introduction

Laboratory and in situ testing have shown that the deformability of anisotropic rocks varies with spatial orientation. This characteristic must be included into any theoretical model that is supposed to simulate their behavior. The purpose of this chapter is twofold:

(i) to review the constitutive relations available for anisotropic media and analyze their applicability to rocks,

(ii) to review the in situ and laboratory testing procedures to assess the deformability of anisotropic rocks.

2.2 Constitutive Relations

2.2.1 Generalized Hooke's Law

If we assume that the anisotropic rock material can be described as linear, elastic, homogeneous and continuous, its general constitutive relation relating stress and strain tensors can be written as follows*

$$\tau_{ij} = C_{ijkl}\, \varepsilon_{kl} \tag{2.1}$$

which is known as the Generalized Hooke's law**. In the most

---

* Repeated indices imply summation. Indices i,j,k take the values 1,2,3

** The present discussion is limited to the infinetisemal character of the theory of linear elasticity.

general 3D case, the tensor of elastic constants C_{ijkl} has 81 independent components. However, due to the symmetry of both the strain and stress tensors, there are at most 36 distinct elastic constants. This number is reduced to 21 if the existence of a strain energy function is further assumed (Lekhnitskii, 1963). Eq (2.1) can be rewritten as follows

$$\varepsilon_{ij} = A_{ijkl} \tau_{kl} \tag{2.2}$$

where A_{ijkl} is a tensor of compliances with 21 distinct components.

A_{ijkl} and C_{ijkl} are symmetric in the indices (i,j) and in (k,l) and in the pairs (ij, kl). They are inverse to each other in the sense that

$$C_{ijkl} A_{klrs} = \frac{1}{2} \left(\delta_{ir} \delta_{js} + \delta_{is} \delta_{jr} \right) \tag{2.3}$$

where δ_{ij} is the Kronecker Delta.

In the present discussion the Generalized Hooke's law described by eq (2.2) will be mostly used. If we consider the matrix representation of the tensors ε_{ij}, τ_{kl} and A_{ijkl} in an arbitrary x,y,z coordinate system, eq (2.2) is equivalent to

$$(\varepsilon)_{xyz} = (A)(\sigma)_{xyz} \tag{2.4}$$

or

$$\begin{bmatrix} \varepsilon_x \\ \varepsilon_y \\ \varepsilon_z \\ \gamma_{yz} \\ \gamma_{xz} \\ \gamma_{xy} \end{bmatrix} = \begin{bmatrix} & & \\ & a_{ij} & \\ & i=1,6 & \\ & j=1,6 & \\ & & \end{bmatrix} \begin{bmatrix} \sigma_x \\ \sigma_y \\ \sigma_z \\ \tau_{yz} \\ \tau_{xz} \\ \tau_{xy} \end{bmatrix} \tag{2.5}$$

Coefficients a_{ij} play different roles and have different physical meanings. This appears if eq (2.5) is rewritten with the following notations (Lekhnitskii, 1963, pp.13).

$$\begin{bmatrix} \varepsilon_x \\ \varepsilon_y \\ \varepsilon_z \\ \gamma_{yz} \\ \gamma_{xz} \\ \gamma_{xy} \end{bmatrix} = \begin{bmatrix} \frac{1}{E_x} & \frac{-\nu_{yx}}{E_y} & \frac{-\nu_{zx}}{E_z} & \frac{\eta_{x,yz}}{G_{yz}} & \frac{\eta_{x,xz}}{G_{xz}} & \frac{\eta_{x,xy}}{G_{xy}} \\ \frac{-\nu_{xy}}{E_x} & \frac{1}{E_y} & \frac{-\nu_{zy}}{E_z} & \frac{\eta_{y,yz}}{G_{yz}} & \frac{\eta_{y,xz}}{G_{xz}} & \frac{\eta_{y,xy}}{G_{xy}} \\ \frac{-\nu_{xz}}{E_x} & \frac{-\nu_{yz}}{E_y} & \frac{1}{E_z} & \frac{\eta_{z,yz}}{G_{yz}} & \frac{\eta_{z,xz}}{G_{xz}} & \frac{\eta_{z,xy}}{G_{xy}} \\ \frac{\eta_{yz,x}}{E_x} & \frac{\eta_{yz,y}}{E_y} & \frac{\eta_{yz,z}}{E_z} & \frac{1}{G_{yz}} & \frac{\mu_{yz,xz}}{G_{xz}} & \frac{\mu_{yz,xy}}{G_{xy}} \\ \frac{\eta_{xz,x}}{E_x} & \frac{\eta_{xz,y}}{E_y} & \frac{\eta_{xz,z}}{E_z} & \frac{\mu_{xz,yz}}{G_{yz}} & \frac{1}{G_{xz}} & \frac{\mu_{xz,xy}}{G_{xy}} \\ \frac{\eta_{xy,x}}{E_x} & \frac{\eta_{xy,y}}{E_y} & \frac{\eta_{xy,z}}{E_z} & \frac{\mu_{xy,yz}}{G_{yz}} & \frac{\mu_{xy,xz}}{G_{xz}} & \frac{1}{G_{xy}} \end{bmatrix} \begin{bmatrix} \sigma_x \\ \sigma_y \\ \sigma_z \\ \tau_{yz} \\ \tau_{xz} \\ \tau_{xy} \end{bmatrix} \tag{2.6}$$

where

(i) E_x, E_y, E_z are the Young's moduli with respect to the directions x,y,z,

(ii) G_{yz}, G_{xz}, G_{xy} are shear moduli for planes that are respectively parallel to yOz, xOz and xOy,

(iii) $\nu_{yx}, \nu_{zx}, \nu_{zy}, \nu_{xy}, \nu_{xz}, \nu_{yz}$ are Poisson's ratios. Poisson's ratio ν_{ij} determines the ratio of strain in the j

direction to the strain in the i direction due to a stress acting in the i direction.

Poisson's ratios ν_{ij} and ν_{ji} are such that*

$$\frac{\nu_{ij}}{E_i} = \frac{\nu_{ji}}{E_j} \tag{2.7}$$

(iv) $\mu_{xz,yz}, ------, \mu_{xz,xy}$ are such that $\mu_{ij,kl}$ characterizes the shear in the plane parallel to the one defined by indices ij that induces the tangential stress in the plane parallel to the one defined by indices kl. Furthermore**

$$\frac{\mu_{ik,jk}}{G_{jk}} = \frac{\mu_{jk,ik}}{G_{ik}} \tag{2.8}$$

(v) $\eta_{x,yz}$ -------- $\eta_{z,xy}$ are called the <u>coefficients of mutual influence of the first kind</u>. $\eta_{yz,x}$ ---------- $\eta_{xy,z}$ are called <u>the coefficients of mutual influence of the second kind.</u> Coefficient $\eta_{k,ij}$ characterizes the stretching in the direction parallel to k induced by the shear stress acting within a plane parallel to the one defined by indices ij. Coefficient $\eta_{ij,k}$ characterizes a shear in the plane defined by indices ij under the influence of a normal stress acting in the k direction. Furthermore**

---

* Any index i, j, k or l correspond to x,y or z.

** Commas do not mean to imply differentiation.

$$\frac{\eta_{ij,k}}{E_k} = \frac{\eta_{k,ij}}{G_{ij}} \qquad (2.9)$$

If the internal composition of the rock material possesses symmetry of any kind, then symmetry can be also observed in its elastic properties.[+] The number of independent components of the tensors of compliances or elastic constants is less then 21. In the present analysis four cases of elastic symmetry are of particular interest:

- one plane of elastic symmetry,
- three orthogonal planes of elastic symmetry,
- one axis of elastic symmetry of rotation,
- complete symmetry.

a) One Plane of Elastic Symmetry

A plane of elastic symmetry exists at a point if the elastic constants or compliances have the same values for every pair of coordinate systems that are the reflected image of one another with respect to the plane. For instance, if plane xOy is a plane of elastic symmetry, it can be shown that in eq (2.5)

$$a_{4i} = a_{5i} = a_{46} = a_{56} = 0 \quad (i = 1, 2, 3) \qquad (2.10)$$

Eq (2.6) reduces to the following one*

+ Symmetry under discussion is a directional property.

* One half of matrix (A) is written since it is symmetric

$$\begin{bmatrix} \varepsilon_x \\ \varepsilon_y \\ \varepsilon_z \\ \gamma_{yz} \\ \gamma_{xz} \\ \gamma_{xy} \end{bmatrix} = \begin{bmatrix} \frac{1}{E_x} & \frac{-\nu_{yx}}{E_y} & \frac{-\nu_{zx}}{E_z} & 0 & 0 & \frac{\eta_{x,xy}}{G_{xy}} \\ & \frac{1}{E_y} & \frac{-\nu_{zy}}{E_z} & 0 & 0 & \frac{\eta_{y,xy}}{G_{xy}} \\ & & \frac{1}{E_z} & 0 & 0 & \frac{\eta_{z,xy}}{G_{xy}} \\ & & & \frac{1}{G_{yz}} & \frac{\mu_{yz,xz}}{G_{xz}} & 0 \\ & & & & \frac{1}{G_{xz}} & 0 \\ & & & & & \frac{1}{G_{xy}} \end{bmatrix} \begin{bmatrix} \sigma_x \\ \sigma_y \\ \sigma_z \\ \tau_{yz} \\ \tau_{xz} \\ \tau_{xy} \end{bmatrix} \tag{2.11}$$

The number of independent elastic constants or compliances is reduced to 13.

b) Three Orthogonal Planes of Elastic Symmetry

Let us assume that three orthogonal planes of elastic symmetry pass through each point of the rock, each one being perpendicular to x,y, or z. It can be shown that in addition to eq (2.10) the following conditions also apply

$$a_{16} = a_{26} = a_{36} = a_{45} = 0 \tag{2.12}$$

Eq (2.6) reduces to the following one

$$\begin{bmatrix} \varepsilon_x \\ \varepsilon_y \\ \varepsilon_z \\ \gamma_{yz} \\ \gamma_{xz} \\ \gamma_{xy} \end{bmatrix} = \begin{bmatrix} \frac{1}{E_x} & \frac{-\nu_{yx}}{E_y} & \frac{-\nu_{zx}}{E_z} & 0 & 0 & 0 \\ & \frac{1}{E_y} & \frac{-\nu_{zy}}{E_z} & 0 & 0 & 0 \\ & & \frac{1}{E_z} & 0 & 0 & 0 \\ & & & \frac{1}{G_{yz}} & 0 & 0 \\ & & & & \frac{1}{G_{xz}} & 0 \\ & & & & & \frac{1}{G_{xy}} \end{bmatrix} \begin{bmatrix} \sigma_x \\ \sigma_y \\ \sigma_z \\ \tau_{yz} \\ \tau_{xz} \\ \tau_{xy} \end{bmatrix} \tag{2.13}$$

The number of independent elastic constants or compliances is reduced to 9. There are, for instance, three Young's moduli E_x, E_y, E_z , three shear moduli G_{yz}, G_{xz}, G_{xy} and three Poisson's ratios ν_{yx}, ν_{zx} and ν_{zy}. A rock material that possesses this type of elastic symmetry is called <u>orthotropic</u>.

c) <u>One Axis of Elastic Symmetry of Rotation</u>

An axis of elastic symmetry g of <u>order</u> n exists at a point when there are sets of equivalent elastic directions that can be superimposed by a rotation through an angle of $2\pi/n$. An axis of the second order is equivalent to a plane of elastic symmetry. For an axis of the third or fourth order, the number of independent elastic constants or compliances is reduced to 7. For an axis of order larger or equal to 6, it can be shown that all directions in the planes normal to it are equivalent with respect to the elastic properties. If z coincides with the axis of elastic symmetry g, the rock is isotropic within the xOy plane. The z axis is defined as <u>axis of radial elastic symmetry</u> of <u>axis of elastic symmetry of rotation</u>. A rock that possesses this type of elastic symmetry is called <u>transversely isotropic</u>. Plane xOy and each plane perpendicular to it are planes of elastic symmetry. Eq (2.6) reduces to eq (2.13) with the additional conditions

$$\begin{aligned}
E_x &= E_y = E \\
E_z &= E' \\
\nu_{xy} &= \nu_{yx} = \nu \\
\nu_{zx} &= \nu_{zy} = \nu' \\
G_{yz} &= G_{xz} = G' \\
G_{xy} &= E \,/\, 2(1+\nu)
\end{aligned} \tag{2.14}$$

The number of independent elastic constants or compliances is reduced to 5. According to eqs (2.14) there are two Young's moduli E and E', one shear modulus G' and two Poisson's ratios ν and ν'.

d) Complete Symmetry

If all planes and axes are one of elastic symmetry the rock material is isotropic. Eq (2.6) reduces to eq (2.13) with the additional conditions

$$E_x = E_y = E_z = E$$
$$\nu_{yx} = \nu_{zx} = \nu_{zy} = \nu$$
$$G_{xy} = G_{yz} = G_{xz} = E \,/\, 2\,(1+\nu) \qquad (2.15)$$

The number of independent elastic constants or compliances is reduced to 2: E and ν.

The elastic constants or compliances introduced previously have ranges of possible variation that are limited since thermodynamic considerations require that the strain energy of an elastic material should always be positive. Using eq (2.4) the strain energy per unit volume is equal to

$$\Omega = \frac{1}{2}\,(\sigma)^t_{xyz}\,(A)\,(\sigma)_{xyz} \qquad (2.15a)$$

If this quadratic form is positive definite, the strain energy will be positive as required. For an isotropic material the requirement is satisfied if the Young's modulus E and the Poisson's ratio ν are such that

$$E > 0 \qquad ; \qquad -1 \leqslant \nu \leqslant 0.5^{*} \qquad (2.16)$$

---

* In the usual cases $0 \leqslant \nu \leqslant 0.5$

For a transversely isotropic material with the five elastic constants defined in eqs (2.14), Pickering (1970) has shown that the following conditions must be satisfied

$$E > 0 \; ; \; E' > 0 \; ; \; G' > 0 \; ; \; -1 \leqslant \nu \leqslant 1^{+}$$

$$\frac{E}{E'}(1-\nu) - 2\nu_{xz}^{2} \geqslant 0^{+} \tag{2.17}$$

where according to eq (2.7)

$$\nu_{xz} = \nu' \frac{E}{E'}^{+} \tag{2.18}$$

Combining the conditions for E and ν in eqs (2.17) and the last of eqs (2.14), it appears that the shear modulus G_{xy} within the plane of transverse isotropy must be also positive.

Conditions (2.17) except the one related to G' can be represented in a E/E', ν, ν_{xz} space (Figure 2.1) by a region limited with a paraboloid; any vertical section with constant ν is a parabola, and any horizontal section with constant E/E' is also parabola. $G' > 0$ is an additional condition. Any given transversely isotropic material is represented by a point. As a special case, isotropic materials lie on a line CD with components $E/E' = 1$ and $\nu_{xz} = \nu$. The limiting values of $\nu_{xz} = \nu$ are then -1 and 0.5. However, it must be remembered that all points lying on CD do not represent isotropic materials since $G' > 0$ is an additional independent condition.

[+] In the usual cases Poisson's ratios ν, ν' and ν_{xz} are positive quantities.

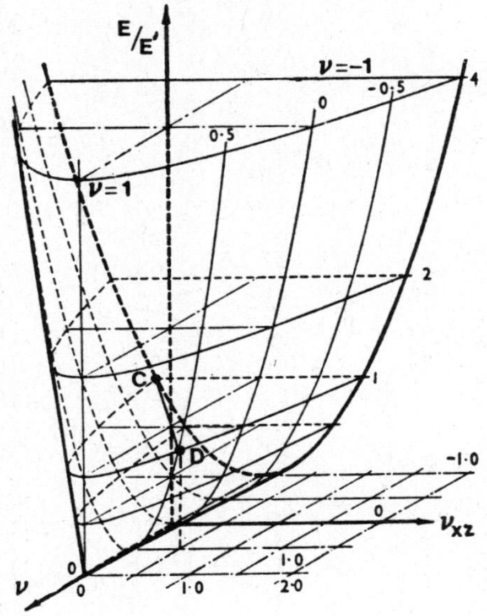

Figure 2.1. Bounds for elastic constants of a transversely isotropic material (Pickering, 1970).

Constitute relations defined by eq (2.1) or (2.2) with the elastic symmetries introduced previously may be used to model the deformability of homogeneous, continuous anisotropic rocks. Transverse isotropy or orthotropy are commonly associated with foliated rocks such as schists, gneisses, slates or bedded rocks such as shales.

2.2.2 Stratified Media

Many rock masses are stratified and clearly non homogeneous. They may be divided into several layers of randomly varying thicknesses and properties. Sometimes, only two types of rock are regularly interlayered. Since it does not seem feasible to take into account the individual properties and geometry of each stratum in any mechanical model of a stratified rock mass, it is more practical to replace the latter by an equivalent homogeneous continuum (Pinto, 1966; Salamon, 1968; Wardle and Gerrard, 1972). A common practice is to assume that both individual stratum and the equivalent continuum consist of a homogeneous linear elastic transversely isotropic material. This concept is valid under certain conditions:

(i) all layers are bounded by parallel planes and no relative displacement takes place on these planes,

(ii) the thickness and elastic properties of the layers vary randomly with respect to the dimension perpendicular to the bounding planes. This randomness is necessary to ensure overall homogeneity of the equivalent material,

(iii) a representative sample of the stratified rock mass on the basis of which the equivalent homogeneous properties are calculated must contain a large number of layers.

Let x,y,z be a coordinate system such that the z axis is perpendicular to the bounding planes of the layers and any plane parallel to them contains the x and y axes. Let n be the number of layers. According to eqs (2.14) the deformability of each layer i can be described by five elastic constants E_i, E'_i, ν_i, ν_{xz_i} and G'_i where

$$\nu_{xz_i} = \nu'_i \frac{E_i}{E'_i} \tag{2.19}$$

Salamon (1968) has shown that the five elastic constants of the equivalent transverse isotropic homogeneous continuum can be expressed in terms of the elastic properties and thickness of each layer. Those five constants can be written as follows

$$\nu = \frac{\sum_{i=1}^{n} \frac{\varphi_i \nu_i E_i}{1-\nu_i^2}}{\sum_{i=1}^{n} \frac{\varphi_i E_i}{1-\nu_i^2}} \quad ; \quad \nu_{xz} = (1-\nu)\sum_{i=1}^{n} \frac{\varphi_i \nu_{xz_i}}{1-\nu_i}$$

$$E = (1-\nu^2)\sum_{i=1}^{n} \frac{\varphi_i E_i}{1-\nu_i^2} \quad ; \quad E' = \frac{1}{\sum_{i=1}^{n} \frac{\varphi_i}{E_i}\left(\frac{E_i}{E'_i} - \frac{2\nu_{xz_i}^2}{1-\nu_i}\right) + \frac{2\nu_{xz}^2}{(1-\nu)E}}$$

$$G' = \frac{1}{\sum_{i=1}^{n} \frac{\varphi_i}{G'_i}} \tag{2.20}$$

where $t_i = \varphi_i L$ is the thickness of the i-th layer and L is the edge dimension of a cube that corresponds to a representative sample of the stratified rock mass on the basis of which the equivalent homogeneous properties are calculated. If there are n layers, $\sum_{i=1}^{n} \varphi_i = 1$.

The shear modulus G_{xy} can be expressed in terms of E and ν defined in eqs (2.20) as follows

$$G_{xy} = \frac{E}{2(1+\nu)} \tag{2.21}$$

If each layer i is isotropic, its elastic constants E_i and ν_i must satisfy conditions (2.16). Wardle and Gerrard (1972) have shown that these conditions impose undue restrictions on the elastic properties of the equivalent material. The elastic constants defined in eqs (2.20) and (2.21) must be such that

$$E > 0 \;;\; E' > 0 \;;\; G' > 0 \;;\; -1 \leqslant \nu \leqslant 0.5^{*}$$

$$-1 \leqslant \nu_{xz} < 2^{*} \qquad G_{xy}/G' \geqslant 1 \tag{2.22}$$

In comparison to conditions (2.17) for a homogeneous single layered transversely isotropic material, conditions (2.22) are more restrictive.

2.2.3 Regularly Jointed Rocks

In the case of anisotropy derived from regular discontinuities, the theory of linear elasticity is invalid. However,

---

* In the usual cases where $\nu_i > 0$, ν, ν_{xz} are such that $0 \leqslant \nu \leqslant 0.5$ and $0 \leqslant \nu_{xz} \leqslant 0.5$

numerical techniques such as the Finite Element Method (Goodman et al, 1968) or the Boundary Element Methods (Crouch and Starfield) can be used to incorporate the discontinuous character of rock masses when modelling their deformability.

Another procedure is to replace the regularly jointed rock by a homogeneous, anisotropic and continuous medium, the behavior of which is equivalent to the behavior of the jointed rock (Duncan and Goodman, 1968; Singh, 1973; Morland, 1976; Eissa, 1980). This procedure can be regarded as a special case of the one presented previously for stratified rock masses. It can also be used when rock anisotropy is derived from regular discontinuities and the intrinsic character of the intact rock

The concept of an "equivalent medium" can be also used to describe the non linear behavior of a discontinuous, homogeneous and anisotropic body or rock containing up to three orthogonal joint sets (Amadei and Goodman, 1981a; Appendix 2.1). The intact rock is assumed to behave in a linear elastic manner with up to three orthogonal planes of elastic symmetry, each one being parallel to one of the joint sets. The joints are modelled in a non linear inelastic fashion in compression and decompression (tension) and in a linear or non linear elastic fashion in shear.

The applicability of the concept of an 'equivalent medium' for modelling the deformability of a stratified or a regularly jointed rock mass depends upon two conditions:

(i) a representative sample of the rock mass on the basis of which the equivalent homogeneous properties are calculated must contain a large number of layers or joints,

(ii) that representative sample must be sufficiently small to be exposed to a homogeneous stress distribution in the equivalent medium.

Those two conditions can only be satisfied if the size of the problem to be dealt with is much larger than the average layer thickness or the joint set spacing. For example, when dealing with a borehole, the appropriate joint set spacings may be of the order of a few centimeters. However, for large cavities they may be as large as a few decimeters.

2.2.4 Effective Stress Laws

The constitutive relations described previously have been derived for a dry continuous or discontinuous anisotropic rock. Several effective stress laws have been proposed for linear elastic, isotropic homogeneous and porous materials (Skempton, 1960; Nur and Byerlee, 1971). Carroll (1979) derived analytically an effective stress law to describe the effect of pore fluid pressure on the linearly elastic response of saturated porous rock that exhibit anisotropy. The anisotropy may be either structural (anisotropic pore geometry) or intrinsic (anisotropic solid material) or both. Using the notations of eqs (2.1), (2.2) and the relation (2.3), the effective stress law takes the form

$$\hat{\tau}_{ij} = \tau_{ij} - P_p \left(\delta_{ij} - C_{ijkl} A^{(s)}_{klmm} \right) \tag{2.23}$$

where

τ_{ij} is the applied stress tensor,

$\hat{\tau}_{ij}$ is the effective stress tensor,

P_p is the pore fluid pressure,

C_{ijkl} is the tensor of elastic constants of the porous rock,

$A^{(s)}_{ijkl}$ is the tensor of compliances for the solid material.

Eq (2.2) reduces to the following

$$\varepsilon_{ij} = A_{ijkl}\,\tau_{kl} - P_p\left(A_{ijkk} - A^{(s)}_{ijkk}\right) \tag{2.24}$$

According to eq (2.23), it is noteworthy that the modification to the applied stress is not hydrostatic as for isotropic rocks. Instead, examination of the last term in eq (2.23) shows that it involves the stress state which corresponds in the porous rock to the strains associated with hydrostatic stress in the solid material.

This effective stress law has been used to evaluate the effect of pore fluid pressure on the response of a regularly anisotropic fractured rock described as an equivalent anisotropic continuum (Appendix 2.1). Joint sets were considered as the only voids in the anisotropic rock.

2.2.5 Transformation of Elastic Constants

In eq (2.4) the components of matrix (A) depend upon the choice of the coordinate system x,y,z. Only in the case of an isotropic body are those components invariant. Let X,Y,Z be an arbitrary fixed global coordinate system and x',y', z' a coordinate system attached to the rectilinear anisotropy. The orientation of that system with respect to the global one may be defined by two angles β and ψ as shown in Figure 1 of Appendix 4.1. If we know the constitutive relation of an anisotropic material in the x',y',z' coordinate system, it is shown in Appendix 4.1 how to calculate its constitutive relation in the global system.

In particular, consider an orthotropic material with one of its three planes of elastic symmetry striking parallel to the Z axis and dipping at angle $\psi = \theta$ (Figure 2.2). Let x',y',z' be a coordinate system attached to that plane with the x' axis directed parallel to its normal. According to eq (2.13) there are nine independent elastic constants to describe the deformability of the anisotropic material in the x',y',z' coordinate system: three Young's moduli $E_{x'}, E_{y'}, E_{z'}$, three shear moduli $G_{y'z'}, G_{x'z'}, G_{x'y'}$ and three Poisson's ratios $\nu_{y'x'}, \nu_{z'x'}$ and $\nu_{z'y'}$. Young's moduli and Poisson's ratios must satisfy eq (2.7). Using eq (9) of Appendix 4.1, the constitutive relation of the anisotropic material in the global coordinate system takes the form

$$(\varepsilon)_{xyz} = (K)(\sigma)_{xyz} \tag{2.25}$$

or

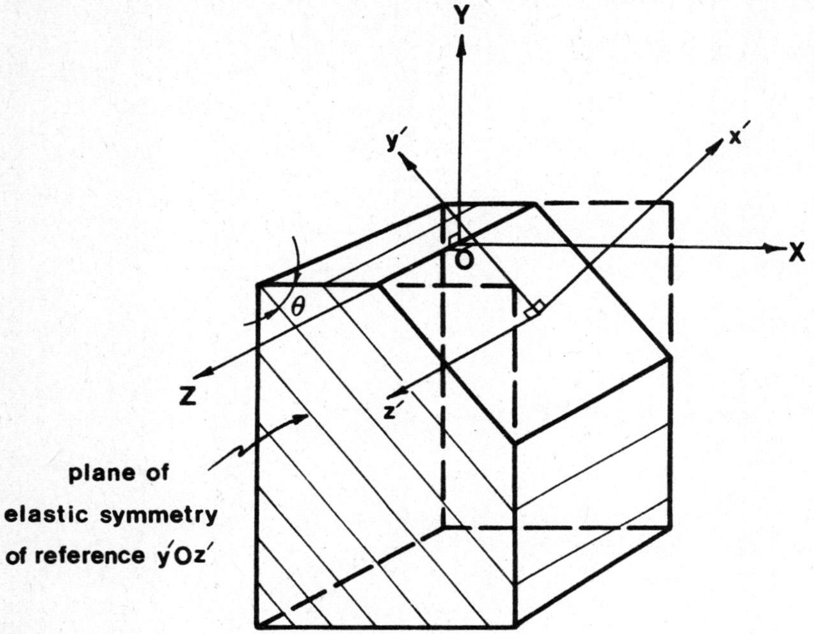

Figure 2.2. Geometry of the problem.
Definition of the coordinate systems.

$$\begin{bmatrix} \varepsilon_x \\ \varepsilon_y \\ \varepsilon_z \\ \gamma_{yz} \\ \gamma_{xz} \\ \gamma_{xy} \end{bmatrix} = \begin{bmatrix} K_{11} & K_{12} & K_{13} & 0 & 0 & K_{16} \\ & K_{22} & K_{23} & 0 & 0 & K_{26} \\ & & K_{33} & 0 & 0 & K_{36} \\ & & & K_{44} & K_{45} & 0 \\ & & & & K_{55} & 0 \\ & & & & & K_{66} \end{bmatrix} \begin{bmatrix} \sigma_x \\ \sigma_y \\ \sigma_z \\ \tau_{yz} \\ \tau_{xz} \\ \tau_{xy} \end{bmatrix} \qquad (2.26)$$

where(K) is a symmetric matrix.

Comparing eq (2.11) with eq (2.26) it can be seen that the latter represents the constitutive relation of an anisotropic material with one plane of elastic symmetry perpendicular to the Z axis. In the global coordinate system, the deformability of the anisotropic material can be defined by 13 elastic constants: three Young's moduli E_x, E_y, E_z ; three shear moduli G_{yz}, G_{xz}, G_{xy}; three Poisson's ratios ν_{yx}, ν_{zx} and ν_{zy} ; one coefficient $\mu_{yz,xz}$ and three coefficients of mutual influence of the first kind $\eta_{x,xy}$, $\eta_{y,xy}$, $\eta_{z,xy}$. Those elastic constants are related to those defined in the x',y',z' coordinate system as follows:

$$\frac{1}{E_x} = K_{11} = \frac{\sin^4\theta}{E_{x'}} + \frac{\cos^4\theta}{E_{y'}} + \frac{\sin^2 2\theta}{4}\left(\frac{1}{G_{x'y'}} - \frac{2\nu_{y'x'}}{E_{y'}}\right)$$

$$\frac{-\nu_{yx}}{E_y} = K_{12} = \frac{-\nu_{y'x'}\sin^4\theta}{E_{y'}} - \frac{\nu_{y'x'}\cos^4\theta}{E_{y'}} + \frac{\sin^2 2\theta}{4}\left(\frac{1}{E_{y'}} + \frac{1}{E_{x'}} - \frac{1}{G_{x'y'}}\right)$$

$$\frac{-\nu_{zx}}{E_z} = K_{13} = -\frac{\nu_{z'x'}\sin^2\theta}{E_{z'}} - \frac{\nu_{z'y'}\cos^2\theta}{E_{z'}}$$

$$\frac{1}{E_y} = K_{22} = \frac{\cos^4\theta}{E_{x'}} + \frac{\sin^4\theta}{E_{y'}} + \frac{\sin^2 2\theta}{4}\left(\frac{1}{G_{x'y'}} - \frac{2\nu_{y'x'}}{E_{y'}}\right)$$

$$-\frac{\nu_{zy}}{E_z} = K_{23} = -\frac{\nu_{z'x'}}{E_{z'}}\cos^2\theta - \frac{\nu_{z'y'}}{E_{z'}}\sin^2\theta$$

$$\frac{1}{E_z} = K_{33} = \frac{1}{E_{z'}}$$

$$\frac{1}{G_{yz}} = K_{44} = \frac{\sin^2\theta}{G_{y'z'}} + \frac{\cos^2\theta}{G_{x'z'}}$$

$$\frac{1}{G_{xz}} = K_{55} = \frac{\cos^2\theta}{G_{y'z'}} + \frac{\sin^2\theta}{G_{x'z'}}$$

$$\frac{1}{G_{xy}} = K_{66} = \frac{\cos^2 2\theta}{G_{x'y'}} + \sin^2 2\theta\left(\frac{1}{E_{x'}} + \frac{1}{E_{y'}} + \frac{2\nu_{y'x'}}{E_{y'}}\right)$$

$$\frac{\mu_{yz,xz}}{G_{xz}} = K_{45} = \frac{\sin 2\theta}{2}\left(\frac{1}{G_{x'z'}} - \frac{1}{G_{y'z'}}\right)$$

$$\frac{\eta_{x,xy}}{G_{xy}} = K_{16} = \sin^2\theta\sin 2\theta\left(\frac{1}{E_{x'}} + \frac{\nu_{y'x'}}{E_{y'}}\right) - \cos^2\theta\sin 2\theta\left(\frac{1}{E_{y'}} + \frac{\nu_{y'x'}}{E_{y'}}\right) + \frac{\sin 2\theta\cos 2\theta}{2G_{x'y'}}$$

$$\frac{\eta_{y,xy}}{G_{xy}} = K_{26} = \sin 2\theta\cos^2\theta\left(\frac{1}{E_{x'}} + \frac{\nu_{y'x'}}{E_{y'}}\right) - \sin 2\theta\sin^2\theta\left(\frac{1}{E_{y'}} + \frac{\nu_{y'x'}}{E_{y'}}\right) - \frac{\sin 2\theta\cos 2\theta}{2G_{x'y'}}$$

$$\frac{\eta_{z,xy}}{G_{xy}} = K_{36} = \sin 2\theta\left(\frac{\nu_{z'y'}}{E_{z'}} - \frac{\nu_{z'x'}}{E_{z'}}\right)$$

(2.27)

As shown by Lekhnitskii (1963, pp.43) the following quantities are invariant

$$
\begin{aligned}
I_1 &= K_{11} + K_{22} + 2K_{12} = \frac{1}{E_{x'}} + \frac{1}{E_{y'}} - 2\frac{\nu_{y'x'}}{E_{y'}} \\
I_2 &= K_{66} - 4K_{12} = \frac{1}{G_{x'y'}} + 4\frac{\nu_{y'x'}}{E_{y'}} \\
I_3 &= K_{44} + K_{55} = \frac{1}{G_{y'z'}} + \frac{1}{G_{x'z'}} \\
I_4 &= K_{23} + K_{13} = -\frac{\nu_{z'x'}}{E_{z'}} - \frac{\nu_{z'y'}}{E_{z'}}
\end{aligned}
\qquad (2.28)
$$

Eqs(2.26) and (2.27) apply for the geometry conditions of Figure 2.2 and for the following types of anisotropy :

(i) orthotropic intact material,

(ii) orthotropic intact material with up to three orthogonal joint sets each one being perpendicular to a plane of elastic symmetry of the intact material (Appendix 2.1),

(iii) transversely isotropic material with the plane of transverse isotropy striking parallel to the Z axis.

2.3 Testing of Anisotropic Rocks

In order to assess the deformability character of anisotropic rocks, testing is required. It can take place in the laboratory or/and in situ.

2.3.1 Laboratory Testing

Two different procedures are commonly used when dealing with class A or class B anisotropic rocks as defined in Chapter 1. Uniaxial and triaxial compression are the tests commonly performed on rock specimens.

a) Class A rocks

Since by definition, those rocks do not present any

apparent direction of symmetry, the approach consists of drilling samples in different directions within a block extracted from the rock mass of interest (Peres-Rodrigues, 1966; Douglass and Voight, 1969; Peres-Rodrigues and Aires-Barros, 1970). The orientation of the block with respect to the rock mass and its natural features (joints, slopes) must also be recorded.

Each sample is then tested in uniaxial compresion. Its deformability is defined by a modulus of deformation. A representative surface is then defined to permit describing the spatial variation of that modulus. An ellipsoid type quadratic law has been proposed by Peres-Rodrigues (1966) to model the spatial variation of the secant modulus of elasticity within granitic rock masses at several dam sites. A good correlation was found between the orientation of the principal axes of the ellipsoid for the modulus of elasticity and the attitudes of joints and natural slopes at those sites. A correlation seemed to exist also with the orientation of the mineralogical composition of the rock (Peres-Rodrigues and Aires-Barros, 1970) or the orientation of the rock defects such as microfractures (Douglas and Voight, 1969). Spatial variation of deformability was also observed on marble by Blankenship (1981a)*.

* Blankenship, Douglas 1981a. "Anisotropy of Marble: Spatial Variation of Strength and Deformation". Unpublished research report, Univ. of California (Supervisor, R.E. Goodman, Dept. of Civil Engineering).

b) Class B rocks

For rocks that are clearly anisotropic and show apparent directions of symmetry two approaches may be used:

(i) If we assume that the deformability of anisotropic rocks cannot be described by a Generalized Hooke's law type, then a procedure similar to the one proposed for class A rocks can be used. Peres-Rodrigues (1970, 1979) has shown that for anisotropic rocks belonging to the schist or micaschist types, a quartic type law can describe the spatial variation of their deformability. The axis of the quartic surface is assumed to be known and to be normal to the schistosity or foliation.

(ii) Another approach is to assume that the deformability of anisotropic rocks can be described by one of the constitutive relations introduced previously. A usual assumption is to accept the rocks to be either transversely isotropic or orthotropic such that planes and/or axes of elastic symmetry coincide with the apparent directions of rock symmetry. For example, a plane of transverse isotropy is often associated with foliation or schistosity planes, bedding or joint sets.

Rock samples (cylindrical or prismatic) are cut at different angles to the apparent directions of rock symmetry and are tested in uniaxial or triaxial compression. Samples must be close to each other in order to eliminate any experimental scatter due to material differences. Sample preparation and testing are supposed to follow the recommendations suggested by the ISRM (1978a, b). However for weak anisotropic rocks, planes of discontinuity often open parallel to the foliation or schistosity planes. For these materials, the suggested procedure designed for competent rocks is far too rigourous and may not be feasible. This was suggested by the experiments by Blankenship (1981b)[+] on Mariposa slate.

When continuous* anisotropic rocks are tested for deformation behavior, the main purpose is the determination of their elastic constants or compliances. Examples have been reported by several authors (Dayre, 1969; Masure, 1970; Pinto, 1970; Ko and Gerstle, 1972; 1976; Simonson et al, 1976; Cook et al, 1978; Eissa, 1980; Lerau et al, 1981). Strains and displacements are recorded during testing by instrumenting rock samples in different directions. Figure 2.3 shows an example of strain gage distributions on prismatic and cylindrical rock samples. Since the rock is assumed to be linearly elastic, measurements must take place in the linear elastic range of the load deformation curves recorded during testing.

Consider a rock sample with the geometry of Figure 2.2. A uniaxial stress is applied in the <u>Y direction</u> with magnitude σ_y .

[+] Blankenship, Douglas 1981b. "Sample Preparation from an Anisotropic, Jointed Rock". Unpublished research report, Univ. of California (Supervisor, R.E. Goodman, Dept. of Civil Eng.

* Continuity is defined here at the sample scale.

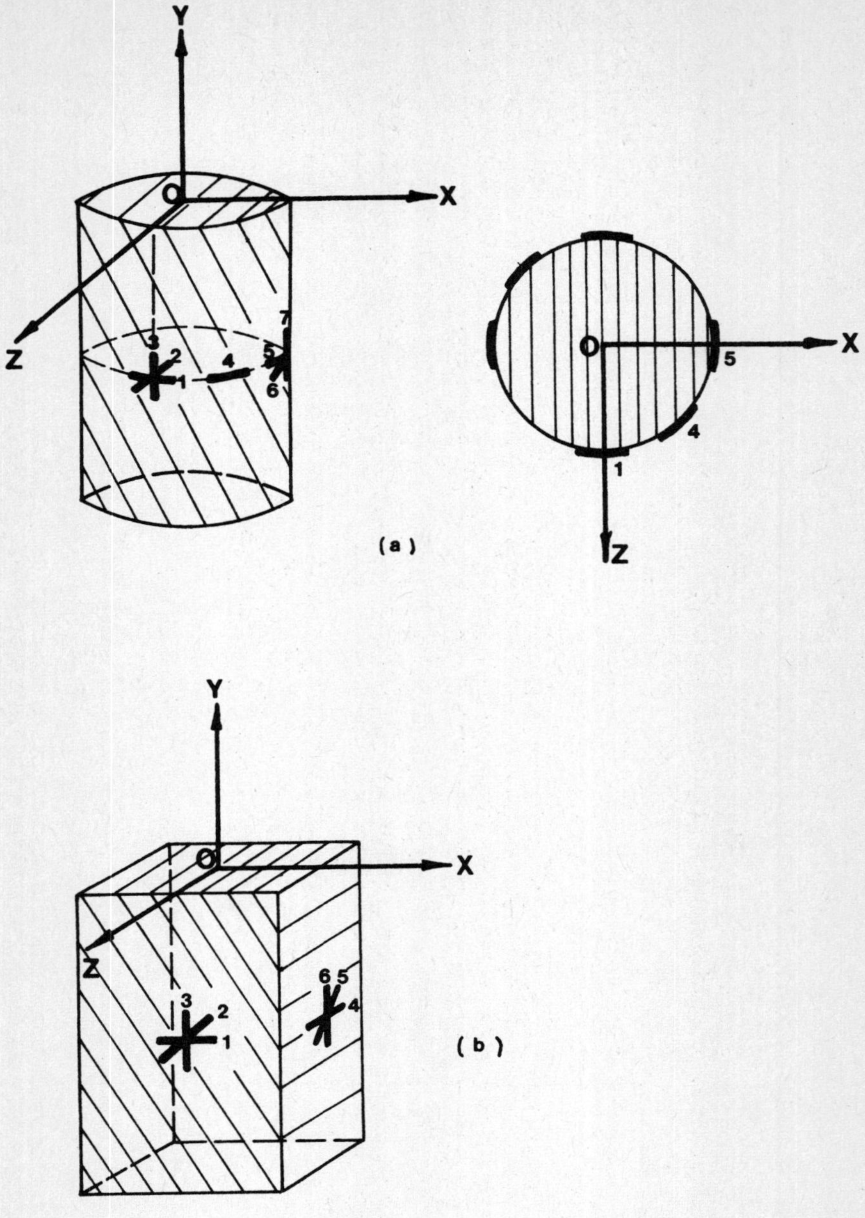

Figure 2.3. Example of strain gage distributions.

If we assume that stresses and strains are uniform within the rock specimen, at any point, the strain components are equal to (eqs 2.27)

$$\varepsilon_x = K_{12}\sigma_y \ ; \ \varepsilon_y = K_{22}\sigma_y \ ; \ \varepsilon_z = K_{23}\sigma_y$$
$$\gamma_{xy} = K_{26}\sigma_y \ ; \ \gamma_{yz} = \gamma_{xz} = 0 \qquad (2.29)$$

If the rock is <u>transversely isotropic</u> in the x',y',z' coordinate system, conditions (2.14)* can be substituted into eqs (2.27). Coefficients K_{12}, K_{22}, K_{23} and K_{26} take the following form

$$\frac{1}{E_y} = K_{22} = \frac{\cos^4\theta}{E'} + \frac{\sin^4\theta}{E} + \frac{\sin^2 2\theta}{4}\left(\frac{1}{G'} - \frac{2\nu'}{E'}\right)$$
$$-\frac{\nu_{yx}}{E_y} = K_{12} = \frac{-\nu'\sin^4\theta}{E'} - \frac{\nu'\cos^4\theta}{E'} + \frac{\sin^2 2\theta}{4}\left(\frac{1}{E} + \frac{1}{E'} - \frac{1}{G'}\right)$$
$$-\frac{\nu_{yz}}{E_y} = K_{23} = -\frac{\nu'}{E'}\cos^2\theta - \frac{\nu}{E}\sin^2\theta$$
$$\frac{\eta_{y,xy}}{G_{xy}} = K_{26} = \sin 2\theta\cos^2\theta\left(\frac{1}{E'} + \frac{\nu'}{E'}\right) - \sin 2\theta\sin^2\theta\left(\frac{1}{E} + \frac{\nu'}{E'}\right) - \frac{\sin 2\theta\cos 2\theta}{2G'} \qquad (2.30)$$

Coefficients E_y, ν_{yx} and ν_{yz} appear as <u>apparent</u> Young's modulus and Poisson's ratios in the global coordinate system X,Y,Z. Figure 2.4 shows an example of variation of those three quantities with the angle θ for a schist tested by Pinto (1970)**. Pinto's experimental results were fitted by a line

---

* with x,y,z replaced respectively by y',z',x'

** The notation used in his paper has been modified to be consistent with the one used in this analysis.

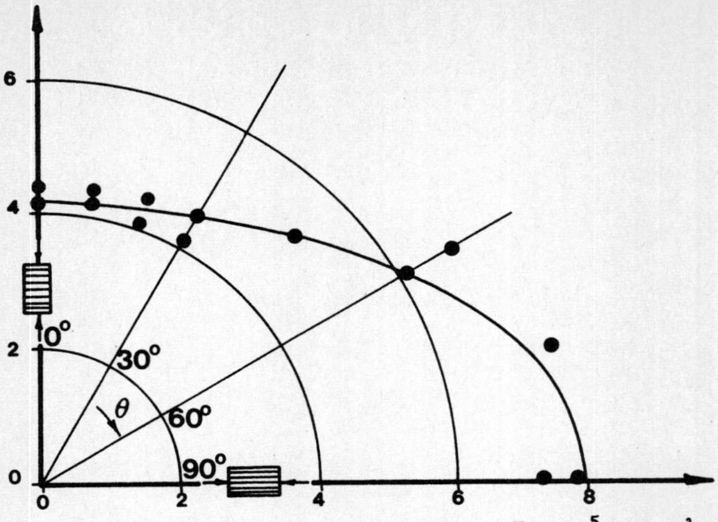

(a) Apparent Young's modulus E_y (10^5 kg/cm^2)

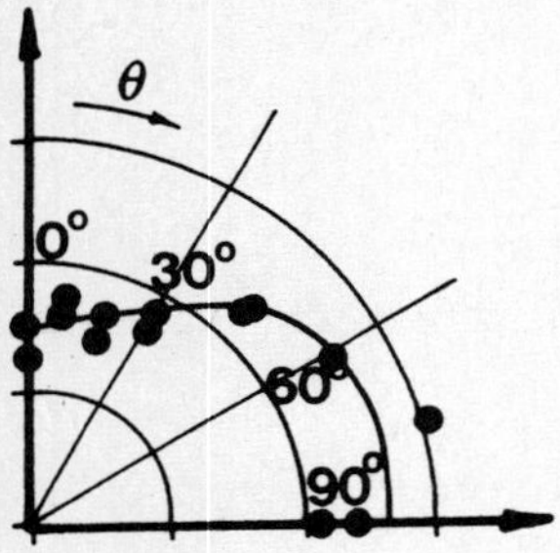

(b) Apparent Poisson's ratio ν_{yx}

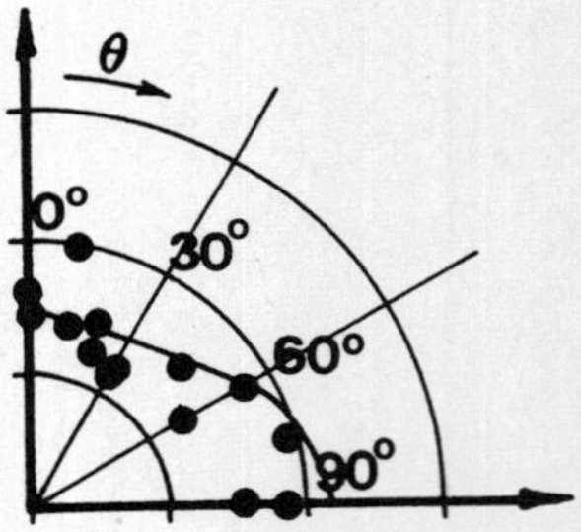

(c) Apparent Poisson's ratio ν_{yz}

Figure 2.4. Variation of the apparent Young's modulus E_y and apparent Poisson's ratios ν_{yx} and ν_{yz} with the angle θ . Comparison between experimental data and theoretical prediction (Pinto,1970).

shown in the figure that accords well with the theoretical prediction of eqs (2.30).

Eqs (2.30) can be used to determine the five elastic constants of the transversely isotropic rock by running three uniaxial compressive tests as shown in Figure 2.5:

- When $\theta = 0$, according to eqs (2.29) and (2.30) we have

$$\varepsilon_x = \varepsilon_z = -\frac{\nu'}{E'}\sigma_y \ ; \quad \varepsilon_y = \frac{\sigma_y}{E'} \quad ; \quad \gamma_{xy} = 0 \tag{2.31a}$$

  Measurements of $\varepsilon_x, \varepsilon_y$ and ε_z determine the values of E' and ν'.

- Similarly when $\theta = \pi/2$

$$\varepsilon_x = -\frac{\nu'}{E'}\sigma_y \ ; \quad \varepsilon_z = -\frac{\nu}{E}\sigma_y \ ; \quad \varepsilon_y = \frac{\sigma_y}{E} \quad ; \quad \gamma_{xy} = 0 \tag{2.31b}$$

  Measurements of ε_z and ε_y determine the values of E and ν.

- A third value of the angle θ is required in order to calculate the value of G'. For instance, when $\theta = \pi/4$ we have

$$\varepsilon_x = \sigma_y\left(\frac{1}{4E} + \frac{1}{4E'} - \frac{\nu'}{2E'} - \frac{1}{4G'}\right)$$

$$\varepsilon_y = \sigma_y\left(\frac{1}{4E} + \frac{1}{4E'} - \frac{\nu'}{2E'} + \frac{1}{4G'}\right) = \frac{\sigma_y}{E_{(45)}}$$

$$\varepsilon_z = \sigma_y\left(-\frac{\nu'}{2E'} - \frac{\nu}{2E}\right)$$

$$\gamma_{xy} = \frac{\sigma_y}{2}\left(\frac{1}{E'} - \frac{1}{E}\right)$$

(2.31c)

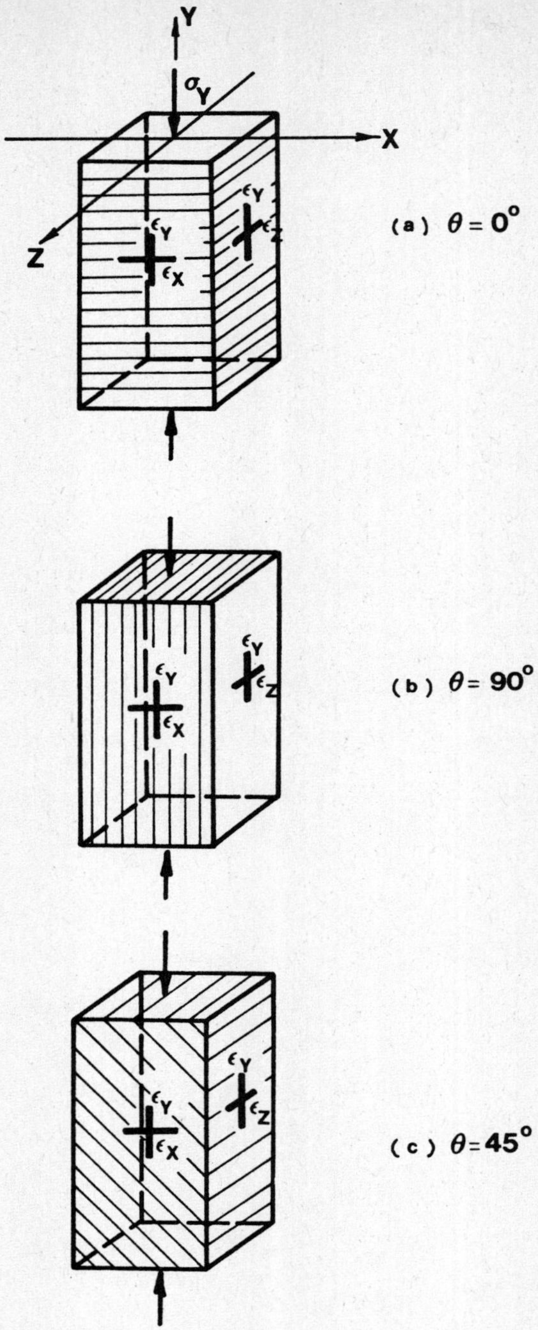

Figure 2.5. Uniaxial compressive tests for three different inclinations of rock anisotropy.

If we define $$\nu_{(45)} = -\frac{\varepsilon_x}{\varepsilon_y} \tag{2.31d}$$

and make use of the two first of eqs (2.31c) we obtain the following relation $$G' = \frac{E_{(45)}}{2(1+\nu_{(45)})} \tag{2.32}$$

If the rock is <u>orthotropic</u> in the x',y',z' coordinate system eqs (2.29) still apply. Coefficients K_{12}, K_{22}, K_{23} and K_{26} are given in eqs (2.27). As for the transversely isotropic rock, an apparent Youngs modulus and two apparent Poisson's ratios can be defined. For the geometry of Figure 2.2, it appears that those three quantities and K_{26} depend upon seven of the nine elastic constants of the rock. They do not depend on $G_{x'z'}$ and $G_{y'z'}$. Therefore, a procedure similar to the one described in Figure 2.5 will be unable to determine those two moduli. However, they can be determined by testing two other rock samples such that the orientation of their planes of elastic symmetry with respect to the global coordinate system has been rotated in comparison to the one in Figure 2.2. This is shown in Figures 2.6.

For the geometry of Figure 2.6a, eqs (2.26) and (2.27) still apply with the following substitution: $1/E_{x'}$, $1/E_{y'}$, $1/E_{z'}$, $\nu_{y'x'}/E_{y'}$, $\nu_{z'x'}/E_{z'}$ $\nu_{z'y'}/E_{z'}$, $1/G_{y'z'}$, $1/G_{x'z'}$, $1/G_{x'y'}$ must be replaced respectively by $1/E_{z'}$, $1/E_{y'}$, $1/E_{x'}$, $\nu_{y'z'}/E_{y'}$, $\nu_{x'z'}/E_{x'}$, $\nu_{x'y'}/E_{x'}$, $1/G_{x'y'}$, $1/G_{x'z'}$, $1/G_{y'z'}$. Using the procedure described in Figure 2.5, the modulus $G_{y'z'}$ can be calculated.

For the geometry of Figure 2.6b, the following substitution must be carried out; $1/E_{x'}$, $1/E_{y'}$, $1/E_{z'}$, $\nu_{y'x'}/E_{y'}$, $\nu_{z'x'}/E_{z'}$, $\nu_{z'y'}/E_{z'}$,

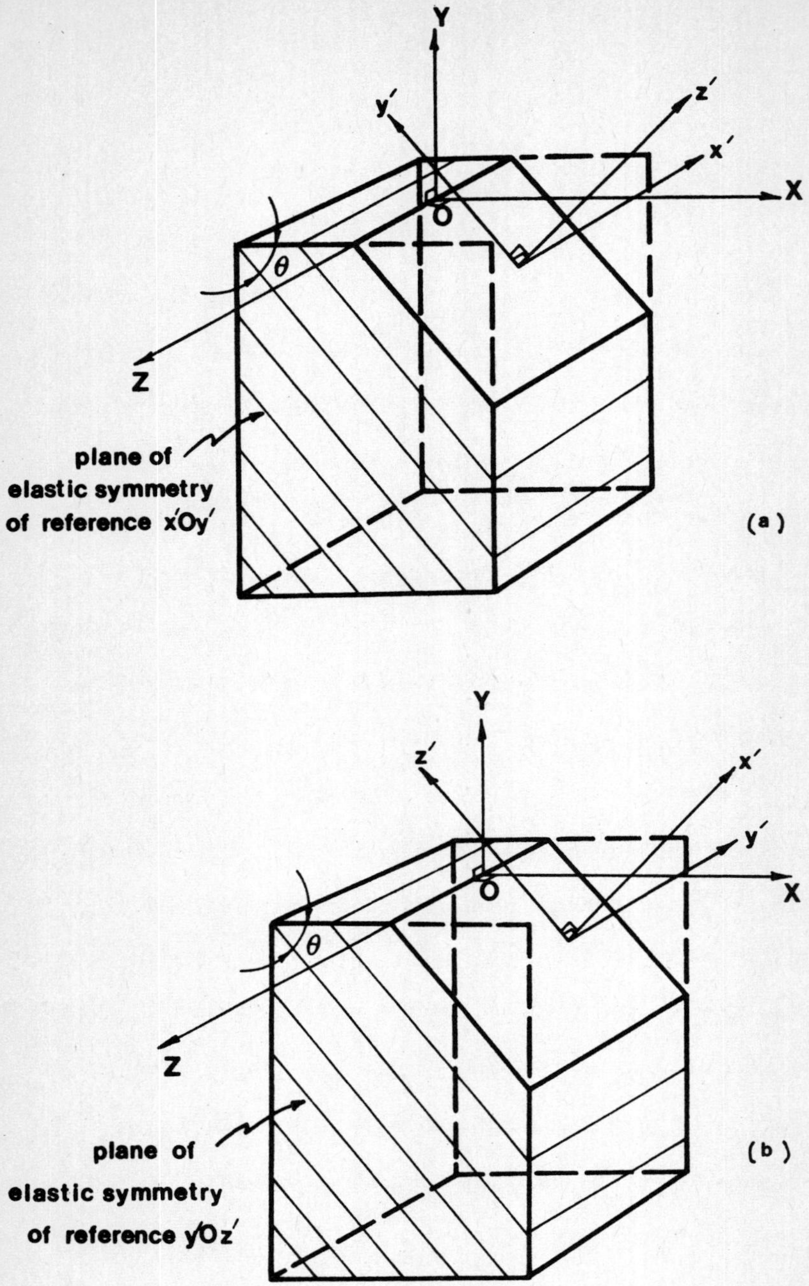

Figure 2.6. Sample orientations for the determination of the elastic constants of an orthotropic rock.

$1/G_{y'z'}, 1/G_{x'z'}, 1/G_{x'y'}$ must be replaced respectively by $1/E_{x'}$, $1/E_{z'}$, $1/E_{y'}$, $\nu_{z'x'}/E_{z'}$, $\nu_{y'x'}/E_{y'}$, $\nu_{z'y'}/E_{z'}$, $1/G_{y'z'}$, $1/G_{x'y'}$, $1/G_{x'z'}$. The modulus $G_{x'z'}$ can be calculated also.

In <u>summary</u>, the deformability of anisotropic rocks described as transversely isotropic or orthogropic can be evaluated by measuring respectively the five and nine elastic constants using a simple uniaxial compression test. The procedure is relatively easy for transversely isotropic materials but more complicated for orthotropic materials. Triaxial $(\sigma_1, \sigma_2 = \sigma_3)$ and multiaxial $(\sigma_1, \sigma_2, \sigma_3)$ tests can also be used (Atkinson and Ko, 1973; Ko and Gerstle, 1972, 1976). This is particularly relevant to study the deformability of orthotropic materials. For such materials it would then be possible to simplify the procedure previously described by inclining the three planes of elastic symmetry with respect to the X,Y,Z directions and the directions of applied stresses $\sigma_1, \sigma_2, \sigma_3$. This will provide additional terms $(K_{4i}, K_{5i}, i = 1,2,3)$ in the (K) matrix in eq (2.25) and the vanishing character of γ_{xz} and γ_{yz} in eq (2.29) will disappear.

Although, the previous procedures are straightforward as long as the directions of symmetry are apparent, quite often the determination of the shear modulus G' or the shear moduli $G_{x'y'}$, $G_{x'z'}$ and $G_{y'z'}$ is avoided. Empirical assumptions are introduced and the following relations are used

$$\frac{1}{G'} = \frac{1}{E} + \frac{1}{E'} + \frac{2\nu'}{E'} \; ; \quad \frac{1}{G_{x'y'}} = \frac{1}{E_{x'}} + \frac{1}{E_{y'}} + \frac{2\nu_{x'y'}}{E_{x'}}$$

$$\frac{1}{G_{x'z'}} = \frac{1}{E_{x'}} + \frac{1}{E_{z'}} + \frac{2\nu_{x'z'}}{E_{x'}} \; ; \quad \frac{1}{G_{z'y'}} = \frac{1}{E_{y'}} + \frac{1}{E_{z'}} + \frac{2\nu_{y'z'}}{E_{y'}} \tag{2.33}$$

Those relations were introduced first by St.-Venant (1863). If they are satisfied the number of independent elastic constants for an orthotropic and a transversely isotropic materials is respectively reduced to 6 and 4. This is the reason why these relations have been used extensively in the literature. However, Martino and Ribacchi (1972) have shown that they are not acceptable for many rocks. This is not surprising since they are purely empirical and have been introduced only to simplify the equation of the surface that represents the variation of elastic moduli with changes of direction*.

It is important to keep in mind that in the determination of the elastic constants, the symmetric character of the rock and its orientation are assumed beforehand. This must be checked during testing by making strain measurements in addition to those required to calculate the elastic constants of the assumed class of symmetry. For example, the vanishing character of the shear strain components γ_{xz} and γ_{yz} in eqs (2.29) may be checked by using strain gages 1,4,5 and 6 in Figure 2.3a or strain gage 5 in Figure 2.3b.

Other methods have been suggested to determine the elastic constants or compliances of anisotropic rocks:

(i) Dynamic methods can be used to determine the elastic constants of orthotropic or transversely isotropic rocks (Duvall, 1965; Liakhovitski and Nevski, 1970; Simonson et al, 1976),

---

* See also Lekhnitskii (1963) pp.51-57.

(ii) Pinto (1979) suggested the use of diametral compression tests. Those tests were carried out on circular cross section specimens, the cross sections being parallel to a plane of elastic symmetry of the rock. In his analysis he assumed that the stresses at the center of the cross sections are equal to those for an isotropic material. This does not seem correct in comparison to the analysis presented by Okubo (1952),

(iii) Becker and Hooker (1967) proposed a procedure by loading hollow cylinders of anisotropic rocks axially and radially. However, this procedure requires several approximations and makes use of eqs (2.33) which effectively reduces the number of elastic constants that must be known.

In order to assess the deformability character of rocks whose anisotropy is induced by regular joint sets, physical models have been used by several authors (Nelson and Hirschfeld, 1968; Einstein and Hirschfeld, 1973; Reik and Zacas, 1978). Triaxial ($\sigma_1, \sigma_2 = \sigma_3$) and multiaxial ($\sigma_1, \sigma_2, \sigma_3$) tests have been carried out on samples built up from blocks of model material. Those blocks are then arranged to reproduce the influence of one, two or three joint sets on the deformability of a fractured rock mass. An example is presented in Figure 2.7. Experiments have shown that the deformability character of the jointed model samples depends on the joint set configuration. In particular, the deformability is strongly influenced by the orientation of the joint sets with respect to the applied loads (Figure 2.8) and the values of their spacing. These

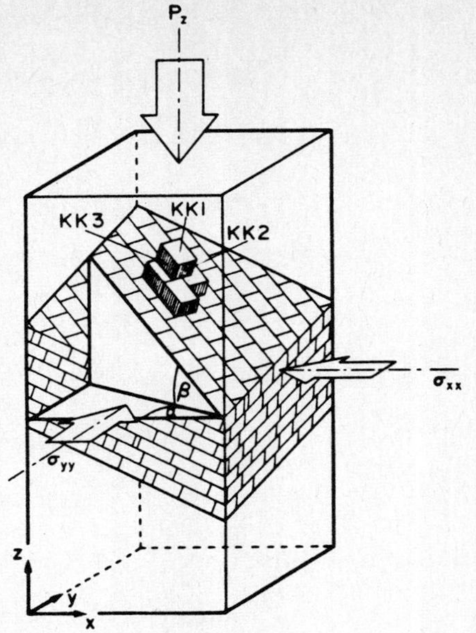

Figure 2.7. Schematic model of jointed rock mass. Three orthogonal joint sets KK1,KK2,KK3 with orientation defined by angles α and β . Multiaxial stress field σ_{xx} , σ_{yy} , P_z (Reik and Zacas,1978).

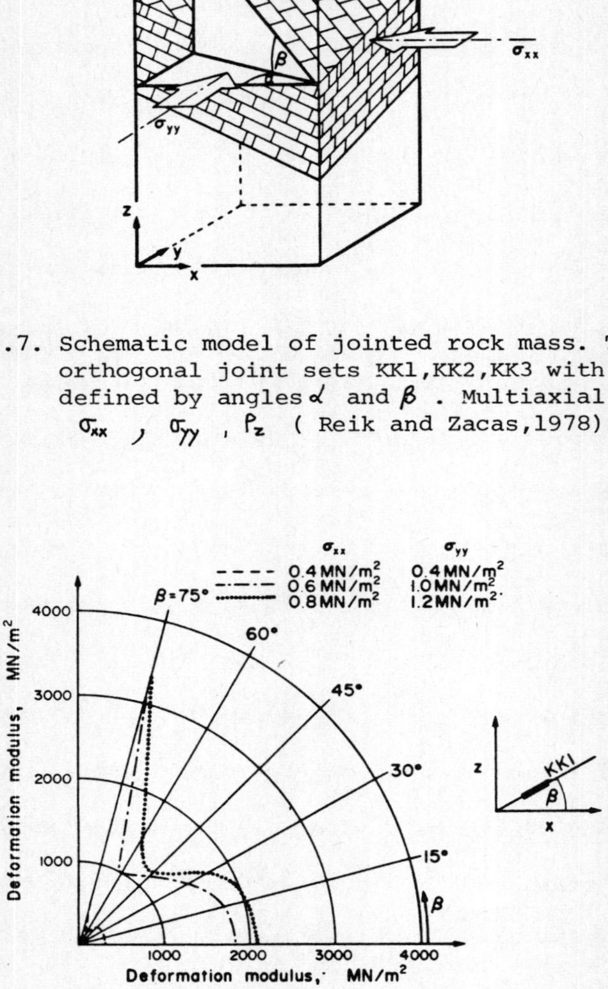

Figure 2.8. Variation of the modulus of deformation (tangent modulus at 50% of max. substainable axial stress) with the angle β ($\alpha = 0°$) (Reik and Zacas,1978).

conclusions have been also suggested by the analytical model proposed by Amadei and Goodman (1981a) (Appendix 2.1).

2.3.2 In Situ Testing

Scale effect is a major problem in rock mechanics (Heuze, 1980). Laboratory tests on small samples are inadequate to measure the deformability of rock masses because of a lack of scaling laws. Therefore, in situ testing is required. In the past 20 years, several techniques have been proposed. They can be separated into two groups: static and dynamic. In the static methods loads are applied at selected surfaces of rock masses and the resulting deformationsare measured. They include plate loading tests, flat jack tests, borehole jack tests and dilatometer tests. In the dynamic methods, the velocity or the frequency of vibrational disturbances is measured. In general, dynamic methods involve larger volumes of rock than the static ones. A number of references review the techniques available and compare the results obtained with several of them (Goodman et al, 1972; Bieniawski, 1978; Heuze et al 1980; Goodman, 1980).

The deformability of a rock mass is usually defined by a so called '<u>modulus of deformation</u>' or '<u>field modulus</u>'. For plate loading tests this is expressed by the slope at an arbitrary point of the applied load settlement curve. For the borehole jack, the modulus of deformation is calculated from an equation derived for an isotropic linear elastic continuum (Goodman et al, 1972). Therefore, it seems incorrect to use this modulus to describe the deformability of anisotropic heterogeneous and discontinuous materials. However, it can be regarded as a way to assess the non isotropic homogeneous continuous character of a rock mass.

If the rock mass is crossed by three orthogonal joint sets, it can be described by an equivalent anisotropic continuum as long as conditions (i) and (ii) of section 2.2.3 and their associated remarks are satisfied. If the joint sets have equal spacing and properties, the modulus of deformation of the rock mass is given by the value of E_n in eq (14) of Appendix 2.1. The same equation can be used to introduce a "reduced modulus of deformation" if the rock mass is highly fractured. This modulus can be related to the degree of rock fracturing (Raphael and Goodman, 1979) or to the RQD* (Heuze, 1971; Kulhawy, 1978; Heuze et al, 1980).

A few cases have been reported in the literature for the in situ determination of the elastic constants of anisotropic rock masses that could be described as orthotropic or transversely isotropic (Oberti et al, 1979; Hata et al, 1979). Additional assumptions are often stated in order to reduce the number of elastic constants that must be known.

Kawamoto (1966) suggested using a pressurized borehole to calculate the elastic constants of an orthotropic rock mass. The hole is drilled perpendicular to one of the planes of elastic symmetry and is pressurized over a length several times

---

* Rock Quality Designation

larger than its diameter. Using eqs (2.33) Kawamoto showed that the number of unknowns is reduced to three; the magnitude of the two principal elastic moduli in the plane perpendicular to the hole axis and their orientation with respect to the coordinate system attached to the hole. Three changes of diameter must be recorded, while pressurizing, in order to determine the three unknowns. Field tests, using this method, have shown that the estimated directions of the principal axes of elastic symmetry coincided well with the direction of stratification.

Consider a hole with radius a and x,y,z a coordinate system attached to it (Figure 2.9). The hole is drilled in an orthctropic rock with three planes of elastic symmetry, each one being perpendicular to one of the coordinate axes. The constitutive relation of the rock material is given by eqs (2.5) and (2.13). Let us consider two cases:

(i) The hole is pressurized over a length several times larger than its diameter,

(ii) The hole is pressurized over a short length.

For each case, it is shown in Appendix 4.8 how to calculate the value of the radial and tangential displacements at any point along the wall of the hole, located at an angle θ from the x axis, induced by a uniform internal pressure q. Assuming that the change in hole diameter Δd is due to the radial component of displacement only we obtain for case (i)

$$\frac{1}{q}\left(\frac{\Delta d}{2a}\right) = \beta_{11}\left(\beta_1+\beta_2-\beta_1\beta_2\right)\cos^2\theta + \frac{\beta_{22}}{\beta_1\beta_2}\left(\beta_1+\beta_2-1\right)\sin^2\theta - \beta_{12} \tag{2.34}$$

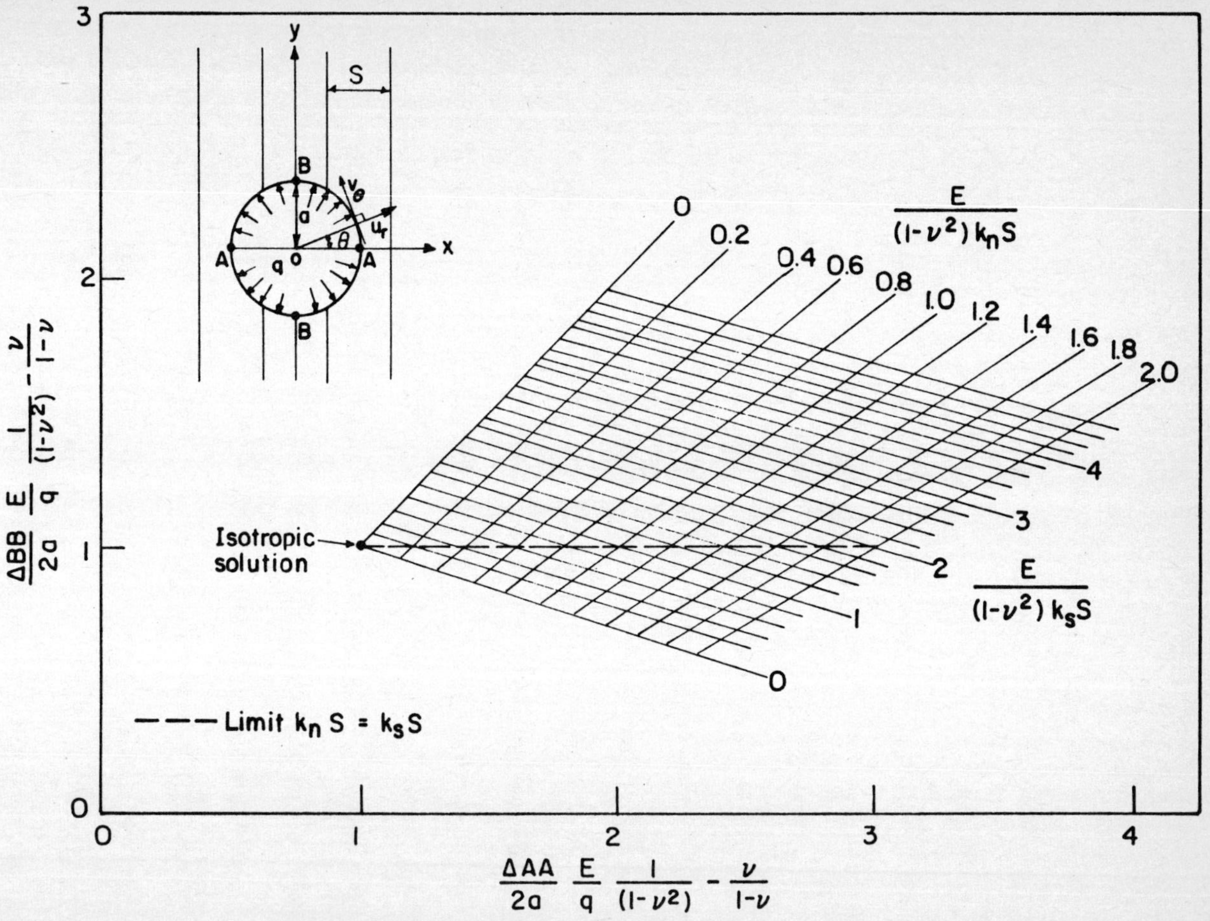

Figure 2.9. Determination of normal and shear stiffnesses of a joint set striking parallel to a borehole.

where $$\beta_{ij} = a_{ij} - \frac{a_{i3}a_{j3}}{a_{33}} \qquad (i,j = 1,2,4,5,6) \tag{2.35}$$

and a_{ij} are defined in eqs (2.5) and (2.13). β_1 and β_2 are such that

$$\beta_1\beta_2 = \sqrt{\frac{\beta_{22}}{\beta_{11}}} \quad ; \quad \beta_1^2 + \beta_2^2 = \frac{2\beta_{12} + \beta_{66}}{\beta_{11}} \tag{2.36}$$

For case (ii) eqs (2.34) and (2.36) still apply but all the coefficients β_{ij} must be replaced by a_{ij}.

Since $\cos^2\theta$ and $\sin^2\theta$ are functions of $\cos 2\theta$ only, eq (2.34) indicates that along the contour of the hole there are at most two independent measurements of change in diameter. Therefore, the nine elastic components of the orthotropic rock can not be determined by diametral measurements in one borehole only. The same conclusions apply if the rock is transversely isotropic or even isotropic. Additional assumptions must be made as far as the value of certain elastic constants is concerned. However, the procedure is simplified for the following example*. The hole is drilled in an isotropic rock cut by one joint set striking parallel to the hole axis and perpendicular to the x axis (Figure 2.9). The joint set has a spacing S and constant normal and shear stiffnesses k_n, k_s. The isotropic rock is defined by a Young's modulus E and a Poisson's ratio ν. We also assume that the jointed material can be

---

* This example has been proposed by Amadei and Goodman (1981b) (Appendix 4.6). The derivations are presented herein.

described as an equivalent anisotropic continuum as defined in Appendix 2.1. Let Δd be equal to ΔAA and ΔBB when θ is respectively equal to 0 and 90 degrees. Using eq (2.34) we can define two quantities X_A and X_B such that

$$X_A = \frac{\Delta AA}{2a}\frac{E}{q}\frac{1}{(1-\nu^2)} - \frac{\nu}{(1-\nu)} = (1+\alpha_n)(\beta_1+\beta_2-\beta_1\beta_2)$$

$$X_B = \frac{\Delta BB}{2a}\frac{E}{q}\frac{1}{(1-\nu^2)} - \frac{\nu}{(1-\nu)} = \sqrt{1+\alpha_n}\,(\beta_1+\beta_2-1) \qquad (2.37)$$

where

$$\alpha_n = \frac{E}{(1-\nu^2)k_n S} \quad ; \quad \alpha_s = \frac{E}{(1-\nu^2)k_s S} \qquad (2.38)$$

and

$$\beta_1\beta_2 = \frac{1}{\sqrt{1+\alpha_n}} \quad ; \quad \beta_1^2+\beta_2^2 = \frac{2+\alpha_s}{1+\alpha_n} \qquad (2.39)$$

X_A and X_B have been plotted in Figure 2.9 for different values of α_n and α_s. When $k_n S$ and $k_s S$ approach infinity, the solution converges to that for the isotropic case and $X_A = X_B = 1$.

Theoretically, by measuring the two changes of diameter ΔAA and ΔBB, it would be possible to calculate the products $k_n S$ and $k_s S$ and the stiffnesses k_n and k_s if the spacing S is known. However, this procedure implies a knowledge of the two intact rock properties E and ν. If we assume that the unloading of a fractured rock mass follows essentially the same path as that of the intact rock (Goodman, 1976, 1980), it would be possible to back-calculate the Young modulus of the intact rock by pressurizing and unpressurizing the hole and by measuring the recoverable part of the radial displacement i.e.,

$$\Delta AA = \Delta BB = 2qa\frac{(1+\nu)}{E} \qquad (2.40)$$

Knowing q, a, ΔAA, ΔBB and assuming a value for ν, it would then be possible to calculate E.

If the hole is pressurized over a short length, a similar procedure can be carried out. Figure 2.9 is still applicable if the quantities $E/(1-\nu^2)$ and $\nu/(1-\nu)$ appearing in the dimensionless quantities are modified to E and ν respectively. Eqs (2.34) can also be used if the joint set is known to strike parallel to the hole axis but its orientation with respect to the x axis is unknown. Three diametral measurements are then required at angles $\theta = \varphi$, $\varphi + \alpha$ and $\varphi + \beta$. φ is determined from those three measurements. A similar procedure can be derived when the intact rock is cut by three orthogonal joint sets with same spacing and properties.

From a practical point of view, changes in diameter can be recorded using devices such as the "tube deformeter" developed by Takano and Shidomoto (1966) or the dilatometer of Rocha et al (1966).

Chapter 3: Strength of Anisotropic Rocks

3.1 Introduction

We have discussed in the previous chapter, how deformability varies with direction. Laboratory and in situ testing have shown that the strength of anisotropic rocks also varies with spatial orientation. Intuitively, both deformability and strength must be affected by the anisotropic character especially if there is symmetry of any kind. The presentation of this chapter follows along the lines of the previous one in focussing on two points:

(i) the strength behavior of anisotropic rocks in the laboratory and in situ,

(ii) the modeling of this behavior for the different types of rock anisotropy mentioned in Chapter 1.

The present chapter is concerned with the study of the peak strength of anisotropic rocks (compressive, tensile, shear) and does not deal with residual strength.

3.2 Experimental Observations

In order to assess the directional character of the strength of anisotropic rocks, rock testing is required. As in the case of deformability tests can be made both in the laboratory and in situ.

3.2.1 Laboratory Testing

Techniques proposed to determine the compressive and tensile strengths of anisotropic rocks include:

(i) uniaxial, triaxial ($\sigma_1, \sigma_2 = \sigma_3$) and multiaxial ($\sigma_1, \sigma_2, \sigma_3$) compression tests,

(ii) direct tensile tests in which samples are subjected to a uniform tensile pull and indirect tensile tests such as the diametral compression of circular discs.

Combination of these test results enables us to determine the shear strength of anisotropic rocks.

a) Compression Tests

For class A anisotropic rocks*, similar conclusions apply for the compressive strength as for the spatial variation of deformability. This holds with regard to the type of variation and its orientation with respect to rock features or properties (Peres - Rodrigues, 1966; Peres - Rodrigues and Aires - Barros, 1970; Douglass and Voight, 1969). For instance, an ellipsoid type quadratic law has been proposed by Peres-Rodrigues to model the spatial variation of the unconfined compressive strength within granitic rock masses.

For certain class B anisotropic rocks belonging to the schist and micaschist type, Peres - Rodrigues (1970, 1979) suggested using a quartic type law to model the spatial variation of the unconfined compressive strength, one axis of the quartic being perpendicular to the schistosity. However, for most class B anisotropic rocks that can be regarded as continuous at the sample scale and show one or several apparent directions of planar or rectilinear anisotropy, the procedure is similar to the one used for the deformability. Samples (cylindrical or prismatic) are cut and tested at different angles to the directions of planar

* Class A and Class B anisotropic rocks are defined in Chapter 1.

anisotropy. For sake of clarity, the present discussion will be limited first to the case of rocks possessing only one apparent direction of anisotropy. Examples of uniaxial and triaxial compression tests on such rocks have been reported by several authors (Donath, 1961, 1964, 1972a,b; Hoek, 1964; Deklotz et al,1966; Mc Lamore and Gray, 1967; Fayed, 1968; Akai et al, 1970; Attewell and Sandford, 1974; Simonson et al, 1976; Allirot and Boehler, 1979). The following general conclusions can be drawn*:

(i) the strength behavior of anisotropic rocks is a function of both the magnitude of the applied stress field and its orientation with respect to the rock anisotropy. An example is shown in Figure 3.1 for Martinsburg slate. The curve of differential stress at failure versus inclination of the slaty cleavage is plotted for different values of the confining pressure,

(ii) in all cases, the minimum value of strength occurs when the planar anisotropy is inclined at an angle varying between 30 and 40 degrees to the direction of the major applied stress σ_1. The maximum value occurs either parallel or perpendicular to the planaranisotropy depending upon the rock type,

(iii) raising the confining pressure may enhance or diminish the anisotropic strength behavior,

(iv) in general failure envelopes are curved and vary with the direction of anisotropy. Often, the plots differential stress at failure versus confining pressure show a moderate curvature. They may be interpolated by straight lines, although with a high

* The experimental results by Donath (1964) on Martinsburg slate are used to illustrate those conclusions.

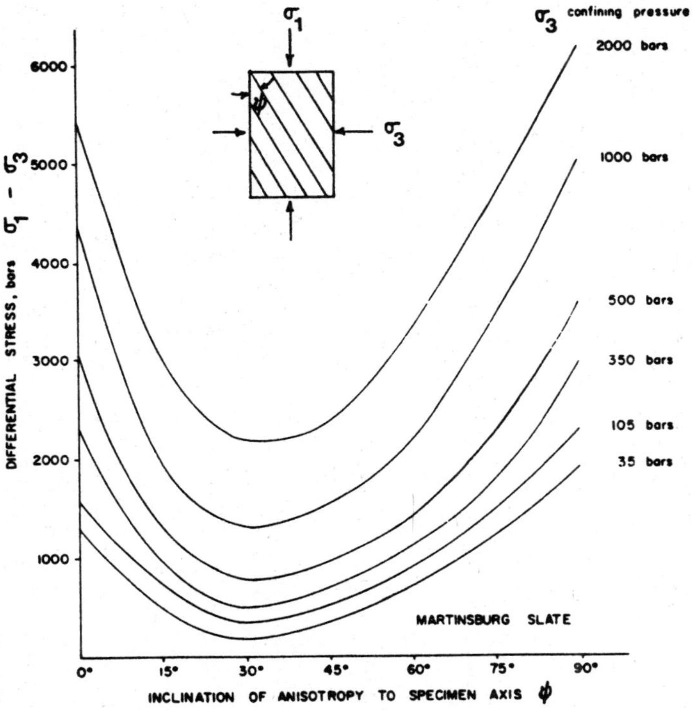

Figure 3.1. Differential stress at failure vs. inclination of cleavages ψ . Martinsburg slate (Donath,1964).

degree of arbitrariness. This occurs for confining pressures less than 70 or 100 MPa. An example is shown in Figure 3.2,

(v) different modes of failure can be observed: brittle failure along or across the planar anisotropy, ductile failure along the planar anisotropy and kinking. The exact nature of the failure and its orientation are dependent upon the confining pressure and the orientation of the anisotropy to the applied stresses. An example is shown in Figure 3.3. In general, failure takes place along the planar anisotropy when the latter is inclined at an angle to the major applied stress varying between 15 and 45 degrees or within the rock for other orientation angles.

A few multiaxial compression tests ($\sigma_1, \sigma_2, \sigma_3$) have been conducted on continuous anisotropic rocks (Jaeger, 1964; Akai et al, 1970). Those tests have emphasized the influence of the intermediate applied stress σ_2 on the strength and the mode of failure. However, they are not numerous enough to draw any major conclusion.

The previous discussion concerned the strength behavior of rocks that present one apparent direction of anisotropy. Very often anisotropic rocks present several directions of planar anisotropy such as bedding, schistosity, cleavage as well as aligment of minerals, microfold axes, and other linear features. Uniaxial and triaxial compression tests on such rocks (Masure, 1970; Dayre, 1970; Pomeroy et al, 1971) have shown strength behavior more complex than previously mentioned particularly if the different directions of anisotropy are coequal in development.

As in the case of deformability, physical models can be used to study the spatial variation of strength and the mode of failure

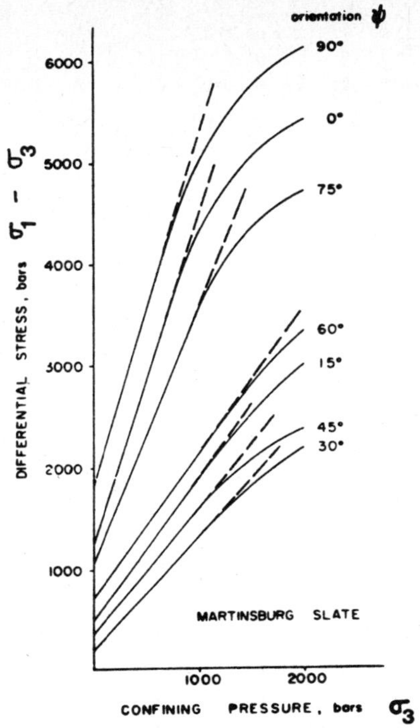

Figure 3.2. Differential stress at failure vs. confining pressure. Martinsburg slate (Donath, 1964).

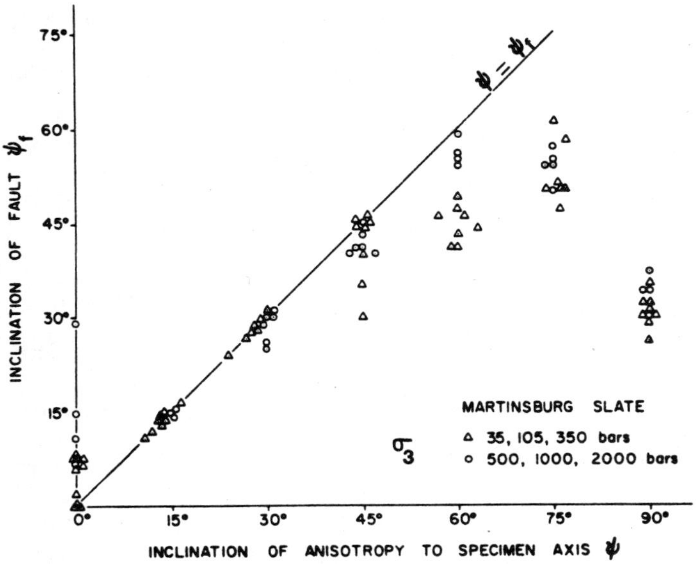

Figure 3.3. Effect of cleavage on the angle of the failure plane.Martinsburg slate (Donath, 1964).

of rocks whose anisotropy is induced by regular joint sets. Uniaxial, biaxial, triaxial and multiaxial compression tests have been carried out on samples built up from blocks of model material (Ladanyi and Archambault, 1972; Einstein and Hirschfeld, 1973*; Reik and Zacas, 1978). The following general conclusions can be drawn:

• the strength behavior of a regularly jointed rock is a function of the magnitude of the applied stress field, its orientation to the joint sets and their properties (spacing, degree of interlocking, surface roughness) . If the rock is cut by one joint set only and is under the influence of a triaxial state of stress ($\sigma_1, \sigma_2 = \sigma_3$) the plot of differential stress at failure <u>versus</u> inclination of the joint set with respect to the major applied stress σ_1 is similar to Figure 3.1. However, this plot often presents flat "shoulders" for orientation of the joint sets near 0 and 90 degrees to σ_1 . For those angles, failure takes place through the intact rock. For other orientation angles, sliding occurs along the joint set,

• the strength of a regularly jointed rock is located between the strength of the intact rock and the strength of the joint sets. This difference is largely reduced as the magnitude of the applied stress field increases,

• the failure pattern of a jointed rock covers a wide range including shear failure through the intact rock, composite shear failure partly through the intact rock and partly along the joint sets, sliding and multiple sliding along pre-existing joint sets, shear zones and kink bands,

---

*See references herein.

• the influence of the intermediate principal stress σ_2 is large when the joint sets are inclined with respect to it, leading in that case to an increase in rock strength. The failure planes may not pass through the direction of σ_2 but are influenced mainly by the joint set orientation.

b) Tensile Tests

Experimental evidence on the behavior of anisotropic rocks when subjected to tensile stresses is rather limited. All the experiments have been conducted on continuous anisotropic rocks presenting a well defined direction of planar anisotropy. Several techniques have been employed to determine the tensile strength of anisotropic rock specimens but none of them have been found to be totally satisfactory. These techniques include the direct uniaxial tensile test (Deklotz et al, 1966; Barla and Goffi, 1973) and the indirect or "Brazilian" test (Hobbs, 1964; Mc Lamore and Gray, 1967; Barron, 1971a).

Experiments have shown that the tensile strength behavior of anisotropic rocks is a function of orientation. The minimum tensile strength occurs when the planar anisotropy is perpendicular to the induced tensile stress within the tested specimens while the maximum strength occurs when the anisotropy is parallel to the induced tensile stress. An illustrative example is shown in Figure 3.4.

c) Qualitative Tests

Beside the quantitative tests previously mentioned, there are qualitative tests that can be used to obtain an initial impression of the anisotropic character of rock strength. They are referred as the "ball test" and the "needle test" by Paulman (1966).

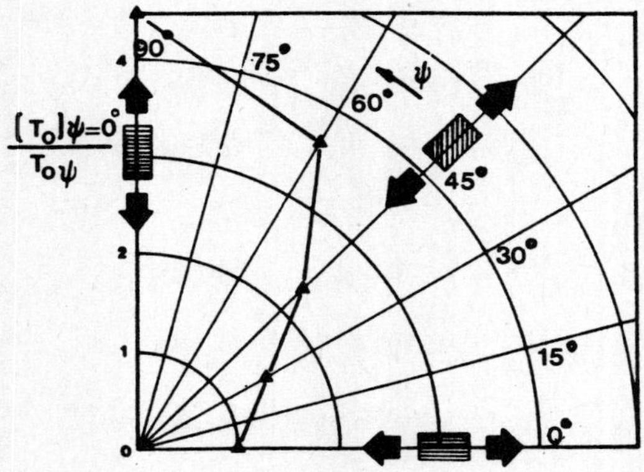

Figure 3.4. Variation of tensile strength $T_{o\psi}$ measured by direct tension ($(T_o)_{\psi=0°} = 7.3$ MPa) for a gneiss (Barla, 1974).

These tests punch or split discs of rock in different directions using a steel ball or a needle. The first technique was used by Dayre (1970) to assess the least strength character of a slate parallel to its lineation.

3.2.2 In Situ Testing

As for deformability, rock strength is scale dependent (Heuze, 1980). Experiments have shown that the laboratory measured strengths are higher than the field measured values and that there is no rational procedure for extrapolating one to another. The in situ techniques to measure rock mass strength are similar to the static techniques mentioned in section 2.3.2 for rock mass deformability. In addition shear tests have been conducted in the field by shearing large rock blocks cut free on their sides and in the natural condition along their base.

Few cases have been reported in the literature for the in situ determination and the spatial variation of the strength of anisotropic rock masses. The only one known by the author is the study reported by Serafim (1964) for the foundations of the Valdecanas dam. In situ shear tests were carried out on blocks of schist cut in different directions with respect to the schistosity.

In situ measurements of rock strength within boreholes has been suggested by Drozd et al (1970) and De la Cruz (1978) by using borehole jack tests. This technique has potential for measuring the strength of anisotropic rock masses, by testing inside boreholes drilled in different directions with respect to the anisotropy.

3.3 Analytical Models

Several analytical models have been proposed to account for the directional character of the strength of anisotropic rocks. They fall broadly into two groups depending upon the continuous, discontinuous character of the corresponding anisotropy.

3.3.1 Discontinuous Models

A theory that describes the shear strength of an isotropic rock cut by a continuous single joint or a joint set has been proposed by Jaeger (1960). A Mohr Coulomb failure criterion is associated to the rock and to the joint set. Let c_j and ϕ_j be the cohesion and friction angle along the planes of weakness. Similarly, let c and ϕ be two parameters that describe the shear strength of the intact rock. c is a shear strength intercept (often called "cohesion") and ϕ (often called "friction angle") is such that $\tan\phi$ describes the rate of increase of peak strength with normal stress.

If the joint set is inclined at an angle ψ to the direction of the major principal stress σ_1 of an applied stress field σ_1, σ_3 * (Figure 3.5), the Mohr Coulomb criterion for the joint set and the intact rock can be stated respectively as follows

$$K_{f_j} = \frac{\sigma_3 + H}{\sigma_1 + H} = \frac{\tan\psi}{\tan(\psi + \phi_j)} \qquad (3.1)$$

---

* The applied stress field can be biaxial (σ_1, σ_3), triaxial ($\sigma_1, \sigma_2 = \sigma_3$) or multiaxial ($\sigma_1, \sigma_2, \sigma_3$) since the Mohr Coulomb criterion does not involve the intermediate principal stress.

and

$$K_{f_r} = \frac{\sigma_3}{\sigma_1} = \frac{1}{\frac{q_u}{\sigma_3} + \tan^2\left(\frac{\pi}{4} + \frac{\phi}{2}\right)} \qquad (3.2)$$

where

$$H = c_j \cot \phi_j \qquad (3.3)$$

and

$$q_u = 2c \tan\left(\frac{\pi}{4} + \frac{\phi}{2}\right) \qquad (3.4)$$

q_u is the expression for the unconfined compressive strength of the intact rock according to the Mohr Coulomb theory.

Using a procedure similar to the one proposed by Bray (1967) and Goodman (1976), σ_1 is assumed to be compressive and the joint set is assumed to have no cohesion ($c_j = H = 0$). In this case, eq (3.1) shows that the principal stress ratio to mobilize the shear strength along the planes of weakness or equivalently to produce slip, varies markedly with the orientation of the joint set with respect to the applied stress field. The maximum value of K_{f_j} occurs when $|\psi| = 45° - \phi_j/2$. This is shown in Figure 3.5 where K_{f_j} has been plotted against ψ within the complete band $-90° < \psi < 90°$ and for $\phi_j = 40°$. The polar representation ($K = \frac{\sigma_3}{\sigma_1}$, ψ) proposed by Bray has been used. Any point located within or at the surface of the shaded area in Figure 3.5 is such that slip occurs along the planes of weakness. This takes place if σ_3 is compressive (K positive) or tensile (K negative).

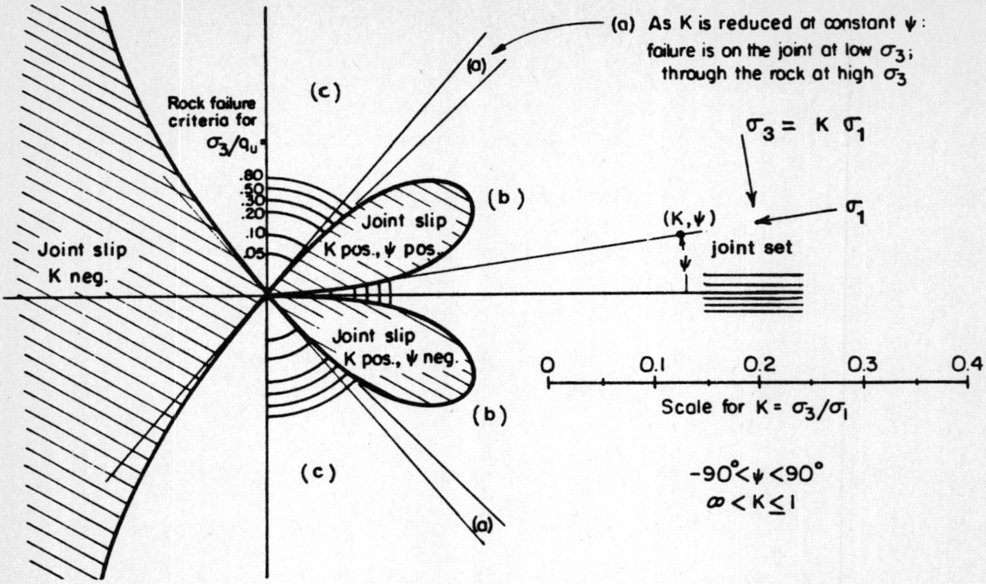

Figure 3.5. Bray's diagram for a joint set with $\phi_j = 40°$ in rock with $\phi = 50°$ (Goodman, 1976).

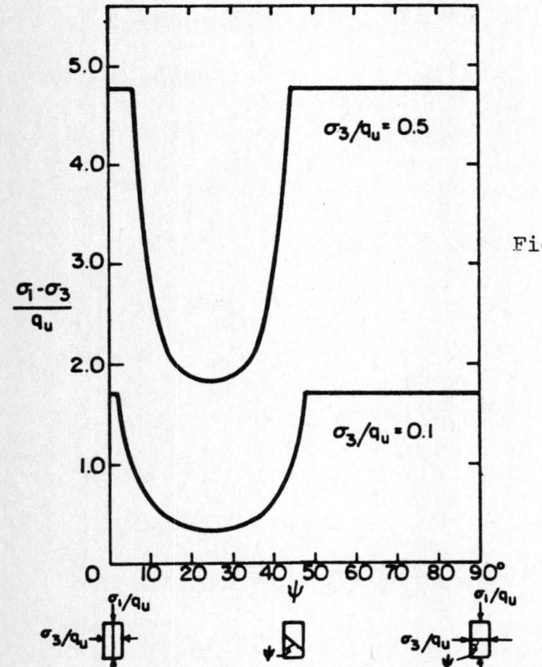

Figure 3.6. Anisotropy of a rock with one joint set: $\phi_j = 40°$, $\phi = 50°$. (Goodman, 1976).

Similarly, the shear strength of the intact rock is represented in Figure 3.5 by circles of radii equal to K_{f_r} (eq 3.2) for each value of the ratio σ_3/q_u and for $\phi = 50°$. It appears that three different modes of failure can take place:

• if the point (K, ψ) is within region (c), the solid rock fails independently of the joint set and the predicted failure plane is inclined at an angle $45° - \phi/2$ to the major principal stress,

• if the point (K, ψ) is within or at the surface of the shaded area in region (b), slip takes place along the planes of weakness,

• if the point (K, ψ) is located in region (a) failure takes place along the joint set at low values of K or within the solid rock at high values of K.

The previous analysis considers sliding only and neglects consideration of joint opening under the applied stress field (σ_1, σ_3). If the planes of weakness have no tensile strength, a reasonable assumption, a joint will open if the normal stress acting across it becomes negative (K negative). This additional constraint must be surimposed on the previous model.

Figure 3.6 is an alternative mode of representaion of the directional character of the strength of a regularly jointed rock that is often used in the literature.

If the intact rock is cut by several joint sets, the graphical representation of Figures 3.5 and 3.6 can be used for each joint set at a time and then superposed. This technique has been suggested by Bray (1967) and Kobayashi (1970). However, this superposition principle is correct as long as there is no interaction between the different joint sets and no stress redistribution. Another approach is to make use of failure criteria that give a "global picture" for the strength of the jointed rock. They may be of the empirical type (Hoek and Brown, 1980) or based on mechanistic or physical models (Ladanyi and Archambault, 1972, 1980).

3.3.2 Continuous Models

In comparison to the experimental observations, Jaegers's theory, described above and often called "plane of weakness" theory, models relatively well the strength behavior of a rock cut by one joint or a single joint set. For anisotropic rocks that can be regarded as continuous at the sample scale, this theory is not applicable since we might expect a continuous variation of their mechanical properties.

Several theories have been proposed to model the directional character of the strength of anisotropic continuous rocks. They fall into two groups: those that neglect the influence of the intermediate principal stress σ_2 and those that do not. These theories will be briefly discussed in the following.

For the theories of the first group, consider a rock that presents a direction of planar anisotropy inclined at an angle ψ to the direction of the major principal stress σ_1 of an applied stress field σ_1, σ_3 (Figure 3.7a). Jaeger (1960) assumed that the shear strength of an anisotropic rock can be described by a modified Mohr Coulomb criterion. On any plane inclined at an angle α to the direction of σ_1 (Figure 3.7b) the resistance to shearing is given by the following expression

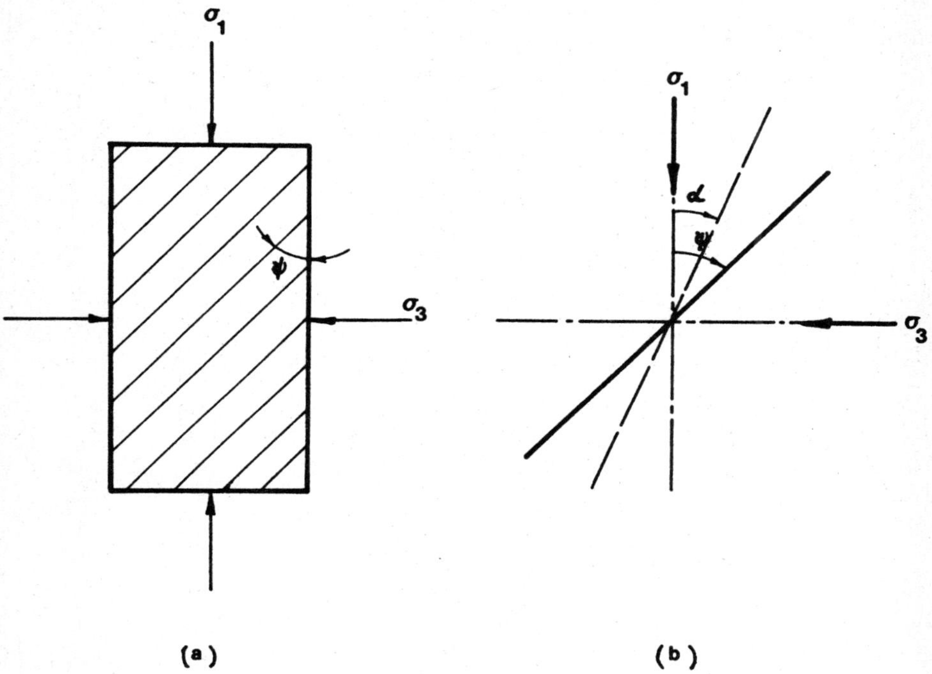

Figure 3.7. Jaeger's theory.

$$\tau = c_\alpha + \sigma_n \tan \phi \tag{3.5}$$

where σ_n and τ are respectively the normal and shear stresses acting along that plane. The shear strength intercept c_α (or "cohesion") is expressed as a function of orientation of the form

$$c_\alpha = S_1 - S_2 \cos 2 (\alpha - \psi) \tag{3.6}$$

According to eq (3.6) the "cohesion" varies continuously between $S_1 - S_2$ and $S_1 + S_2$. The same form of orientation dependence has previously been proposed by Casagrande and Carrillo (1944). These authors suggested also the following variation of $\tan \phi$ with orientation

$$\tan \phi_\alpha = T_1 - T_2 \cos 2 (\alpha - \psi) \tag{3.7}$$

S_1, S_2, T_1 and T_2 are constants that must be determined by curve fitting.

When applied to the prediction of the directional character of the strength of anisotropic rocks tested in <u>triaxial</u> and <u>uniaxial compression</u>, the following conclusions have been suggested by Jaeger regarding his theory:

(i) compared with the "plane of weakness" theory, the present theory develops a broader minimum in the variation of strength with orientation and an absence of "shoulders" close to the extreme orientations,

(ii) the predicted differential stress at failure $\sigma_1 - \sigma_3$ has a non linear dependence on confining pressure σ_3 .

(iii) the predicted angle of failure depends upon the angle ψ and the level of the applied stresses themselves. In place of two planes of failure as suggested by the classical Mohr Coulomb formulation, the theory predicts only one that lies between the direction of planar anisotropy and the nearest to it of the two Mohr Coulomb directions. Failure takes place along a plane inclined at $45°-\phi/2$ to the direction of σ_1 only if $\psi = 45°-\phi/2$.

Mc Lamore and Gray (1967) also analyzed the behavior of shales and slates in terms of variable "cohesion" and "friction angle". Although they presented their theory in the line of Jaeger's one, their approach is different. Samples of shale and slate were tested in triaxial compression for different orientations of the anisotropy with respect to the applied stress field. For each orientation angle, the plot of differential stress at failure versus confining pressure was found moderately curved. An example is shown in Figure 3.8a. Each plot was interpolated by a straight line then, a "cohesion" c_ψ and a "friction angle" ϕ_ψ were determined assuming a Mohr Coulomb strength criterion. Figure 3.8b shows an example of variation of those two quantities with the angle ψ for the plots in Figure 3.8a. Mc Lamore and Gray proposed that both c_ψ and $\tan \phi_\psi$ could be described as continuous functions of direction according to

$$c_\psi = S_1 - S_2 \left(\cos 2 \left(\psi_{min,c} - \psi\right)\right)^n$$

$$\tan \phi_\psi = T_1 - T_2 \left(\cos 2 \left(\psi_{min,\phi} - \psi\right)\right)^m \qquad (3.8)$$

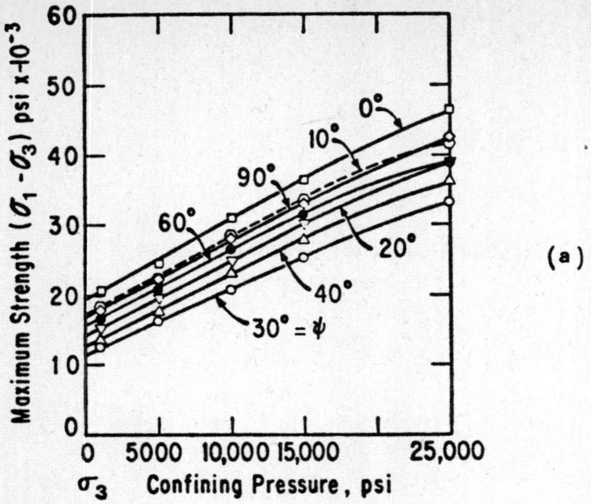

Figure 3.8. (a) Differential stress at failure vs. confining pressure. (Green River Shale-2; Mc Lamore and Gray, 1967).

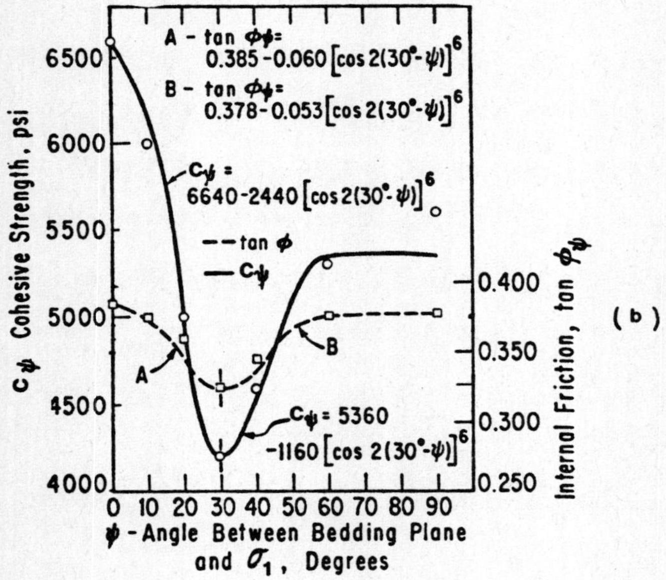

Figure 3.8. (b) Variation of c_ψ and $\tan\phi_\psi$ with respect to ψ (Green River Shale-2; Mc Lamore and Gray, 1967).

where S_1, S_2, T_1, T_2, n, m are constants that must be determined by curve fitting,

ψ is defined in Figure 3.7a,

$\psi_{min,c}$ and $\psi_{min,\phi}$ are the values of ψ corresponding to minima in c and $\tan\phi$ respectively. In the Figure 3.8b, $\psi_{min,c} = \psi_{min,\phi} = 30°$.

Very often one set of constants does not fit well the variation of c_ψ and $\tan\phi_\psi$ over the entire range of ψ from 0 to 90°. In that case Mc Lamore and Gray suggested using different sets of constants for different ranges of orientation angles as shown in Figure 3.8b. Finally, since a Mohr Coulomb envelope was assumed for each orientation angle, failure must take place at an angle $45° - \phi_\psi/2$ to the σ_1 direction, ϕ_ψ being defined in eq (3.8).

In the theories discussed previously, the rock material was assumed to fail in shear under a compressive applied stress field. Another group of theories assumes that the rock is composed of two populations of cracks; long cracks parallel to the planar anisotropy superimposed on a homogeneous, randomly distributed array of smaller cracks. Walsh and Brace (1964) assumed that failure may occur through the growth of either the long or the small cracks depending upon the orientation of the long crack system to the applied stress field. They made use of the Griffith's theory and its modification by McClintock and Walsh (1962) to predict respectively the tensile and shear strengths of rocks with any type of anisotropy. Comparison between experimental results and those predicted by the Walsh and Brace's theory (Mc Lamore and Gray, 1967) is not good for the extreme values of the orientation angle ψ.

The theory produces "shoulders" in the plot differential stress at failure versus confining pressure as in Figure 3.6. Other theories similar to the one previously mentioned have been proposed by Hoek (1964) and Barron (1971b).

The theories of the first group are purely empirical and require a large amount of curve fitting. They also have the conceptual disadvantage of being applicable only to the prediction of the directional character of strength in triaxial conditions. Furthermore they are independent of the type of rock symmetry. In the theories of the second group, the influence of the intermediate principal stresss σ_2 and the type of rock symmetry are taken into account. However, the applicability of these theories as far as the influence of σ_2 is concerned has not yet been verified experimentally since, as shown in section 3.2.1 there is a lack of experimental data from multiaxial tests.

Nova (1980) proposed a failure criterion similar to the Mohr Coulomb criterion but expressed in tensorial form. He showed that if the rock is transversely isotropic, the material parameters necessary to determine completely the failure condition and the failure plane are reduced to four: two cohesions and two friction coefficients. Those quantities are determined experimentally. When applied to a standard triaxial compression test, experimental results are fitted relatively well by the theoretical prediction. However, one of the main assumptions of this theory is that the available shear strength along any plane within a rock is the sum of two terms, one related to the cohesion and the other one related to friction. This additive character is often subject to controversy for a solid material (Handin, 1969).

Another theory has been proposed by Boehler and Sawczuck (1977) and Allirot and Boehler (1979). Yielding and failure of oriented solids is studied within the framework of the theory of representation for invariant tensor functions. This theory has been successfully applied to predict the directional character of strength and failure of anisotropic materials such as sand, clay or rock.

A reasonably general yield criterion expressed as a quadratic function, has been proposed by von Mises (1928) for anisotropic metals. It can be written as follows

$$\frac{1}{2}\left(K_{11}\sigma_x^2 + K_{22}\sigma_y^2 + K_{33}\sigma_z^2 + K_{44}\tau_{yz}^2 + K_{55}\tau_{xz}^2 + K_{66}\tau_{xy}^2\right) + K_{12}\sigma_x\sigma_y + \text{------} + K_{14}\sigma_x\tau_{yz} + \text{------} + K_{56}\tau_{xz}\tau_{xy} = \text{const.} \quad (3.9)$$

Eq (3.9) contains 21 independent constants of the form K_{ij} that depend on the choice of reference axes x,y,z and their orientation with respect to the anisotropy. If the material is orthotropic and the axes x,y,z are each one perpendicular to a plane of elastic symmetry, Hill (1950) showed that eq (3.9) reduces to the following

$$F'(\sigma_y-\sigma_z)^2 + G'(\sigma_z-\sigma_x)^2 + H'(\sigma_x-\sigma_y)^2 + 2L'\tau_{yz}^2 + 2M'\tau_{xz}^2 + 2N'\tau_{xy}^2 = 1 \quad (3.10)$$

where F', G'------, N' are constants expressed in terms of K_{ij}. If the material is transversely isotropic additional relations exist between these 6 constants. For instance if the z axis is one of elastic symmetry of rotation, Hill (1950) showed that $F' = G'$, $L' = M'$ and $N' = F' + 2H'$, the number of independent constants being reduced to 3. For a complete symmetry $L' = M' = N' = 3F' = 3G' = 3H'$

and eq (3.10) reduces to the von Mises criterion.

In general, yield or failure criteria for metals do not apply to geological materials such as rocks since hydrostatic stress may influence yielding. Pariseau (1972) extended Hill's theory to such materials for two types of elastic symmetry: orthotropy and transverse isotropy. For an orthotropic material eq (3.10) is re-written as follows

$$\left| F(\sigma_y - \sigma_z)^2 + G(\sigma_z - \sigma_x)^2 + H(\sigma_x - \sigma_y)^2 + L\tau_{yz}^2 + M\tau_{xz}^2 + N\tau_{xy}^2 \right|^{\frac{n}{2}}$$

$$-\left(U\sigma_x + V\sigma_y + W\sigma_z\right) = 1 \qquad (3.11)$$

where the 9 coefficients $F, G, -----, W$ and n ($n \geq 1$) are material constants and the axes x,y,z are each directed perpendicular to a plane of elastic symmetry. If the material is transversely isotropic within a plane, say yOz, the following relations apply also

$$G = H \;;\; M = N \;;\; V = W \;;\; L = 2G + 4F \qquad (3.12)$$

The number of independent coefficients of the stresses in eq (3.11) is reduced to 5. For a complete symmetry $L = M = N = 6F = 6G = 6H$ and $U = V = W$. In that case there are 2 independent coefficients. It is noteworthy that according to Pariseau's theory the number of coefficients to model the strength and deformability (section 2.2.1) is the same for geological materials that can be described as orthotropic, transversely isotropic or even isotropic.

In order to discuss the applicability of Pariseau's model, the present analysis will be limited to a transversely isotropic material with $n = 1$. The geometry of the problem is shown in Figure 3.9. Let X,Y,Z be a global coordinate system and x,y,z a local coordinate

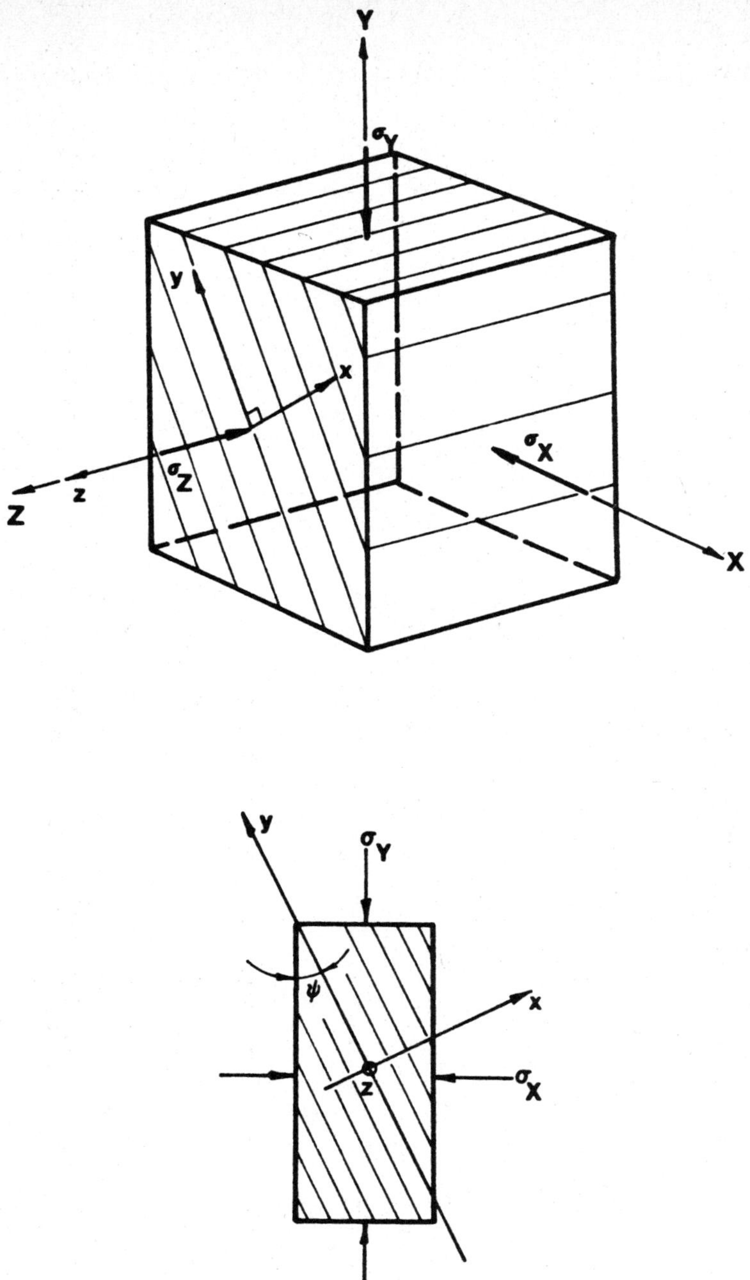

Figure 3.9. Geometry of the problem.

system attached to the plane of transverse isotropy with the x axis directed parallel to its normal. $\sigma_x, \sigma_y, \sigma_z$ are applied principal stresses. The planar anisotropy strikes parallel to the Z axis and is inclined at an angle ψ to the Y axis. Let $\sigma_1, \sigma_3, \sigma_2$ be respectively the major, minor and intermediate principal stresses among σ_x, σ_y and σ_z.

Pariseau (1972) showed that the five independent coefficients F, G, M, U and V can be expressed in terms of the unconfined compressive, tensile and torsional strengths measured parallel or perpendicular to the anisotropy as follows

$$2U = \frac{1}{T_{ox}} - \frac{1}{C_{ox}} \quad ; \quad 2V = \frac{1}{T_{oy}} - \frac{1}{C_{oy}}$$

$$2G = \frac{1}{4}\left(\frac{1}{C_{ox}} + \frac{1}{T_{ox}}\right)^2 \quad ; \quad 2F = \frac{1}{2}\left(\frac{1}{C_{oy}} + \frac{1}{T_{oy}}\right) - 2G$$

$$M = 1/S_{xy}^2 \tag{3.13}$$

where*

T_{ox}, T_{oy} are the absolute values of the tensile strength in the x,y directions,

C_{ox}, C_{oy} are the unconfined compressive strengths in the x,y directions,

S_{xy} is the torsional strength measured in the x,y plane.

Combining eqs (3.11) and (3.12) and making use of the transformation law for stresses, eq (3.11) can be expressed in terms of the three applied principal stresses σ_x, σ_y and σ_z and takes the general form

* Owing to the axial symmetry $T_{oz} = T_{oy}$; $C_{oz} = C_{oy}$; $S_{xz} = S_{xy}$.

$$f\left(\sigma_x, \sigma_y, \sigma_z, \psi, n, F, G, M, U, V\right) = 1 \qquad (3.14)$$

According to eq (3.14) the failure criterion can be expressed in terms of the three principal applied stresses, one parameter of orientation ψ , one arbitrary coefficient n and five coefficients depending upon the mechanical properties of the material.

Consider the following loading condition $\sigma_y = \sigma_1$; $\sigma_x = \sigma_z = \sigma_3$. It can be shown that eq (3.14) reduces to the following

$$\sigma_1 - \sigma_3 = \frac{2}{(A' - B)} + \frac{2B\,\sigma_3}{(A' - B)} \qquad (3.15)$$

where

$$B = U + 2V$$
$$A' = (U - V)\cos 2\psi + V + \left|4F\cos^4\psi + 4G\left(\sin^4\psi + \cos^2 2\psi\right) + M\sin^2 2\psi\right|^{1/2} \qquad (3.16)$$

According to eq (3.15), for each value of the orientation angle ψ , the differential stress at failure is a linear function of the confining pressure. In particular when $\sigma_3 = 0$, the unconfined compressive strength $\sigma_1 = C_{o\psi y}$ is equal to*

$$C_{o\psi y} = 2/(A' - B) \qquad (3.17)$$

Substituting eq (3.17) into eq (3.15) the failure criterion reduces to

$$1/K_\psi = \frac{\sigma_1}{\sigma_3} = \frac{C_{o\psi y}}{\sigma_3} + \frac{A' + B}{A' - B} \qquad (3.18)$$

Following the line of other authors assume that eq (3.15) represents a Mohr Coulomb criterion expressed in terms of prin-

---

* $C_o(\psi = 0°)_y = C_{oy}$; $C_o(\psi = 90°)_y = C_{ox}$

cipal stresses σ_1, σ_3 . Thus, two coefficients C_ψ and $\tan \phi_\psi$ can be introduced for each value of the inclination angle ψ such that

$$C_\psi = \frac{1}{\sqrt{A'^2 - B^2}} \quad ; \quad \tan \phi_\psi = B\, C_\psi \tag{3.19}$$

The second of eqs (3.19) implies that for a given material and a given orientation angle ψ , the ratio $\tan \phi_\psi / C_\psi$ must be constant. Furthermore, as ψ varies, the minima and maxima of $\tan \phi_\psi$ and C_ψ must occur for the same values of ψ .

The tensile strength in the Y direction denoted $T_{o\psi y}$ can be calculated using the following loading condition $\sigma_y = \sigma_3 = - T_{o\psi y}$ and $\sigma_x = \sigma_z = \sigma_1 = 0$. It can be shown that*

$$T_{o\psi y} = 2 \;/\; (B - A'') \tag{3.20}$$

where $$A'' = 2\,(U - V) \cos 2\psi + 2V - A' \tag{3.21}$$

A procdure similar to the one previously mentioned can be carried out to predict the unconfined compressive and tensile strengths in the X direction denoted $C_{o\psi x}$ and $T_{o\psi x}$. In fact, due to the geometrical symmetry of the problem the following relations apply

$$C_{o\psi x} = C_{o(90°-\psi)y} \quad ; \quad T_{o\psi x} = T_{o(90°-\psi)y} \tag{3.22}$$

Since the plane of transverse isotropy strikes parallel to the Z axis, the unconfined compressive and tensile strengths in the Z direction are independent of the angle ψ .

Pariseau's theory requires knowledge of the torsional strength

---

* $T_{o(\psi=0)y} = T_{oy}$; $T_{o(\psi=90°)y} = T_{ox}$

in a plane perpendicular to the plane of transverse isotropy in order to calculate coefficient M in eq (3.13). However, torsion tests are not common practice in rock mechanics. Another method can be used by running an additional unconfined compressive test beside those parallel to the x and y directions. For instance, if a test is run on a sample whose anisotropy is inclined at an angle $\psi = 45°$, combining eqs (3.16) and (3.17) and solving for M, we obtain

$$M = \left(\frac{2}{C_{o(\psi=45)y}} + (U+V) \right)^2 - (F+G) \qquad (3.23)$$

where $C_{o(\psi=45)y}$ is the unconfined compressive strength measured in the Y direction when $\psi = 45°$. Similarly, if tensile tests cannot be run or their results are considered doubtful, an alternative way to calculate the five coefficients U, V, F, G and M would be to carry out triaxial compression tests and to use eq (3.15) directly.

The applicability of Pariseau's theory is illustrated by the following numerical example: $C_{ox} = 280$ MPa; $C_{oy} = 175$ MPa; $C_{o(\psi=45)y} = 140$ MPa; $T_{ox} = 28$ MPa; $T_{oy} = 56$ MPa. The loading is such that $\sigma_y = \sigma_1$, $\sigma_x = \sigma_2 = \sigma_3$. Figure 3.10 shows the plot of differential stress at failure $(\sigma_1 - \sigma_3)$ versus inclination of the anisotropy ψ for different values of the confining pressure σ_3 as predicted by eq (3.15). Figures 3.11a and 3.11b show respectively the variation of the coefficients C_ψ and $\tan \phi_\psi$ defined in eq (3.19) with the orientation angle ψ. If we assume that eq (3.15) represents a Mohr Coulomb criterion expressed in terms of σ_1 and σ_3, failure must take place at an angle $45° - \phi_\psi/2$ to the direction of σ_1.

For anisotropic rocks tested in <u>triaxial compression</u>, Figures 3.10 and 3.11 show that the directional character of the strength

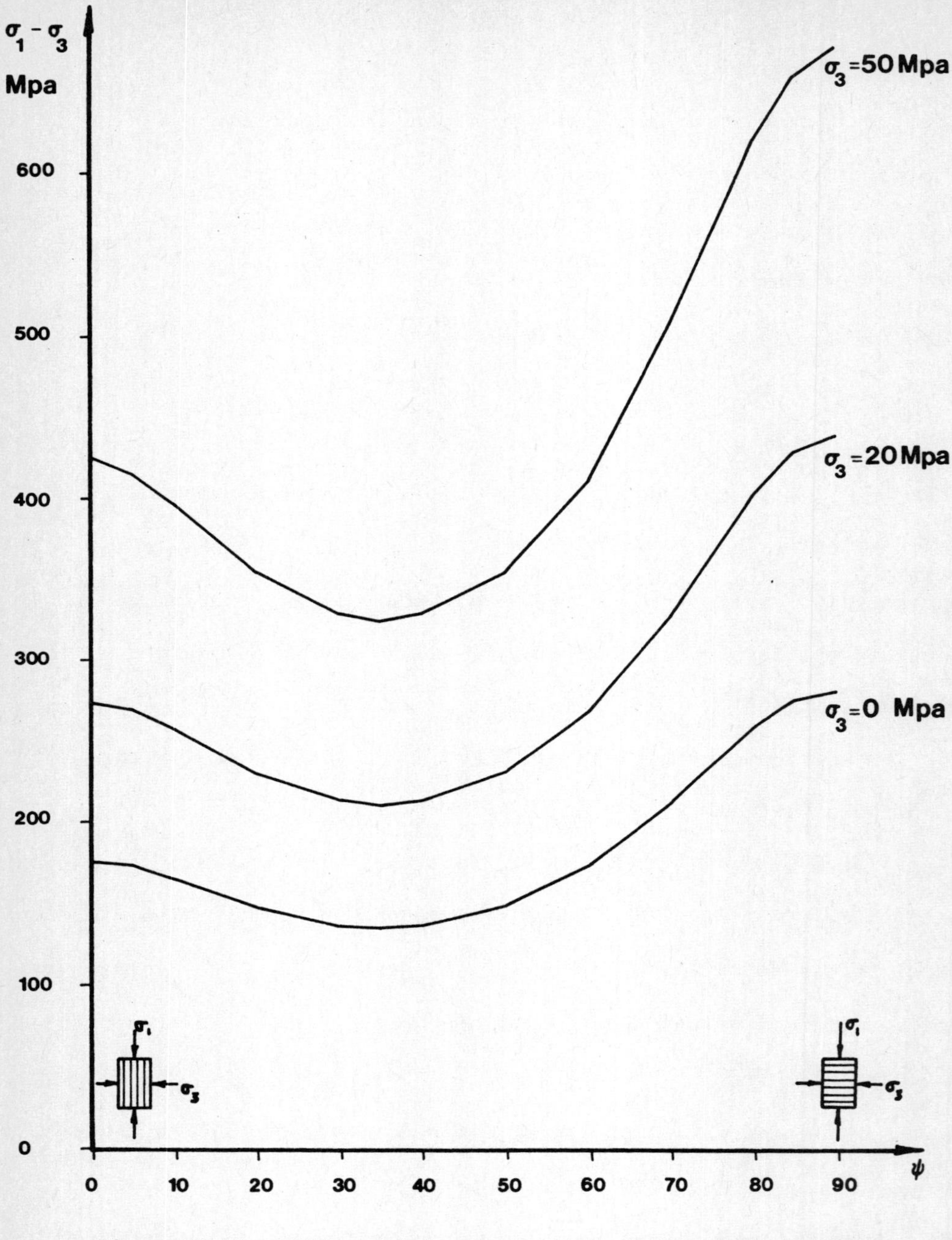

Figure 3.10. Differential stress at failure vs. inclination of anisotropy (theoretical example).

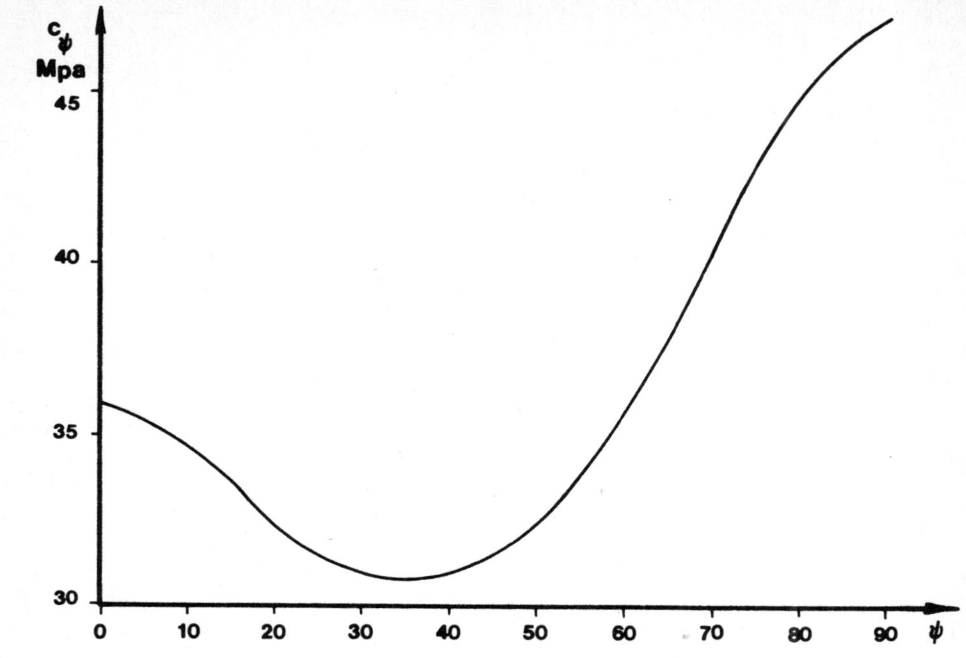

Figure 3.11. (a) Variation of c_ψ with inclination of anisotropy ψ.

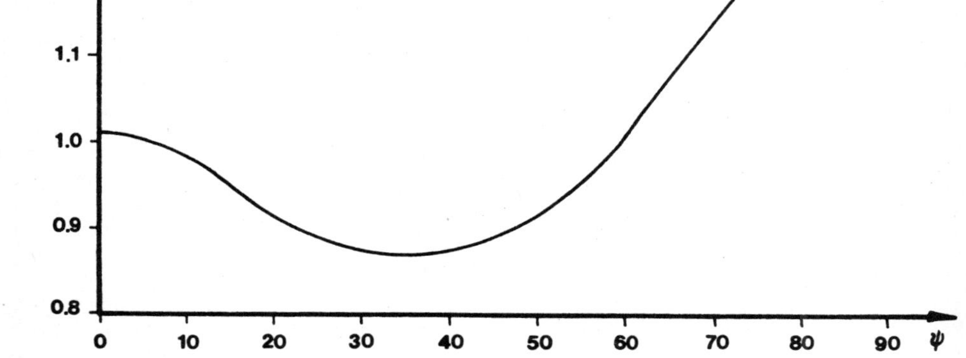

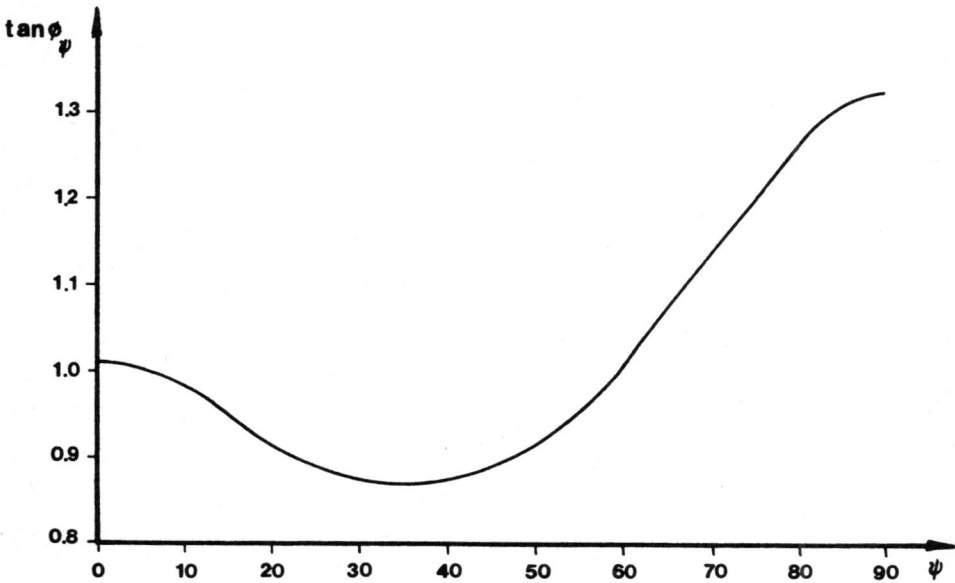

Figure 3.11. (b) Variation of $\tan\phi_\psi$ with inclination of anisotropy ψ.

predicted by the Pariseau's theory follows essentially the same trend as the one observed experimentally. This remark suggests that the applicability of this theory should be further investigated and compared to experimental results since the example presented above was imaginary. It suggests also that a possible approach to model the spatial variation of the strength of anisotropic rocks would be to follow the ideas used for deformability; that is, keep the expression for the failure criterion general and simplify it according to the results of laboratory tests and apparent material symmetries.

As far as the prediction of the tensile strength is concerned, yield or failure criteria that consider the rock as an isotropic material are unsuccessful. "Cut off" surfaces are often overlaid on the yield surfaces in order to solve this problem. For anisotropic rocks we might expect the same conclusions and testing is needed. This is further enhanced since from a practical point of view, testing techniques are far from totally satisfactory. Furthermore, when indirect methods are used, a few theoretical investigations have been reported for the interpretation of test results. The value of the tensile strength is usually calculated using formulas applicable to isotropic materials (Hobbs, 1964; Mc Lamore and Gray, 1967; Barron, 1971a) even though similar formulas exist for anisotropic materials (Okubo, 1952).

Two types of model were considered previously, discontinuous and continuous, each one being associated with a different type of rock anisotropy. A third model can be imagined to predict the strength of rocks that are both anisotropic and possess one or several planes of weakness. A method similar to the one proposed in section 3.3.1 can be used. In order to illustrate the method let us assume that the rock can be described as a transversely isotropic material with the geometry of Figure 3.9 and is cut by a joint set parallel to the plane of transverse isotropy. Let c_j and ϕ_j be the cohesion and friction angle along the planes of weakness. The Mohr Coulomb failure criterion for the joint set can be expressed by the value of the ratio Kf_j in eq (3.1). Similarly, assuming Pariseau's theory, the failure criterion for the intact rock can be expressed in terms of the ratio K_ψ in eq (3.18). Both Kf_j and K_ψ depend upon the orientation angle ψ. Their variation within the band $0^\circ < \psi < 90^\circ$ has been plotted in Figure 3.12 for the following numerical example: the rock has the same properties as those used to draw Figures 3.10 and 3.11 and the joint set is such that $c_j = 0$ and $\phi_j = 30^\circ$. In comparison to Figure 3.5, the shear strength of the intact rock is now represented by curved lines instead of circles. Three modes of failure can take place as in Figure 3.5.

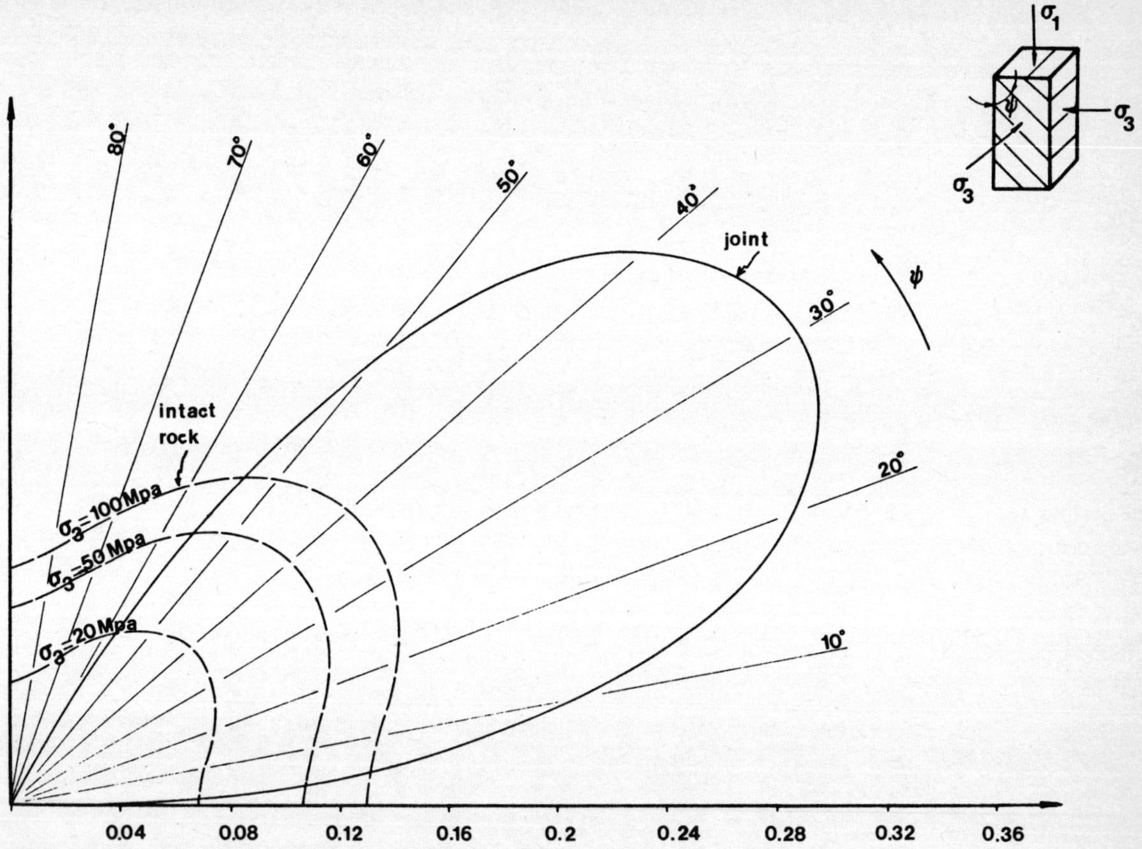

Figure 3.12. Bray's diagram for a joint set with $\phi_j = 30°$ in a transversely isotropic rock.

Chapter 4: Elastic Equilibrium of an Anisotropic Homogeneous Body Bounded Internally by a Cylindrical Surface of Arbitrary Cross Section

4.1 Introduction

The purpose of this chapter is to study the elastic equilibrium of an anisotropic, homogeneous and continuous medium bounded internally by a cylindrical surface of arbitrary cross section. The medium is subjected to body and surface forces. The work of Lekhnitskii (1963) is heavily referenced and used as a guide line for the present analysis. His model is extended to include the influence of a boundary stress component parallel to the hole axis.

The general expressions for the components of stress and displacement at any point in the anisotropic medium are derived first. Then, the plane strain and plane stress formulations available in the literature and their restrictions for anisotropic materials are reviewed. Finally, several closed form solutions are presented.

4.2 Geometry and Definition of the Problem

Let us consider the elastic equilibrium of an anisotropic, homogeneous and continuous body; the body possesses rectilinear anisotropy of a general form. It is bounded internally by a cylindrical surface of arbitrary cross section; the region outside the cylindrical surface being infinite (Figure 4.1).

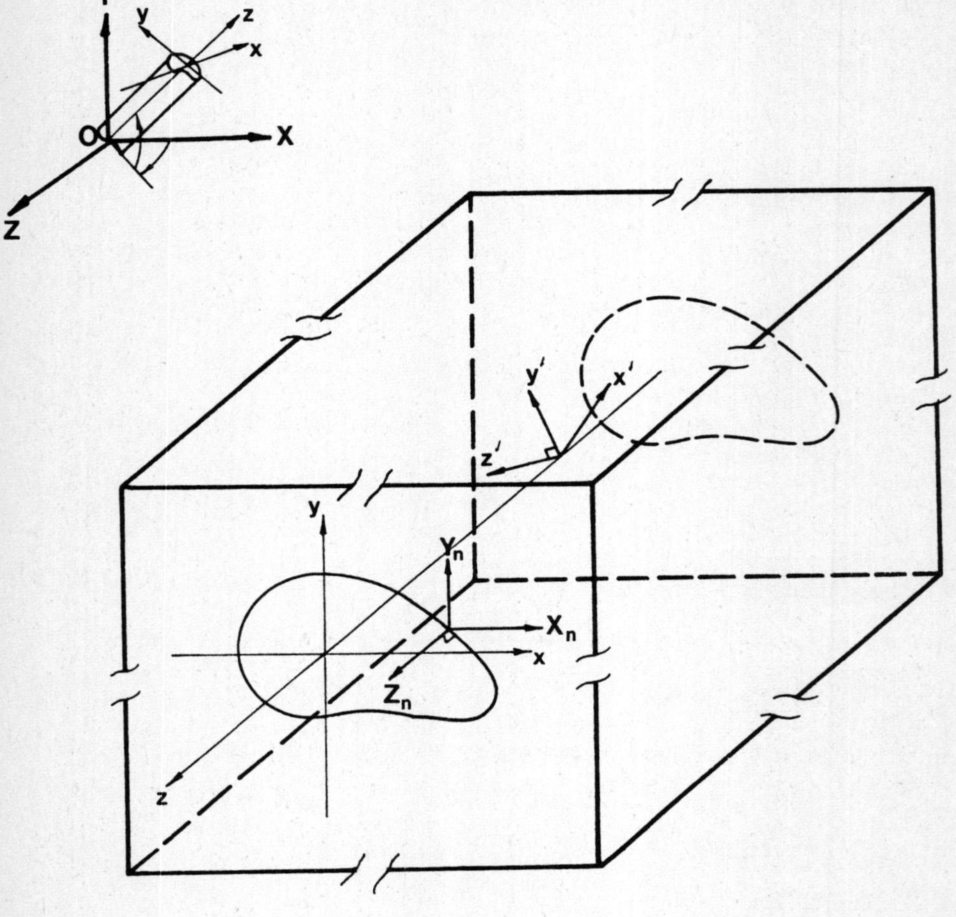

Figure 4.1. Geometry of the problem.

Let x,y,z be a system of cartesian coordinates with the z axis defining the longitudinal axis of the hole. The orientation of the hole is defined with respect to a fixed global coordinate system X,Y,Z. The body has planes and/or axes of elastic symmetry with respect to directions independent of the x,y,z directions. Thus, let x',y',z' be a system of cartesian coordinates attached to the rectilinear anisotropy. Its orientation with respect to the global coordinate system is assumed to be known.

Let X_n, Y_n, Z_n be the components in the x,y,z directions of the surface forces per unit area acting along the cylindrical surface and $X, Y, (Z=0)$ the components of the body forces per unit volume acting within the body. Surface and body forces are assumed not to vary along the generator of the hole.

The problem is to determine around the hole the distribution of stresses, strains and displacements induced by the surface and body forces. This is often known as the <u>first fundamental problem</u> of the statics of an elastic body (Muskhelishvili, 1953).

4.3 Formulation of the Problem

4.3.1 Basic Equations

At any point around the hole, the state of stress in the x,y,z coordinate system is defined by six independent components: $\sigma_x, \sigma_y, \sigma_z, \tau_{yz}, \tau_{xz}, \tau_{xy}$. Similarly, the state of strain is defined by $\varepsilon_x, \varepsilon_y, \varepsilon_z, \gamma_{yz}, \gamma_{xz}, \gamma_{xy}$. There are three components of displacement u, v, w respectively in the x,y,z directions. For any problem of elastostatics stress, strain and displacement components must satisfy the following equations: equations of equilibrium, equations of compatibility for strains, strain displacement relations, constitutive relations and the boundary conditions.

(i) Equations of Equilibrium

Let us assume that the body forces are derivable from a potential $\bar{U}$ such that

$$X = -\frac{\partial \bar{U}}{\partial x} \quad ; \quad Y = -\frac{\partial \bar{U}}{\partial y} \tag{4.1}$$

Furthermore, if the surface and body forces do not vary along the generator the components of stress will depend only on the coordinates x and y. Thus, the equations of equilibrium will have the following form

$$\begin{aligned} \frac{\partial \sigma_x}{\partial x} + \frac{\partial \tau_{xy}}{\partial y} + \frac{\partial \tau_{xz}}{\partial z}^{\,0} - \frac{\partial \bar{U}}{\partial x} &= 0 \\ \frac{\partial \tau_{xy}}{\partial x} + \frac{\partial \sigma_y}{\partial y} + \frac{\partial \tau_{yz}}{\partial z}^{\,0} - \frac{\partial \bar{U}}{\partial y} &= 0 \\ \frac{\partial \tau_{xz}}{\partial x} + \frac{\partial \tau_{yz}}{\partial y} + \frac{\partial \sigma_z}{\partial z}^{\,0} &= 0 \end{aligned} \tag{4.2}$$

(ii) Strain Displacement Relations

Infinitesimal strains and displacements are related through the following equations

$$\varepsilon_x = -\frac{\partial u}{\partial x} \; ; \; \varepsilon_y = -\frac{\partial v}{\partial y} \; ; \; \varepsilon_z = -\frac{\partial w}{\partial z} \; ; \; \gamma_{yz} = -\left(\frac{\partial w}{\partial y} + \frac{\partial v}{\partial z}\right)$$

$$\gamma_{xz} = -\left(\frac{\partial w}{\partial x} + \frac{\partial u}{\partial z}\right) \; ; \; \gamma_{xy} = -\left(\frac{\partial u}{\partial y} + \frac{\partial v}{\partial x}\right) \tag{4.3}$$

(iii) Equations of Compatibility for Strains

Strains are compatible if they result in displacements that produce no separation between small elements of the body. By using eqs (4.3) it can be shown that there exist six distinct relations between the strain components:

$$\frac{\partial^2 \varepsilon_x}{\partial y^2} + \frac{\partial^2 \varepsilon_y}{\partial x^2} = \frac{\partial^2 \gamma_{xy}}{\partial x \partial y}$$

$$\frac{\partial^2 \varepsilon_y}{\partial z^2} + \frac{\partial^2 \varepsilon_z}{\partial y^2} = \frac{\partial^2 \gamma_{yz}}{\partial y \partial z}$$

$$\frac{\partial^2 \varepsilon_z}{\partial x^2} + \frac{\partial^2 \varepsilon_x}{\partial z^2} = \frac{\partial^2 \gamma_{xz}}{\partial x \partial z}$$

$$\frac{2\,\partial^2 \varepsilon_x}{\partial y \partial z} = \frac{\partial}{\partial x}\left(\frac{\partial \gamma_{xz}}{\partial y} + \frac{\partial \gamma_{xy}}{\partial z} - \frac{\partial \gamma_{yz}}{\partial x}\right)$$

$$\frac{2\,\partial^2 \varepsilon_y}{\partial x \partial z} = \frac{\partial}{\partial y}\left(\frac{\partial \gamma_{yz}}{\partial x} + \frac{\partial \gamma_{xy}}{\partial z} - \frac{\partial \gamma_{xz}}{\partial y}\right)$$

$$\frac{2\,\partial^2 \varepsilon_z}{\partial x \partial y} = \frac{\partial}{\partial z}\left(\frac{\partial \gamma_{yz}}{\partial x} + \frac{\partial \gamma_{xz}}{\partial y} - \frac{\partial \gamma_{xy}}{\partial z}\right)$$

(4.4)

(iv) Constitutive Relation

The strain components are related to the stress components through the constitutive relation of the anisotropic body (see Chapter 2). If the constitutive relation is known in an x',y',z'coordinate system attached to the rectilinear anisotropy (Figure 4.1), one can calculate the constitutive relation in the x,y,z coordinate system attached to the hole. (See Appendix 4.1). A general form can be written as follows

$$\begin{bmatrix} \varepsilon_x \\ \varepsilon_y \\ \varepsilon_z \\ \gamma_{yz} \\ \gamma_{xz} \\ \gamma_{xy} \end{bmatrix} = \begin{bmatrix} a_{11} & a_{12} & a_{13} & a_{14} & a_{15} & a_{16} \\ a_{21} & a_{22} & a_{23} & a_{24} & a_{25} & a_{26} \\ a_{31} & a_{32} & a_{33} & a_{34} & a_{35} & a_{36} \\ a_{41} & a_{42} & a_{43} & a_{44} & a_{45} & a_{46} \\ a_{51} & a_{52} & a_{53} & a_{54} & a_{55} & a_{56} \\ a_{61} & a_{62} & a_{63} & a_{64} & a_{65} & a_{66} \end{bmatrix} \begin{bmatrix} \sigma_x \\ \sigma_y \\ \sigma_z \\ \tau_{yz} \\ \tau_{xz} \\ \tau_{xy} \end{bmatrix} \qquad (4.5)$$

or in a matrix form

$$(\varepsilon)_{xyz} = (A)(\sigma)_{xyz} \qquad (4.6)$$

(v) Boundary Conditions

For the first fundamental problem, the boundary condition along the contour of the hole can be written as follows*

$$\begin{aligned} \sigma_x \cos(\underset{\sim}{n},x) + \tau_{xy} \cos(\underset{\sim}{n},y) &= X_n \\ \tau_{xy} \cos(\underset{\sim}{n},x) + \sigma_y \cos(\underset{\sim}{n},y) &= Y_n \\ \tau_{xz} \cos(\underset{\sim}{n},x) + \tau_{yz} \cos(\underset{\sim}{n},y) &= Z_n \end{aligned} \qquad (4.7)$$

where $\underset{\sim}{n}$ is the unit vector normal to the contour (Figure 4.2).

---

* In the model of Lekhnitskii (1963), Z_n is automatically set to zero.

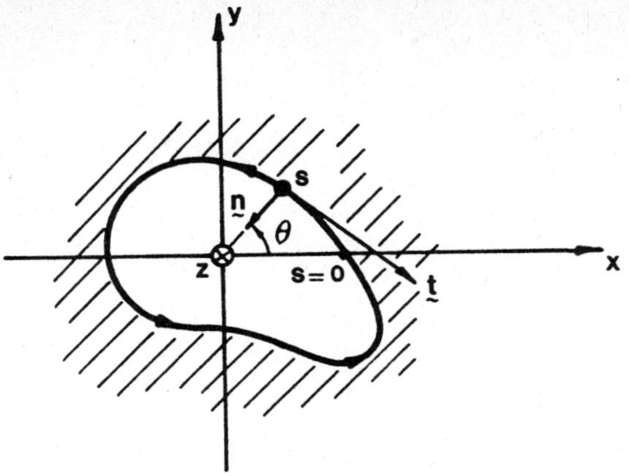

Figure 4.2. Orientation of the coordinate system attached to the contour of the hole.

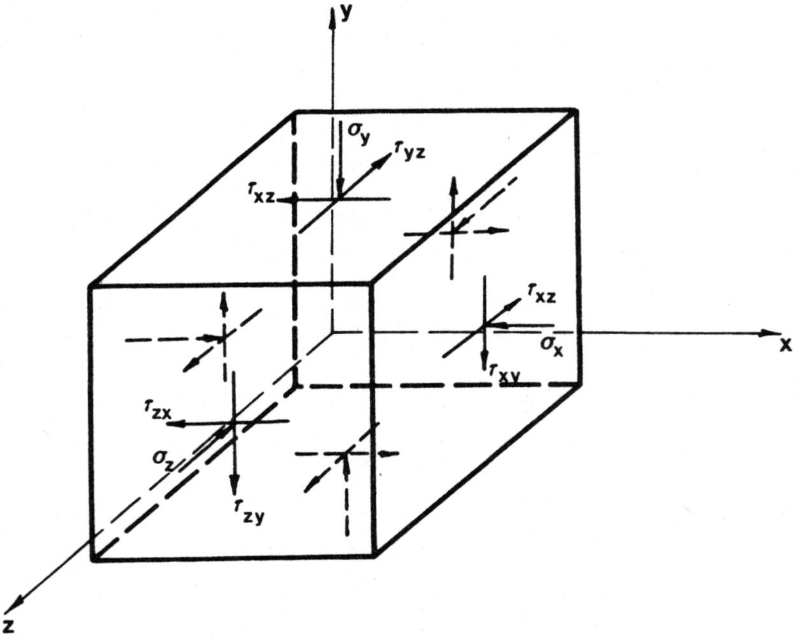

Figure 4.3. Sign convention for stresses. Orientation of the positive shear stresses.

Similarly let $\underset{\sim}{t}$ be the tangential unit vector in a clockwise direction on the contour. $\cos(\underset{\sim}{n},x)$ and $\cos(\underset{\sim}{n},y)$ can be expressed in terms of an arc length s defined from an arbitrary initial point $s = 0$ as follows

$$\cos(\underset{\sim}{n},x) = -\frac{dy}{ds} \quad ; \quad \cos(\underset{\sim}{n},y) = \frac{dx}{ds} \qquad (4.8)$$

4.3.2 Sign Conventions

In the following analysis and in Chapters 5 and 6 we adopt the following sign conventions for stress, strain and displacement components:

- The displacement components u,v,w respectively in the x,y,z directions are taken positive if directed in the positive directions of the coordinate axes,

- Compressive stresses are taken as positive. On any plane the sense of positive shear stresses is defined by the sense of the positive normal stress (see Goodman 1980, pp.341); if the normal compression is directed parallel to a positive coordinate axis, the shear stresses are positive parallel to the other positive coordinate directions and vice versa. As an example, Figure 4.3 shows the orientation of the positive shear stresses on three faces perpendicular to a coordinate system x,y,z ,

- Contractile longitudinal strains in the x,y,z directions are taken as positive. A positive shear strain increases beyond 90° the angle between two lines originally at right angle to each other.

4.3.3 Beltrami Michell Equations of Compatibility

The equations of equilibrium (eqs 4.2) can be satisfied by introducing two stress functions F (x,y) and Ψ (x,y) such that

$$\sigma_x = \frac{\partial^2 F}{\partial y^2} + \overline{U} \quad ; \quad \sigma_y = \frac{\partial^2 F}{\partial x^2} + \overline{U}$$

$$\tau_{xy} = -\frac{\partial^2 F}{\partial x \partial y} \quad ; \quad \tau_{xz} = \frac{\partial \Psi}{\partial y} \quad ; \quad \tau_{yz} = -\frac{\partial \Psi}{\partial x}$$

(4.9)

If the stress components are assumed to depend only on the coordinates x and y, the strain components are themselves also independent of z through the constitutive relation of the anisotropic body. It is then possible to simplify the equations of compatibility for strains (eqs 4.4) as follows:

$$\frac{\partial^2 \varepsilon_x}{\partial y^2} + \frac{\partial^2 \varepsilon_y}{\partial x^2} = \frac{\partial^2 \gamma_{xy}}{\partial x \partial y}$$

$$\frac{\partial \gamma_{xz}}{\partial y} - \frac{\partial \gamma_{yz}}{\partial x} = k$$

$$\frac{\partial^2 \varepsilon_z}{\partial y^2} = \frac{\partial^2 \varepsilon_z}{\partial x^2} = \frac{\partial^2 \varepsilon_z}{\partial x \partial y} = 0$$

(4.10)

where k is an arbitrary constant which appears as a result of the integration of the fourth and fifth equations of compatibility (eqs 4.4). It is easily verified that the three last conditions of compatibility (4.10) are satisfied only if ε_z is a linear function of x and y. Then,

$$\varepsilon_z = a_{33}\left(Ax + By + C\right) \tag{4.11}$$

where A, B, C are arbitrary constants. Combining eq (4.11) and the expression for ε_z in eq (4.5) we obtain

$$\sigma_z = Ax + By + C - \frac{1}{a_{33}}\left(a_{31}\sigma_x + a_{32}\sigma_y + a_{34}\tau_{yz} + a_{35}\tau_{xz} + a_{36}\tau_{xy}\right) \tag{4.12}$$

The first two conditions of compatibility (4.10) may be combined with the constitutive relation (eq 4.5) and the expression for σ_z (eq 4.12) to produce two relations between the stress components and the function $\overline{U}$. Those relations are often called the Beltrami Michell equations of compatibility. If we make use of eqs (4.9) we can rewrite those two relations as a system of two differential equations that the stress functions must satisfy:

$$L_4 F + L_3 \Psi = -(\beta_{12} + \beta_{22})\frac{\partial^2 \overline{U}}{\partial x^2} + (\beta_{16} + \beta_{26})\frac{\partial^2 \overline{U}}{\partial x \partial y} - (\beta_{11} + \beta_{12})\frac{\partial^2 \overline{U}}{\partial y^2}$$

$$L_3 F + L_2 \Psi = k + A a_{34} - B a_{35} + (\beta_{14} + \beta_{24})\frac{\partial \overline{U}}{\partial x} - (\beta_{15} + \beta_{25})\frac{\partial \overline{U}}{\partial y} \tag{4.13}$$

where

$$\beta_{ij} = a_{ij} - \frac{a_{i3} a_{j3}}{a_{33}} \qquad (i, j = 1, 2, 4, 5, 6) \tag{4.14}$$

L_2, L_3, L_4 are differential operators of the second, third and fourth orders which have the form

$$L_2 = \beta_{44}\frac{\partial^2}{\partial x^2} - 2\beta_{45}\frac{\partial^2}{\partial x \partial y} + \beta_{55}\frac{\partial^2}{\partial y^2}$$

$$L_3 = -\beta_{24}\frac{\partial^3}{\partial x^3} + (\beta_{25}+\beta_{46})\frac{\partial^3}{\partial x^2 \partial y} - (\beta_{14}+\beta_{56})\frac{\partial^3}{\partial x \partial y^2} + \beta_{15}\frac{\partial^3}{\partial y^3}$$

$$L_4 = \beta_{22}\frac{\partial^4}{\partial x^4} - 2\beta_{26}\frac{\partial^4}{\partial x^3 \partial y} + (2\beta_{12}+\beta_{66})\frac{\partial^4}{\partial x^2 \partial y^2} - 2\beta_{16}\frac{\partial^4}{\partial x \partial y^3} + \beta_{11}\frac{\partial^4}{\partial y^4}$$

(4.15)

Combining the constitutive relation of the anisotropic material and the strain displacement relations, Lekhnitskii (1963) proposed the following general expressions for the displacement components*

$$u(x,y,z) = \frac{1}{2} z^2 a_{33} A + \bar{\theta} yz - U(x,y) + \omega_2 z - \omega_3 y + u_0$$

$$v(x,y,z) = \frac{1}{2} z^2 a_{33} B - \bar{\theta} xz - V(x,y) + \omega_3 x - \omega_1 z + v_0$$

$$w(x,y,z) = -z(Ax + By + C) a_{33} - W(x,y) + \omega_1 y - \omega_2 x + w_0$$

(4.16)

where A, B, C are the arbitrary constants appearing in eqs (4.12). $\bar{\theta}$ is another arbitrary constant such that

$$k = -2\bar{\theta} \qquad (4.17)$$

---

* The expressions proposed in his book (eqs 17.12 and 17.13 pp.107-108) have been modified to satisfy the sign convention used in the present analysis.

$\omega_1, \omega_2, \omega_3$ characterize the rigid body rotations around the coordinate axes that are not accompanied by deformation and are such that

$$\omega_1 = \omega_{yz} \; ; \quad \omega_2 = -\omega_{xz} \; ; \quad \omega_3 = \omega_{xy} \tag{4.18}$$

u_o, v_o, w_o characterize the rigid body translations parallel to the coordinate axes that are not accompanied by deformation.

U, V, W are three functions of x and y defined as follows.

$$\frac{\partial U}{\partial x} = \beta_{11}\sigma_x + \beta_{12}\sigma_y + \beta_{14}\tau_{yz} + \beta_{15}\tau_{xz} + \beta_{16}\tau_{xy} + a_{13}(Ax+By+C)$$

$$\frac{\partial V}{\partial y} = \beta_{12}\sigma_x + \beta_{22}\sigma_y + \beta_{24}\tau_{yz} + \beta_{25}\tau_{xz} + \beta_{26}\tau_{xy} + a_{23}(Ax+By+C)$$

$$\frac{\partial U}{\partial y} + \frac{\partial V}{\partial x} = \beta_{16}\sigma_x + \beta_{26}\sigma_y + \beta_{46}\tau_{yz} + \beta_{56}\tau_{xz} + \beta_{66}\tau_{xy} + a_{36}(Ax+By+C)$$

$$\frac{\partial W}{\partial x} = \beta_{15}\sigma_x + \beta_{25}\sigma_y + \beta_{45}\tau_{yz} + \beta_{55}\tau_{xz} + \beta_{56}\tau_{xy} + a_{35}(Ax+By+C) + \bar{\theta}y$$

$$\frac{\partial W}{\partial y} = \beta_{14}\sigma_x + \beta_{24}\sigma_y + \beta_{44}\tau_{yz} + \beta_{45}\tau_{xz} + \beta_{46}\tau_{xy} + a_{34}(Ax+By+C) - \bar{\theta}x \tag{4.19}$$

4.3.4 Boundary Conditions

The two stress functions F and Ψ are solutions of the system (4.13) and must satisfy the boundary conditions along the contour of the hole. If we express the stress components of eqs (4.7) in terms of F and Ψ using eqs (4.9) and express $\cos(\underset{\sim}{n}, x)$ and $\cos(\underset{\sim}{n}, y)$ in terms of an arc length s along the contour (eqs 4.8) we can rewrite the boundary conditions as follows:

$$\frac{\partial}{\partial y}\left(\frac{\partial F}{\partial y}\right)\frac{dy}{ds} + \frac{\partial}{\partial x}\left(\frac{\partial F}{\partial y}\right)\frac{dx}{ds} = -X_n - \bar{U}\frac{dy}{ds}$$

$$\frac{\partial}{\partial y}\left(\frac{\partial F}{\partial x}\right)\frac{dy}{ds} + \frac{\partial}{\partial x}\left(\frac{\partial F}{\partial x}\right)\frac{dx}{ds} = Y_n - \overline{U}\frac{dx}{ds}$$

$$\frac{\partial \Psi}{\partial y}\frac{dy}{ds} + \frac{\partial \Psi}{\partial x}\frac{dx}{ds} = -Z_n \tag{4.20}$$

By integrating eqs (4.20) with respect to the arc length s from a certain initial point ($s=0$) to the variable point s, we obtain:

$$\frac{\partial F}{\partial y} = \int_0^s \left(-X_n - \overline{U}\frac{dy}{ds}\right) ds + C_1$$

$$\frac{\partial F}{\partial x} = \int_0^s \left(Y_n - \overline{U}\frac{dx}{ds}\right) ds + C_2$$

$$\Psi = -\int_0^s \left(Z_n\right) ds + C_3 \tag{4.21}$$

where C_1, C_2, C_3 are constants of integration that can be set equal to zero for a simply connected region.

4.3.5 <u>The General Expressions for the Stress Functions</u>

Lekhnitskii (1963) proposed a solution of the system of equations (4.13) that satisfies the boundary conditions (eqs 4.21). The general solution can be written as follows

$$F = F^* + F_0 \quad ; \quad \Psi = \Psi^* + \Psi_0 \tag{4.22}$$

where F_0 and Ψ_0 are particular solutions of the nonhomogeneous system (4.13) and F^* and Ψ^* are the solutions of the following homogeneous system

$$L_4 F^* + L_3 \Psi^* = 0$$

$$L_3 F^* + L_2 \Psi^* = 0 \tag{4.23}$$

Solving the system (4.23) in terms of F^* we are left with a sixth order differential equation

$$\left(L_4 L_2 - L_3^2\right) F^* = 0 \tag{4.24}$$

This sixth order differential equation makes sense only if L_3 is different from zero. The special case $L_3 = 0$ will be studied in section 4.4. The algebraic equation that corresponds to eq (4.24) can be written as follows

$$\ell_4(\mu)\,\ell_2(\mu) - \ell_3^2(\mu) = 0 \tag{4.25}$$

where

$$\begin{aligned} \ell_2(\mu) &= \beta_{55}\mu^2 - 2\beta_{45}\mu + \beta_{44} \\ \ell_3(\mu) &= \beta_{15}\mu^3 - (\beta_{14} + \beta_{56})\mu^2 + (\beta_{25} + \beta_{46})\mu - \beta_{24} \\ \ell_4(\mu) &= \beta_{11}\mu^4 - 2\beta_{16}\mu^3 + (2\beta_{12} + \beta_{66})\mu^2 - 2\beta_{26}\mu + \beta_{22} \end{aligned} \tag{4.26}$$

Eq (4.25) always has six roots μ_k (k = 1 to 6). Lekhnitskii (1963) has proved that the roots are always complex or purely imaginary*, three of them being the conjugate of the three others. Let μ_1, μ_2, μ_3 be those roots and $\bar{\mu}_1, \bar{\mu}_2, \bar{\mu}_3$ their respective conjugates.

If we define three complex numbers $\lambda_1, \lambda_2, \lambda_3$ equal to

$$\lambda_1 = -\frac{\ell_3(\mu_1)}{\ell_2(\mu_1)} \; ; \quad \lambda_2 = -\frac{\ell_3(\mu_2)}{\ell_2(\mu_2)} \; ; \quad \lambda_3 = -\frac{\ell_3(\mu_3)}{\ell_4(\mu_3)} \tag{4.27}$$

Lekhnitskii (1963) has shown that the general expressions for the stress functions F and Ψ take the following form

---

* Similar theorems have been proposed for the equations $\ell_4(\mu) = 0$ and $\ell_2(\mu) = 0$

$$F = 2\,\mathrm{Re}\left(F_1(z_1) + F_2(z_2) + F_3(z_3)\right) + F_0$$

$$\Psi = 2\,\mathrm{Re}\left(\lambda_1 F_1'(z_1) + \lambda_2 F_2'(z_2) + \frac{1}{\lambda_3} F_3'(z_3)\right) + \Psi_0 \qquad (4.28)$$

where

(i) Re is the notation for the real part of the complex expressions in the brackets.

(ii) $F_k(z_k)$ (k=1,2,3) are analytic functions of the complex variables $z_k = x+\mu_k y$.(x,y) are the coordinates of the point within the body where stress, strain and displacement components must be determined.

(iii) A prime indicates a derivation with respect to the complex varibables z_k (k=1,2,3).

4.3.6 The General Expressions for the Components of Stress and Displacement

Lekhnitskii (1963) introduced three new analytic functions ϕ_k (k=1,2,3) of the complex variables z_k equal to

$$\phi_1(z_1) = F_1'(z_1)\,;\quad \phi_2(z_2) = F_2'(z_2)\,;\quad \phi_3(z_3) = \frac{1}{\lambda_3} F_3'(z_3) \qquad (4.29)$$

Using eqs (4.28) and (4.29) we can write the expressions for Ψ and the first derivatives of F with respect to x and y in the following way

$$\frac{\partial F}{\partial x} = 2\,\mathrm{Re}\left(\phi_1(z_1) + \phi_2(z_2) + \lambda_3\phi_3(z_3)\right) + \frac{\partial F_0}{\partial x}$$

$$\frac{\partial F}{\partial y} = 2\,\mathrm{Re}\left(\mu_1\phi_1(z_1) + \mu_2\phi_2(z_2) + \lambda_3\mu_3\phi_3(z_3)\right) + \frac{\partial F_0}{\partial y}$$

$$\Psi = 2\,\mathrm{Re}\left(\lambda_1\phi_1(z_1) + \lambda_2\phi_2(z_2) + \phi_3(z_3)\right) + \Psi_0 \qquad (4.30)$$

Combining eqs (4.9) and (4.30) we obtain the general

expressions for the components of stress

$$\sigma_x = 2\,\mathrm{Re}\left(\mu_1^2 \phi_1'(z_1) + \mu_2^2 \phi_2'(z_2) + \lambda_3 \mu_3^2 \phi_3'(z_3)\right) + \frac{\partial^2 F_0}{\partial y^2} + \bar{U}$$

$$\sigma_y = 2\,\mathrm{Re}\left(\phi_1'(z_1) + \phi_2'(z_2) + \lambda_3 \phi_3'(z_3)\right) + \frac{\partial^2 F_0}{\partial x^2} + \bar{U}$$

$$\tau_{xy} = -2\,\mathrm{Re}\left(\mu_1 \phi_1'(z_1) + \mu_2 \phi_2'(z_2) + \lambda_3 \mu_3 \phi_3'(z_3)\right) - \frac{\partial^2 F_0}{\partial x \partial y}$$

$$\tau_{xz} = 2\,\mathrm{Re}\left(\lambda_1 \mu_1 \phi_1'(z_1) + \lambda_2 \mu_2 \phi_2'(z_2) + \mu_3 \phi_3'(z_3)\right) + \frac{\partial \Psi_0}{\partial y}$$

$$\tau_{yz} = -2\,\mathrm{Re}\left(\lambda_1 \phi_1'(z_1) + \lambda_2 \phi_2'(z_2) + \phi_3'(z_3)\right) - \frac{\partial \Psi_0}{\partial x}$$

(4.31)

where $\phi_k'(z_k)$ denotes the derivative of $\phi_k(z_k)$ with respect to z_k. From eq (4.12)

$$\sigma_z = Ax + By + C - \frac{1}{a_{33}}\left(a_{31}\sigma_x + a_{32}\sigma_y + a_{34}\tau_{yz} + a_{35}\tau_{xz} + a_{36}\tau_{xy}\right)$$

Substituting eqs (4.31) into eqs (4.19) and integrating we obtain

$$u = 2\,\mathrm{Re}\left(\sum_{k=1}^{3} p_k \phi_k(z_k)\right) + u^\circ$$

$$v = 2\,\mathrm{Re}\left(\sum_{k=1}^{3} q_k \phi_k(z_k)\right) + v^\circ$$

$$w = 2\,\mathrm{Re}\left(\sum_{k=1}^{3} r_k \phi_k(z_k)\right) + w^\circ$$

(4.32)

where

$$p_k = \beta_{11}\mu_k^2 + \beta_{12} - \beta_{16}\mu_k + \lambda_k\left(\beta_{15}\mu_k - \beta_{14}\right)$$

$$q_k = \beta_{12}\mu_k + \frac{\beta_{22}}{\mu_k} - \beta_{26} + \lambda_k\left(\beta_{25} - \frac{\beta_{24}}{\mu_k}\right)$$

$$r_k = \beta_{14}\mu_k + \frac{\beta_{24}}{\mu_k} - \beta_{46} + \lambda_k\left(\beta_{45} - \frac{\beta_{44}}{\mu_k}\right)$$

$(k = 1, 2)$

(4.33a)

$$p_3 = \lambda_3 \left(\beta_{11}\mu_3^2 + \beta_{12} - \beta_{16}\mu_3\right) + \beta_{15}\mu_3 - \beta_{14}$$

$$q_3 = \lambda_3 \left(\beta_{12}\mu_3 + \frac{\beta_{22}}{\mu_3} - \beta_{26}\right) + \beta_{25} - \frac{\beta_{24}}{\mu_3}$$

$$r_3 = \lambda_3 \left(\beta_{14}\mu_3 + \frac{\beta_{24}}{\mu_3} - \beta_{46}\right) + \beta_{45} - \frac{\beta_{44}}{\mu_3} \qquad (4.33b)$$

U°, V°, W° are solutions of eqs (4.19) that correspond to the functions F_0, Ψ_0, $\bar{U}$ and to the linear functions $a_{ij}(Ax+By+C)$, $\bar{\theta}y$ and $-\bar{\theta}x$ that contain the arbitrary constants $\bar{\theta}$, A, B, C.

Then, eqs (4.32) can be substitute into eqs (4.16) to obtain the general expressions for the components of displacement.

If we combine eqs (4.21) and eqs (4.30) the boundary conditions take the form

$$2\,\mathrm{Re}\left(\mu_1\phi_1(z_1) + \mu_2\phi_2(z_2) + \lambda_3\mu_3\phi_3(z_3)\right) = \int_0^s \left(-X_n - \bar{U}\frac{dy}{ds}\right)ds - \frac{\partial F_0}{\partial y} + c_1$$

$$2\,\mathrm{Re}\left(\phi_1(z_1) + \phi_2(z_2) + \lambda_3\phi_3(z_3)\right) = \int_0^s \left(Y_n - \bar{U}\frac{dx}{ds}\right)ds - \frac{\partial F_0}{\partial x} + c_2$$

$$2\,\mathrm{Re}\left(\lambda_1\phi_1(z_1) + \lambda_2\phi_2(z_2) + \phi_3(z_3)\right) = -\int_0^s Z_n\,ds - \Psi_0 + c_3 \qquad (4.34)$$

<u>In summary</u>, the problem of the elastic equilibrium of an anisotropic homogeneous and continuous medium bounded internally by a cylindrical surface of arbitrary cross section is reduced to the determination of three functions ϕ_k of the complex variables $z_k = x + \mu_k y$ (k=1,2,3). These functions must be such that the stress and displacement components (eqs 4.31 and 4.16) defined by them are single valued continuous functions of the coordinates x and y up to the internal contour. Furthermore, the three functions $\phi_k(z_k)$ must satisfy the boundary conditions on the contour of the hole (eqs 4.34).

4.4 Special Case of Anisotropy: A Plane of Elastic Symmetry Perpendicular to the Hole Axis

Consider the special case when there is a plane of elastic symmetry perpendicular to the hole axis. This occurs, for instance, for the following conditions of anisotropy:

(i) orthotropic body with one plane of elastic symmetry perpendicular to the hole axis and the two other planes parallel to the hole axis,

(ii) transverse isotropy in a plane striking parallel to the hole axis,

(iii) transverse isotropy in a plane perpendicular to the hole axis, and

(iv) isotropy.

Conditions (ii), (iii), (iv) can be derived from condition (i).

As far as the constitutive relation of the anisotropic material is concerned (see Chapter 2), this is satisfied when

$$a_{4i} = a_{5i} = a_{46} = a_{56} = 0 \qquad (i = 1, 2, 3) \tag{4.35}$$

or using eqs (4.14) when

$$\beta_{4i} = \beta_{5i} = \beta_{46} = \beta_{56} = 0 \qquad (i = 1, 2) \tag{4.36}$$

Replacing those equalities into eqs (4.15) and (4.26) we obtain

$$L_3 = 0 \quad ; \quad \ell_3(\mu) = 0 \tag{4.37}$$

Then, from eqs (4.27), (4.33a) and (4.33b)

$$\lambda_1 = \lambda_2 = \lambda_3 = 0 \quad ; \quad p_3 = q_3 = r_1 = r_2 = 0 \tag{4.38}$$

Therefore F and Ψ are solutions of two <u>independent</u> differential equations.

$$L_4 F = -(\beta_{12}+\beta_{22})\frac{\partial^2 \overline{u}}{\partial x^2} + (\beta_{16}+\beta_{26})\frac{\partial^2 \overline{u}}{\partial x \partial y} - (\beta_{11}+\beta_{12})\frac{\partial^2 \overline{u}}{\partial y^2}$$

$$L_2 \Psi = -2\overline{\theta}$$

(4.39)

For the anisotropic condition (i)

$$L_2 = \beta_{44}\frac{\partial^2}{\partial x^2} - 2\beta_{45}\frac{\partial^2}{\partial x \partial y} + \beta_{55}\frac{\partial^2}{\partial y^2}$$

$$L_4 = \beta_{22}\frac{\partial^4}{\partial x^4} - 2\beta_{26}\frac{\partial^4}{\partial x^3 \partial y} + (2\beta_{12}+\beta_{66})\frac{\partial^4}{\partial x^2 \partial y^2} - 2\beta_{16}\frac{\partial^4}{\partial x \partial y^3} + \beta_{11}\frac{\partial^4}{\partial y^4}$$

(4.40)

If the two planes of elastic symmetry of the orthotropic material of condition (i) parallel to the hole axis, are also parallel to planes x0z and y0z then, a_{45}, a_{16}, a_{26}, a_{36}, β_{45}, β_{16}, β_{26}, β_{36} are equal to zero and eqs (4.40) reduce to

$$L_2 = \beta_{44}\frac{\partial^2}{\partial x^2} + \beta_{55}\frac{\partial^2}{\partial y^2} = \beta_{44}\left(\frac{\partial^2}{\partial x^2} + \frac{\partial^2}{\partial Y^2}\right) \quad \text{with } Y = y\sqrt{\frac{\beta_{44}}{\beta_{55}}}$$

$$L_4 = \beta_{22}\frac{\partial^4}{\partial x^4} + (2\beta_{12}+\beta_{66})\frac{\partial^4}{\partial x^2 \partial y^2} + \beta_{11}\frac{\partial^4}{\partial y^4} \qquad (4.41)$$

For a material that possesses a transverse isotropy in a plane perpendicular to the hole axis or for an isotropic material then, $\beta_{44} = \beta_{55}$ and $\beta_{11} = \beta_{22} = (2\beta_{12}+\beta_{66})/2$ and eqs (4.40) reduce to

$$L_2 = \beta_{44}\left(\frac{\partial^2}{\partial x^2} + \frac{\partial^2}{\partial y^2}\right) = \beta_{44}\nabla^2$$

$$L_4 = \beta_{11}\left(\frac{\partial^4}{\partial x^4} + 2\frac{\partial^4}{\partial x^2 \partial y^2} + \frac{\partial^4}{\partial y^4}\right) = \beta_{11}\nabla^4$$

(4.42)

where ∇^2 and ∇^4 are respectively the Laplace and Biharmonic operators.

Combining eqs (4.31) and (4.32) with the conditions (4.35) and (4.38) we obtain the general expressions for the stress and displacement components when there is one plane of elastic symmetry perpendicular to the hole axis:

$$\sigma_x = 2\,\mathrm{Re}\,(\mu_1^2 \phi_1'(z_1) + \mu_2^2 \phi_2'(z_2)) + \frac{\partial^2 F_0}{\partial y^2} + \bar{U}$$

$$\sigma_y = 2\,\mathrm{Re}\,(\phi_1'(z_1) + \phi_2'(z_2)) + \frac{\partial^2 F_0}{\partial x^2} + \bar{U}$$

$$\tau_{xy} = -2\,\mathrm{Re}\,(\mu_1 \phi_1'(z_1) + \mu_2 \phi_2'(z_2)) - \frac{\partial^2 F_0}{\partial x \partial y}$$

$$\tau_{xz} = 2\,\mathrm{Re}\,(\mu_3 \phi_3'(z_3)) + \frac{\partial \Psi_0}{\partial y}$$

$$\tau_{yz} = -2\,\mathrm{Re}\,(\phi_3'(z_3)) - \frac{\partial \Psi_0}{\partial x}$$

$$\sigma_z = Ax + By + C - \frac{1}{a_{33}}(a_{31}\sigma_x + a_{32}\sigma_y + a_{36}\tau_{xy}) \tag{4.43}$$

and

$$u = 2\,\mathrm{Re}\,(p_1 \phi_1(z_1) + p_2 \phi_2(z_2)) + u^\circ$$

$$v = 2\,\mathrm{Re}\,(q_1 \phi_1(z_1) + q_2 \phi_2(z_2)) + v^\circ$$

$$w = 2\,\mathrm{Re}\,(r_3 \phi_3(z_3)) + w^\circ \tag{4.44}$$

μ_1, μ_2 are solutions of the equation $l_4(\mu) = 0$ and μ_3 is solution of the equation $l_2(\mu) = 0$.

Similarly, the boundary conditions (eqs (4.34)) take the form

$$2\,\mathrm{Re}\,(\mu_1 \phi_1(z_1) + \mu_2 \phi_2(z_2)) = \int_0^s \left(-X_n - \bar{U}\frac{dy}{ds}\right) ds - \frac{\partial F_0}{\partial y} + c_1$$

$$2\,\mathrm{Re}\,(\phi_1(z_1) + \phi_2(z_2)) = \int_0^s \left(Y_n - \bar{U}\frac{dx}{ds}\right) ds - \frac{\partial F_0}{\partial x} + c_2$$

$$2\,\mathrm{Re}\,(\phi_3(z_3)) = -\int_0^s Z_n\, ds - \Psi_0 + c_3 \tag{4.45}$$

Therefore, for this special case of anisotropy, the problem of the elastic equilibrium of an anisotropic body bounded internally by a cylindrical surface of arbitrary cross section can be considered as the sum of two problems:

(i) a plane problem involving stresses and displacements in the plane xOy only. The stress function F must satisfy the first of eqs (4.39). This problem is reduced to the determination of two functions $\phi_1(z_1)$ and $\phi_2(z_2)$ that satisfy along the contour of the hole the two first of eqs (4.45),

(ii) a longitudinal problem or antiplane problem involving τ_{xz}, τ_{yz} and the displacement component w. The stress function ψ must satisfy the second of eqs (4.39). This problem is reduced to the determination of a function ϕ_3 (z_3) that satisfies along the contour of the hole the last of eqs (4.45).

The concepts of plane and antiplane problems were first introduced by Filon (1937) for isotropic materials and were generalized by Milne-Thomson (1962) for anisotropic materials.

4.5 Plane Strain and Plane Stress Formulations

4.5.1 Plane Problems of the Theory of Elasticity

Until the last section, the solution of the problem of the elastic equilibrium of an anisotropic medium bounded internally by a cylindrical surface of arbitrary cross section loaded by surface and body forces has been kept general. The only main assumption was that stresses, strains, body and surfaces forces do not vary along the generator of the hole. From a practical point of view this solution is complex. The different constants appearing from integration such as $\bar{\theta}, A, B, C$ must be calculated. For an anisotropic body of finite length and finite cross section, Lekhnitskii (1963) has shown how to calculate them from the moments and forces applied at the ends of the body. For a long body with an infinite cross section the method cannot be applied.

From an engineering point of view, we would like to simplify the general solution by making some assumptions that would be <u>approximately</u> satisfied in the "real world".

A first simplification is possible when the dimension of the anisotropic body is very large in the longitudinal direction of the hole (approximately infinite). This is referred to as <u>plane strain formulation</u>. A solution can be proposed for the distributions of stresses, strains and displacements in any cross section far enough from the ends of the hole. Three dimensional numerical methods such as the Boundary Integral Equation Method (Hocking, 1976) have been used for isotropic materials to investigate how far from the ends of a hole the plane strain formulation is applicable. A distance equal to two or three times the hole diameter seems reasonable. Plane strain formulations are frequently used in rock mechanics when dealing with underground structures or stress measurements in boreholes.

A second simplification is when the dimensions of the anisotropic body are large in two directions only. Certain mine pillars may satisfy that condition. This is referred to as <u>plane stress formulation</u>. The analogy between the body and a thin plate is often assumed.

Several formulations of plane strain and plane stress have been proposed in the literature. The purpose of this section is to review these formulations* carefully and critically and to determine their applicability and restrictions in anisotropic materials

---

* See also Teodorescu (1964)

4.5.2 Plane Strain Formulations

(i) Usual Plane Strain Formulation and its Restrictions

In the usual plane strain formulation, a body with a hole is assumed to be approximately infinite in length and to be subjected to the action of body and surface forces directed normal to the axis of the hole. Furthermore, the displacement components induced by those forces must be such that u and v are functions of x and y alone and w vanishes (Love, 1920; Sokolnikoff, 1941; Timoshenko and Goodier, 1970). Therefore, according to eqs (4.3)

$$\varepsilon_z = \gamma_{xz} = \gamma_{yz} = 0 \tag{4.46}$$

Also, the boundary stress component Z_n must be zero. Then, among the six components of strain in eqs (4.3) only ε_x, ε_y and γ_{xy} are non zero.

Since by hypothesis, the induced components of displacement u, v must be independent of z, and w must vanish at any point x,y,z in the anisotropic medium, according to eqs (4.16)

$$A = B = C = \bar{\theta} = 0 \tag{4.47}$$

Combining eqs (4.47), the third of eqs (4.16) and eqs (4.32), the induced component of displacement w is equal to zero if and only if

$$2\,\mathrm{Re}\left(r_1 \phi_1(z_1) + r_2 \phi_2(z_2) + r_3 \phi_3(z_3)\right) + W^\circ = 0 \tag{4.48}$$

Let us consider the special case when there is one plane of elastic symmetry perpendicular to the hole axis and $\phi_3(z_3) = 0$. Then, from eqs (4.33a) r_1 and r_2 are zero. Since $\bar{\theta}$ = o, the particular solution ψ_0 of the second of eqs (4.39) can be set equal to zero. Substituting those conditions into the fourth and fifth of eqs (4.43) the stress components τ_{xz} and τ_{yz} become zero.

Combining eqs (4.47), the expressions for $\partial W/\partial x$ adn $\partial W/\partial y$ in eqs (4.19) and the previous remarks, we can set W° to zero. Therefore, we arrive to the conclusion that if the following conditions are satisfied

(a) $A = B = C = \bar{\theta} = 0$

<u>and</u> (b) the material has one plane of elastic symmetry perpendicular to the hole axis,

<u>and</u> (c) the loading is such that $\phi_3(z_3) = 0$

(4.49)

then eq (4.48) is satisfied and the classical plane strain formulation is applicable. Conditions (4.49) also imply that the long axis of the hole must coincide with a principal stress direction.

Since $\tau_{xz} = \tau_{yz} = 0$, the state of stress at any point is specified by σ_x, σ_y and τ_{xy} only. Only one stress function F remains that satisfies the first of eqs (4.39).

It is very important to note that conditions (4.49) are <u>sufficient</u> but not necessary. Since ϕ_1, ϕ_2, ϕ_3 are determined from the boundary conditions (eqs 4.34) and the single valued character of the stress and displacement components and since

r_1, r_2, r_3 are functions of the material properties, there may be a combination of loading conditions, anisotropy and body forces that satisfy eqs (4.48). Indeed, the problem described by the conditions (4.49) is very important in rock mechanics. As will be shown in section 4.6 for the case of a long hole whose longitudinal axis is perpendicular to a plane of elastic symmetry and under a three dimensional stress field applied at infinity, the analytic function $\phi_3(z_3)$ is related to the components $\tau_{xz,o}$ and $\tau_{yz,o}$ of the stress field. In that case, the condition $\phi_3(z_3) = 0$ restricts the hole to be in a principal stress direction.

(ii) Generalized or Complete Plane Strain Formulation

Because of the restrictions on the applicability of the usual plane strain method for anisotropic materials, a more general formulation has been proposed. The essential notion in this more general plane strain concept is that all components of stress, strain, displacement, body and surface forces are to be identical in all planes perpendicular to the hole axis. In other words,

$$\frac{\partial u}{\partial z} = \frac{\partial v}{\partial z} = \frac{\partial w}{\partial z} = 0 \qquad (4.50)$$

and the boundary stress component Z_n is non zero. The displacement components* u, v, w are functions of x and y alone.

---

* In the generalized plane strain formulation, rigid body rotations and translations are included into the displacement components (eqs 4.16).

Then, among the six components of strain in eqs (4.3) only ε_z is zero. This plane strain concept is called generalized plane strain (Lekhnitskii, 1963; Milne-Thomson, 1962). It has been also called complete plane strain (Brady and Bray, 1978) for isotropic materials. Combining eqs (4.16) and (4.50) the following conditions must be satisfied+

$$A = B = C = \overline{\theta} = 0 \quad ; \quad \omega_1 = \omega_2 = 0 \tag{4.51}$$

Indeed, the classical plane strain formulation appears as a special case of the generalized concept when, for instance, conditions (4.49(b)(c)) are also satisfied.

The generalized plane strain concept will be applied frequently in the following chapters. Therefore, it is worthwhile to rewrite the basic equations derived in section 4.3 with the conditions (4.51).

(i) The equations of compatibility for strains are reduced to two. The three last of eqs (4.10) are automatically satisfied. In terms of the two stress functions F and Ψ, they take the form

$$L_4 F + L_3 \Psi = -(\beta_{12} + \beta_{22}) \frac{\partial^2 \overline{U}}{\partial x^2} + (\beta_{16} + \beta_{26}) \frac{\partial^2 \overline{U}}{\partial x \partial y} - (\beta_{11} + \beta_{12}) \frac{\partial^2 \overline{U}}{\partial y^2}$$

$$L_3 F + L_2 \Psi = (\beta_{14} + \beta_{24}) \frac{\partial \overline{U}}{\partial x} - (\beta_{15} + \beta_{25}) \frac{\partial \overline{U}}{\partial y} \tag{4.52}$$

(ii) The general expression for the displacement compo-

+Conditions (4.51) are necessary and sufficient to satisfy eqs (4.50).

nents can be written as follows

$$u(x,y) = -2\,\mathrm{Re}\left(\sum_{k=1}^{3} p_k \phi_k(z_k)\right) - U^\circ - \omega_3 y + u_o$$
$$v(x,y) = -2\,\mathrm{Re}\left(\sum_{k=1}^{3} q_k \phi_k(z_k)\right) - V^\circ + \omega_3 x + v_o$$
$$w(x,y) = -2\,\mathrm{Re}\left(\sum_{k=1}^{3} r_k \phi_k(z_k)\right) - W^\circ + w_o \tag{4.53}$$

(iii) The general expression for the components of stress (eqs 4.31) and the boundary conditions (eqs 4.34) do not change except that

$$\sigma_z = -\frac{1}{a_{33}}\left(a_{31}\sigma_x + a_{32}\sigma_y + a_{34}\tau_{yz} + a_{35}\tau_{xz} + a_{36}\tau_{xy}\right) \tag{4.54}$$

When body forces are absent then

$$\bar{U} = F_o = \Psi_o = U^\circ = V^\circ = W^\circ = 0 \tag{4.55}$$

4.5.3 <u>Plane Stress Formulations</u>

(i) <u>General Formulation with the Condition $\sigma_z = 0$</u>

Let us introduce the condition $\sigma_z = 0$ into the general formulation of section 4.3. Then, eq (4.12) is replaced by the following one

$$\varepsilon_z = a_{33}\left(Ax + By + C\right) = a_{31}\sigma_x + a_{32}\sigma_y + a_{34}\tau_{yz} + a_{35}\tau_{xz} + a_{36}\tau_{xy} \tag{4.56}$$

The two first equations of compatibility (eqs 4.10) may be combined with the constitutive relation (eqs 4.5), the condition $\sigma_z = 0$ and eqs (4.9) to produce two differential equations that the stress functions F and Ψ must satisfy

$$L_4^* F + L_3^* \Psi = -(a_{12} + a_{22}) \frac{\partial^2 \bar{U}}{\partial x^2} + (a_{16} + a_{26}) \frac{\partial^2 \bar{U}}{\partial x \partial y} - (a_{11} + a_{12}) \frac{\partial^2 \bar{U}}{\partial y^2}$$

$$L_3^* F + L_2^* \Psi = -2\bar{\theta} + (a_{14} + a_{24}) \frac{\partial \bar{U}}{\partial x} - (a_{15} + a_{25}) \frac{\partial \bar{U}}{\partial y}$$

(4.57)

L_2^*, L_3^*, L_4^* are differential operators of the second, third and fourth orders. They have the same form as L_2, L_3, L_4 in eqs (4.15) but all the coefficients β_{ij} are replaced by the coefficients a_{ij} with $i, j = 1, 2, 4, 5, 6$.

Using eqs (4.9), we can rewrite eq (4.56) in terms of the two stress functions F and Ψ as follows

$$a_{31} \frac{\partial^2 F}{\partial y^2} + a_{32} \frac{\partial^2 F}{\partial x^2} - a_{36} \frac{\partial^2 F}{\partial x \partial y} + a_{35} \frac{\partial \Psi}{\partial y} - a_{34} \frac{\partial \Psi}{\partial x} + (a_{31} + a_{32}) \bar{U}$$
$$= a_{33} (Ax + By + C) \qquad (4.58)$$

Thus, our problem is to find two stress functions F and Ψ that satisfy eqs (4.57) with the boundary conditions (eqs 4.21). F and Ψ must be such that ε_z is also a linear function of x and y (eq 4.56).

This problem is more complex than the one presented when σ_z is different from zero. In section 4.3, the linear character of ε_z with respect to the variables x and y (eq 4.11) has been included in the system of differential equations (eqs 4.13) through the expression for σ_z. When $\sigma_z = 0$, it produces an additional equation (eq 4.56). Indeed, no simple solution has been proposed for anisotropic materials that includes this additional equation. A possible approach is to solve the system (4.57) with the boundary conditions (eqs 4.21), obtain a solution and see if the linear character of ε_z with respect to x and y is approximately satisfied. The system (4.57) can be solved the same way as the system (4.13). However, all the coefficients β_{ij} entering into the algebraic equation (eq 4.25) must be replaced by the coefficients a_{ij} for $i,j = 1,2,4,5,6$. The general expressions for the components of stress (eqs 4.31) and displacement (eqs 4.16 and 4.32) and the boundary conditions (eqs 4.34) do not change. However, in eqs (4.19) and (4.33) all the coefficients β_{ij} are replaced by the coefficients a_{ij} for $i,j = 1,2,4,5,6$. Furthermore, the quantities $u°$, $v°$, $w°$ appearing in eqs (4.32) are solutions of eqs (4.19) that now correspond to the functions $F_0, \Psi_0, \bar{u}$ and to the linear functions $\bar{\theta}y$ and $-\bar{\theta}x$.

(ii) <u>Usual Plane Stress Formulation and its Restrictions</u>

In the usual plane stress formulation, the body is assumed to be analogous to a thin plate subjected to the action of body and surface forces. Those forces are applied parallel to the plane of the plate and distributed uniformly over its thickness. Furthermore, the stress components σ_z, τ_{xz} and τ_{yz} are zero on both faces of the plate and are assumed to be also zero within the plate (Love, 1920; Sokolnikoff, 1941; Timoshenko and Goodier, 1970).

Since $\sigma_z = \tau_{xz} = \tau_{yz} = 0$, the state of stress at any point within the anisotropic material is specified by σ_x, σ_y and τ_{xy} only. Then, from eqs (4.9) the stress function Ψ can be set equal to zero. According to the last of eqs (4.30) this implies

$$\Psi = 2\,\mathrm{Re}\left(\lambda_1 \phi_1(z_1) + \lambda_2 \phi_2(z_2) + \phi_3(z_3)\right) + \Psi_0 = 0 \tag{4.59}$$

This also implies that the long axis of the hole must coincide with a principal stress direction (the principal stress in that direction being equal to zero). Indeed, the usual plane stress formulation appears as a special case of the general one derived in (i).

In order to solve eq (4.59), let us consider the special case when there is one plane of elastic symmetry perpendicular to the hole axis and $\phi_3(z_3) = 0$. Then, from eqs (4.27) λ_1 and λ_2 are zero. If we also assume that $\overline{\theta} = 0$, then the particular solution Ψ_0 of the second of eqs (4.57) can be set equal to zero.

Therefore, we arrive to the conclusion that if the following conditions are satisfied

(a) $\overline{\theta} = 0$,

and (b) the material has one plane of elastic symmetry perpendicular to the hole axis,

and (c) the loading is such that $\phi_3(z_3) = 0$ (4.60)

then eq (4.59) is satisfied and the classical plane stress formulation is applicable.

As for conditions (4.49) and for the same reasons, conditions (4.60) are sufficient but not necessary.

Only one stress function remains that satisfies the equation

$$L_4^* F = -(a_{12} + a_{22})\frac{\partial^2 \bar{u}}{\partial x^2} + (a_{16} + a_{26})\frac{\partial^2 \bar{u}}{\partial x \partial y} - (a_{11} + a_{12})\frac{\partial^2 \bar{u}}{\partial y^2} \tag{4.61a}$$

When $\Psi = 0$, eq (4.58) takes for form

$$a_{31}\frac{\partial^2 F}{\partial y^2} + a_{32}\frac{\partial^2 F}{\partial x^2} - a_{36}\frac{\partial^2 F}{\partial x \partial y} + (a_{31} + a_{32})\bar{u} = a_{33}(Ax + By + C) \tag{4.61b}$$

If we compare the applicability of the usual plane strain and plane stress formulation for anisotropic materials both of them require similar conditions to be satisfied (4.49 and 4.60) • one condition about the orientation of the anisotropy,

- one condition about the function $\phi_3(z_3)$,
- one condition about the constants A, B, C or $\bar{\theta}$.

Note that for those conditions and since Z_n must be equal to zero, the third boundary condition of eqs (4.34) is automatically satisfied. Only one stress function remains that satisfies the first of eqs (4.39) or eq (4.61a).

Other plane stress formulations have been proposed in the literature. They make use of an averaging process; if the plate is thin enough, the determination of the average values of the components of stress, strain and displacement taken over the thickness of the plate may lead to knowledge nearly as useful as that of the actual values at each point.

(ii) Complete Plane Stress Formulation

Let us assume that the anisotropic body is now analogous to a plate of thickness $2h$ (Figure 4.4), the middle plane of which will be taken to coincide with the xOy plane of the coordinate

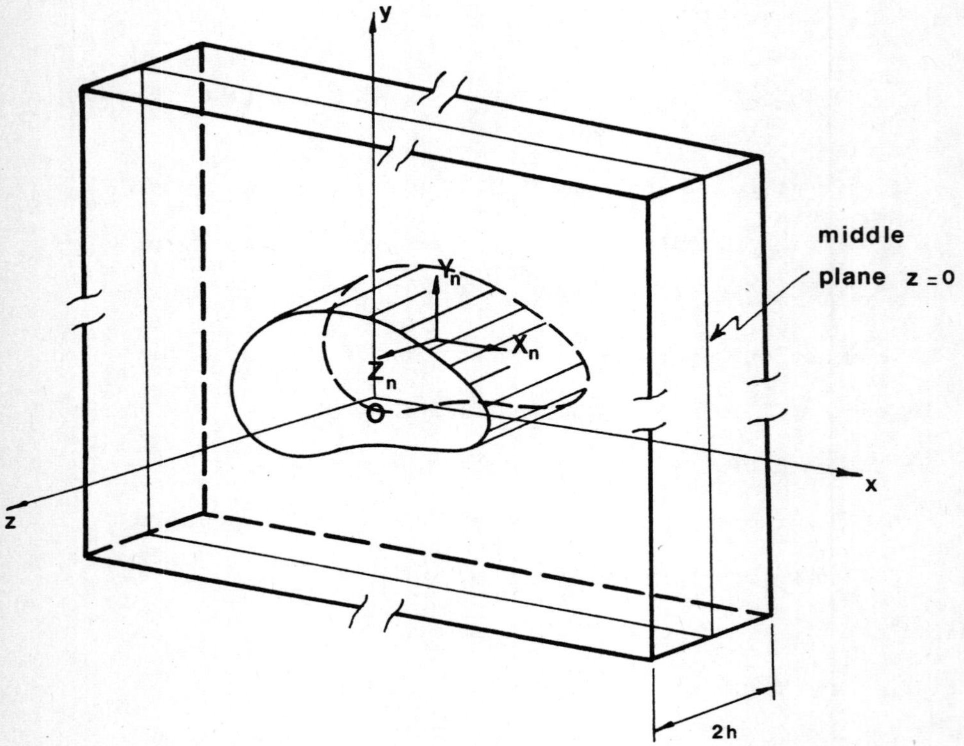

Figure 4.4. Plate theory. Geometry of the problem .

system. The body is bounded internally by a cylindrical surface of arbitrary cross section and is subjected to the same body and surface forces as those mentioned at the beginning of this chapter. It also possesses a rectilinear anisotropy of a general form.

The faces of the plate are assumed to be free of forces. Body and surface forces are also assumed to be symmetrically distributed with respect to the middle plane $z = 0$.

Let the components of stress, strain, displacement, body and surface forces be functions of x,y,z. Their mean values taken over the thickness of the plate can be calculated as shown in Appendix 4.2 and therefore are functions of x and y only.

The concept of complete plane stress has been introduced by Brady and Bray (1978) for isotropic plates subjected to zero body forces and to surface forces resulting from a three dimensional stress field applied at infinity ($\sigma_{x,o}$, $\sigma_{y,o}$, $\sigma_{z,o} = 0$, $\tau_{yz,o}$, $\tau_{xz,o}$, $\tau_{xy,o}$)*. They have shown that if the equations of equilibrium, constitutive relations, strain displacement relations, equations of compatibility for strains and boundary conditions are expressed in terms of the average values of stress, strain and displacement across the plate thickness, then they are in all respects identical to those for complete plane strain.

---

*Commas do not mean to imply differentiation.

In Appendix 4.2, those basic equations have been rederived in terms of average values of stress, strain and displacement for anisotropic plates subjected to body and surface forces of a general form.

It is shown that the same conclusions apply if the components of displacement u, v, w satisfy the following conditions

$$
\begin{aligned}
f_u &= u(x,y,h) - u(x,y,-h) = 0 \\
f_v &= v(x,y,h) - v(x,y,-h) = 0 \\
f_w &= w(x,y,h) - w(x,y,-h) = 0
\end{aligned}
\qquad (4.62)
$$

and*

$$
\begin{aligned}
\left.\frac{\partial u}{\partial z}\right|_{+h} - \left.\frac{\partial u}{\partial z}\right|_{-h} &= 0 \\
\left.\frac{\partial v}{\partial z}\right|_{+h} - \left.\frac{\partial v}{\partial z}\right|_{-h} &= 0
\end{aligned}
\qquad (4.63)
$$

f_u, f_v, and f_w appear as a result of integration of the strain components γ_{xz}, γ_{yz} and ε_z with respect to the variable z (eqs (9) in Appendix 4.2). If we adjust the derivations of Brady and Bray (eqs (10)) to the present coordinate system, it can be seen that they have automatically set the quantities f_u, f_v, f_w to zero. Therefore, their conclusion is not at all surprising since, as shown in Appendix 4.2, eqs (4.62) and (4.63) are obtained respectively by integration of eqs (4.50) and

---

* $f\big|_{\pm h}$ indicates the value of the function f at $z = h$ or $z = -h$.

$$\frac{\partial}{\partial z}\left(\frac{\partial u}{\partial z}\right) = \frac{\partial}{\partial z}\left(\frac{\partial v}{\partial z}\right) = 0 \qquad (4.64)$$

Since the complete plane stress formulation requires eqs (4.62) and (4.63) to be satisfied, it cannot be used in general. However, for thin plates, we should not except the components of displacement to vary substantially across the plate. In that case, the complete plane stress formulation may be thought as an approximation.

(iv) Generalized Plane Stress Formulation

In order to introduce this formulation let us recall first in the left hand side of Table 4.1 the basic conditions to be satisfied by the stress, strain and displacement components in the classical plane strain formulation when conditions (4.49) are applied.

Then, let us consider the elastic equilibrium of the anisotropic plate used in the previous formulation with the following additional conditions:

- the plate is thin and has at each point a plane of elastic symmetry parallel to its middle plane,

- the faces of the plate are free of external forces. Surface and body forces are also assumed to be symmetrically distributed with respect to the middle plane $z = 0$ and parallel to it ($Z_n = 0$).

For reasons of symmetry, it is obvious that the points in

Table 4.1

Classical Plane Strain Formulation	Generalized Plane Stress Formulation
• $\frac{\partial \sigma_x}{\partial x} + \frac{\partial \tau_{xy}}{\partial y} - \frac{\partial \overline{U}}{\partial x} = 0$ $\frac{\partial \tau_{xy}}{\partial x} + \frac{\partial \sigma_y}{\partial y} - \frac{\partial \overline{U}}{\partial y} = 0$	• $\frac{\partial \overline{\sigma}_x}{\partial x} + \frac{\partial \overline{\tau}_{xy}}{\partial y} - \frac{\partial \overline{U}}{\partial x} = 0$ $\frac{\partial \overline{\tau}_{xy}}{\partial x} + \frac{\partial \overline{\sigma}_y}{\partial y} - \frac{\partial \overline{U}}{\partial y} = 0$
• $\sigma_x = \frac{\partial^2 F}{\partial y^2} + \overline{U}$; $\sigma_y = \frac{\partial^2 F}{\partial x^2} + \overline{U}$ $\tau_{xy} = -\frac{\partial^2 F}{\partial x \partial y}$	• $\overline{\sigma}_x = \frac{\partial^2 F}{\partial y^2} + \overline{U}$; $\overline{\sigma}_y = \frac{\partial^2 F}{\partial x^2} + \overline{U}$ $\overline{\tau}_{xy} = -\frac{\partial^2 F}{\partial x \partial y}$
• $\frac{\partial^2 \varepsilon_x}{\partial y^2} + \frac{\partial^2 \varepsilon_y}{\partial x^2} = \frac{\partial^2 \gamma_{xy}}{\partial x \partial y}$	• $\frac{\partial^2 \overline{\varepsilon}_x}{\partial y^2} + \frac{\partial^2 \overline{\varepsilon}_y}{\partial x^2} = \frac{\partial^2 \overline{\gamma}_{xy}}{\partial x \partial y}$
• $\varepsilon_x = -\frac{\partial u}{\partial x}$; $\varepsilon_y = -\frac{\partial v}{\partial y}$ $\gamma_{xy} = -\left(\frac{\partial u}{\partial y} + \frac{\partial v}{\partial x}\right)$	• $\overline{\varepsilon}_x = -\frac{\partial \overline{u}}{\partial x}$; $\overline{\varepsilon}_y = -\frac{\partial \overline{v}}{\partial y}$ $\overline{\gamma}_{xy} = -\left(\frac{\partial \overline{u}}{\partial y} + \frac{\partial \overline{v}}{\partial x}\right)$
• $\varepsilon_x = \beta_{11}\sigma_x + \beta_{12}\sigma_y + \beta_{16}\tau_{xy}$ $\varepsilon_y = \beta_{21}\sigma_x + \beta_{22}\sigma_y + \beta_{26}\tau_{xy}$ $\gamma_{xy} = \beta_{61}\sigma_x + \beta_{62}\sigma_y + \beta_{66}\tau_{xy}$	• $\overline{\varepsilon}_x = a_{11}\overline{\sigma}_x + a_{12}\overline{\sigma}_y + a_{16}\overline{\tau}_{xy}$ $\overline{\varepsilon}_y = a_{21}\overline{\sigma}_x + a_{22}\overline{\sigma}_y + a_{26}\overline{\tau}_{xy}$ $\overline{\gamma}_{xy} = a_{61}\overline{\sigma}_x + a_{62}\overline{\sigma}_y + a_{66}\overline{\tau}_{xy}$
• $\sigma_x \cos(\underset{\sim}{n}, x) + \tau_{xy} \cos(\underset{\sim}{n}, y) = X_n$ $\tau_{xy} \cos(\underset{\sim}{n}, x) + \sigma_y \cos(\underset{\sim}{n}, y) = Y_n$	• $\overline{\sigma}_x \cos(\underset{\sim}{n}, x) + \overline{\tau}_{xy} \cos(\underset{\sim}{n}, y) = \overline{X}_n$ $\overline{\tau}_{xy} \cos(\underset{\sim}{n}, x) + \overline{\sigma}_y \cos(\underset{\sim}{n}, y) = \overline{Y}_n$

the middle plane z = 0 will remain in it after deformation, that the displacement component w will be small and that the variations of the components u and v over the thickness of the plate will be insignificant. Consequently, satisfactory results are obtained if we use instead, the mean values of the displacement components over the thickness of the plate.

Since it is assumed that $\sigma_z = \tau_{xz} = \tau_{yz} = 0$ on both faces of the plate, it is shown in Appendix 4.2 (eq (8)) that $\partial\sigma_z/\partial z = 0$ at $z = \pm h$. Hence, σ_z is not only zero for $z = \pm h$ but also its derivative with respect to z vanishes for these values. Therefore, since the plate is thin, σ_z is small throughout the plate and one may assume as a good approximation that $\sigma_z = 0$ everywhere.

Let us recall in the right hand side of Table 4.1 the conditions to be satisfied by the mean values of the displacement components u, v the strain components $\varepsilon_x, \varepsilon_y, \gamma_{xy}$ and the stress components $\sigma_x, \sigma_y, \tau_{xy}$ as obtained in Appendix 4.2.

Comparison between the left and right hand sides of Table 4.1 shows that the mean values of the displacement components u, v the strain components $\varepsilon_x, \varepsilon_y, \gamma_{xy}$, the stress components $\sigma_x, \sigma_y, \tau_{xy}$, the body and surface force components satisfy the same equations which govern the classical formulation of plane strain, the only difference being that all the coefficients β_{ij} appearing in the constitutive relation are replaced by the coefficients a_{ij} $(i, j = 1,2,6)$. Then, by comparison with the first of eqs (4.39), the stress function F appearing in the generalized plane stress formulation satisfies the equation

$$a_{22}\frac{\partial^4 F}{\partial x^4} - 2a_{26}\frac{\partial^4 F}{\partial x^3 \partial y} + (2a_{12}+a_{66})\frac{\partial^4 F}{\partial x^2 \partial y^2} - 2a_{16}\frac{\partial^4 F}{\partial x \partial y^3} + a_{11}\frac{\partial^4 F}{\partial y^4} =$$

$$-(a_{12}+a_{22})\frac{\partial^2 \bar{U}}{\partial x^2} + (a_{16}+a_{26})\frac{\partial^2 \bar{U}}{\partial x \partial y} - (a_{11}+a_{12})\frac{\partial^2 \bar{U}}{\partial y^2} \tag{4.65}$$

Eq (4.65) is the same as eq (4.61a) for the classical plane stress formulation.

The algebraic equation that corresponds to the homogeneous part of eq (4.65) takes the form

$$a_{11}\mu^4 - 2a_{16}\mu^3 + (2a_{12}+a_{66})\mu^2 - 2a_{26}\mu + a_{22} = 0 \tag{4.66}$$

If μ_1, μ_2 are the roots of that equation, then according to eqs (4.43) and (4 .53) the average values of the components of stress and displacement are equal to

$$\begin{aligned}
\bar{\sigma}_x &= 2\,\mathrm{Re}\,(\mu_1^2\phi_1'(z_1) + \mu_2^2\phi_2'(z_2)) + \frac{\partial^2 F_0}{\partial y^2} + \bar{U} \\
\bar{\sigma}_y &= 2\,\mathrm{Re}\,(\phi_1'(z_1) + \phi_2'(z_2)) + \frac{\partial^2 F_0}{\partial x^2} + \bar{U} \\
\bar{\tau}_{xy} &= -2\,\mathrm{Re}\,(\mu_1\phi_1'(z_1) + \mu_2\phi_2'(z_2)) - \frac{\partial^2 F_0}{\partial x \partial y} \\
\bar{u} &= -2\,\mathrm{Re}\,(p_1\phi_1(z_1) + p_2\phi_2(z_2)) - u^\circ - \omega_3 y + u_0 \\
\bar{v} &= -2\,\mathrm{Re}\,(q_1\phi_1(z_1) + q_2\phi_2(z_2)) - v^\circ + \omega_3 x + v_0
\end{aligned} \tag{4.67}$$

The stress functions ϕ_1 and ϕ_2 must also satisfy the two first of eqs (4.45) with body and surface force components replaced by their mean values.

The stressed state of a plate for which $\sigma_z = 0$ everywhere and τ_{xz}, τ_{yz} vanish on its faces has been first considered by Filon (1903) and called generalized plane stress by Love (1920). This formulation has been first stated for isotropic material. For such materials, it has been shown that there are relations between the stress and displacement components in the classical plane stress theory and those of the generalized plane stress theory (Filon, 1930). The generalized plane stress formulation was then extended by Lekhnitskii (1957), Green and Zerna (1968) to anisotropic plates with one plane of elastic symmetry parallel to their middle plane.

4.5.4 Remarks

As shown throughout this section, plane solutions bring simplifications of calculations as compared to the general formulation developed at the beginning of this chapter. This is very useful especially from a practical and technical viewpoint. In any cases, the anisotropic character of the material complicates the derivations.

The classical formulations of plane stress and plane strain commonly used for isotropic materials no longer apply to anisotropic materials except for certain orientations of anisotropy and loading conditions. More general formulations must be introduced.

There is however, an important difference between plane strain and plane stress formulations:

(i) the maintenance of a state of plane strain requires the application of a stress in the z direction (eq 4.54 for instance),

(ii) the maintenance of a state of plane stress does not require the application of a stress in the z direction (since $\sigma_z = 0$) but instead requires the stress components to satisfy an additional equation (eq 4.56). This equation has been found to have a physical meaning when dealing with an isotropic plate and the usual plane stress formulation. Love (1920)*[1], Timoshenko and Goodier (1970)*[2] have shown that the additional equation requires the stress and surface forces to vary in a parabolic fashion within the plate (zero body forces were assumed in their analysis). They have also shown that as the plate gets thinner the influence of the additional equation becomes negligible. No such derivations have been proposed for anisotropic materials.

When dealing with thin plates, the generalized plane stress formulation can be introduced. It has been shown that solutions of problem of plane strain may be used to study the elastic equilibrium of thin plates deformed in their own plane by forces applied in the plane. The average values across the plate thickness of stress and displacement components are determined instead of their actual values; mean values of the displacement components being independent of z. Furthermore, this formulation is limited to plates with one plane of elastic symmetry parallel to their middle planes. No such

---

*[1] Love (1920) Article 145

*[2] Timoshenko and Goodier (1970) Article 98.

solution exists for any inclined anisotropy.

Those few remarks show us the approximate character of the plane stress formulations.

4.6 Particular Solution for an Infinite Cylinder With a Circular Cross Section

As a special case of the general theory developed in section 4.3, let us consider the elastic equilibrium of an anisotropic, homogeneous body bounded internally by an infinite cylinder with a circular cross section of radius a. Boundary stresses with components X_n, Y_n, Z_n are applied on the surface of the hole and are assumed not to vary along its generator. Body forces are assumed to be absent. The generalized plane strain formulation is used to calculate the distributions of stresses, strains and displacements induced by those forces in the body.

4.6.1 Fourier Series Boundary Conditions

We assume that X_n, Y_n, Z_n can be expanded as Fourier series in $\cos m\theta$ and $\sin m\theta$ as follows

$$X_n = a_{o_x} + \sum_{m=1}^{\infty} \left(a_{m_x} \cos m\theta + b_{m_x} \sin m\theta \right) \tag{4.68}$$

and similar expressions for Y_n, Z_n where $a_{o_x}, a_{m_x}, b_{m_x}$ are replaced respectively by $a_{o_y}, a_{m_y}, b_{m_y}$ and $a_{o_z}, a_{m_z}, b_{m_z}$. θ is an angle (see Figure 4.2) that assumes all values from zero to 2π for a complete circuit along the circular contour. Furthermore,

$$a_{o_x} = \frac{1}{2\pi}\int_{-\pi}^{\pi} X_n \, d\theta \; ; \quad a_{o_y} = \frac{1}{2\pi}\int_{-\pi}^{\pi} Y_n \, d\theta \; ; \quad a_{o_z} = \frac{1}{2\pi}\int_{-\pi}^{\pi} Z_n \, d\theta \tag{4.69}$$

Let P_x, P_y and P_z be respectively the resultant forces (per unit length) of X_n, Y_n, Z_n along the circular contour of radius a, such that

$$P_x = \int_{-\pi}^{\pi} X_n a\, d\theta \;; \quad P_y = \int_{-\pi}^{\pi} Y_n a\, d\theta \;; \quad P_z = \int_{-\pi}^{\pi} Z_n a\, d\theta \tag{4.70}$$

Combining eqs (4.69) and (4.70), we obtain the following relations

$$a_{o_x} = \frac{P_x}{2\pi a} \;; \quad a_{o_y} = \frac{P_y}{2\pi a} \;; \quad a_{o_z} = \frac{P_z}{2\pi a} \tag{4.71}$$

Since the body forces are absent eqs (4.55) are satisfied. Substituting eqs (4.71) into eqs (4.68) and the expressions for Y_n and Z_n, then integrating the results with respect to the variable θ, we can rewrite eqs (4.34) as follows

$$\begin{aligned}
2\,\mathrm{Re}\left(\mu_1 \phi_1(z_1) + \mu_2 \phi_2(z_2) + \lambda_3\mu_3 \phi_3(z_3)\right) &= -\frac{P_x\theta}{2\pi} + \sum_{m=1}^{\infty}\left(b_m e^{im\theta} + \bar{b}_m e^{-im\theta}\right) \\
2\,\mathrm{Re}\left(\phi_1(z_1) + \phi_2(z_2) + \lambda_3 \phi_3(z_3)\right) &= \frac{P_y\theta}{2\pi} + \sum_{m=1}^{\infty}\left(a_m e^{im\theta} + \bar{a}_m e^{-im\theta}\right) \\
2\,\mathrm{Re}\left(\lambda_1 \phi_1(z_1) + \lambda_2 \phi_2(z_2) + \phi_3(z_3)\right) &= -\frac{P_z\theta}{2\pi} + \sum_{m=1}^{\infty}\left(c_m e^{im\theta} + \bar{c}_m e^{-im\theta}\right)
\end{aligned} \tag{4.72}$$

where

$$\begin{aligned}
a_m &= -\frac{a}{2m}\left(b_{m_y} + i a_{m_y}\right) \;; & \bar{a}_m &= -\frac{a}{2m}\left(b_{m_y} - i a_{m_y}\right) \\
b_m &= \frac{a}{2m}\left(b_{m_x} + i a_{m_x}\right) \;; & \bar{b}_m &= \frac{a}{2m}\left(b_{m_x} - i a_{m_x}\right) \\
c_m &= \frac{a}{2m}\left(b_{m_z} + i a_{m_z}\right) \;; & \bar{c}_m &= \frac{a}{2m}\left(b_{m_z} - i a_{m_z}\right)
\end{aligned} \tag{4.73}$$

4.6.2 General Expressions for the Analytic Functions $\phi_k(z_k)$ (k = 1,2,3)

Let S be the infinite region around the hole defined in the x0y plane. As shown in section 4.3, $\phi_k(z_k)$ is a function of the variable $z_k = x + \mu_k y$. Since μ_k is always complex or imaginary, one can write

$$z_k = x + \mu_k y = x + (\alpha_k + i\beta_k)y = (x + \alpha_k y) + i\beta_k y = x_k + iy_k \tag{4.74}$$

where* $x_k = x + \alpha_k y$; $y_k = \beta_k y$

(k = 1,2,3)

Therefore, the functions $\phi_k(z_k)$ (k = 1,2,3) must be determined not in the region with cross section S but in the regions S_1, S_2, S_3 obtained from S by means of the affine transformations 4.74**. The solution of our problem is obtained by mapping conformally the regions S_k onto the exterior of unit circles $|\xi_k|=1$ in the ξ_k planes (k = 1,2,3) such that any point on the contour of the region S_k is now located on the contour of the corresponding unit circle. Lekhinitskii (1963) proposed the following mapping functions+

$$\frac{z_k}{a} = \frac{1}{2}(1 - i\mu_k)\xi_k + \frac{1}{2}(1 + i\mu_k)\frac{1}{\xi_k} \tag{4.75a}$$

---

* β_k is always positive (Lehnitskii, 1963, pp 122).

**When the body is isotropic, it can be shown from $l_4(\mu)=0$ and $l_2(\mu)=0$ that $\mu_1 = \mu_2 = \mu_3 = i$ and therefore S_1, S_2, S_3 are identical to S

\+ Dimensionless quantities have been introduced in the analysis.

whose function inverse is

$$\xi_k = \frac{\frac{z_k}{a} + \sqrt{\left(\frac{z_k}{a}\right)^2 - 1 - \mu_k^2}}{1 - i\mu_k} \tag{4.75b}$$

(for k = 1,2,3).

It can be shown that if $\xi_k = e^{i\theta}$, then $\frac{z_k}{a} = \cos\theta + \mu_k \sin\theta$ and vice versa.

A general expression for the functions ϕ_k can be proposed as follows (Lekhnitskii, 1963)

$$\phi_k(z_k) = A_k \ln \xi_k + \sum_{m=1}^{\infty} A_{km} \xi_k^{-m} \tag{4.76}$$

If we combine eqs (4.76) and eqs (4.72), it is shown in Appendix 4.3 how to obtain the general expressions for the functions $\phi_k(z_k)$ and $\phi_k'(z_k)$ in terms of the coefficients defined in eqs (4.73).

4.6.3 Special Cases of Fourier Distributions

The equation of the contour of the circular hole in parametric form is

$$x = a\cos\theta \quad ; \quad y = a\sin\theta \tag{4.77}$$

Combining eqs (4.77) with eqs (4.8), we can rewrite eqs (4.7) as such

$$\begin{aligned} -\sigma_x \cos\theta - \tau_{xy} \sin\theta &= X_n \\ -\tau_{xy}\cos\theta - \sigma_y \sin\theta &= Y_n \\ -\tau_{xz}\cos\theta - \tau_{yz}\sin\theta &= Z_n \end{aligned} \tag{4.78}$$

(i) Internal Pressure q Distributed Uniformly along the Hole

Eqs (4.78) can be rewritten in terms of σ_r, $\tau_{r\theta}$ and τ_{rz}

using the equations of Appendix 4.4. Since $\tau_{r\theta} = \tau_{rz} = 0$ and $\sigma_r = q$ along the contour of the hole then, eqs (4.78) take the form

$$X_n = -q\cos\theta \; ; \quad Y_n = -q\sin\theta \; ; \quad Z_n = 0 \tag{4.79}$$

Comparing eqs (4.79) with the general expressions for X_n, Y_n, Z_n (eqs 4.68) and using eqs (4.73) we obtain

$$\bar{a}_1 = \frac{qa}{2} \; ; \quad \bar{b}_1 = \frac{iqa}{2} \; ; \quad \bar{c}_1 = 0 \tag{4.80}$$

and all other coefficients $\bar{a}_m, \bar{b}_m, \bar{c}_m$ are zero*.

(ii) Three Dimensional Stress Field Applied at Infinity

The boundary conditions of the problem are such that:

- the contour of the hole is free from surface forces i.e. $X_n = Y_n = Z_n = 0$ and
- the stress components must be equal at infinity to the stress field components $\sigma_{x,o}$; $\sigma_{y,o}$; $\sigma_{z,o}$; $\tau_{yz,o}$; $\tau_{xz,o}$; $\tau_{xy,o}$**.

Since we are dealing with a linear elastic material, the present problem can be decomposed into two:

a) find the distribution of stress, strain and displacement components in the medium without any hole under the influence of a uniform stress field. Along a fictitious contour of radius a, the surface stress components are then equal to:

$$\begin{aligned} X'_n &= -\sigma_{x,o}\cos\theta - \tau_{xy,o}\sin\theta \\ Y'_n &= -\tau_{xy,o}\cos\theta - \sigma_{y,o}\sin\theta \\ Z'_n &= -\tau_{xz,o}\cos\theta - \tau_{yz,o}\sin\theta \end{aligned} \tag{4.81}$$

---

* of course $a_{ox} = a_{oy} = a_{oz} = 0$ since $P_x = P_y = P_z = 0$.

** Commas do not mean to imply differentiation.

b) find the distribution of stress, strain and displacement components in the medium with a hole. There is no stress field applied at infinity and the contour of the hole is under the influence of surface stresses with components X_n, Y_n, Z_n such that

$$X_n + X'_n = 0 \ ; \quad Y_n + Y'_n = 0 \ ; \quad Z_n + Z'_n = 0 \tag{4.82}$$

Combining eqs (4.82) with eqs (4.81), then comparing the results with the general expressions for X_n, Y_n, Z_n (eqs 4.68) and finally using eqs (4.73), the solution of problem b) is obtained by setting

$$\begin{aligned} \bar{a}_1 &= -\frac{a}{2}\left(\sigma_{y,o} - i\,\tau_{xy,o}\right) \\ \bar{b}_1 &= \frac{a}{2}\left(\tau_{xy,o} - i\,\sigma_{x,o}\right) \\ \bar{c}_1 &= \frac{a}{2}\left(\tau_{yz,o} - i\,\tau_{xz,o}\right) \end{aligned} \tag{4.83}$$

and setting all the other coefficients $\bar{a}_m, \bar{b}_m, \bar{c}_m$ to zero. Again, for that loading condition $a_{o_x} = a_{o_y} = a_{o_z} = 0$ since $P_x = P_y = P_z = 0$.

(iii) Other Fourier Distributions

The present analysis is not limited to the two special cases of loading described in the previous section as long as the coefficients of eqs (4.73) are known. It can be also used for more complicated loading conditions such as the borehole jack for instance (Goodman et al, 1972; De la Cruz, 1978) in relation to the measurement of rock deformability and/or strength.

4.6.4 Closed Form Solutions

For the two special cases of Fourier distributions presented in section 4.6.3, it is now possible to substitute the coefficients $\bar{a}_1, \bar{b}_1, \bar{c}_1$ defined in eqs (4.80) and (4.83) into the general expressions for the functions $\phi_k(z_k)$ and $\phi_k'(z_k)$ (k = 1,2,3) in Appendix 4.3. Then, using eqs (4.31), (4.54), (4.5) and (4.53) we can calculate at at any point around the hole the stress, strain and displacement components induced by those Fourier distributions.

(i) In Appendix 4.5, it is shown how to determine the distribution of total stresses, strains and displacements around a circular hole drilled in an anisotropic body initially subjected to a 3D stress field applied at infinity. The influence of an internal pressure (if any), uniformly distributed along the hole, and acting after it is drilled, is also investigated.

As far as the 3D stress field applied at infinity is concerned, the problem is decomposed into two as suggested in parts (ii) a) and (ii) b) of section 4.6.3. This decomposition is indeed very useful in practice. The solution of subproblem (ii) b) is used to determine the distributions of stresses, strains and displacements induced by drilling the hole in the anisotropic medium. If a generalized plane strain formulation is assumed then, according to eq (4.50), the process of drilling the hole must induce zero longitudinal strain. General expressions for the induced components of stress, strain and displacement at any point of the medium in terms of the 3D stress field components are proposed in eqs (15), (18a), (19a) and (25). For subproblem (ii) a), the application of a uniform

3D stress field on a stress free body will produce displacements and strains. If we assume a plane strain formulation with a constant longitudinal strain component equal to

$$\varepsilon_{z_0} = a_{31}\sigma_{x,0} + a_{32}\sigma_{y,0} + a_{33}\sigma_{z,0} + a_{34}\tau_{yz,0} + a_{35}\tau_{xz,0} + a_{36}\tau_{xy,0} \qquad (4.84)$$

then eqs (26) and (35a) give the general expressions for the initial components of strain and displacement. Therefore, if we add the solutions of the two subproblems, we obtain at any point in the medium the total components of stress, strain and displacement in terms of the 3D stress field components (eqs (16), (18b), (19b) and (35c)). All the total stress, strain and displacement components are independent of z except the axial displacement since the total longitudinal strain is non zero but equal to the initial one before the hole was drilled.

Furthermore, since the material is linear elastic and therefore path independent, the distributions of stresses, strains and displacements expressed in terms of total quantities are the same if instead the hole is drilled first and the 3D stress field is applied thereafter with a constant longitudinal strain ε_{z_0} equal to eq (4.84).

Two general expressions are proposed for the total radial displacement component u_r at any point along the contour $r = a$. It is shown that for a general class of anisotropy

$$\frac{u_r}{a} = f_{11u}\sigma_{x,0} + f_{12u}\sigma_{y,0} + f_{13u}\sigma_{z,0} + f_{14u}\tau_{yz,0} + f_{15u}\tau_{xz,0} + f_{16u}\tau_{xy,0} \qquad (4.85)$$

or

$$\frac{u_r}{a} = M_1 + M_2 \cos 2\theta + M_3 \sin 2\theta \tag{4.86}$$

The coefficients f_{1i_u} (i = 1 to 6) are functions of the material properties, the orientation of the hole with respect to the directions of anisotropy and the angle θ . Instead, coefficients M_1, M_2 and M_3 are functions of the six stress field components, the material properties and the orientation of the hole with respect to the directions of anisotropy. Eq (4.86) indicates that along the circular contour, there are at most <u>three</u> independent values of u_r .

When there is <u>one plane of elastic symmetry</u> perpendicular to the hole axis, the general form of eq (4.86)* remains unchanged and eq (4.85) becomes

$$\frac{u_r}{a} = f_{11_u} \sigma_{x,0} + f_{12_u} \sigma_{y,0} + f_{13_u} \sigma_{z,0} + f_{16_u} \tau_{xy,0} \tag{4.87}$$

It is also shown that when there is one plane of elastic symmetry perpendicular to the hole axis, the equation $\phi_3(z_3)=0$ of conditions (4.49) is satisfied at any point x,y <u>if and only if</u> $\tau_{xz,0}$ and $\tau_{yz,0}$ vanish, in other words when the longitudinal axis of the hole is parallel to a principal stress field direction.

As mentioned in section 4.5.2, the classical plane strain

---

* M_1, M_2, M_3 are functions of $\sigma_{x,0}, \sigma_{y,0}, \sigma_{z,0}$ and $\tau_{xy,0}$ and independent of $\tau_{xz,0}$ and $\tau_{yz,0}$.

formulation appears as a special case of the generalized one. Combining conditions (4.49) and the remarks previously mentioned for a hole drilled in a 3D stress field, it is found that if there is one plane of elastic symmetry perpendicular to the hole axis and if the hole is parallel to a principal stress field direction, then the generalized plane strain formulation reduces to the classical one. The simultaneous character of those two conditions must be underlined. Therefore, the choice between classical and generalized plane strain formulations depends upon two parameters:

- the orientation of the directions of anisotropy with respect to the geometrical symmetry of the problem and,
- the values of the field stresses.

Table 4.2 summarizes the different types of plane strain formulation. The existence of a plane of elastic symmetry perpendicular to the hole axis is defined in the table by the condition $L_3 = 0$ (eq 4.37).

When the material is isotropic, eqs (4.86) and (4.87) are still valid. The expressions for the coefficients M_1, M_2, M_3 and $f_{1i_u} (i=1,6)$ are equal to those proposed by Hiramatsu and Oka (1968) or Fairhurst (1967).

Finally, if after drilling the hole, an internal pressure is uniformly distributed along its contour, its contribution to the total stress, strain and displacement components is calculated using a generalized plane strain formulation (eqs (58) to (61)). Then, according to eqs (4.50) the internal pressure induces also zero longitudinal strain.

	Orientation of the directions of anisotropy with respect to the geometrical symmetry of the problem	Field stresses applied at infinity	Type of plane strain formulation
$L_3 \neq 0$	Any orientation with $L_3 \neq 0$	$\sigma_{x,o}$; $\sigma_{y,o}$; $\sigma_{z,o}$; $\tau_{yz,o}$; $\tau_{xz,o}$; $\tau_{xy,o}$	Generalized plane strain*
$L_3 = 0$	• Orthotropy with one plane of elastic symmetry perpendicular to the hole axis • Transverse isotropy parallel or perpendicular to the hole axis • Isotropy	$\tau_{xz,o} = 0$ $\tau_{yz,o} = 0$	Plane strain
		$\tau_{xz,o} \neq 0$ $\tau_{yz,o} \neq 0$	Generalized plane strain = plane problem +antiplane problem

* τ_{xz}, τ_{yz}, γ_{xz}, γ_{yz} can be induced even though $\tau_{xz,o}$ and/or $\tau_{yz,o}$ are zero.

Table 4.2. Types of plane strain formulation .

(ii) In Appendix 4.7, the anisotropic body is now analogous to a thin plate with one plane of elastic symmetry parallel to its middle plane. A generalized plane stress formulation is used to determine the distribution of average stresses, strains and displacements (across the plate thickness), induced by drilling a circular hole in a 2D stress field parallel to the plate with average components $\overline{\sigma}_{x,o}, \overline{\sigma}_{y,o}$ and $\overline{\tau}_{xy,o}$. The influence of an internal pressure, uniformly distributed along the hole is also investigated.

In particular, it is shown that at any point around the circular hole, the average radial induced displacement can be expressed as follows

$$\frac{\overline{u}_{rh}}{a} = f_{11}^{(h)}\overline{\sigma}_{x,o} + f_{12}^{(h)}\overline{\sigma}_{y,o} + f_{13}^{(h)}\overline{\tau}_{xy,o} \qquad (4.88)$$

An expression similar to eq (4.86) is also proposed along the contour $r = a$. Then coefficients M_1, M_2 and M_3 are now functions of $\overline{\sigma}_{x,o}, \overline{\sigma}_{y,o}, \overline{\tau}_{xy,o}$ and the material properties.

(iii) In Appendix 4.8, the anisotropic body possesses three orthogonal planes of elastic symmetry, each one being perpendicular to one of the coordinate axes attached to the hole. It is shown how to calculate the distribution of the displacement components induced by an internal pressure uniformly distributed along its contour. Two formulations are proposed.

If the hole is very long, a classical plane strain formulation is used. In particular, at any point along the contour, the radial displacement is found to be a function of $\cos 2\theta$ only. It implies that along the contour there are at most two independent values of u_r .

If the hole is pressurized over a short length a generalized plane stress formulation is used. Similar conclusions apply now to the induced radial displacement expressed in terms of an average quantity across the pressurized zone.

4.6.5 Remarks

The general expressions for the distribution of stresses, strains and displacements around a long circular hole drilled in an anisotropic body subjected to a 3D stress field applied at infinity have been proposed by several authors. Those derivations were mostly related to the influence of rock anisotropy on stress measurements.

Berry and Fairhurst (1966), Fairhurst (1967) and Berry (1968)* assumed the rock material to be transversely isotropic only. Stress, strain and displacement components were calculated at any point along the contour of the circular hole. Additional assumptions were also used as far as the orientation of the anisotropy and the principal stress field with respect to the geometrical symmetry of the hole were concerned. Berry and Fairhurst (1966) assumed the hole to be in a principal stress field direction and to have its axis parallel or perpendicular to the axis of elastic symmetry of the anisotropy. For Fair-

---

* Their model is an extension of the one proposed by Milne – Thomson (1962) to allow for a constant axial strain to occur.

hurst (1967) and Berry (1968) the hole was inclined with respect to both the anisotropy and the principal stress field. However, the orientation of the anisotropy was defined by one angle only.

Lekhnitskii (1963) proposed general expressions for the distributions of stresses, strains and displacements at any point of a body with any type of anisotropy. However, he assumed the hole to be loaded by boundary stresses with components $X_n, Y_n, Z_n = 0$. Therefore, according to eqs (4.81) and (4.82) it implies that $\tau_{xz,o}$ and $\tau_{yz,o}$ must vanish. As shown in this chapter, the influence of the boundary stress component Z_n and therefore $\tau_{xz,o}$ and $\tau_{yz,o}$ can be introduced in Lekhnitskii's general theory through the boundary conditions (eqs 4.34) and the general expressions for the functions $\phi_k(z_k)$ (k = 1,2,3) in Appendix 4.3. A similar method was used by Niwa et al (1970).

In the whole section 4.6, we were mostly concerned about a long cylinder with a circular cross section. A similar analysis can be derived for a long cylinder of arbitrary cross section loaded by surface stresses with components X_n, Y_n, Z_n. The contour of the cross section must be approximated by Fourier series (Leknitskii, 1957; Niwa and Hirashima, 1970) and a new mapping function (eq 4.75a) must be introduced.

The generalized plane strain formulation can be also incorporated into numerical models. Eissa (1980) proposed an indirect formulation of the boundary element method to calculate the distribution of stresses, strains and displacements around cavities of any shape in anisotropic bodies. However, the model does not apply when the directions of anisotropy do not correspond with the geometrical symmetry of the cavity.

4.6.6 Numerical Examples

In order to illustrate the closed form solutions of section 4.6.4, several numerical examples are presented. A computer program called Berni 2 (see Appendix 6.3) is used to calculate the distribution of total stresses, strains and displacements around an infinitely long circular hole drilled in an infinite medium with a general anisotropic character. The medium is subjected to a 3D stress field applied at infinity. The equilibrium of the anisotropic body is such that, at any point around the hole, the total longitudinal strain is constant and defined by eq (4.84). The influence of an internal pressure, uniformly distributed along the hole and acting after it is drilled is also taken into account. The computer program calculates also the distributions of stresses, strains and displacements induced by the hole drilling and/or by the application of an internal pressure. For that purpose, a generalized plane strain formulation is used.

As a first example, consider the following 3D stress field components $\sigma_{x,o} = 7.\ \text{MPa}$; $\sigma_{y,o} = 2.5\ \text{MPa}$; $\sigma_{z,o} = 5.8\ \text{MPa}$; $\tau_{yz,o} = -1.8\ \text{MPa}$; $\tau_{xz,o} = 1.\ \text{MPa}$; $\tau_{xy,o} = 2.6$. Let us introduce two types of material properties

(i) an isotropic material with a Young's Modulus $E = 40\,000\ \text{MPa}$ and a Poisson's ratio $\nu = 0.25$ and

(ii) a transversely isotropic material with the following properties* $E_1 = 2\ 10^4$ MPa ; $E_2 = E_3 = 4\ 10^4$ MPa ; $G_{12} = G_{13} = 4\ 10^3$ MPa; $\nu_{12} = \nu_{13} = 0.2$; $\nu_{23} = \nu_{32} = 0.25$

The hole and the plane of transverse isotropy have the following orientations with respect to the global coordinate system X,Y,Z (Figures 1,2, Appendix 4.1):

(i) the coordinate system x,y,z is parallel to the global coordinate one ($\beta_h = 90°$; $\delta_h = 0°$) and (ii) the plane of transverse isotropy strikes at different angles β varying from 0 to 90 degrees and dips at an angle ψ of 30 degrees.

Figures 4.5a, 4.5b, and 4.5c show respectively the polar distributions of total tangential stresses, induced radial and tangential displacements** along the wall of the hole (θ varying from 0 to 90 degrees) for the isotropic material and the transversely isotropic material when β is equal to 0, 40 and 90 degrees.

As a second example, the hole is now pressurized over its whole length with an internal pressure of 1 MPa. Material properties, hole and transverse isotropy orientations are similar to those for the previous example. Figures 4.6a and 4.6b show respectively the polar distributions of tangential

---

* Directions 1, 2, 3 correspond respectively to x', y', z', in Figure 1 of Appendix 4.1.

** Displacement components are expressed in terms of dimensionless quantities $\frac{u_r}{a}$, $\frac{v_\theta}{a}$

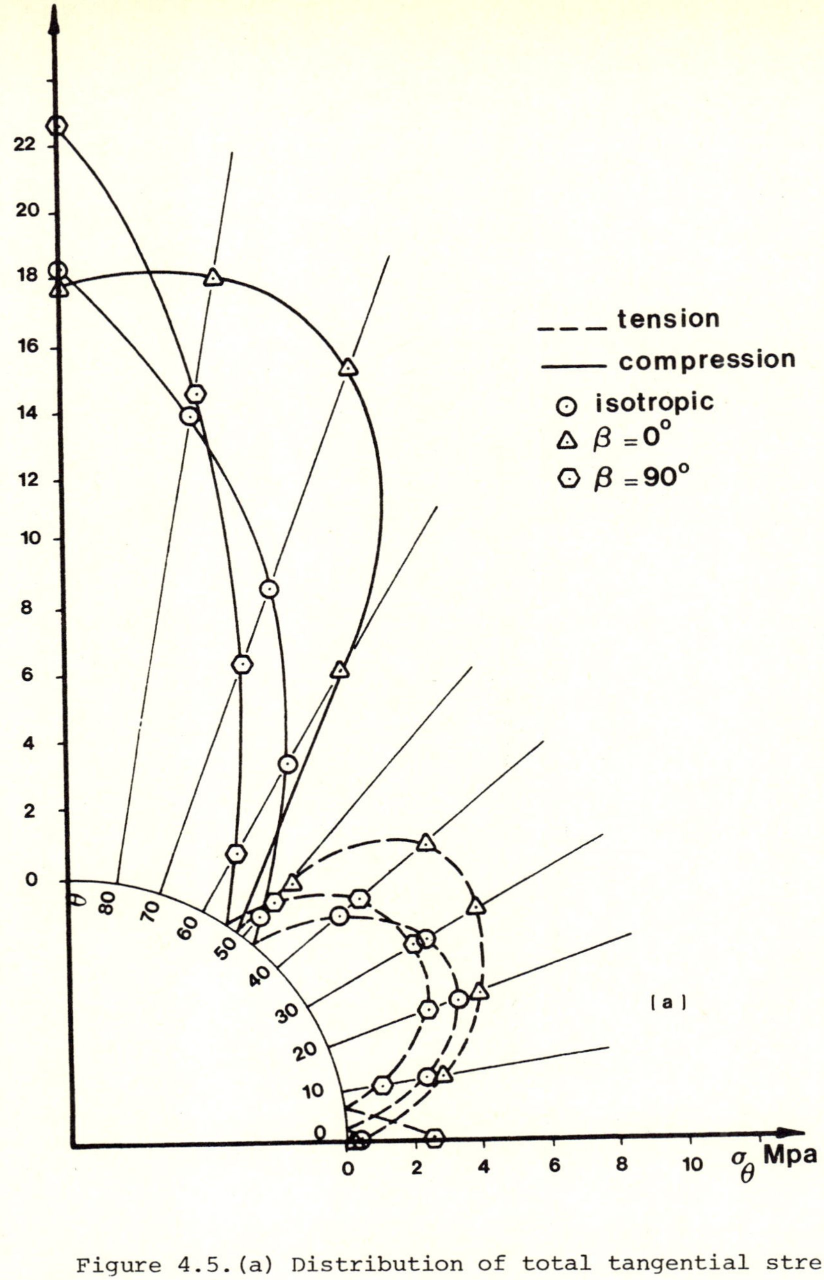

Figure 4.5.(a) Distribution of total tangential stresses.

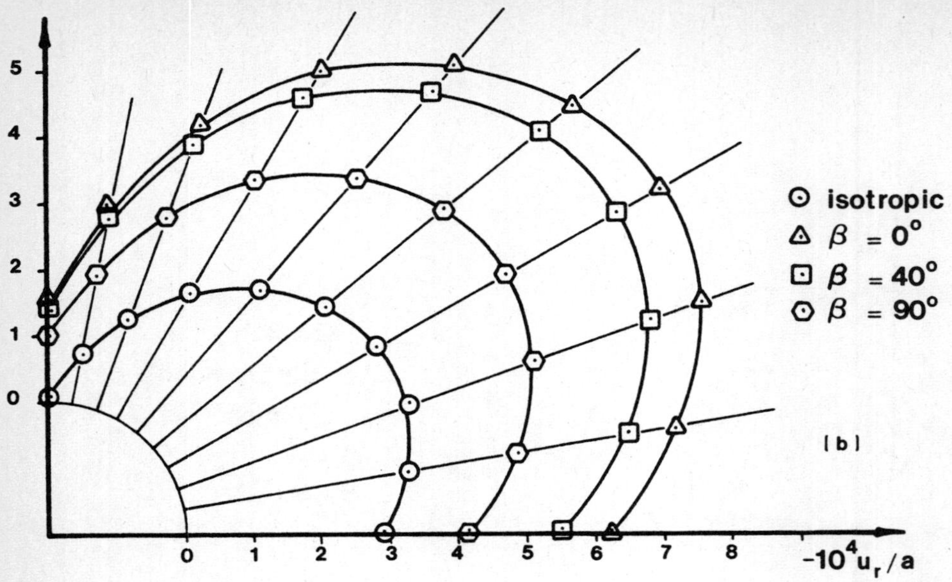

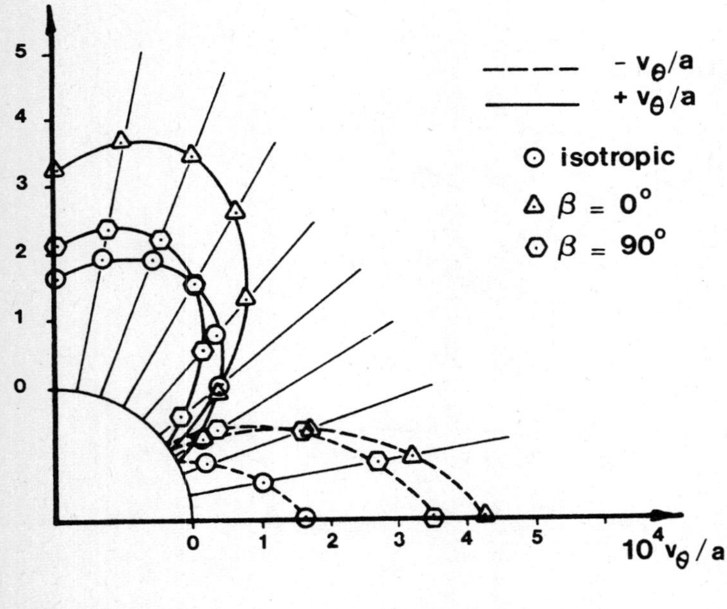

Figure 4.5. (b) Distribution of induced radial displacements.
(c) Distribution of induced tangential displacements.

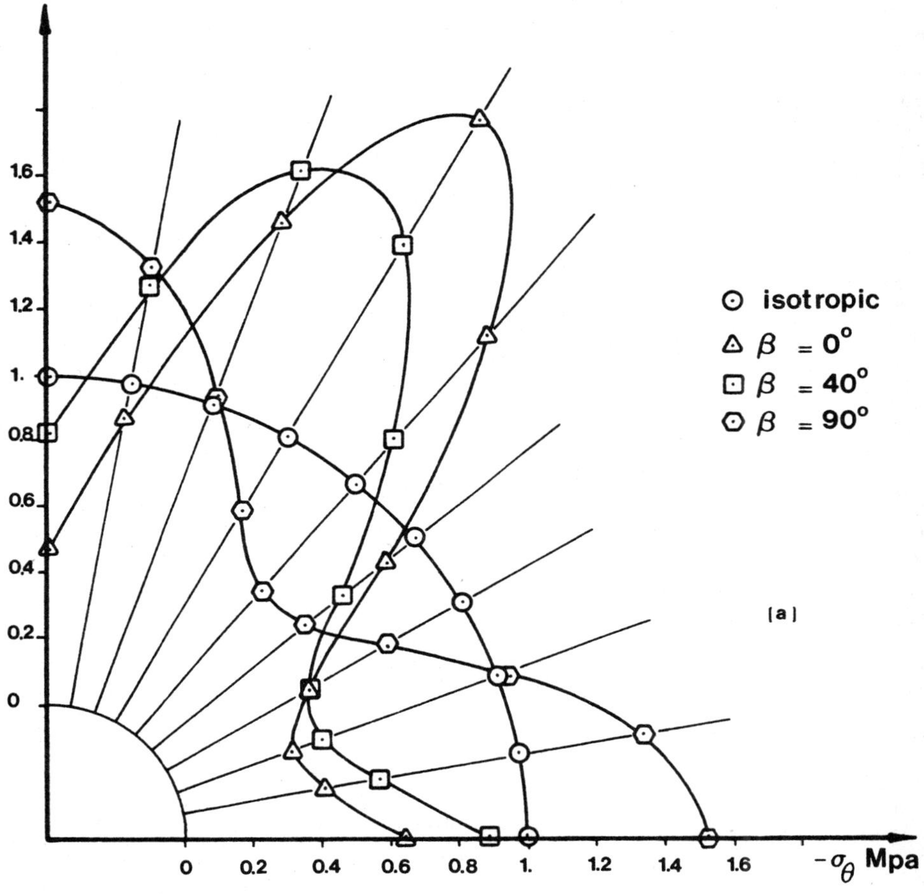

Figure 4.6. (a) Distribution of induced tangential stresses .

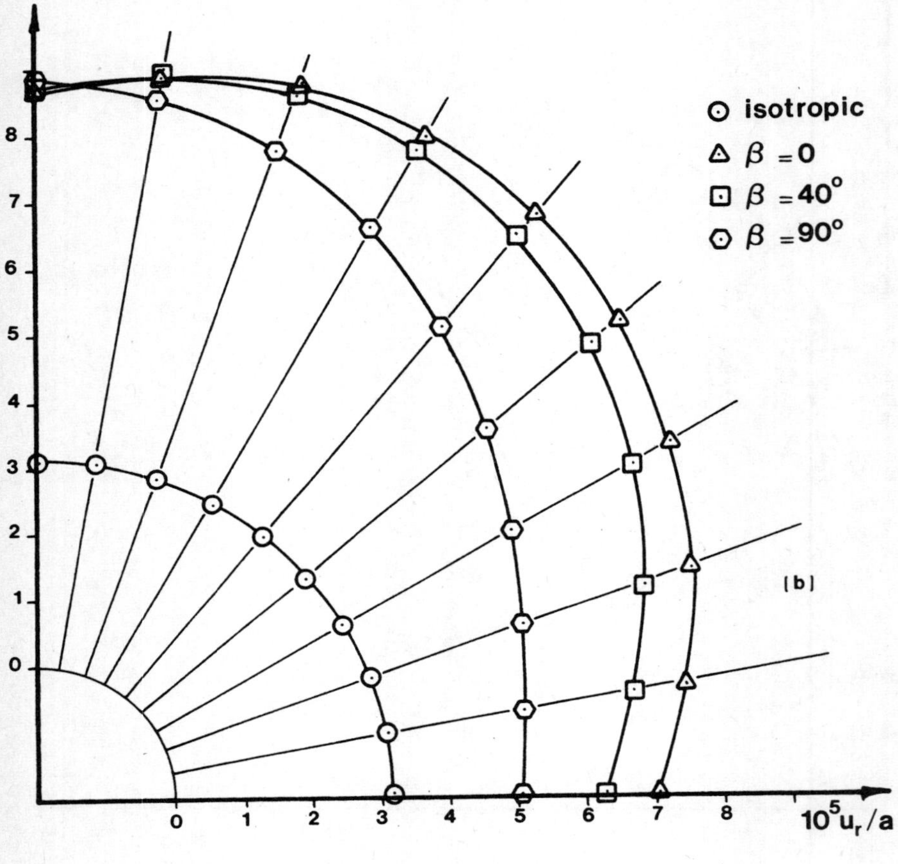

Figure 4.6. (b) Distribution of induced radial displacements.

stresses and radial displacements induced by the internal pressure along the wall of the hole (θ varying from 0 to 90 degrees) for the isotropic material and the transversely isotropic material when β is equal to 0, 40 to 90 degrees.

The two previous examples were concerned with an intact material, isotropic within a plane and strongly anisotropic in a direction perpendicular to it. It appears that the stress and displacement distributions are largely influenced by the anisotropic character of the material and its orientation with respect to the applied stress field. This influence is the largest when the anisotropy runs parallel to the hole axis and decreases appreciably as the anisotropy becomes perpendicular to it.

Similar conclusions were obtained by Berry and Fairhurst (1966) for a transversely isotropic rock and a hole drilled in a principal stress field direction. Both strongly and moderately anisotropic materials were investigated. The same conclusions also apply when the intact rock is isotropic or anisotropic but is cut by one joint set. The jointed material is then replaced by an equivalent anisotropic continuum (Appendix 4.6).

When the anisotropy is described by more than 5 coefficients, it is not only difficult to measure the material properties, as shown in section 2.3, but it is also difficult to draw general conclusions on the influence of anisotropy on the stress and displacement distributions around a hole. The influence exists and can be investigated for a given type and orientation of anisotropy using the models described in section 4.6.5.

Chapter 5: Elastic Equilibrium of an Anisotropic Homogeneous Body Bounded Internally by an Isotropic Hollow Inclusion of Circular Cross Section

5.1 Introduction

Consider an infinite, anisotropic, homogeneous and continuous medium bounded internally by a cylindrical surface of circular cross section. The hole contains a hollow inclusion assumed to be linear elastic, isotropic, continuous, homogeneous and perfectly bonded to the anisotropic body. Theoretically, both the hole and the inclusion are assumed to be infinitely long. The purpose of this chapter is twofold:

(i) to obtain relations between the components of a 3D stress field applied at infinity and the stress, strain and displacement components at each point of any cross section of the inclusion located far enough from its ends,

(ii) to study how far within the anisotropic body the stress field is disturbed by the presence of the hollow inclusion.

In the present analysis, body forces are assumed to be absent and allowance is made for a constant axial strain.

A similar analysis has been proposed for a full inclusion. Eshelby (1957) has shown that if an infinite elastic matrix (isotropic or anisotropic) contains a single elliptical or ellipsoidal inclusion, imposition of stresses at infinity produces uniform stress and strain fields within the inclusion. This fact has been used by several authors (Lekhnitskii, 1957;

Berry, 1970; Niwa and Hirashima, 1971). As far as the hollow inclusion is concerned, Duncan Fama and Pender (1980) proposed exact closed form solutions for the distributions of stresses and displacements within and around a circular isotropic inclusion in an isotropic body when a 3D stress field is applied at infinity. No such solutions exist when the body is anisotropic. However, anisotropy of a general character can be handled by the present model.

5.2 Geometry and Definition of the Problem

The geometry of the present problem is similar to the one of Chapter 4. In addition, let a and b be respectively the outer and inner radii of the hollow inclusion (Figure 5.1).

The boundary conditions can be defined in the cylindrical coordinate system (r,θ,z) as follows

(i) at $r = b$, $\sigma_r = \tau_{r\theta} = \tau_{rz} = 0$

(ii) at $r = a$ the stress components $\sigma_r, \tau_{r\theta}, \tau_{rz}$ and the displacement components u_r, v_θ, w are continuous,

(iii) as $r \longrightarrow \infty$, the stress components must approach the stress field components. Let (σ_o) be defined such that

$$(\sigma_o)^t = (\sigma_{x,o} \; \sigma_{y,o} \; \sigma_{z,o} \; \tau_{yz,o} \; \tau_{xz,o} \; \tau_{xy,o}) \tag{5.1}$$

Since the inclusion is infinitely long, we assume all the components of stress, strain and displacement, in any cross sectior of the inclusion located far enough from its ends, to be independent of z. However, allowance for an axial strain ε_{z_o} requires the longitudinal displacement component to be a function of z.

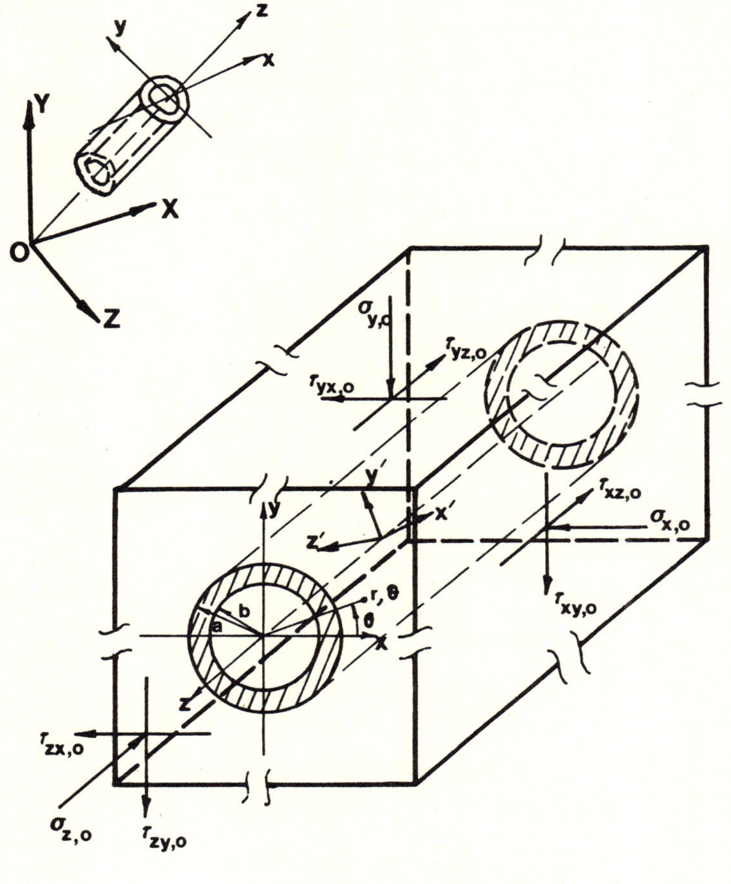

Figure 5.1. Geometry of the problem.

The present problem can be decomposed into two, referred to as problems (A) and (B) in this chapter (Figure 5.2). In order to simulate the perfect bonding between the inclusion and the anisotropic medium we introduce surface stress components X'_n, Y'_n, Z'_n in problem (A) and X_n, Y_n, Z_n in problem (B) such that

$$X'_n = -X_n \quad ; \quad Y'_n = -Y_n \quad ; \quad Z'_n = -Z_n \tag{5.2}$$

Furthermore, perfect bonding also implies that the axial strain ε_{z_0} must be the same both in the inclusion and in the anisotropic body.

Problem (B) can in turn be decomposed into two subproblems since the medium is assumed to be linear elastic:

(i) subproblem (B_1) which is the elastic equilibrium of an infinite anisotropic, homogeneous body bounded internally by a hole of circular cross section. A 3D stress field is applied at infinity. The deformation of the body takes place with a constant longitudinal strain ε_{z_0}.

(ii) subproblem (B_2) which is the elastic equilibrium of the same body loaded along its internal contour by stresses with components X_n, Y_n, Z_n. The deformation of the body takes place under zero axial strain.

In sections 5.3 and 5.4 problems (A) and (B) will be solved separately assuming Fourier series types boundary stresses X_n, Y_n, Z_n. Then, in section 5.5, the condition of continuity for the displacements u_r, v_θ, w at $r = a$ will allow us to express the coefficients of those Fourier series in terms of the stress field components.

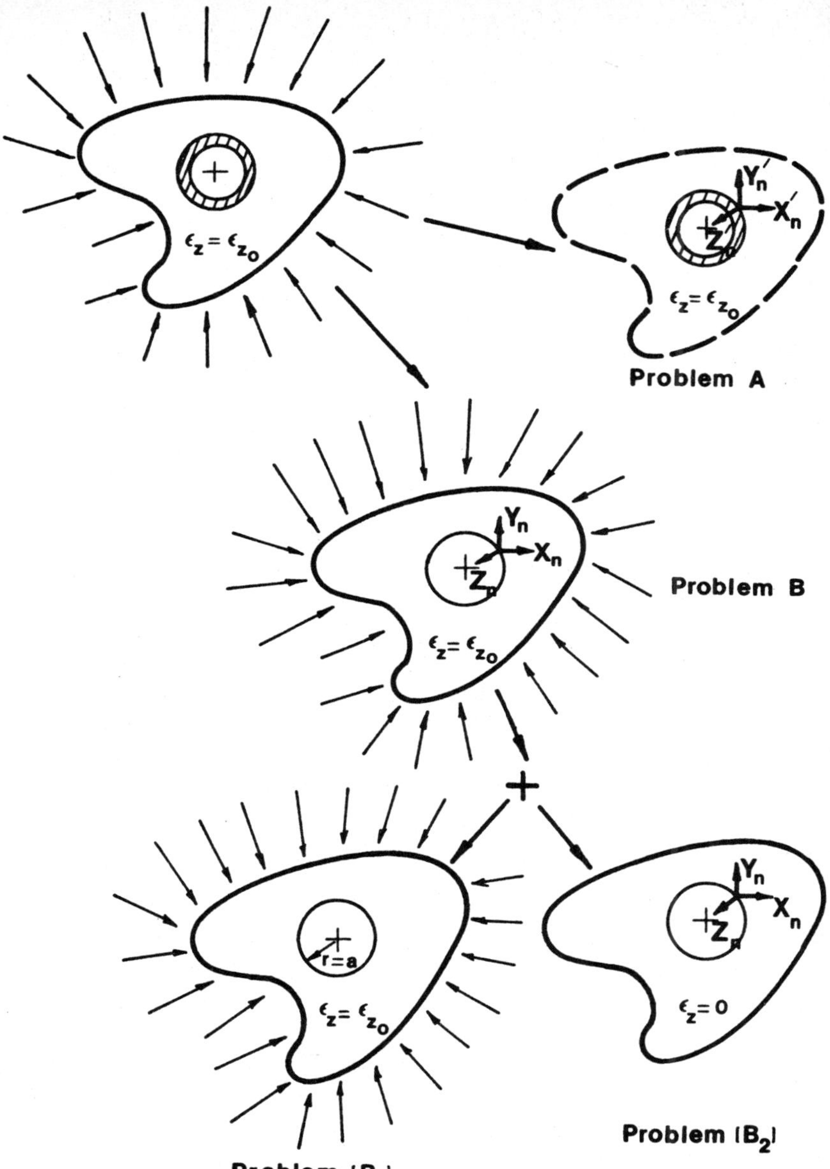

Figure 5.2. Decomposition of the problem.

5.3 Formulation of Problem (A)

5.3.1 Basic Equations

As shown in Figure 5.2, problem (A) is related to the equilibrium of a linear elastic, isotropic annulus loaded on its external boundary by surface stresses with components X'_n, Y'_n, Z'_n. At any point within the annulus, stress, strain and displacement components must satisfy the basic equations presented in section 4.3.1. For the present problem, they can be rewritten in the r, θ, z coordinate system as follows:

(i) Equations of Equilibrium

$$\begin{aligned} \frac{\partial \sigma_r}{\partial r} + \frac{1}{r}\frac{\partial \tau_{r\theta}}{\partial \theta} + \frac{\sigma_r - \sigma_\theta}{r} &= 0 \\ \frac{\partial \tau_{r\theta}}{\partial r} + \frac{1}{r}\frac{\partial \sigma_\theta}{\partial \theta} + \frac{2\tau_{r\theta}}{r} &= 0 \\ \frac{\partial \tau_{rz}}{\partial r} + \frac{1}{r}\frac{\partial \tau_{\theta z}}{\partial \theta} + \frac{\tau_{rz}}{r} &= 0 \end{aligned} \tag{5.3}$$

(ii) Strain Displacement Relations

$$\varepsilon_r = -\frac{\partial u_r}{\partial r} \;;\quad \varepsilon_\theta = -\frac{u_r}{r} - \frac{1}{r}\frac{\partial v_\theta}{\partial \theta} \;;\quad \varepsilon_z = -\frac{\partial w}{\partial z}$$

$$\gamma_{\theta z} = -\frac{1}{r}\frac{\partial w}{\partial \theta} \;;\quad \gamma_{rz} = -\frac{\partial w}{\partial r} \;;\quad \gamma_{r\theta} = -\frac{1}{r}\frac{\partial u_r}{\partial \theta} - \frac{\partial v_\theta}{\partial r} + \frac{v_\theta}{r} \tag{5.4}$$

(iii) Equations of Compatibility for Strains

Since by hypothesis the displacement components u, v and the strain components are independent of z and for a constant axial strain, eqs (4.4) reduce to the two first of eqs (4.10) with the constant k equal to zero. All the other compatibility equations are automatically satisfied.

(iv) Constitutive Relation

The elastic constants are reduced to two: a Youngs modulus E and a Poisson's ratio ν. The constitutive relation takes the form

$$\varepsilon_r = \frac{1}{E}\left(\sigma_r\left(1-\nu^2\right) - \nu\left(1+\nu\right)\sigma_\theta\right) - \nu\,\varepsilon_{z_0}$$

$$\varepsilon_\theta = \frac{1}{E}\left(\sigma_\theta\left(1-\nu^2\right) - \nu\left(1+\nu\right)\sigma_r\right) - \nu\,\varepsilon_{z_0}$$

$$\varepsilon_z = \varepsilon_{z_0}$$

$$\gamma_{\theta z} = \frac{1}{G}\tau_{\theta z}$$

$$\gamma_{rz} = \frac{1}{G}\tau_{rz}$$

$$\gamma_{r\theta} = \frac{1}{G}\tau_{r\theta} \qquad (5.5)$$

(v) <u>Boundary Conditions</u>

Along the external contour eqs (4.8) are replaced by the following

$$\cos(\underset{\sim}{n},x) = \frac{dy}{ds} \quad ; \quad \cos(\underset{\sim}{n},y) = -\frac{dx}{ds} \qquad (5.6)$$

Combining eqs (4.77) and (5.6), then substituting the result into eqs (4.7) and using the equations of <u>Appendix 4.4</u>, the boundary conditions on the outer contour of the annulus can be expressed in terms of σ_r, $\tau_{r\theta}$ and τ_{rz} as follows

$$\sigma_r = X_n' \cos\theta + Y_n' \sin\theta$$
$$\tau_{r\theta} = -X_n' \sin\theta + Y_n' \cos\theta$$
$$\tau_{rz} = Z_n' \qquad (5.7a)$$

On the inner contour $\sigma_r = \tau_{rz} = \tau_{r\theta} = 0$ (5.7b)

5.3.2 Beltrami Michell Equations of Compatibility

Eqs (5.3) are satisfied if we introduce two stress functions $\phi(r,\theta)$ and $\psi(r,\theta)$ such that

$$\sigma_r = \frac{1}{r}\frac{\partial \phi}{\partial r} + \frac{1}{r^2}\frac{\partial^2 \phi}{\partial \theta^2}$$

$$\sigma_\theta = \frac{\partial^2 \phi}{\partial r^2}$$

$$\tau_{r\theta} = \frac{1}{r^2}\frac{\partial \phi}{\partial \theta} - \frac{1}{r}\frac{\partial^2 \phi}{\partial r \partial \theta}$$

$$\tau_{rz} = \frac{1}{r}\frac{\partial \psi}{\partial \theta}$$

$$\tau_{\theta z} = -\frac{\partial \psi}{\partial r} \qquad (5.8)$$

If we combine the constitutive relation (eqs 5.5) with the two remaining equations of compatibility for strains expressed in the r, θ, z coordinate system, then ϕ and ψ must satisfy respectively a Biharmonic and a Laplace equations i.e.

$$\nabla^2\nabla^2\phi = 0 \qquad ; \qquad \nabla^2\psi = 0 \qquad (5.9)$$

where

$$\nabla^2 = \frac{\partial^2}{\partial r^2} + \frac{1}{r}\frac{\partial}{\partial r} + \frac{1}{r^2}\frac{\partial^2}{\partial \theta^2}$$

Comparing eqs (5.3) to (5.9), problem (A) can be decomposed into a plane and an antiplane problems as defined in section 4.4:

(i) a plane problem involving stress, strain and displacement components in the r, θ plane only. The stress function ϕ satisfies a Biharmonic equation and must be such that the stress components σ_r and $\tau_{r\theta}$ defined in eqs (5.8) satisfy the boundary conditions (eqs 5.7a,b),

(ii) an antiplane problem involving $\tau_{rz}, \tau_{\theta z}$ and the displacement w. The stress function Ψ satisfies a Laplace equation and must be such that the stress component τ_{rz} defined in eqs (5.8) satisfies also the boundary conditions (eqs 5.7a,b).

5.3.3 Fourier Series Boundary Conditions

Let us assume that X'_n, Y'_n, Z'_n can be expanded as Fourier series in $\cos m\theta$ and $\sin m\theta$ as follows

$$X'_n = -a_{o_x} - \sum_{m=1}^{N} \left(a_{m_x} \cos m\theta + b_{m_x} \sin m\theta \right) \tag{5.10}$$

and similar expressions for Y'_n and Z'_n where $a_{o_x}, a_{m_x}, b_{m_x}$ are replaced respectively by a_{oy}, a_{my}, b_{my} and a_{oz}, a_{mz}, b_{mz}. N is an arbitrary number whose values will be specified later in this chapter.

Since the annulus must be in equilibrium, the resultant forces (per unit length) of X'_n, Y'_n, Z'_n along the circular contour of radius $r = a$ must vanish. Then, according to eqs (4.71) a_{o_x}, a_{oy}, a_{oz} must be set equal to zero.

Combining eqs (5.7a) and the expressions for X'_n, Y'_n, Z'_n (eq 5.10), the boundary conditions on the outer contour of the annulus take the form

$$\begin{aligned} \sigma_r &= A_o + A_1 \cos\theta + B_1 \sin\theta + \sum_{n=2}^{N+1} \left(A_n \cos n\theta + B_n \sin n\theta \right) \\ \tau_{r\theta} &= C_o + C_1 \cos\theta + D_1 \sin\theta + \sum_{n=2}^{N+1} \left(C_n \cos n\theta + D_n \sin n\theta \right) \\ \tau_{rz} &= \sum_{n=1}^{N} \left(E_n \cos n\theta + F_n \sin n\theta \right) \end{aligned} \tag{5.11}$$

where

$$A_o = -\frac{(a_{1x} + b_{1y})}{2} \; ; \quad A_1 = -\frac{(a_{2x} + b_{2y})}{2} \; ; \quad B_1 = -\frac{(b_{2x} - a_{2y})}{2}$$

$$\left.\begin{aligned} A_n &= -\left(\frac{(a_{n+1_x} + b_{n+1_y})}{2} + \frac{(a_{n-1_x} - b_{n-1_y})}{2}\right) \\ B_n &= -\left(\frac{(b_{n+1_x} - a_{n+1_y})}{2} + \frac{(b_{n-1_x} + a_{n-1_y})}{2}\right) \end{aligned}\right\} n = 2, N-1$$

$$A_N = -\frac{(a_{N-1_x} - b_{N-1_y})}{2} \quad ; \quad B_N = -\frac{(b_{N-1_x} + a_{N-1_y})}{2}$$

$$A_{N+1} = -\frac{(a_{N_x} - b_{N_y})}{2} \quad ; \quad B_{N+1} = -\frac{(b_{N_x} + a_{N_y})}{2}$$

$$C_o = -\frac{(a_{1_y} - b_{1_x})}{2} \quad ; \quad C_1 = -B_1 \quad ; \quad D_1 = A_1$$

$$\left.\begin{aligned} C_n &= -\left(\frac{(a_{n+1_y} - b_{n+1_x})}{2} + \frac{(a_{n-1_y} + b_{n-1_x})}{2}\right) \\ D_n &= -\left(\frac{(b_{n+1_y} + a_{n+1_x})}{2} + \frac{(b_{n-1_y} - a_{n-1_x})}{2}\right) \end{aligned}\right\} n = 2, N-1$$

$$C_N = B_N \; ; \; D_N = -A_N \; ; \; C_{N+1} = B_{N+1} \; ; \; D_{N+1} = -A_{N+1}$$

$$E_n = -a_{n_z} \; ; \quad F_n = -b_{n_z} \quad ; \; n = 1, N \tag{5.12}$$

Note that in eqs (5.11) a constant term C_o appears in the expression of the shear stress $\tau_{r\theta}$. Equilibrium of the annulus prohibits any unbalancing moment. Then, C_o must vanish, i.e.

$$a_{1_y} - b_{1_x} = 0 \tag{5.13}$$

5.3.4 <u>Formulation of the Plane Problem</u>

Michell (1899) sought a general solution of the Biharmonic equation that included the θ dependence and was periodic in nature. The general solution is

$$\begin{aligned} \phi = {} & a_o + b_o \ln r + c_o r^2 + d_o r^2 \ln r + (A_o^* + B_o \ln r + C_o r^2 + D_o r^2 \ln r)\theta \\ & + \left(a_1 r + b_1 r \ln r + \frac{c_1}{r} + d_1 r^3\right)^{\sin\theta}_{\cos\theta} + (A_1 r + B_1 r \ln r)\theta^{\sin\theta}_{\cos\theta} \\ & + \sum_{n=2}^{\infty} \left(a_n r^n + b_n r^{2+n} + c_n r^{-n} + d_n r^{2-n}\right)^{\sin n\theta}_{\cos n\theta} \end{aligned} \tag{5.14}$$

$a_o, b_o, c_o, d_o, \text{--------}, a_n, b_n, c_n, d_n$ are coefficients that must be determined from the boundary conditions of the problem.

The notation for $\sin n\theta$ and $\cos n\theta$ is meant to imply, for instance, that $b_n r^{2+n} \sin n\theta$ is one term and $b'_n r^{2+n} \cos n\theta$ is another independent term available to satisfy the boundary conditions.

For the <u>annulus problem</u> with the present boundary conditions and the requirement that displacements and stresses must be single valued functions of the angle θ, it is shown in Little (1973)* that ϕ reduces to

$$\phi = a_o + b_o \ln r + c_o r^2 + A_o^* \theta + \left(a_1 r + \frac{c_1}{r} + d_1 r^3\right)_{\cos\theta}^{\sin\theta} + \sum_{n=2}^{\infty} \left(a_n r^n + b_n r^{2+n} + c_n r^{-n} + d_n r^{2-n}\right)_{\cos n\theta}^{\sin n\theta} \tag{5.15}$$

A <u>full inclusion</u> may be considered as a special hollow inclusion since $b = 0$. For that problem ϕ reduces to (Little, 1973)

$$\phi = c_o r^2 + \left(d_1 r^3\right)_{\cos\theta}^{\sin\theta} + \sum_{n=2}^{\infty} \left(a_n r^n + b_n r^{2+n}\right)_{\cos n\theta}^{\sin n\theta} \tag{5.16}$$

It is shown in <u>Appendix 5.1</u> how to calculate the general expressions for the stress, strain and displacement components at any point in the r, θ plane using eq (5.15) or (5.16) and the boundary conditions (5.11) and (5.7b). Conditions (5.7b) disappear when dealing with a full inclusion. The general expressions for the displacement components u_r/a and v_θ/a along the external contour of the inclusion are shown to be

---

* See also Timoshenko and Goodier (1970) pp 132-136.

equal to

$$\frac{u_r}{a} = (g_o A_o + \nu \varepsilon_{z_o}) + g_1 B_1 \sin\theta + g_1 A_1 \cos\theta + \sum_{n=2}^{N+1} (h_n B_n + k_n C_n) \sin n\theta$$
$$+ \sum_{n=2}^{N+1} (h_n A_n - k_n D_n) \cos n\theta + \frac{s_1^+}{a} \cos\theta + \frac{s_2^+}{a} \sin\theta$$

$$\frac{v_\theta}{a} = r_1 B_1 \cos\theta - r_1 A_1 \sin\theta + \sum_{n=2}^{N+1} (s_n B_n + t_n C_n) \cos n\theta$$
$$+ \sum_{n=2}^{N+1} (-s_n A_n + t_n D_n) \sin n\theta + \frac{s_2^+}{a} \cos\theta - \frac{s_1^+}{a} \sin\theta + \varpi_1^+$$

(5.17)

where

s_1^+, s_2^+, ϖ_1^+ are rigid body translations in the x,y directions and a rigid body rotation around the z axis,

$A_o, A_1, B_1, B_n, C_n, A_n, D_n$ are coefficients appearing in the boundary conditions (eqs 5.11),

$g_o, g_1, r_1, h_n, k_n, s_n, t_n$ are functions of the elastic properties of the inclusion, its geometry defined by the ratio a/b and the n values varying between 2 and $N+1$. Those quantities are defined in eqs (24b), (25b) of <u>Appendix 5.1</u> for a hollow inclusion and in eqs (28) for a full inclusion.

For any value of n, k_n is found to be equal to s_n. However, h_n is equal to t_n for a full inclusion only. In order to illustrate these remarks, Figure 5.3 shows the variations of the quantities $E(h_n + k_n)$ and $E(s_n + t_n)$ with n and a/b for an inclusion with a Poisson's ratio equal to 0.3. The case $a/b = \infty$ corresponds to a full inclusion.

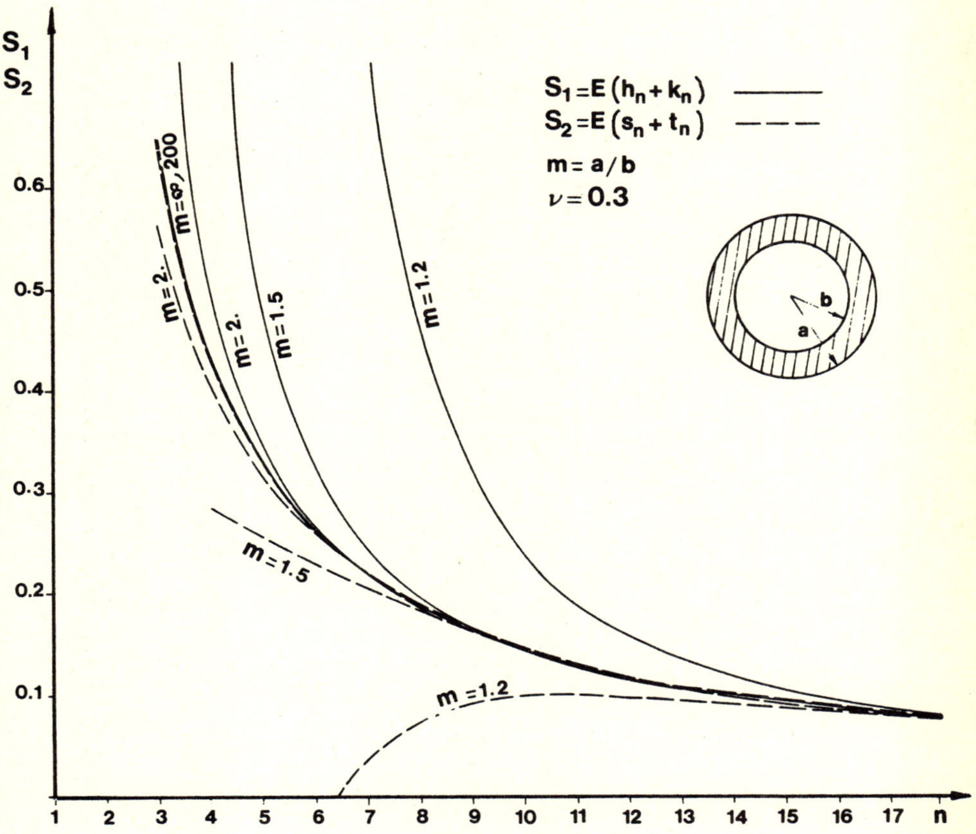

Figure 5.3. Variation of $E(h_n+k_n)$ and $E(s_n+t_n)$ with n .

5.3.5 Formulation of the Antiplane Problem

In Appendix 5.2, it is shown that a general solution of the Laplace equation can be written as follows

$$\Psi = a_o^* + b_o^* \ln r + c_o^* \theta \ln r + d_o^* \theta + \sum_{n=1}^{\infty} \left(\left(a_n^* r^n + \frac{b_n^*}{r^n} \right) \cos n\theta + \left(c_n^* r^n + \frac{d_n^*}{r^n} \right) \sin n\theta \right) \tag{5.18}$$

$a_o^*, b_o^*, c_o^*, d_o^*, a_n^*, b_n^*, c_n^*, d_n^*$ are coefficients that must be determined from the boundary conditions of the problem. The general expressions for the stress and displacement components $\tau_{rz}, \tau_{\theta z}, w$ at any point r, θ, z within a hollow inclusion are presented. In particular the displacement component w/a along the external contour of the inclusion is found to be equal to

$$\frac{w}{a} = -\frac{z}{a} \varepsilon_{z_o} + \sum_{n=1}^{N} \left(u_n E_n \cos n\theta + u_n F_n \sin n\theta \right) + \frac{s_3^+}{a} \tag{5.19}$$

where

s_3^+ is a rigid body translation in the z direction,

E_n, F_n are coefficients appearing in the boundary conditions (eq 5.11), u_n is function of the elastic properties of the inclusion, its geometry defined by the ratio a/b and the n values varying between 1 and N.

For a full inclusion, the general expressions for the stress and displacement components can be obtained directly by setting $a/b \to \infty$ into the expressions for a hollow inclusion.

5.4 Formulation of Problem (B)

As shown in Figure 5.2, problem (B) is equivalent to the

elastic equilibrium of an infinite, anisotropic and homogeneous body bounded internally by a hole of circular cross section. A 3D stress field is applied at infinity. Boundary stresses are also applied along the circular contour. Their components X_n, Y_n, Z_n are obtained by substituting eqs (5.10) into eq (5.2). $a_{o_x}, a_{o_y}, a_{o_z}$ must vanish and condition (5.13) must be satisfied.

If the constant axial strain ε_{z_o} is defined by eq (4.84), problem (B) can be solved using the general theory developed in Chapter 4. In particular, a closed form solution of subproblem (B_1) has been already derived in section 4.6.4 and in Appendix 4.5. Subproblem (B_2) corresponds to a body with Fourier series types boundary stresses X_n, Y_n, Z_n on its internal contour and deforming under zero axial strain. The solution to that problem has been also proposed in sections 4.6.1, 4.6.2 and in Appendix 4.3.

It is shown in Appendix 5.3 that for problem (B) the total displacement components $u_r/a, v_\theta/a$ and w/a along the contour of the hole can be written as follows

$$\frac{u_r}{a} = \left(M_{u_1}^{(o)} + M_{u_1}^{(h)}\right) + \left(\frac{u_t^{(o)} + u^{(o)}}{a}\right)\cos\theta + \left(\frac{v_t^{(o)} + v^{(o)}}{a}\right)\sin\theta + \left(M_{u_2}^{(o)} + M_{u_2}^{(h)}\right)\cos 2\theta + \left(M_{u_3}^{(o)} + M_{u_3}^{(h)}\right)\sin 2\theta + \sum_{m=1}^{N}\left(\frac{X_m^{(1)} + Y_m^{(2)}}{2}\right)\cos(m+1)\theta + \left(\frac{X_m^{(2)} - Y_m^{(1)}}{2}\right)\sin(m+1)\theta + \sum_{m=1}^{N}\left(\frac{X_m^{(1)} - Y_m^{(2)}}{2}\right)\cos(m-1)\theta + \left(\frac{-Y_m^{(1)} - X_m^{(2)}}{2}\right)\sin(m-1)\theta$$

$$\frac{v_\theta}{a} = \left(\omega + \omega_{xy}^{(o)}\right) + M_{v_1}^{(h)} + \left(\frac{v_t^{(o)} + v^{(o)}}{a}\right)\cos\theta - \left(\frac{u_t^{(o)} + u^{(o)}}{a}\right)\sin\theta +$$

$$\left(M_{v_2}^{(o)}+M_{v_2}^{(h)}\right)\cos 2\theta+\left(M_{v_3}^{(o)}+M_{v_3}^{(h)}\right)\sin 2\theta+\sum_{m=1}^{N}\left(\frac{X_m^{(2)}-Y_m^{(1)}}{2}\right)\cos(m+1)\theta+$$

$$\left(\frac{-X_m^{(1)}-Y_m^{(2)}}{2}\right)\sin(m+1)\theta+\sum_{m=1}^{N}\left(\frac{X_m^{(2)}+Y_m^{(1)}}{2}\right)\cos(m-1)\theta+\left(\frac{X_m^{(1)}-Y_m^{(2)}}{2}\right)\sin(m-1)\theta$$

$$\frac{w}{a}=\left(\frac{w_t^{(o)}+w^{(o)}}{a}\right)-\frac{z}{a}\varepsilon_{z_o}+\left(M_{w_2}^{(o)}+M_{w_2}^{(h)}\right)\cos\theta+$$

$$\left(M_{w_3}^{(o)}+M_{w_3}^{(h)}\right)\sin\theta+$$

$$\sum_{m=1}^{N}\left(X_m^{(3)}\cos m\theta-Y_m^{(3)}\sin m\theta\right) \tag{5.20}$$

where $(u_t^{(o)}+u^{(o)})$, $(v_t^{(o)}+v^{(o)})$, $(w_t^{(o)}+w^{(o)})$ are rigid body translations in the x,y,z directions and $(\omega+\omega_{xy}^{(o)})$ is a rigid body rotation around the z axis, $M_{u_i}^{(o)}$, $M_{v_i}^{(o)}$, $M_{w_i}^{(o)}$, $M_{u_i}^{(h)}$, $M_{v_i}^{(h)}$, $M_{w_i}^{(h)}$, $(i=1,2,3)$ are functions of the 3D stress field components, the anisotropic properties of the body and the orientation of the hole with respect to the directions of anisotropy, $X_m^{(1)}, X_m^{(2)}, X_m^{(3)}, Y_m^{(1)}, Y_m^{(2)}, Y_m^{(3)}$ $(m=1,N)$ depend upon the coefficients $a_{m_x}, b_{m_x}, a_{m_y}, b_{m_y}, a_{m_z}, b_{m_z}$ entering into the expressions for the boundary stresses. They also depend on the anisotropic properties of the body and the orientation of the hole with respect to the directions of anisotropy.

5.5 Condition of Continuity

In sections 5.3 and 5.4 we calculated the general expressions for the displacement components along the contour $r = a$ that is common to both the anisotropic body and the isotropic inclusion. Displacement components were expressed in terms of dimensionless ratios u_r/a, v_θ/a and w/a. Furthermore, both hollow and full inclusions were considered.

Since, by hypothesis, the inclusion and the anisotropic body are perfectly bonded, it is now possible to express the condition of continuity across the contour $r = a$ by equating the displacement components derived in eqs (5.17) and (5.19) to those in eqs (5.20)*. Furthermore, let the rigid body translations be such that

$$\begin{aligned} s_1^+ &= u^{(0)} + u_t^{(0)} \\ s_2^+ &= v^{(0)} + v_t^{(0)} \\ s_3^+ &= w^{(0)} + w_t^{(0)} \end{aligned} \qquad (5.21)$$

and ϖ' be defined as follows:

$$\varpi' = \varpi_1^+ - \omega_{xy}^{(0)} \qquad (5.22)$$

Since the displacement components can be expressed in terms of Fourier series in $\sin n\theta$ and $\cos n\theta$, the continuity condition will be satisfied if and only if the corresponding coefficients of sin and cos appearing on both sides of the equality sign are equal. This, plus conditions (5.13) and (5.21) allow us to write the following system of $6N+1$ equations whose unknowns are the $6N$ coefficients $a_{m_x}, b_{m_x}, a_{m_y}, b_{m_y}, a_{m_z}, b_{m_z}$ $(m = 1, N)$ and the rigid body rotation $\omega - \varpi'$

---

* Note that the condition of continuity can be also expressed in terms of strains since ε_θ, ε_z and $\gamma_{\theta z}$ must be continuous across the boundary $r = a$ according to eqs (5.4).

$$\frac{(X_1^{(1)} - Y_1^{(2)})}{2} - g_o A_o = \nu E_{z_o} - M_{u_1}^{(o)} - M_{u_1}^{(h)} \qquad (a)$$

$$(\omega - \varpi') + \frac{(X_1^{(2)} + Y_1^{(1)})}{2} = - M_{v_1}^{(h)} \qquad (b)$$

$$X_2^{(1)} - Y_2^{(2)} - (g_1 - r_1) A_1 = 0 \qquad (c)$$

$$X_2^{(2)} + Y_2^{(1)} + (g_1 - r_1) B_1 = 0 \qquad (d)$$

$$\frac{(X_3^{(1)} - Y_3^{(2)})}{2} + \frac{(X_1^{(1)} + Y_1^{(2)})}{2} - (h_2 A_2 - k_2 D_2) = - M_{u_2}^{(o)} - M_{u_2}^{(h)} \qquad (e)$$

$$\frac{(X_3^{(2)} + Y_3^{(1)})}{2} + \frac{(X_1^{(2)} - Y_1^{(1)})}{2} - (s_2 B_2 + t_2 C_2) = - M_{v_2}^{(o)} - M_{v_2}^{(h)} \qquad (f)$$

$$\frac{(-Y_3^{(1)} - X_3^{(2)})}{2} + \frac{(X_1^{(2)} - Y_1^{(1)})}{2} - (h_2 B_2 + k_2 C_2) = - M_{u_3}^{(o)} - M_{u_3}^{(h)} \qquad (g)$$

$$\frac{(X_3^{(1)} - Y_3^{(2)})}{2} + \frac{(-X_1^{(1)} - Y_1^{(2)})}{2} - (-s_2 A_2 + t_2 D_2) = - M_{v_3}^{(o)} - M_{v_3}^{(h)} \qquad (h)$$

$$\left.\begin{aligned}
&(X_{n+1}^{(1)} - Y_{n+1}^{(2)}) + (X_{n-1}^{(1)} + Y_{n-1}^{(2)}) - 2(h_n A_n - k_n D_n) = 0 && (i)\\
&(X_{n+1}^{(2)} + Y_{n+1}^{(1)}) + (X_{n-1}^{(2)} - Y_{n-1}^{(1)}) - 2(s_n B_n + t_n C_n) = 0 && (j)\\
&(-Y_{n+1}^{(1)} - X_{n+1}^{(2)}) + (X_{n-1}^{(2)} - Y_{n-1}^{(1)}) - 2(h_n B_n + k_n C_n) = 0 && (k)\\
&(X_{n+1}^{(1)} - Y_{n+1}^{(2)}) + (-X_{n-1}^{(1)} - Y_{n-1}^{(2)}) - 2(-s_n A_n + t_n D_n) = 0 && (l)
\end{aligned}\right\} n : 3, N-1$$

$$(X_{N-1}^{(1)} + Y_{N-1}^{(2)}) - 2(h_N + k_N) A_N = 0 \qquad (m)$$

$$(X_{N-1}^{(2)} - Y_{N-1}^{(1)}) - 2(h_N + k_N) B_N = 0 \qquad (o)$$

$$(X_N^{(1)} + Y_N^{(2)}) - 2(h_{N+1} + k_{N+1}) A_{N+1} = 0 \qquad (p)$$

$$(X_N^{(2)} - Y_N^{(1)}) - 2(h_{N+1} + k_{N+1}) B_{N+1} = 0 \qquad (q)$$

$$X_1^{(3)} - u_1 E_1 = - M_{w_2}^{(o)} - M_{w_2}^{(h)} \qquad (r)$$

$$-Y_1^{(3)} - u_1 F_1 = - M_{w_3}^{(o)} - M_{w_3}^{(h)} \qquad (s)$$

$$\left.\begin{array}{l} X_n^{(3)} - u_n E_n = 0 \\ Y_n^{(3)} + u_n F_n = 0 \end{array}\right\} n: 2, N \qquad \begin{array}{r}(t)\\(u)\end{array}$$

$$a_{1y} - b_{1x} = 0 \qquad (v)$$

(5.23)*

Combining eqs (5.23) with eqs (5.12) and eqs (8) in Appendix 5.3, the left hand side of system (5.23) can be rewritten in terms of the $6N + 1$ unknowns.

However, it can be shown that the continuity condition implies also that six equations must be satisfied by the unknowns in addition to eqs (5.23). Those equations take the form

$$A_1 = 0 \qquad (a)$$

$$B_1 = 0 \qquad (b)$$

$$X_{N-1}^{(1)} + Y_{N-1}^{(2)} - 2(\lambda_N + t_N) A_N = 0 \qquad (c)$$

$$X_{N-1}^{(2)} - Y_{N-1}^{(1)} - 2(\lambda_N + t_N) B_N = 0 \qquad (d)$$

$$X_N^{(2)} - Y_N^{(1)} - 2(\lambda_{N+1} + t_{N+1}) B_{N+1} = 0 \qquad (e)$$

$$X_N^{(1)} + Y_N^{(2)} - 2(\lambda_{N+1} + t_{N+1}) A_{N+1} = 0 \qquad (f)$$

(5.24)*

Let us gather in one system, the equations among the $6N+1$ (eqs 5.23) that involve the coefficients $a_{mx}, b_{mx}, a_{my}, b_{my}, a_{mz}, b_{mz}$ with even values of m only. It is found that those quantities are solution of a homogeneous system. Unless the matrix of the coefficients of that system is singular, the system has the trivial solution. Eqs (5.24) are then reduced in number to two:

---

* In writing eqs (5.23 m,o,p,q) and eqs (5.24 c,d,e,f) use has been made of the relations $B_N = C_N$; $D_N = -A_N$; $B_{N+1} = C_{N+1}$; $D_{N+1} = -A_{N+1}$ in eqs (5.12).

$$X^{(1)}_{N-1} + Y^{(2)}_{N-1} - 2\,(s_N + t_N)\,A_N = 0$$

$$X^{(2)}_{N-1} - Y^{(1)}_{N-1} - 2\,(s_N + t_N)\,B_N = 0$$

(5.25a)

when N is <u>even</u> or,

$$X^{(2)}_{N} - Y^{(1)}_{N} - 2\,(s_{N+1} + t_{N+1})\,B_{N+1} = 0$$

$$X^{(1)}_{N} + Y^{(2)}_{N} - 2\,(s_{N+1} + t_{N+1})\,A_{N+1} = 0$$

(5.25b)

when N is <u>odd</u>.

Note that eqs (5.25a) or (5.25b) are automatically satisfied for a <u>full inclusion</u> since $h_n + k_n = s_n + t_n$ for any value of n. Then, eqs (5.24c,d) or (5.24e,f) are respectively identical to eqs (5.23m,o) and (5.23p,q).

For a <u>hollow inclusion</u>, those two extra equations will be approximately satisfied if the left hand sides of eqs (5.25a) or (5.25b) are infinetisimal quantities or if the values of N or $N + 1$ are such that $s_N + t_N$ or $s_{N+1} + t_{N+1}$ are approximately equal to $h_N + k_N$ or $h_{N+1} + k_{N+1}$ as shown in Figure 5.3. Numerical examples presented in section 5.8 show that this approximate character is indeed satisfied in practice.

Since all the unknowns $a_{m_x}, b_{m_x}, a_{m_y}, b_{m_y}, a_{m_z}, b_{m_z}$ vanish for any even value of m, combining eqs (5.12) with eqs (19) to (23) in <u>Appendix 5.1</u> and eqs (16) in <u>Appendix 5.2</u>, it appears that for a given radius r within the inclusion, $\sigma_r, \sigma_\theta, \tau_{r\theta}, u_r$ and v_θ are trigonometric functions of $\sin n\theta$ and $\cos n\theta$ with even values of n only.* However, τ_{rz}, $\tau_{\theta z}$ and w are similar functions with odd values of n only. This could have been expected since the system must be in equilibrium and present a circular symmetry.

5.6 Closed Form Solutions

Let us define the $(6N+1, 1)$ column matrix (X) such that

$$(X)^t = (\omega - \varpi' \ a_{1x} \ b_{1x} \ a_{1y} \ b_{1y} \ a_{1z} \ b_{1z} \text{------} a_{Nx} \ b_{Nx} \ a_{Ny} \ b_{Ny} \ a_{Nz} \ b_{Nz}) \tag{5.26}$$

If we substitute eqs (5.12) and eqs (8) of Appendix 5.3 into the left hand side of system (5.23) and if we substitute eqs (2) and (4) of Appendix 5.3 in the right hand side of the same system, then, eqs (5.23) can be written in matrix form as follows

$$(A_x)(X) = (C_x)(\sigma_o) \tag{5.27}$$

where

(A_x) and (C_x) are respectively $(6N+1, 6N+1)$ and $(6N+1, 6)$ matrices whose components are functions of the anisotropic properties of the body, the orientation of the hole with respect to the directions of anisotropy, the elastic properties of the inclusion and its geometry i.e. the ratio a/b (for a hollow inclusion only), (σ_o) is the stress field component matrix as defined in eq (5.1).

---

* a similar conclusion applies also to σ_z according to eq (5) Appendix 5.1.

Therefore, as long as (A_x) is non singular, eq (5.27) can be solved for matrix (X).

Let us define the $(6N+7, 1)$ column matrix (Y) such that

$$(Y)^t = (A_0\ A_1\ B_1\ C_1\ D_1\ E_1\ F_1 ------ A_{N+1}\ B_{N+1}\ C_{N+1}\ D_{N+1}\ C_0\ \omega-\varpi') \tag{5.28}$$

Eqs (5.12) can be rewritten in matrix form as follows

$$(Y) = (D_x)(X) \tag{5.29}$$

where (D_x) is a $(6N+7, 6N+1)$ matrix whose components are equal to $0.5, -0.5, 1$ or -1.

At any point $(r/a, \theta)$ within the inclusion, the components of the state of stress are defined by eqs (5), (19), (20), and (21) in Appendix 5.1 and by eqs (16) in Appendix 5.2. Using matrix (Y) they can be rewritten in matrix form as follows

$$\begin{bmatrix} \sigma_r \\ \sigma_\theta \\ \sigma_z \\ \tau_{\theta z} \\ \tau_{rz} \\ \tau_{r\theta} \end{bmatrix} = (E_x)(Y) + (F_x)(\sigma_0) \tag{5.30}$$

or $$(\sigma)_{r\theta z} = (E_x)(Y) + (F_x)(\sigma_0) \tag{5.31}$$

(E_x) is a $(6, 6N+7)$ matrix whose components are functions of r/a, θ and the geometry of the inclusion (for a hollow inclusion only), (F_x) is a $(6, 6)$ matrix whose components are all equal to 0 except those of the third line where

$$f_{x_{3i}} = E\, a_{3i} \qquad (i = 1, 6) \tag{5.32}$$

Similarly, the displacement components $u_r/a, v_\theta/a$ and w/a at any point $(r/a, \theta)$ within the inclusion defined by eqs (22) and (23) in Appendix 5.1 and by eqs (16) in Appendix 5.2 can be rewritten in matrix form as follows*

$$\begin{bmatrix} \frac{u_r}{a} \\ \frac{v_\theta}{a} \\ \frac{w}{a} \end{bmatrix} = (E_{xu})(Y) + (F_{xu})(\sigma_o) \tag{5.33}$$

or

$$(u)_{r\theta z} = (E_{xu})(Y) + (F_{xu})(\sigma_o) \tag{5.34}$$

where

(E_{xu}) is a $(3, 6N+7)$ matrix whose components are functions of r/a, θ, the elastic properties of the inclusion and its geometry (for a hollow inclusion only).

(F_{xu}) is a $(3, 6)$ matrix whose components are equal to zero except those of the first and third lines where

$$f_{xu_{1i}} = \nu\left(\frac{r}{a}\right) a_{3i} \quad ; \quad f_{xu_{3i}} = -\left(\frac{z}{a}\right) a_{3i} \quad (i = 1, 6) \tag{5.35}$$

Substituting eqs (5.27) and (5.29) into eqs (5.31) and (5.34), it follows that

---

* In writing eq (5.33) rigid body translations and rotation $\delta_1^+, \delta_2^+, \delta_3^+, \varpi_1^+$ were arbitrarily chosen to be equal to $\delta_1^+ = \delta_2^+ = \delta_3^+ = 0$; $\varpi_1^+ = -(\omega - \varpi')$. According to eqs (5.21) and (5.22) this is equivalent to say that within the anisotropic body $u^{(o)} + u_+^{(o)} = v^{(o)} + v_+^{(o)} = w^{(o)} + w_+^{(o)} = \omega + \omega_{xy}^{(o)} = 0$.

$$(\sigma)_{r\theta z} = \left((E_x)(D_x)(A_x)^{-1}(C_x) + (F_x)\right)(\sigma_o) \tag{5.36}$$

$$(U)_{r\theta z} = \left((E_{xu})(D_x)(A_x)^{-1}(C_x) + (F_{xu})\right)(\sigma_o) \tag{5.37}$$

If we call (G_x) the compliance matrix for the inclusion whose components are defined by eqs (5.5), then at any point $(r/a, \theta)$ within the inclusion the strain components are equal to

$$(\varepsilon)_{r\theta z} = (Q)(\sigma_o) \tag{5.38}$$

where

$$(Q) = (G_x)\left[(E_x)(D_x)(A_x)^{-1}(C_x) + (F_x)\right] \tag{5.39}$$

Similarly, at any point $(r/a, z/a, \theta)$ within the inclusion eq (5.37) can be rewritten as

$$(U)_{r\theta z} = (Q_u)(\sigma_o) \tag{5.40}$$

where

$$(Q_u) = (E_{xu})(D_x)(A_x)^{-1}(C_x) + (F_{xu}) \tag{5.41}$$

In conclusion, if we know the following parameters:

(i) the elastic properties of the inclusion defined by a Young modulus E and a Poisson's ratio ν,

(ii) the geometry of the inclusion defined by the ratio

a/b *,

(iii) the anisotropic properties of the body and the orientation of the anisotropy with respect to the hole axis,

then, the components of matrices (A_x) and (C_x) can be calculated. Similarly, at any point $(r/a, z/a, \theta)$ within the inclusion the components of matrices (Q) and (Q_u) can be determined.+

Therefore, at any point within the inclusion the strain and displacement components can be related to the stress field components applied at infinity through eqs (5.38) and (5.40). This is always possible as long as matrix (A_x) is non singular.

5.7 Remarks

5.7.1 Rigid Body Translations

In solving system (5.23) with the six additional equations (5.24) we assumed eqs (5.21) to be satisfied. If we do not make those assumptions eqs (5.23) and (5.24) remain the same except

---

* This is only valid for a hollow inclusion. For a full inclusion the geometric property of the inclusion does not enter in the present dimensionless analysis .

+ The general expressions for the components of matrices (A_x), (C_x), (D_x), (E_x), (F_x), (E_{xu}), (F_{xu}), (Q) and (Q_u) are not presented in the present analysis. However, they have been computerized and included into programs Berni 1 and Berni 3 (see Appendices 6.2 and 6.4).

eqs (5.24a,b) that now take the form

$$\left(\frac{u^{(0)} + u_t^{(0)} - \Delta_1^+}{a}\right) - \left(\frac{r_1 + g_1}{2}\right) A_1 = 0$$

$$\left(\frac{v^{(0)} + v_t^{(0)} - \Delta_2^+}{a}\right) - \left(\frac{r_1 + g_1}{2}\right) B_1 = 0 \tag{5.42}$$

Furthermore, it can be shown that there is an other equation in addition to eqs (5.24) that must be satisfied. It originates from the constant terms in the displacement component w and takes the form

$$w^{(0)} + w_t^{(0)} - \Delta_3^+ = 0 \tag{5.43}$$

Since all the coefficients $a_{mx}, b_{mx}, a_{my}, b_{my}, a_{mz}, b_{mz}$ with even values of m vanish, it implies that A_1, B_1 must be zero. Therefore, eqs (5.42) and (5.43) reduce to eqs (5.21).

5.7.2 Variations in Length of Oblique Distances

In eq (5.40) we were able to relate the displacement components at any point $(r/a, z/a, \theta)$ within the inclusion to the 3D stress field components applied at infinity. Another quantity that is of special interest is the change in length between two points located in different cross sections of the inclusion. Especially, it is shown in Appendix 5.4 that for two points P_1 and P_2 located on the inner surface of a hollow inclusion with the geometry defined in Figure 5.4, the following relation applies*

---

* as long as u_r, v_θ and w are small in comparison to L, a and b.

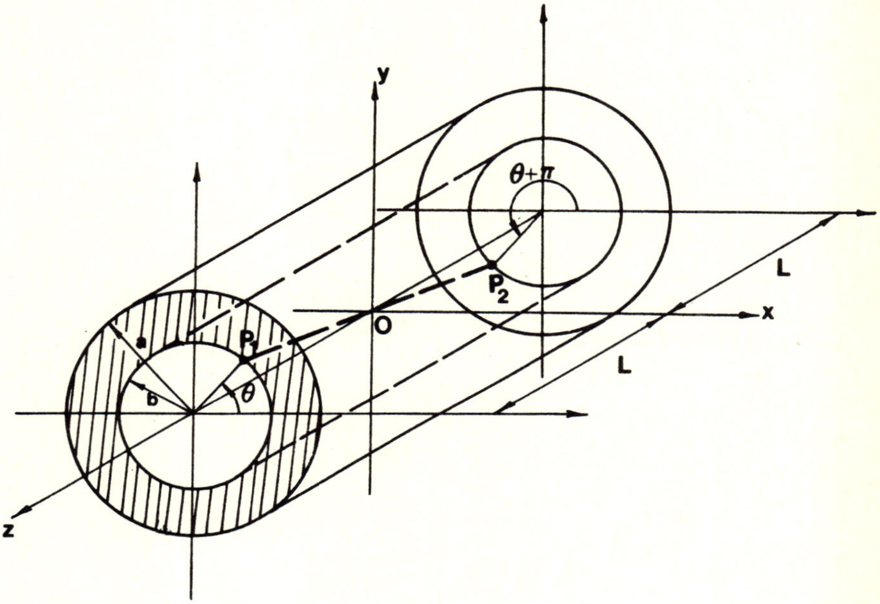

Figure 5.4. Variation in length of the oblique distance P_1P_2 .
Geometry of the problem.

$$\left(\frac{\Delta s_\theta}{2a}\right)\left(\frac{s_\theta}{2a}\right) \simeq \left(\frac{b}{a} \quad 0 \quad \frac{L}{a}\right)\begin{bmatrix}\frac{u_r}{a}\\ \frac{v_\theta}{a}\\ \frac{w}{a}\end{bmatrix} \tag{5.44}$$

where

s_θ is the initial length between points P_1 and P_2 located in two cross sections with a distance $2L$ apart,

Δs_θ is the change in length between those two points,

$\frac{u_r}{a}, \frac{v_\theta}{a}, \frac{w}{a}$ are the displacement components calculated at point P_1 with coordinates $(b/a, L/a, \theta)$

If L is equal to zero, then $s_\theta = 2b$ and Δs_θ is now reduced to a variation of the inner diameter of the inclusion with

$$\Delta s_\theta \simeq 2u_r \tag{5.45}$$

If eq (5.40) is substituted into eq (5.44) we are able to relate the change in length Δs_θ to the 3D stress field components as follows

$$\left(\frac{\Delta s_\theta}{2a}\right)\left(\frac{s_\theta}{2a}\right) \simeq \left(\frac{b}{a} \quad 0 \quad \frac{L}{a}\right)(Q_u)(\sigma_o) \tag{5.46}$$

where the components of matrix (Q_u) are calculated at point $(b/a, L/a, \theta)$.*

---

* Note that eqs (5.44), (5.45) and (5.46) still apply if the inclusion is solid and points P_1, P_2 are located on a circle with radius b within the inclusion .

5.7.3 Induced Stress Field Within the Anisotropic Body

As mentioned in Appendix 5.3 it is possible to calculate the stress components at any point in the anisotropic body induced by the presence of an inclusion or equivalently by the boundary stresses X_n, Y_n, Z_n in subproblem (B_2).

For an inclusion with given elastic and geometric properties, for a body anisotropy with given properties and orientation and for a given 3D stress field applied at infinity, eq (5.27) can be solved for matrix (X). Then, coefficients $\bar{a}_m, \bar{b}_m, \bar{c}_m$ $(m=1,N)$ in eqs (4.73) may be calculated and substituted into the expressions for $\phi_k(z_k)$ and $\phi_k'(z_k)$ $(k=1,2,3)$ in eqs (12) and (16) of Appendix 4.3 with $A_1 = A_2 = A_3 = 0$. Finally, the induced stress components are obtained using eqs (4.31) and (4.54).

5.7.4 Limiting Cases

The elastic equilibrium of a full inclusion and the one for a hole without an inclusion can be regarded as two limiting cases of the general theory developed for a hollow inclusion. In particular, it was shown in Appendices 5.1 and 5.2 how to handle the first limiting case. The second case takes place when the ratio a/b is equal to 1 and when the boundary stresses X_n, Y_n, Z_n vanish. Our problem is then reduced to problem (B_1) of Figure 5.2. The solution of that problem was the object of Chapter 4*. In particular, eq (5.44) still applies for two points located on the surface

---

* Section 4.6.4 and Appendix 4.5.

of the hole. Adding eqs (1) and (3) of Appendix 5.3, setting the rigid body translations to zero and substituting the result into eq (5.44) with $a/b = 1$, we obtain

$$\left(\frac{\Delta s_\theta}{2a}\right)\left(\frac{s_\theta}{2a}\right) = \left(M_{u_1}^{(o)} + M_{u_1}^{(h)}\right) + \left(M_{u_2}^{(o)} + M_{u_2}^{(h)}\right)\cos 2\theta + \left(M_{u_3}^{(o)} + M_{u_3}^{(h)}\right)\sin 2\theta$$
$$- \left(\frac{L}{a}\right)^2 \varepsilon_{z_o} + \left(\frac{L}{a}\right)\left(M_{w_2}^{(o)} + M_{w_2}^{(h)}\right)\cos\theta + \left(\frac{L}{a}\right)\left(M_{w_3}^{(o)} + M_{w_3}^{(h)}\right)\sin\theta \quad (5.47)$$

If L is equal to zero then $s_\theta = 2a$ and eq (5.47) reduces to eq (4.86). Furthermore, eq (5.47) indicates that along the circular contour there are at most five independent values of Δs_θ when L is not zero. They reduce to three when L vanishes.

5.8 Numerical Examples

In order to illustrate the closed form solutions of section 5.6, numerical examples are presented. A computer program called Berni 1 (see Appendix 6.2) is used to calculate the distributions of stresses, strains and displacements within an isotropic inclusion perfectly bonded to an infinite body with an anisotropy of a general form. The body is subjected to a 3D stress field applied at infinity. At any point within and around the inclusion, the longitudinal strain is constant and defined by eq (4.84). The computer program calculates also the distribution of stresses induced by the inclusion within the anisotropic body. Both solid and hollow inclusion are considered.

As an example, consider the following 3D stress field

components*: $\sigma_{x,0} = 7.\ MPa$; $\sigma_{y,0} = 2.5\ MPa$; $\sigma_{z,0} = 5.8\ MPa$; $\tau_{yz,0} = -1.8\ MPa$; $\tau_{xz,0} = 1.\ MPa$; $\tau_{xy,0} = 2.6$. Two types of material properties are considered; isotropic and transversely isotropic. The elastic properties of the inclusion are defined by a Poisson's ratio $\nu = 0.3$ and a Youngs modulus E_{inc}. In order to study the influence of the latter upon the stress, strain and displacement distributions, high and low modulus inclusions are considered with Youngs moduli respectively equal to 2.10^4 MPa and $2\ 10^3$ MPa.

5.8.1 Isotropic Solution

Let the elastic properties of the isotropic body be defined by a Youngs modulus $E = 4\ 10^4$ MPa and a Poisson's ratio $\nu = 0.25$. The elastic modulus of the inclusion can be also defined with respect to the one of the isotropic body by the ratio E_{inc}/E respectively equal to 0.5 and 0.05 for the high and the low modulus inclusions.

Closed form solutions for the stress, strain and displacement distributions were proposed by Duncan-Fama and Pender (1980) and used as a verification for the present model. It is found through the numerical examples that the variable N appearing in the summation term in eqs (5.10) can be set equal to 3. In other words A_n, B_n, C_n, D_n are equal to zero for $n = 4, 6, 8, \ldots$ and E_n, F_n are equal to zero for $n = 3, 5, 7 \ldots$. Therefore $\sigma_r, \sigma_\theta, \tau_{r\theta}, \sigma_z, u_r, v_\theta$ are trigonometric functions of $\cos 2\theta$ and $\sin 2\theta$, but $\tau_{rz}, \tau_{\theta z}$ and w are functions of $\cos\theta$ and $\sin\theta$.

* In terms of principal stresses $\sigma_1 = 8.2$ MPa; $\sigma_2 = 6.6$ MPa; $\sigma_3 = 0.5$ MPa with orientations as shown in Figures 5.6 and 5.10.

This was predicted by the closed form solution of Duncan-Fama and Pender (1980).

For the present elastic properties and stress field components the elastic equilibrium of the inclusion takes place under a constant longitudinal strain $\varepsilon_{z_0} = 8.562\ 10^{-5}$.

Figures 5.5 show the distributions of the major principal stress σ_1 along the contour $r/a = 1$ for different values of the ratio a/b, for the high modulus inclusion (Figure 5.5a) and the low modulus inclusion (Figure 5.5b). Similar plots can be obtained for σ_2 and σ_3. It can be seen that as the inclusion gets thicker, the principal stress field within the inclusion becomes more uniform. Indeed, Figure 5.6 shows the orientation and magnitude of that stress field within the corresponding solid inclusion. It appears that the value of the modulus of the inclusion influences slightly the orientation of the principal stresses however, their magnitudes are largely affected.

Figures 5.7a and 5.7b show respectively the distribution of the radial stress component σ_r at the contact inclusion, isotropic body for the high and the low modulus inclusions and for different values of the ratio a/b. σ_r can vary between high and low values especially for the solid inclusion. It can be as large as the stress components applied at infinity and is largely reduced with the ratio a/b and/or by using the low modulus inclusion. The limiting case of the hole without any inclusion corresponds of course to

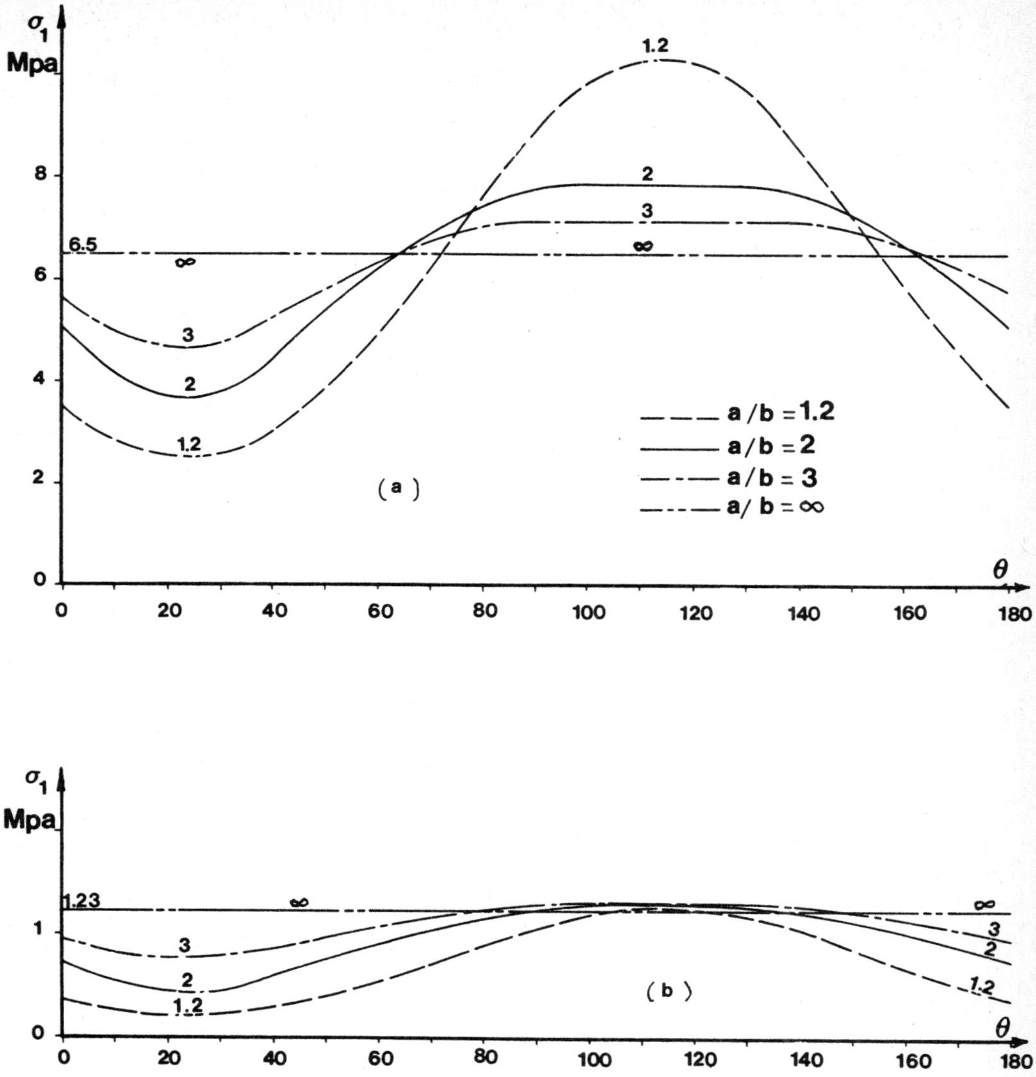

Figure 5.5. Distribution of the principal stress σ_1 along the contour $r/a = 1$. (a) $E_{inc} = 2.10^4$ MPa (b) $E_{inc} = 2\ 10^3$ MPa. Isotropic solution.

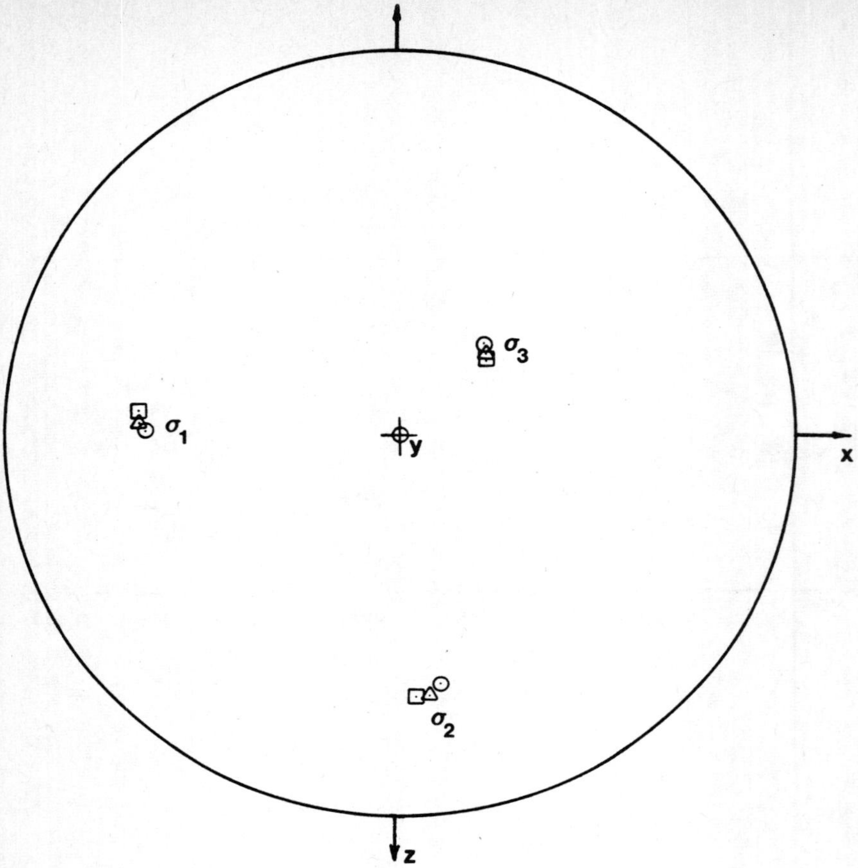

⊡ Applied stress field.

principal stresses (Mpa)	Einc = 2.10⁴ Mpa △	Einc = 2.10³ Mpa ⊙
σ_3	0.88	0.24
σ_2	4.64	0.73
σ_1	6.50	1.23

Figure 5.6. Orientation and magnitude of the principal stresses within a solid inclusion for $E_{inc} = 2\ 10^4$ and $E_{inc} = 2\ 10^3$ MPa .Lower hemisphere stereographic projection. Isotropic solution.

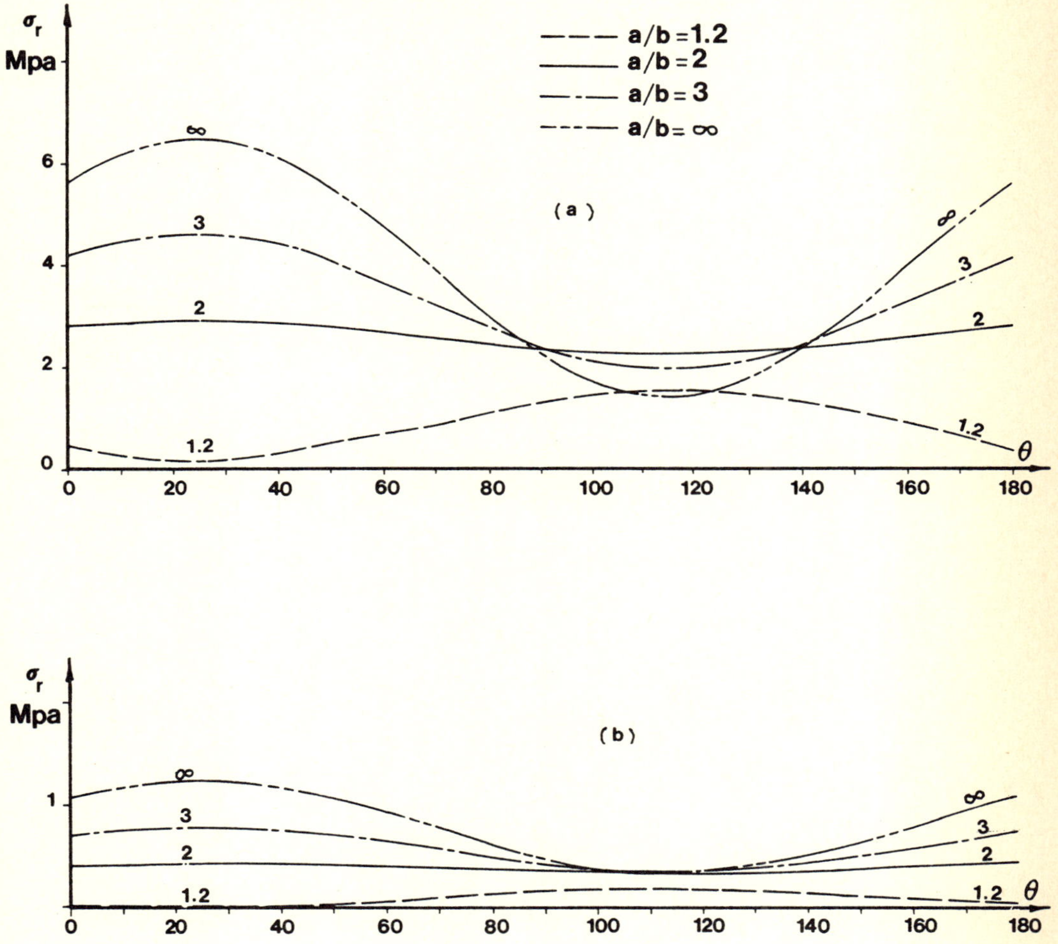

Figure 5.7. Distribution of the radial stress σ_r along the contour $r/a = 1$. (a) $E_{inc} = 2\ 10^4$ MPa (b) $E_{inc} = 2\ 10^3$ MPa. Isotropic solution.

the condition $\sigma_r = 0$.

Figure 5.8a shows the distribution of the radial stress induced by the inclusion within the infinite body along the x axis ($\theta = 0^\circ$).

The magnitude of σ_r decays rapidly but is still non negligible at a distance of five times the radius of the hole unless a/b is less than 2 and/or the low modulus inclusion is considered. Similarly, Figures 5.8b and 5.8c show respectively the distributions of the induced stress components $\tau_{r\theta}$ and τ_{rz} along the x axis.

Finally, Figure 5.9 represents the distribution of the radial displacement u_r/a along the contour $r/a = 1$ for the high and low modulus inclusions and $a/b = 2.$. Similar variations apply also to v_θ/a and w/a .

5.8.2 Anisotropic Solution

Let the infinite body be transversely isotropic with elastic properties equal to those used in the numerical examples of section 4.6. The hole and the plane of transverse isotropy have the following orientations with respect to the global coordinate system X,Y,Z (Figures 1,2; Appendix 4.1):

(i) the coordinate system x,y,z is parallel to the global one,

(ii) the plane of transverse isotropy strikes at different angles β varying from 0 to 90 degrees and dips at an angle ψ of 30 degrees. The special case of a horizontal anisotropy $\psi = 0^\circ$ is also studied.

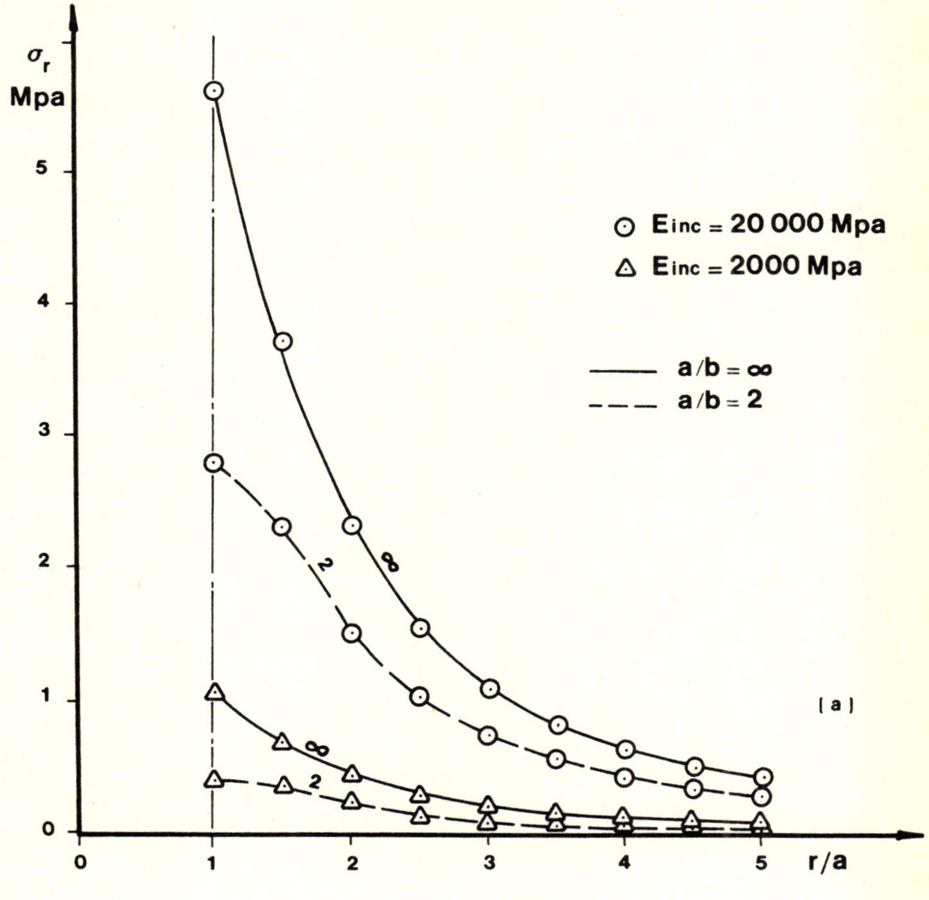

Figure 5.8. (a) Radial stress distribution induced by an inclusion within an isotropic body (θ = 0°).

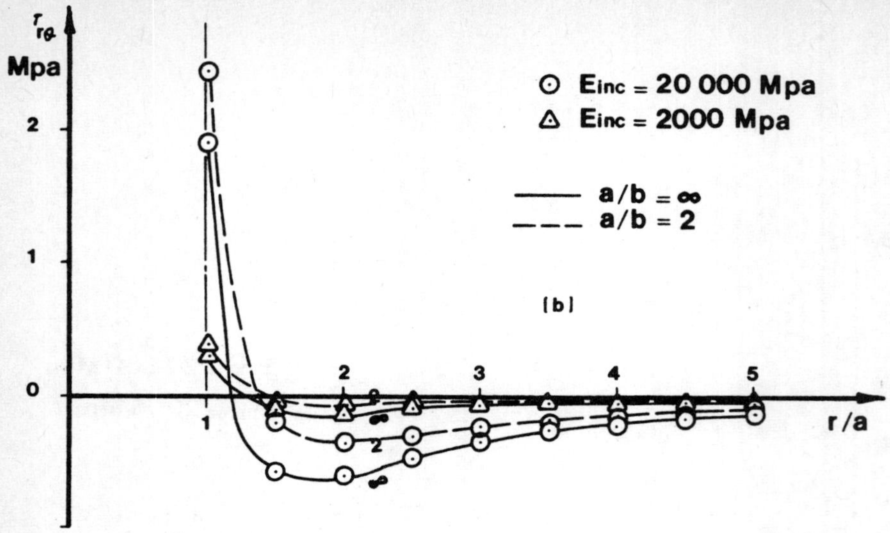

Figure 5.8. (b) Distribution of tangential stress $\tau_{r\theta}$ induced by an inclusion within an isotropic body ($\theta = 0°$).

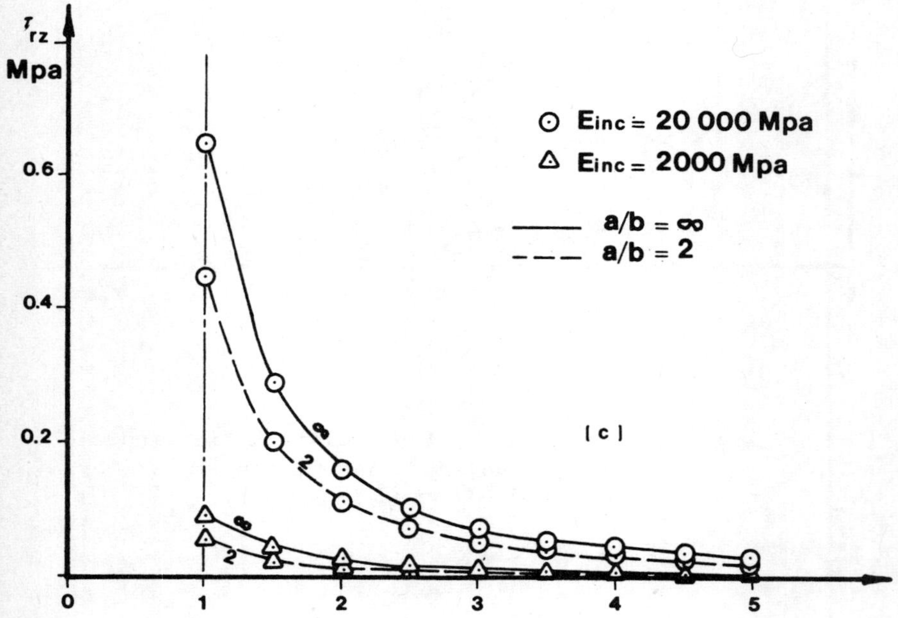

Figure 5.8. (c) Distribution of tangential stress τ_{rz} induced by an inclusion within an isotropic body ($\theta = 0°$).

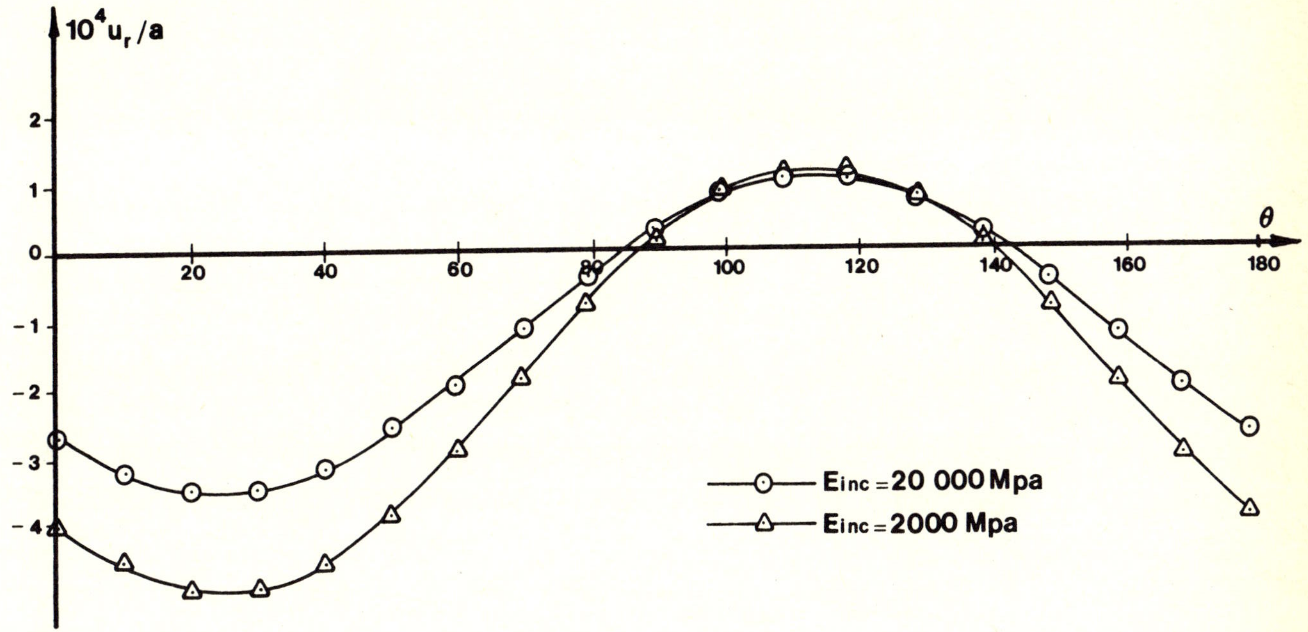

Figure 5.9. Distribution of the radial displacement u_r/a along the contour $r/a = 1$.
Hollow inclusion with $a/b = 2$.
Isotropic solution.

For those conditions, the elastic equilibrium takes place under a constant longitudinal strain ε_{z_0} whose value varies between $6.359\ 10^{-5}$ for $\beta = 0^\circ$ and $1.22\ 10^{-4}$ for $\beta = 90^\circ$. For a horizontal anisotropy $\varepsilon_{z_0} = 7.625\ 10^{-5}$. The elastic modulus of the inclusion can be also defined with respect to those for the anisotropic body by the two ratios E_{inc}/E_1, E_{inc}/E_2 respectively equal to 1, 0.5 and 0.1, 0.05 for the high and the low modulus inclusions. It is found through the numerical examples that N must, at least, be set equal to 5 in eqs (5.10) for a hollow inclusion. For those values the left hand sides of eqs (5.25a) or (5.25b) become infinitesimal quantities. For a full inclusion, N can be reduced to 3 as in the isotropic solution. Eqs (5.25b) are now exactly satisfied.

As for the isotropic solution, the principal stress field within an hollow inclusion is found to become more uniform as the inclusion gets thicker. Indeed, Figure 5.10 shows the orientation and magnitude of the uniform stress field within the corresponding solid inclusion, when the anisotropy strikes parallel or perpendicular to the hole axis. For a given orientation of anisotropy the value of the modulus of the inclusion influences mostly the magnitude of the principal stress components as in the isotropic case. For a given inclusion modulus, rotation of the plane of transverse isotropy produces a rotation of σ_1 and σ_2 with axis approximately parallel to the σ_3 direction.

Finally, Figures 5.11 and 5.12 show the distribution of the radial stresses induced respectively by a solid and a

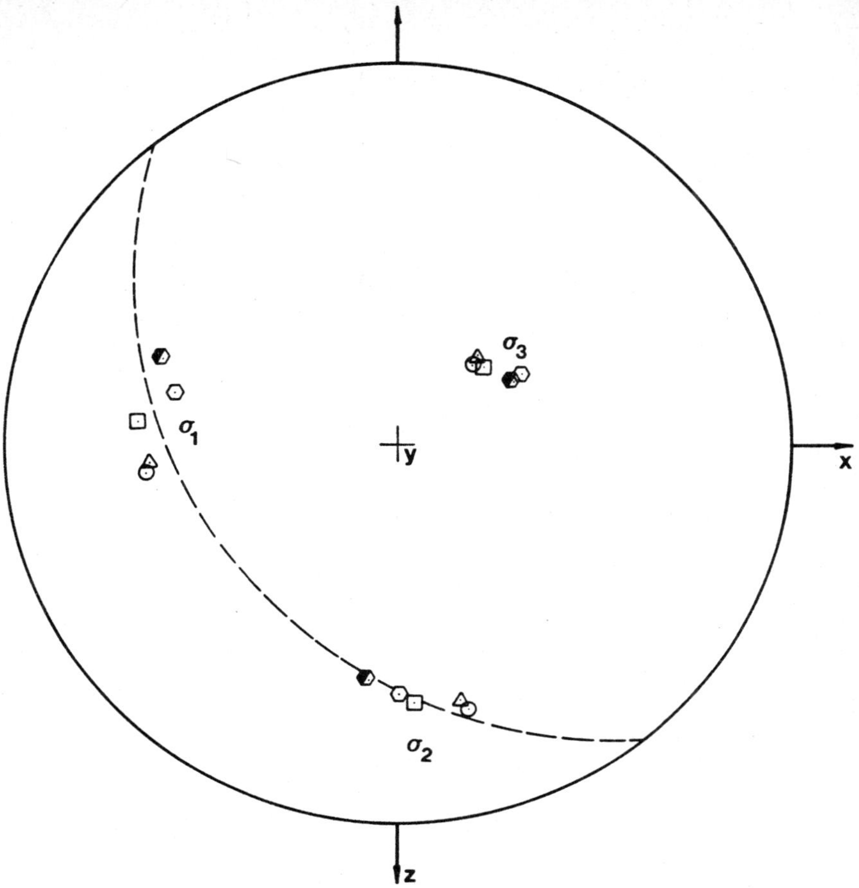

⊡ Applied stress field.

principal stresses (Mpa)	$\beta = 0°$		$\beta = 90°$	
	Einc=2.10⁴ ⊙	Einc=2.10³ △	Einc=2.10⁴ (filled hexagon)	Einc=2.10³ (hexagon)
σ_3	-0.66	0.23	-0.56	0.05
σ_2	4.84	1.17	4.94	0.87
σ_1	9.20	2.59	8.08	1.88

Figure 5.10. Orientation and magnitude of the principal stress field within a solid inclusion for $E_{inc} = 2\ 10^4$ and $E_{inc} = 2\ 10^3$ MPa . Lower hemisphere stereographic projection. Anisotropic solution.

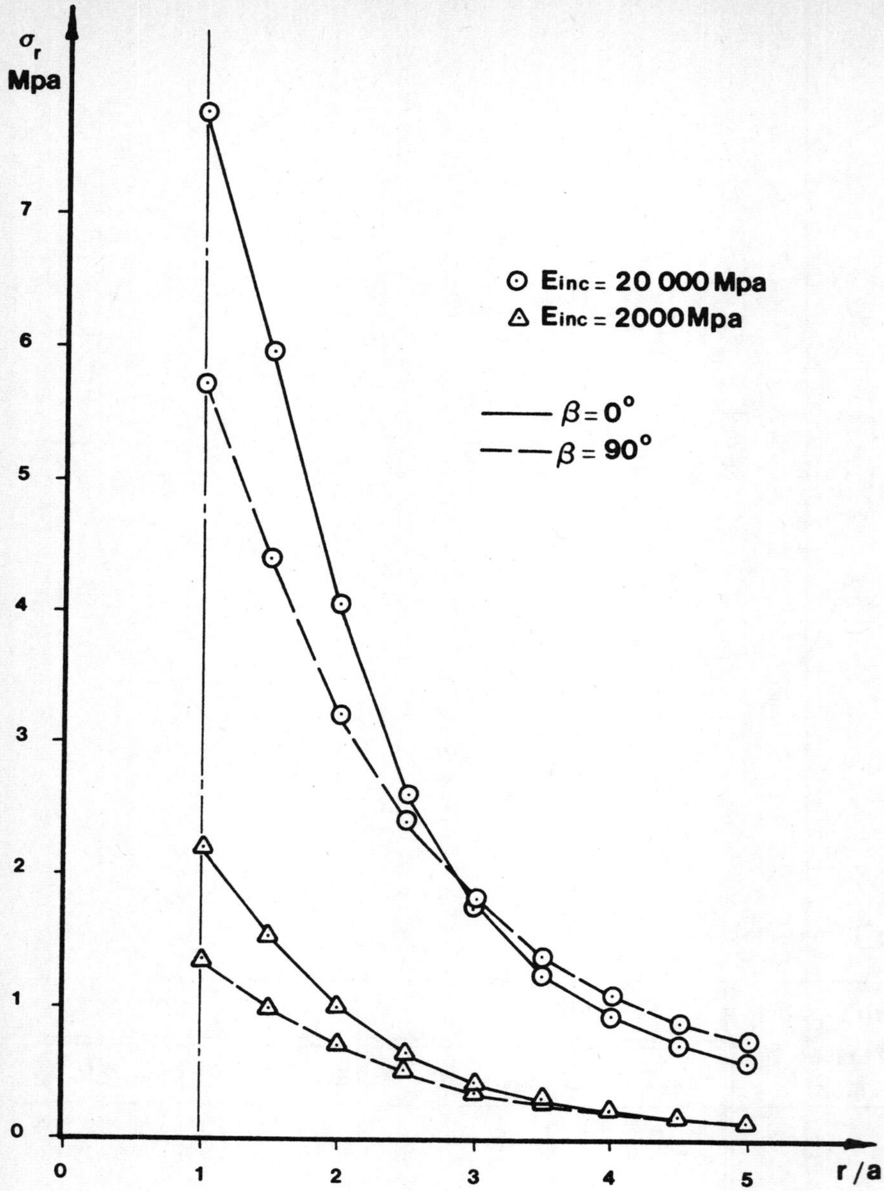

Figure 5.11. Radial stress distribution induced by a solid inclusion within an anisotropic body ($\theta = 0°$).

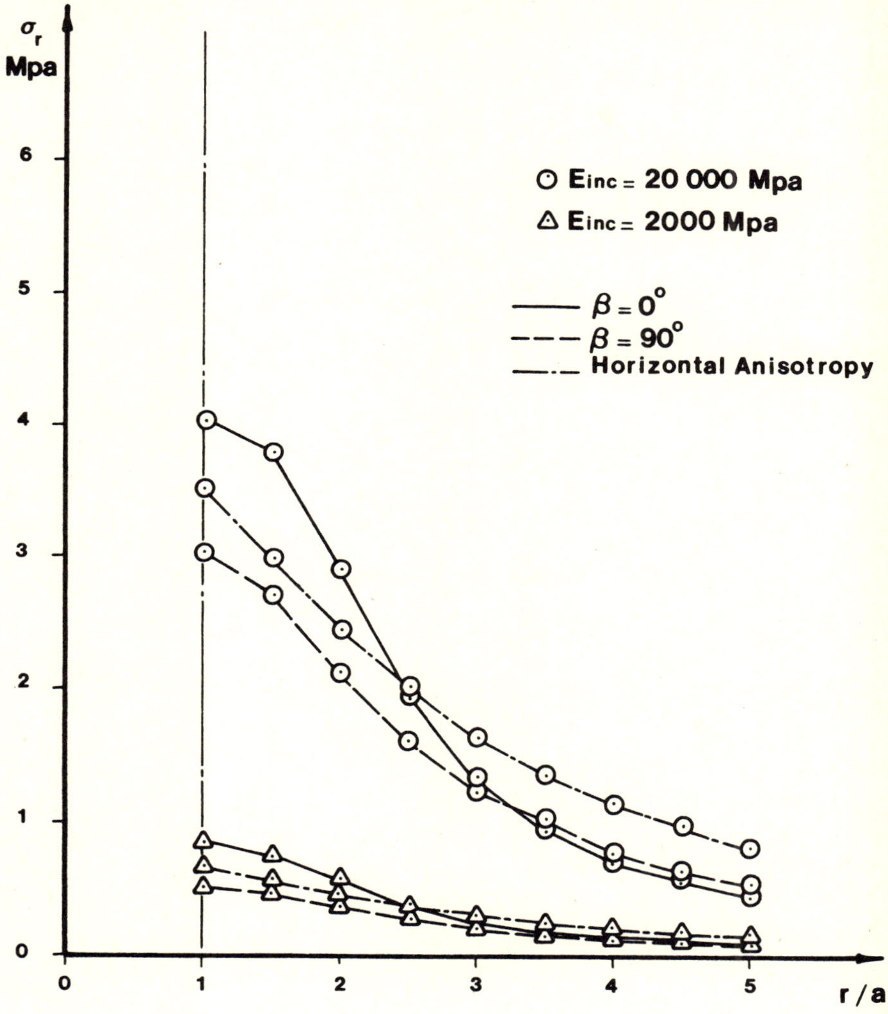

Figure 5.12. Radial stress distribution induced by a hollow inclusion ($a/b = 2.$) within an anisotropic body ($\theta = 0°$).

hollow inclusions within the transversely isotropic body. In reference to the isotropic solution, it appears that for both cases, the radial stress distribution is largely influenced by the anisotropic character of the body and its orientation as least closed to the contact $r/a = 1$. This influence is the largest for $\beta = 0°$ and decreases appreciably for $\beta = 90°$. Cases with intermediate values of β are contained between those two. Note also that the influence of the anisotropy takes place regardless of the value of E_{inc}. Figure 5.12 shows also the distribution of the radial stresses induced by a hollow inclusion within a body whose plane of transverse isotropy is horizontal ($\psi = 0°$). Conclusions similar to those for the radial stress component apply also for $\tau_{r\theta}$ (Figure 5.13a). In comparison to the isotropic solution, the distribution of the tangential stress τ_{rz} induced within the infinite body, seems to be highly affected by the anisotropy for any value of β (Figure 5.13b).

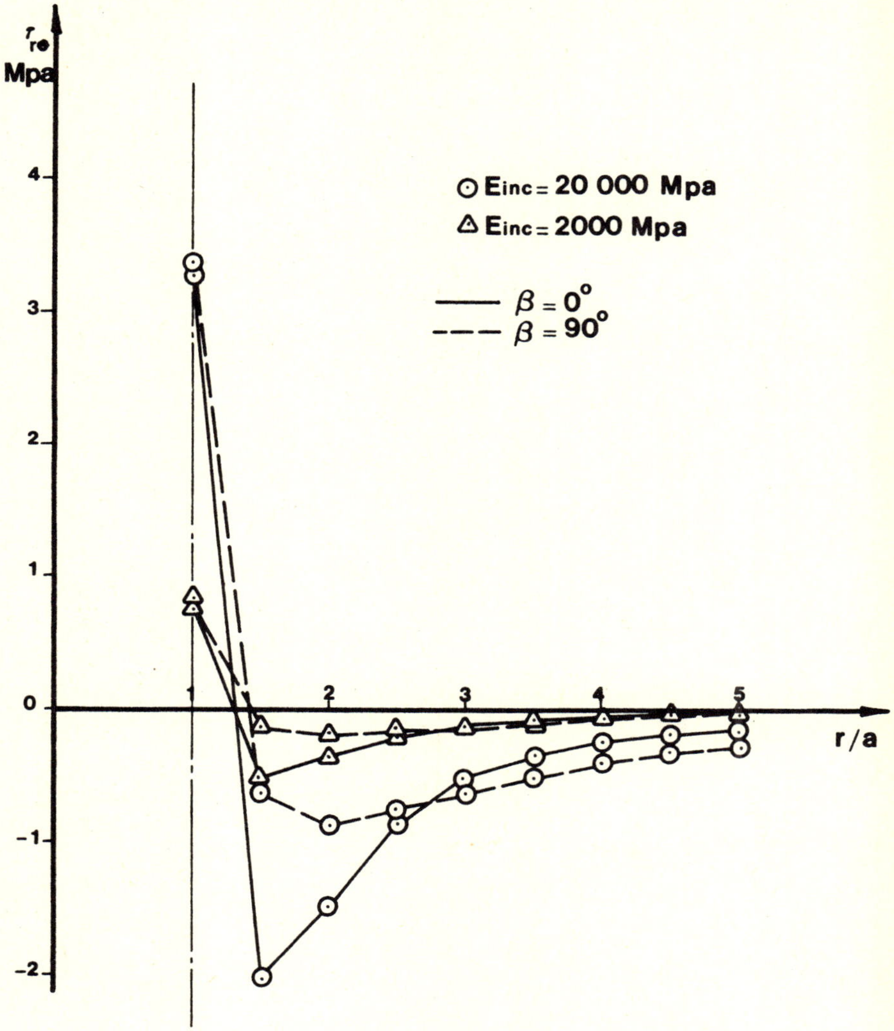

Figure 5.13. (a) Distribution of tangential stress $\tau_{r\theta}$ induced by a solid inclusion within an anisotropic body ($\theta = 0°$).

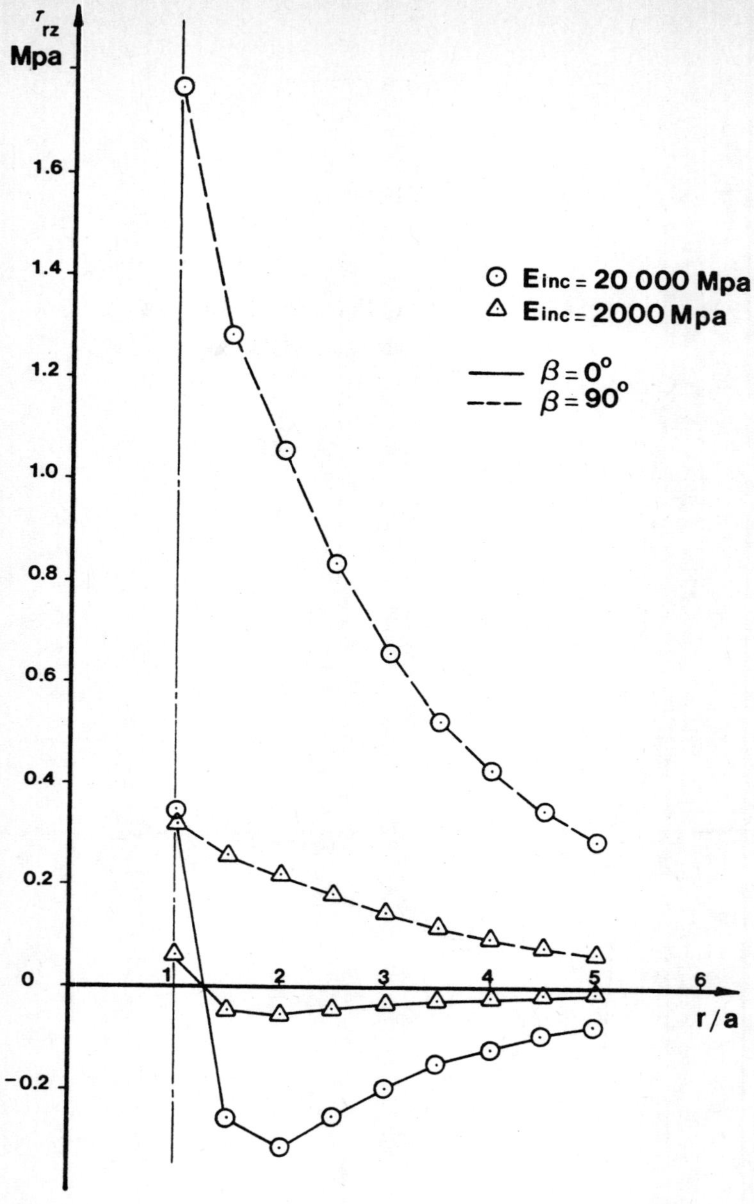

Figure 5.13. (b) Distribution of tangential stress τ_{rz} induced by a solid inclusion within an anisotropic body ($\theta = 0°$).

Chapter 6: Influence of Rock Anisotropy on Stress Measurements by Overcoring Techniques

6.1 Introduction

Knowledge of the in situ state of stress is important in the application of rock mechanics principles. A number of references review the techniques available for measuring in situ stress in rocks and compare their advantages and disadvantages (Leeman, 1964; Fairhurst, 1967; Obert and Duvall, 1967, pp 409-459; McGarr and Gay, 1978; Goodman, 1980). Stresses in situ can be measured in boreholes, on outcrops or on the walls of underground galleries. Except for a few techniques that attempt to measure the state of stress directly, most involve the measurement of strains or displacements resulting from disturbing the state of stress in the rock. Strains or displacements are related to stresses through assumptions about material behavior.

In situ stress determinations are seldom accurate. Apart from measurement inaccurracies (Schmidt and Frenk, 1972) that can be minimized by proper care, some inaccuracies can be worsened by the way the mechanical behavior of the rock is modelled. Strains and displacements recorded during testing are usually related to the rock stresses on the assumption that the rock behaves as a homogeneous, isotropic, continuous and linearly elastic medium, a situation that is not even approximately correct in many practical cases. Additional assumptions are also postulated concerning (i) the orientation of the principal stress field with respect to the directions of measurement, (ii) the mode of deformation (plane strain, plane stress) when disturbing the state of stress in the rock and (iii) what is actually measured. The final assumptions are so restrictive that one should not be surprised to arrive sometimes to erroneous conclusions.

The present chapter draws attention to the influence of rock anisotropy on stress measurements by overcoring techniques. The rock material is described as linear elastic, homogeneous and continuous as defined in Chapter 1. No restrictions are made on the type or the orientation of the anisotropy, the orientation of the principal stress field, or the mode of deformation of the rock. The closed-form solutions derived in Chapters 4 and 5 are applied to calculate in situ stress components from measurements of strains or displacements using several existing overcoring techniques*. New techniques are proposed. The general formulas for the influence of rock anisotropy on stress measurements by undercoring techniques are also derived.

6.2 In Situ Determination of Stress by Relief Techniques

The overcoring and undercoring techniques can be classified as relief techniques. In general, the latter are defined as

---

* The present chapter does not deal with the technical part of these techniques.

methods or procedures that wholly or partially isolate a rock specimen from the stress field in the surrounding rock. Strain and/or displacement measurements on the isolated specimen or on the surrounding rock associated with the relief are recorded. For the overcoring and undercoring techniques those measurements take place respectively in boreholes and on exposed rock surfaces in the vicinity of the point at which the state of stress has to be determined. This requires the stress field to be homogeneous throughout the zone of interest before the measurements are performed. Indeed, this is a reasonable assumption in the absence of heterogenities or major geological features.

Figure 6.1 illustrates the three steps commonly involved in any overcoring technique:

(i) in Figure 6.1a, a large diameter hole is drilled in the volume of rock where stresses have to be determined. The hole is drilled to a distance sufficiently far from the investigation chamber that the effect of the chamber itself on the stress measurements can be neglected. A distance at least equal to one chamber diameter is required,

(ii) in Figure 6.1b, a small pilot hole is drilled at the end of the previous one. An instrumented device is inserted into the hole. The corresponding device must be able to measure displacements, strains or both if required. The pilot hole must be long enough not only to neglect the effects of its own ends on the measurements but also to neglect the disturbance in stress caused by the larger hole. The instrumented device can be also

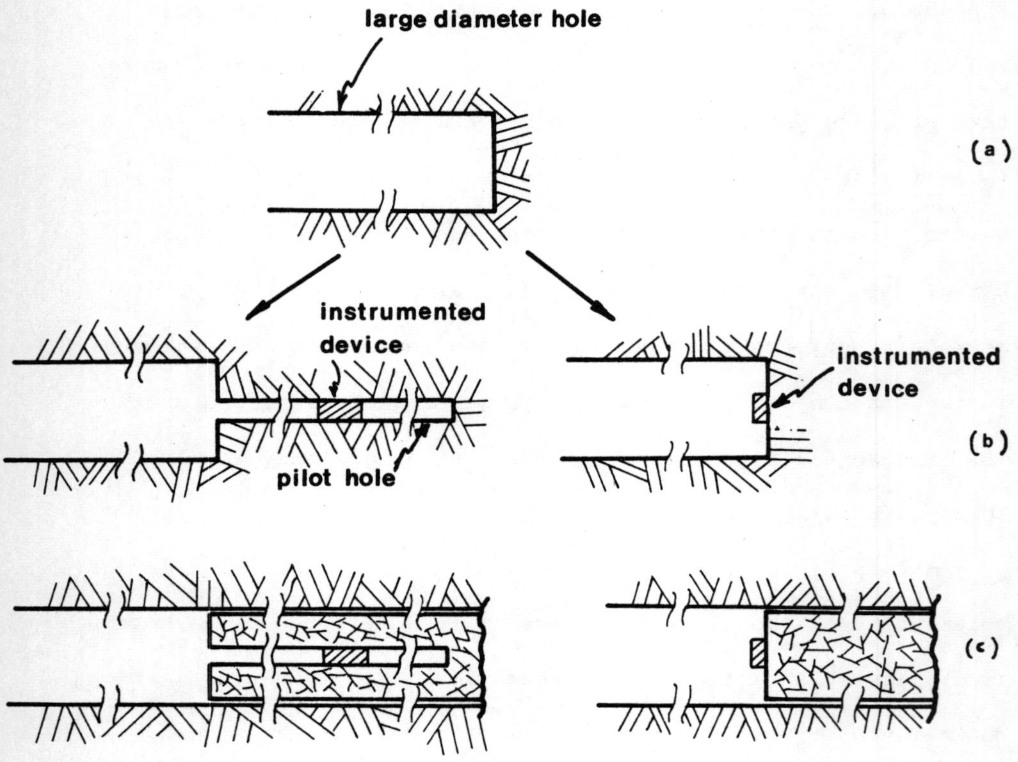

Figure 6.1. Steps commonly involved in any overcoring technique (see explanations of the procedure in the text).

positioned directly onto a flat end of the large diameter hole and no pilot hole is needed,

(iii) in Figure 6.1c, the large diameter hole is resumed, partially or totally relieving stresses and strains within the cylinder of rock that is formed. The resulting changes of displacement or strain within the instrumented device are then recorded.

Instrumented devices that can measure the changes of one or several diameters of the pilot hole have been developed by several authors (Merrill and Peterson, 1961; Obert et al 1962; Merrill, 1967; Crouch and Fairhurst, 1967; Royea, 1969). The best known of these is the U.S. Bureau of Mines (USBM) six arms deformation gage that can monitor the hole diameter changes along three diameters simultaneously. That gage was subsequently improved by Hooker et al (1974). Bonnechere and Cornet (1977) proposed a gage than can measure both diametral and longitudinal displacements.

Changes in strains can be measured by bonding strain rosettes directly to the surface of the rock. With the CSIR 'doorstopper' (Leeman, 1971), three strain gages are glued onto a flat end of the large diameter hole in prescribed directions. The CSIR triaxial strain cell (Leeman and Hayes, 1966) consists of three triple-gage strain rosettes glued to the wall of the pilot hole at known orientations and positions. Improvements of that cell were realized by Hiltscher et al (1979) for deep water-filled boreholes. Strain gages can also be embedded into solid epoxy probes (Rocha and Silverio, 1969; Blackwood, 1977) or thin-walled epoxy probes (Rocha et al, 1974; Worotnicki and Walton, 1976). Both probes must be perfectly bonded into the pilot holes.

The undercoring technique, also called stress relief technique by center hole drilling is a name applied by Duvall (1974) to a procedure for measuring stresses on an exposed rock surface by monitoring radial displacements of points around a center hole (Figure 6.2). The procedure is as follows:

(i) six measuring pins, 60 degrees apart, are positioned on the rock surface along a circle. The distance across the three diameters is measured,

(ii) a hole is then drilled in the center of the measuring pins so that they are close to the surface of the hole. After drilling, the distance across the three diameters is measured again.

The changes in distance across the three diameters induced by the hole drilling are the measurements that are used to calculate the stresses within the rock surface.

6.3 Information Obtained From Measuring Techniques

It is important to know and understand what is measured by in situ stress determination instruments.

The terminology currently used to describe in situ stresses shows a great diversity. Voight (1966) classified in situ stresses into two groups; gravitational and tectonic. Tectonic stresses are themselves decomposed into current and residual components. Obert (1966) split in situ stresses into

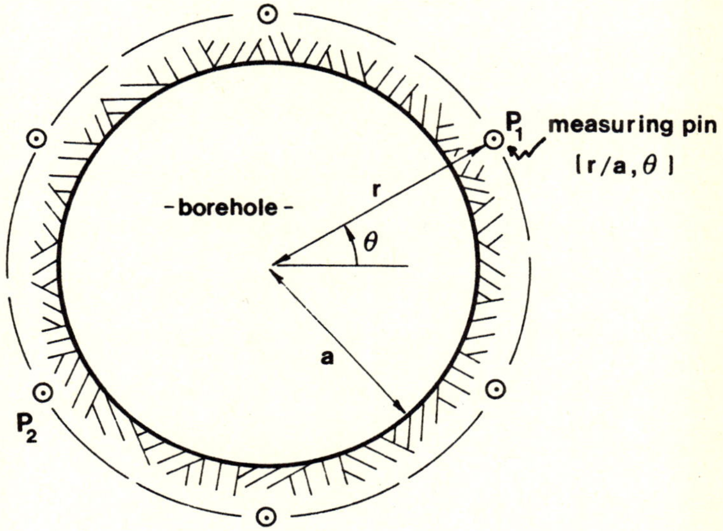

Figure 6.2. Set up for the undercoring technique.

external stresses composed of gravitational and tectonic stresses and internal stresses composed of residual stresses. External stresses have been also called regional stresses by Fairhurst (1967). Bielenstein and Barron (1971) suggested the terminology illustrated in Figure 6.3, more complete than the two previous ones and defined as follows:

"Induced stresses are man made stress components due to removal or addition of material. They are superimposed on natural stresses which exist prior to any excavation. The natural stress field can be composed of gravitational stresses (due to the mass of overburden), tectonic stresses and residual stresses. Tectonic stresses may be active tectonic stresses (due to the active present day straining of the earth's crust) and remanent tectonic stresses (due to past tectonic events which have only been partially relieved by natural processes)".

Bielenstein and Barron separated also the relief techniques into the categories of short term (relief within two hours) and long term techniques (relief occuring after two hours). The present analysis is limited to the first category for which the rock response can be considered to be essentially elastic. We also assume that strains and displacements measured by overcoring and undercoring are related to the external stresses only and that measuring devices do not sense the influence of residual stresses. This last assumption is not always approximately correct in practice since several authors have shown that residual stresses present in some rocks are only

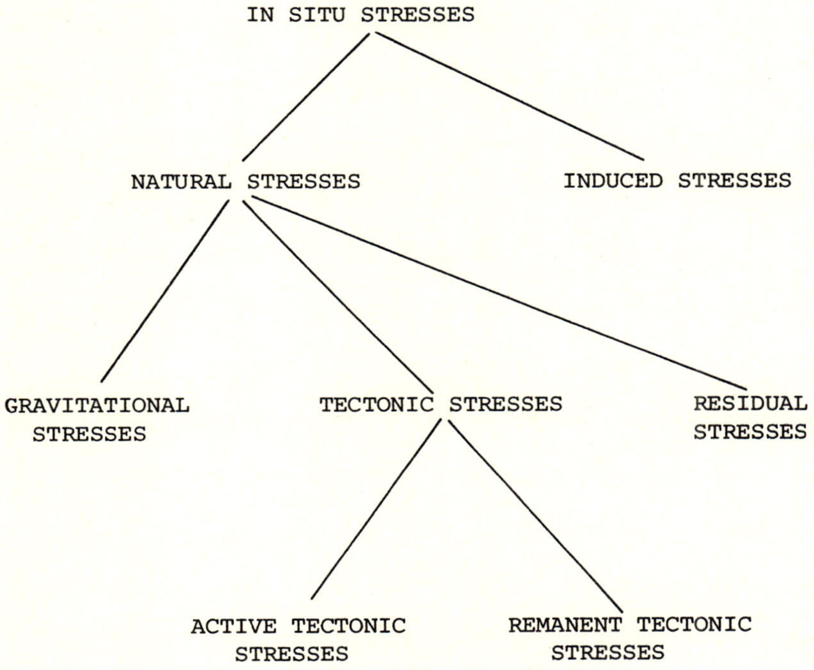

Figure 6.3. Terminology (Bielenstein and Barron, 1971).

partially released by stress measurements (Voight, 1966; Nichols and Savage, 1976*; Hiltscher, 1976; Hiltscher et al, 1979). Residual stresses may then be viewed as a possible source of inaccuracies in stress measurements, the influence of which has been disregarded; there are also a possible source of deformation of engineering structures. However, the influence of residuals can be neglected if the external stresses to be measured are expected to be substantially much larger.

Another prerequisite for the application of relief, techniques is that the rock stresses to be measured must not be too great in relation to the strength of the rock. If this should be the case, core discing, flaking of borehole walls and end can take place and measurements will not be possible (Hast, 1979).

The scale of the tests is also important. Most of the data obtained from stress measurements fit into the _mesoscopic_ scale as defined by Turner and Weiss (1963) i.e. a scale to cover rock bodies from hand-specimen size up to directly observable exposures, for example engineering excavations. Each measurement site can be regarded as a 'sample'. Extrapolation to a larger scale requires 'samples' at a number of locales.

6.4 General Formulas for Overcoring and Undercoring Techniques

---

* see also references therein.

6.4.1 Overcoring Techniques

Recalling the general procedure described in Figure 6.1, the process of overcoring can be seen as cancelling the components of initial stress acting across a cylindrical surface in the rock. Hence, this is equivalent to add tractions equal in magnitude but opposite in sign to those that existed across the cylindrical overcore surface before it was drilled. In other words, σ_r , $\tau_{r\theta}$ and τ_{rz} vanish along the outer contour of the overcored sample.

If the instrumented device in contact with the rock surface does not interfere with the deformation of the rock, then, the sample will be completely free of strains and stresses after completion of overcoring. This is true regardless of its size and shape. Overcoring releases strains, displacements and stresses not only due to the in situ stress state but also those induced by the introduction of the pilot hole and the large diameter hole. Devices that permit the strain relief without interference include the USBM deformation gage, the CSIR 'doorstopper' and the CSIR triaxial strain cell.

If the instrumented device does interfere with the deformation of the rock, overcoring does not produce a total relief since the presence of an inclusion results in the retention of 'residual stresses'* and strains within the inclusion and the rock. This is the case with solid or hollow inclusion probes. The partial relief implies that the size

* Not the same residual stresses as discussed in Section 6.3.

of the overcoring diameter must influence the distribution of stresses and strains within the epoxy inclusion probe. Furthermore, imposition of negative stresses at the overcore surface will not produce, in general, a uniform state of stress within a solid probe as was shown for an infinite medium.

For a solid inclusion in an isotropic medium, Duncan-Fama (1979) has shown that the errors involved in neglecting the finiteness of the overcoring diameter are not significant for values of the ratio between the shear modulus of the epoxy to that of the rock less than 0.05 and overcoring diameters greater than three times the pilot hole diameter. If the inclusion is hollow, it must be also thin walled for the approximation to be valid (Duncan-Fama and Pender, 1980).

As shown in Chapter 5, the distributions and magnitudes of the stress components σ_r, $\tau_{r\theta}$ and τ_{rz} induced by an inclusion within an infinite anisotropic body with a 3D stress field applied at infinity seem to depend upon the elastic and geometric properties of the inclusion as well as the elastic properties and the orientation of the anisotropy especially close to the contact between the inclusion and the anisotropic body. Although it is very difficult to draw general conclusions for anisotropic media, it seems that for <u>very low modulus solid inclusions</u> and <u>low modulus thin walled hollow inclusions</u>, stresses and strains in the rock are disturbed in a region very close to the inclusion. In particular, for those conditions the induced stress components σ_r, $\tau_{r\theta}$ and τ_{rz} decay rapidly and become negligible at distances less than three to four times the radius of the pilot hole*. This seems to apply regardless of the type and orientation of the anisotropy. Hence, provided we overcore beyond those limits, the errors involved in neglecting the finiteness of the overcore diameter become insignificant and the overcoring diameter can be set equal to infinity.

For a complete relief techniques and the partial relief techniques with the additional restrictions mentioned above and since the rock material is linearly elastic, the changes in strains and displacements recorded by an instrumented device during overcoring can be assumed to be equal in magnitude and opposite in sign to those that would be developed if the pre-existing in situ stresses were applied at a large distance from a central hole already drilled and instrumented in the rock.

Let the in situ stress field (σ_o) be defined by eq (5.1) in the x,y,z coordinate system attached to the hole where the measurements take place. We can describe its removal by overcoring as the application of a 3D stress field - (σ_o) applied at infinity. Therefore, all the closed form solutions derived in chapters 4+ and 5 between total strain and displacement components and the components of a 3D stress field

---

* The limiting case of a pilot hole without any inclusion corresponds of course to the condition of zero induced stresses σ_r, $\tau_{r\theta}$ and τ_{rz} within the rock .

+ section 4.6.4 and Appendix 4.5

applied at infinity are valid for overcoring techniques. In all those closed form solutions (σ_o) must be replaced by - (σ_o). However, it must be pointed out that this is only applicable if the following conditions are satisfied:

(i) the hole must be very long (theoretically infinite),

(ii) measurements must take place in sections remote from the ends of the hole and remote from the ends of the measuring devices if one uses instrumented solid or hollow probes. The latter must be also perfectly bonded to the walls of the hole,

(iii) allowance must be made for a constant axial strain to occur during stress relief.

Therefore, the applicability of the closed form solutions derived in the previous chapters is limited to the overcoring techniques mentioned in section 6.2 that do not involve measurements close to or at the ends of a hole. The CSIR 'doorstopper' for instance, would not be suitable.

Since measurements of changes in length or strain can take place in several boreholes, it is more appropriate to rewrite the closed form solutions of the previous chapters in terms of the in situ stress field $(\sigma_o)_{XYZ}$ expressed in a fixed global coordinate system X,Y,Z *. $(\sigma_o)_{XYZ}$ is related to (σ_o) as follows

$$(\sigma_o) = (T_\sigma)(\sigma_o)_{XYZ} \tag{6.1}$$

---

* see Figure 2, <u>Appendix</u> 4.1.

where (T_σ) is the transformation matrix defined in eq (11) of Appendix 4.1.

If we substitute eq (6.1) into eqs (5.38), we obtain the general expression between the six components of strain at any point $(r/a, \theta)$ within an inclusion (hollow or solid) induced during overcoring and the six components of the in situ stress field applied at infinity as follows

$$-(\varepsilon)_{r\theta z} = (Q)(T_\sigma)(\sigma_o)_{xyz} \tag{6.2}$$

Similarly, let P_1 and P_2 be two points located on the inner surface of a hollow inclusion (Figure 5.4). Point P_1 has coordinates $(b/a, L/a, \theta)$. If s_θ is the initial length between those points and Δs_θ is the change in length induced during overcoring, then substituting eq (6.1) into eq (5.46) we obtain

$$-\left(\frac{\Delta s_\theta}{2a}\right)\left(\frac{s_\theta}{2a}\right) \simeq \left(\frac{b}{a} \quad 0 \quad \frac{L}{a}\right)(Q_u)(T_\sigma)(\sigma_o)_{xyz} \tag{6.3}$$

If L is equal to zero and $s_\theta = 2b$, then Δs_θ is now reduced to the change of an inner diameter of the inclusion with

$$-\left(\frac{\Delta s_\theta}{2a}\right) \simeq (q_{u_{11}} \; q_{u_{12}} \; q_{u_{13}} \; q_{u_{14}} \; q_{u_{15}} \; q_{u_{16}})(T_\sigma)(\sigma_o)_{xyz} \tag{6.4}$$

where $q_{u_{1i}}$ $(i=1,6)$ are the components of the first line of matrix (Q_u). It is important to note that in eqs (6.2), (6.3) and (6.4), the components of matrices (Q) and (Q_u) are defined in the coordinate system x,y,z, attached to the hole where measurements take place and therefore vary from one hole to another. It can also be shown throughout the computaton of the components of matrix (Q_u) that when there is a plane of elastic symmetry perpendicular to a hole axis, components $q_{u_{14}}$ and $q_{u_{15}}$ vanish.

Eqs (6.2) and (6.4) with the remarks just mentioned are indeed generalizations of the closed form solutions derived in section 4.6.4 and Appendix 4.5 when there is no inclusion and measurements are made on the walls of the hole. Eq (6.4) is reduced to eq (4.85) or (4.87) previously combined with eqs (5.1)[+] and (6.1) by setting $a/b \longrightarrow 1$, $r/a \longrightarrow 1$ and the modulus of the inclusion $E_{inc} \longrightarrow 0$. A similar reduction takes place between eq (6.2) and eq (19b) of Appendix 4.5.

For overcoring techniques that do involve measurements close to or at the ends of a hole, eqs (6.1) to (6.4) are not applicable since the ends of the hole may produce stress concentrations at the location of the measurements. For the CSIR 'doorstopper' technique closed form solutions have been proposed between the strain components at the end of a hole and the 3D stress field applied at infinity for isotropic (Gray and Toews, 1968) or anisotropic rocks (Barla and Wane, 1970; Ribacchi, 1977). However, those closed form solutions require the knowledge of stress concentration factors. Those quantities must be determined numerically or using the theory of photo-elasticity. No theoretical solution exists. For isotropic material the concentration factors have been proposed by

---

[+]with (σ_o) replaced by $-(\sigma_o)$.

several authors (Bonnechere, 1967; Van Heerden, 1968; Crouch, 1969; De la Cruz 1969; Hocking, 1976) and seem to depend slightly on the elastic constants of the material at least in a region not too far from the center of the hole. For anisotropic materials, the problem is more complicated since the stress concentration factors depend upon the elastic constants of the material and cannot be determined once and for all. For a given type of anisotropy and orientation with respect to the hole, stress concentration factors can be computed using three dimensional numerical techniques.

For the remaining part of this study, we shall restrict ourselves to overcoring techniques for which conditions (i) to (iii) mentioned previously are satisfied. Furthermore, the applicability of the closed form solutions derived in the previous chapters requires the overcored holes to be circular before overcoring. This is not exactly true in practice since in drilling a hole for the placement of an instrumented device, a hole eccentricity cannot be ruled out. Agarwal (1968) has shown that for a circular hole in an isotropic medium subjected to a uniaxial stress field, a small ellipticity created during the drilling of the hole can be neglected if the measuring device has a precision of $\pm$ 50 μ in. Hole eccentricity will be disregarded for the present analysis but must be investigated, especially for anisotropic materials.

6.4.2 Undercoring Techniques

If we recall the general procedure described in Figure 6.2,

the process of undercoring consists of measuring three changes of distances between reference pins induced by drilling a circular hole in their center. Since the deformation of the anisotropic medium takes place under a plane stress condition, the closed form solutions derived in Chapter 4* for the generalized plane stress formulation are valid if the hole has one plane of elastic symmetry perpendicular to its longitudinal axis.

Let P_1 and P_2 be two points positioned along a circle on the rock surface as shown in Figure 6.2. Point P_1 has coordinates (r/a, θ) and P_2 is diametrically opposite to P_1. The change in length Δs_θ between P_1 and P_2 induced by drilling the center hole is equal to (eq 4.88).

$$\frac{\Delta s_\theta}{2a} = f_{11}^{(h)}\overline{\sigma}_{x,o} + f_{12}^{(h)}\overline{\sigma}_{y,o} + f_{13}^{(h)}\overline{\tau}_{xy,o} \tag{6.5}$$

where $f_{11}^{(h)}$, $f_{12}^{(h)}$ and $f_{13}^{(h)}$ are defined in eq (14) of Appendix 4.7 and $\overline{\sigma}_{x,o}$, $\overline{\sigma}_{y,o}$, $\overline{\tau}_{xy,o}$ are the components of the average in situ stress field in the x0y coordinate system attached to the center hole. The in situ stress field is acting at a large distance within a plate parallel to the rock surface. The plate is assumed to be thin with a thickness equal to the depth of the center hole within the rock.

We can also imagine an undercoring technique where changes in strain instead of distances are recorded. Strain gages are glued on a rock surface at known positions (r/a, θ) and

---

* Section 4.6.4 and Appendix 4.7.

orientation. Using a procedure similar to the one described in section 6.2, the changes in strain induced by drilling a center hole can be related to $\overline{\sigma_{x,0}}$, $\overline{\sigma_{y,0}}$ and $\overline{\tau_{xy,0}}$ through eq (11) of Appendix 4.7.

6.4.3 Absolute Stresses - Changes in Stress

The general formulas derived in sections 6.4.1 and 6.4.2 are applicable for the determination of the absolute state of stress. There are cases where the in situ stress field may vary with time. In order to measure any change of the in situ state of stress during an interval of time, say Δt , two procedures may be used:

(i) the absolute state of stress is determined at time t and at time $t + \Delta t$ and the difference represents the change of the state of stress during Δt,

(ii) if the material is linear elastic, the principle of superposition makes it possible to calculate the change of the state of stress directly from measurements of change in length or strains during the corresponding interval of time. It is then, possible to make use of eqs (5.38), (5.46) and (6.1) and the corresponding equations in section 4.6.4 and Appendix 4.5 where $(\sigma_o)_{xyz}$ is now replaced by its variation $(\Delta\sigma_o)_{xyz}$ such that

$$(\Delta\sigma_o)^t_{xyz} = (\Delta\sigma_{x,o}\ \Delta\sigma_{y,o}\ \Delta\sigma_{z,o}\ \Delta\tau_{yz,o}\ \Delta\tau_{xz,o}\ \Delta\tau_{xy,o}) \qquad (6.6)$$

Furthermore, if solid or hollow inclusion are used as instrumented devices, there are no restrictions as far as their geometry and elastic properties are concerned as was the case for the absolute measurements since no relief is involved.

Similarly, eqs (6.5) or eq (11) of Appendix 4.7 can be used where $\overline{\sigma}_{x,o}$, $\overline{\sigma}_{y,o}$ and $\overline{\tau}_{xy,o}$ are replaced by their variations $\Delta\overline{\sigma}_{x,o}$, $\Delta\overline{\sigma}_{y,o}$ and $\Delta\overline{\tau}_{xy,o}$.

6.5 General Results for Overcoring in Anisotropic Media

6.5.1 Introduction

Eqs (6.2) and (6.3) show that strains and changes in length induced by overcoring are linear functions of the six components of the in situ stress field. Therefore, determination of those components requires that we set up a system of six independent equations from the results of six measurements. However, such measurements will not provide any information as to the precision of the calculated stress values. In order to improve the accuracy of these results, additional measurements must be made and a least square estimate of the stress components carried out (Panek, 1966; Gray and Toews, 1968; 1975). Least square solutions can be treated in the same way as the problem of multilinear regression analysis. A brief summary of the method is presented in Appendix 6.1 using the matrix formulation proposed by Draper and Smith (1966).

Several questions arise concerning those measurements:

(i) how many independent measurements can we make in a single borehole and therefore how many boreholes do we need to obtain the six independent measurements required?

(ii) are the numbers influenced by the anisotropy of the

rock and its orientation with respect to the boreholes?

(iii) how do anisotropy type and orientation influence the determination of the in situ stress field?

The object of this section is to make use of eqs (6.2), (6.3) and (6.4) to answer these questions.

6.5.2 Isotropic Solution

Before we go any further, it is worthwhile to summarize the conclusions that have been drawn for isotropic rocks as far as question (i) is concerned*.

For instrumented devices that measure variations in borehole diameters, there are at most three independent measurements in a single borehole. If each borehole can furnish three independent measurements, it would seem to be possible to determine the six in situ stress components by using two non parallel boreholes as was proposed by Panek (1966) and Leeman (1967). However, it turns out that the six equations derived from those measurements are dependent and consequently only five independent measurements can be obtained. If we set up nine simultaneous equations from the results of diametral measurements made into three non parallel boreholes, the six in situ stress field components can be calculated. Indeed, the three extra equations or any equations resulting from additional measurements will improve the accuracy of the results (Gray and Toews, 1968).

---

* reference is made to the overcoring techniques mentioned in Section 6.2.

If both diametral and longitudinal displacement measurements are performed in a single borehole, the complete state of stress can be calculated as shown by Bonnechere and Cornet (1977). Hiratmatsu and Oka (1968) have also shown that, for a single borehole in an isotropic medium, there are at most five independent measurements of variation in length of oblique distances such as P_1P_2 in Figure 5.4 (with $a = b$). Therefore, the complete state of stress can be also determined in a single borehole by measuring both borehole diameters and oblique distances.

For instrumented devices that measure changes in strains by bonding strain rosettes to the walls of the pilot hole, the complete state of stress can be obtained from six independent measurements in a single borehole. Strain rosettes are commonly positioned as shown in Figure 6.4[+] (Leeman and Hayes, 1966; Leeman, 1971). Each rosette i is located at an angle θ_i from the x axis of the pilot hole. A rosette consists of three strain gages (A_i, B_i and C_i). Let A_i and B_i gages be aligned with the θ and z directions i.e. with the circumferential and axial directions of the pilot hole. C_i gages are inclined at angles ψ_i to the θ axes. If ε_{A_i} , ε_{B_i} and ε_{C_i} are the changes in strain measured

---

[+] If gages A_i, B_i and C_i have a non negligible length, the measured strains are average values over the corresponding gage lengths.

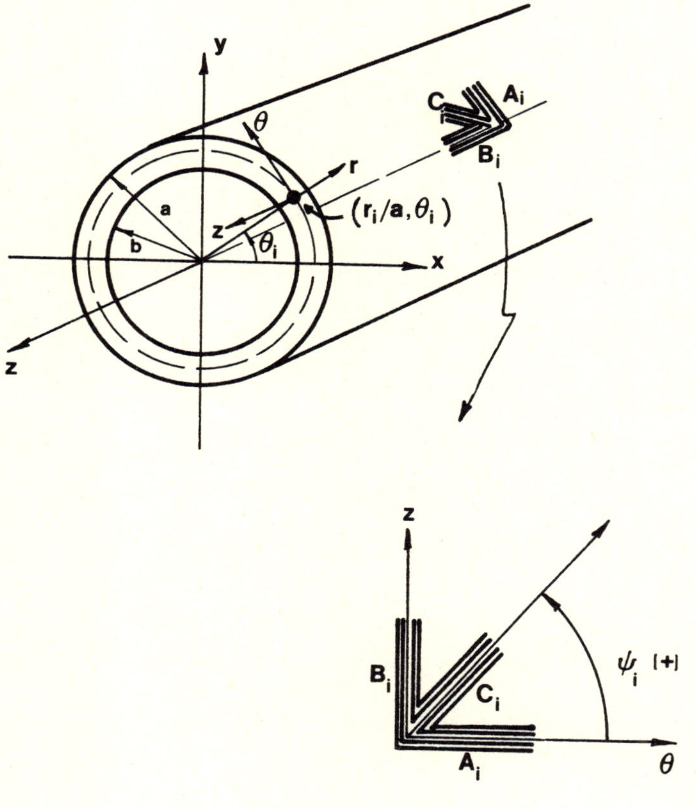

Figure 6.4. Hollow epoxy probe. Rosette configuration
and definition of coordinate systems.

in gages A_i, B_i and C_i respectively*, the strain components in the θz plane at the corresponding point of measurement are equal to

$$\begin{aligned}\varepsilon_{\theta_i} &= \varepsilon_{A_i}\\ \varepsilon_{z_i} &= \varepsilon_{B_i}\\ \gamma_{\theta z_i} &= \left(\varepsilon_{C_i} - \varepsilon_{A_i}\cos^2\psi_i - \varepsilon_{B_i}\sin^2\psi_i\right)\frac{2}{\sin 2\psi_i}\end{aligned} \tag{6.7}$$

If the strain rosettes are oriented as shown in Figure 6.4, three rosettes are required to calculate the complete state of stress. Care must be taken that the difference between any two of the three angles θ_1, θ_2 and θ_3 must not be equal to π. In the CSIR triaxial strain cell $\theta_1 = \pi/2$; $\theta_2 = \pi$; $\theta_3 = 7\pi/4$ and $\psi_i = 5\pi/4$ $(i = 1,2,3)$. Any additional strain rosette will provide measurements that may improve the accuracy of the results.

Measurements of changes in strain can also be carried out by embedding strain gages within solid epoxy probes cemented to the walls of the pilot hole. The complete state of stress can be determined in a single borehole if at least six independent directions of measurement are used within the probe (Rocha and Silverio, 1969; Blackwood, 1977). This method has several advantages with respect to the previous one (i) strain gages are protected against water, (ii) their deformation is less affected by rock heterogeneities and discontinuities, (iii) measurements can take place in weathered or

* with $a = b$ and $r_i/a = 1$

jointed rocks and (iv) no surface preparation is required. The main disadvantage is that low modulus epoxy must be used in order to reduce the tensile stresses at the epoxy, rock interface which may be sufficiently large to break the bond between the epoxy probe and the rock during overcoring*. This also reduces the effect of ignoring the overcoring diameter as discussed in section 6.4.1.

In order to overcome the main disadvantage, another technique is to have the strain gages embedded within a thin-walled soft epoxy probe (Rocha et al, 1974; Worotnicki and Walton, 1976; Pender 1977). Strain rosettes are commonly positioned as shown in Figure 6.4. For that configuration, at least three strain rosettes are required to obtain the complete state of stress from measurements in one borehole.+

6.5.3 Anisotropic Solution; Literature Review

The influence of rock anisotropy on stress measurements has been investigated by several authors.

Berry and Fairhurst (1966) presented the analytical expression for the radial displacement component at the surface of a circular hole in an infinite transversely isotropic rock. The hole was supposed to be in a principal stress direction and to lie perpendicular or parallel to the axis of elastic symmetry of the anisotropy. Both moderately and markedly anisotropic materials were studied. It was found that the radial displacement distribution along the contour of the

* See for instance Figure 5.8a with σ_r replaced by $-\sigma_r$.

+ Eqs (6.7) still apply for each rosette i with coordinates $r_i/a, \theta_i$.

circular hole in the anisotropic medium differs appreciably from that predicted by the isotropic solution when the hole is drilled perpendicular to the axis of elastic symmetry. The influence of the anisotropy is largely reduced when the hole is drilled parallel to the axis of elastic symmetry. This conclusion suggested that neglecting rock anisotropy when drilling parallel to the bedding for instance could lead to highly erroneous conclusions (over 50% of error).

This model was generalized by Berry (1968). An analytical expression for both strain and displacement components at the surface of a circular hole in an infinite transversely isotropic rock was proposed. The hole was allowed to be inclined with respect to both anisotropy and principal in situ stresses. However, the orientation of the anisotropy with respect to the hole was defined by one angle only. This solution permitted the influence of rock anisotropy to be considered in the overcoring techniques for which diametral measurements or changes in strain along the walls of a borehole are involved. It was shown that if diametral measurements are used, there are at most three independent measurements in a single borehole and two boreholes can be used to determine the complete state of stress. However, when one of the borehole is within the plane of transverse isotropy and the other one is perpendicular to it or when the rock is weakly anisotropic, three boreholes are required. If strain measurements are used, the complete state of stress can be obtained from measurements in one borehole only.

Becker and Hooker (1967) and Becker (1968) proposed an analytical expression for the radial displacement component at the surface of a circular hole in an orthotropic medium. The hole was perpendicular to a plane of elastic symmetry and in a principal stress direction. Restrictive approximations and simplifications proposed by Kawamoto (1963) were also used. It was shown that the orientation and magnitude of the in situ stress field perpendicular to the hole axis can be calculated from the results of three changes in diameter during overcoring. This procedure was used by Hooker and Johnson (1969) to investigate near surface horizontal stresses in several rock quarries having an orthotropic anisotropy. They concluded that isotropic versus anisotropic stress determinaton can present differences as large as 25% in magnitude and 25 degrees in orientation.

Hirashima and Koga (1977) proposed a model based on Niwa et al (1970)'s closed form solutions. This model permitted the influence of any rock anisotropy type to be considered in the overcoring techniques that are based upon diametral measurements or changes in strain along the walls of a borehole. If diametral measurements are used, they suggested that at least three boreholes are required to calculate the complete state of stress. If strain measurements are used, one borehole is enough.

They also showed that when boreholes are drilled within

a plane of transverse isotropy, the value of the ratio between the modulus parallel and the one perpendicular to that plane incluences the determination of the in situ stress field. However, the expression for σ_z that they proposed in their paper (eq (3)) does not seem correct.

Berry (1970) proposed a closed form solution to calculate stress changes in a transversely isotropic rock by making six independent measurements within an instrumented solid inclusion. It was suggested that this instrumented device be used only for overcoring with low modulus inclusions. A similar closed form solution was also derived by Niwa and Hirashima (1971) for the measurement of the absolute in situ state of stress.

6.5.4 Anisotropic Solution; Present Analysis

The previous literature review shows that questions (i) to (iii) of section 6.5.1 have never been completely answered for anisotropic media. Instead of focusing our analysis to a specific method of overcoring, let us assume that strain and displacement measurements can always take place respectively within or at the inner surface of a hollow inclusion perfectly bonded to the surface of a pilot hole. This approach is very general since measurements on the rock surface or within a solid inclusion can then be considered as two limiting cases.

The inclusion is also assumed to be soft for two reasons:

• to reduce the effect of ignoring the overcoring diameter as discussed in section 6.4.1,

• to reduce the tensile stresses at the contact rock,

inclusion since, if the rock was under a compressive state of stress before overcoring, the radial stress σ_r induced by the inclusion within the rock becomes tensile after overcoring*. It can then be large enough to break the bond at the contact rock inclusion.

For given geometry and elastic properties of the inclusion, given elastic properties of the rock anisotropy, given orientation of the anisotropy and the pilot hole where measurements take place with respect to a fixed arbitrary global coordinate system, the components of matrices (Q) and (Q_u) appearing in eqs (6.2), (6.3) and (6.4) can be calculated respectively at any measuring point within or at the inner surface of the inclusion. Similarly, the components of matrix (T_σ) can be also determined.

Figures 6.5 show three types of measurement that can be associated with a hollow inclusion. In Figure 6.5a, the change of three inner diameters (or more if needed) of the hollow inclusion are recorded. Eq (6.4) can then be used to relate each measurement to the 3D stress field applied at infinity. This suggests the following measuring techniques:

(i) After drilling the pilot hole, a hollow inclusion is cemented onto the walls of that hole. Then, an instrumented device such as the USBM deformation gage is inserted into the hollow inclusion. After overcoring, the measurements of diameter take place at the inner surface of the hollow inclusion

---

* see for instance Figure 5.12 with σ_r replaced by $-\sigma_r$

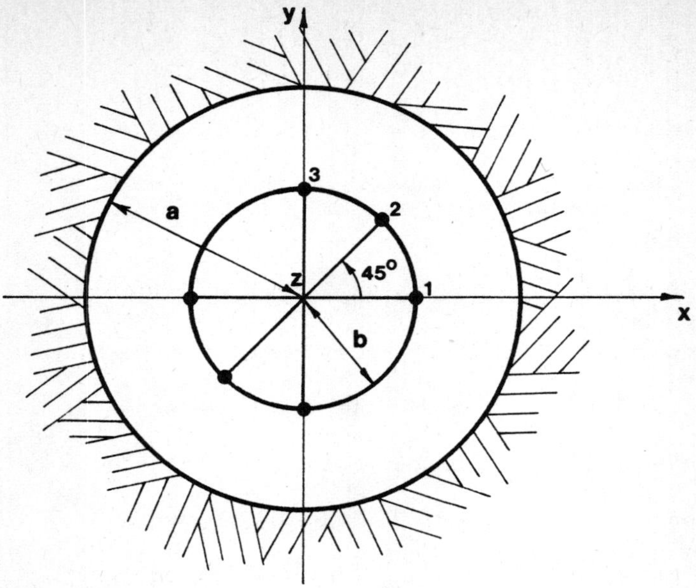

Figure 6.5. (a) Diametral measurements.

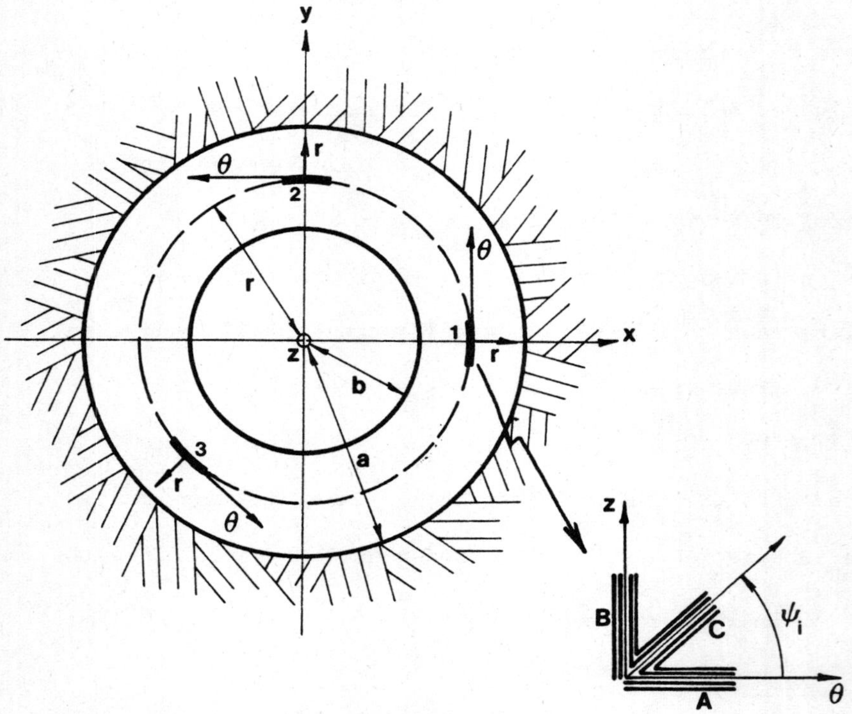

Figure 6.5. (b) Strain measurements.

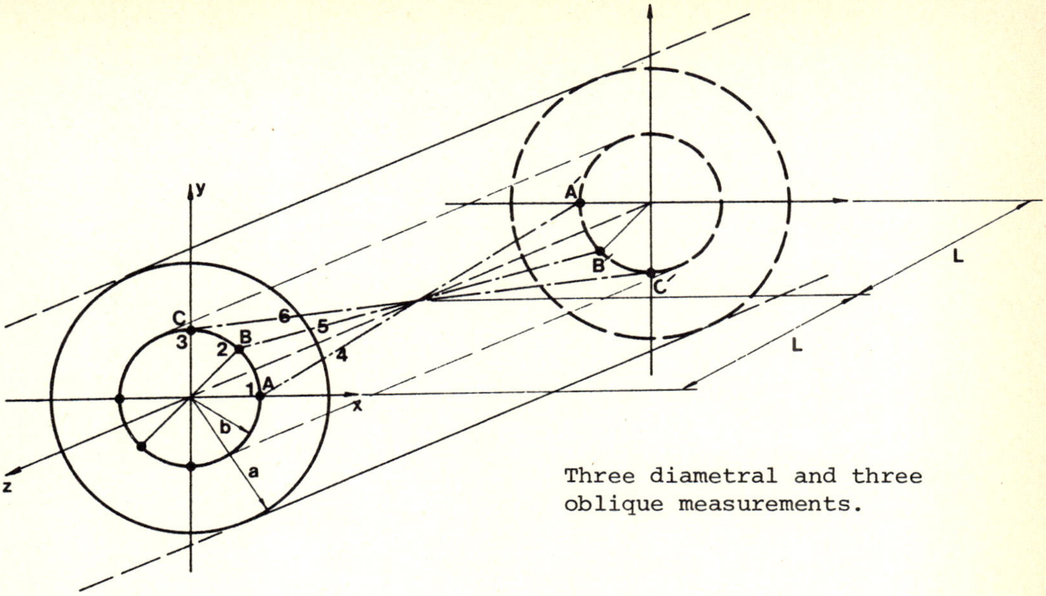

Three diametral and three oblique measurements.

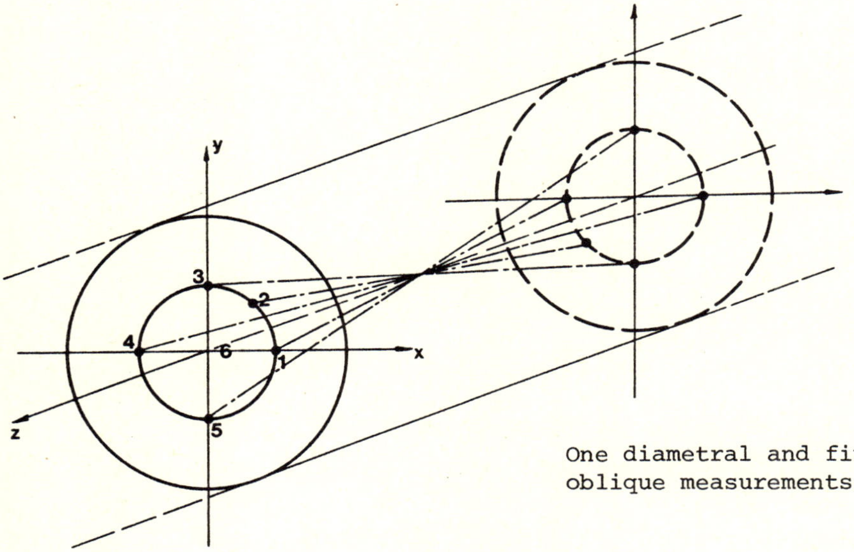

One diametral and five oblique measurements.

Figure 6.5. (c) Combination of measurements of changes in diameter and changes in length of oblique distances within an instrumented hollow inclusion.

instead at the rock surface in the classical procedure. The advantages of this technique with respect to the classical one are such that the diametral measurements are less affected by rock heterogeneities, discontinuities or wall irregularities. Measurements can also take place in weathered or jointed rocks.

(ii) After drilling the pilot hole, an instrumented hollow inclusion already containing gages that can measure changes in diameter is cemented onto the walls of the hole. Additional gages can be inserted to measure changes in length of oblique distances as shown in Figure 6.5c. Eq (6.3) can then be used to relate each measurement to the 3D stress field applied at infinity.

In Figure 6.5b, three strain rosettes (or more if needed) are embedded within a hollow inclusion along a circle of radius r. Each rosette is oriented as shown in Figure 6.4. For each rosette i with coordinates r/a, θ_i, the strain components $\varepsilon_{\theta_i}, \varepsilon_{z_i}$ and $\gamma_{\theta z_i}$ can be calculated from the measured values ε_{A_i}, ε_{B_i} and ε_{C_i} using eqs (6.7) and related to the 3D stress field applied at infinity through eq (6.2). This is valid as long as the length of the measuring gages is small. If this is not the case, the components of matrices $(\varepsilon)_{r\theta z}$ and $(Q)(T_\sigma)$ must be replaced by their average values calculated over the corresponding gage length.

The present analysis is not limited to the types of measurement and orientations shown in Figures 6.5. For instance measurements of strain and displacement may be combined within the same instrumented hollow inclusion. Similarly, strain gages can be inclined with respect to the θ, z axes and additional ones used.

In order to answer questions (i), (ii) and (iii) of section 6.5.1, three computer programs were developed. There are named Berni 1 Berni 2 and Berni 3. They are presented respectively in Appendices 6.2, 6.3 and 6.4. As far as stress measurements are concerned, those programs make use of the closed form solutions derived previously and the statistical analysis of Appendix 6.1. They calculate the least square estimates of the principal in situ stress components and their orientation with respect to an arbitrary fixed global coordinate system from measurements of change in strain within solid or hollow inclusions, change in diameter and change in length of oblique distances at the inner surface of hollow inclusions. Those measurements take place in one or several boreholes.

a) Number of Independent Measurements in a Single Borehole

When there is no inclusion and measurements take place directly on the rock surface, eqs (4.86), (5.47) and the remarks associated with them imply that for a circular hole, there at most three independent measurements of change in diameter and five independent measurements of change in length of oblique distances.

When the measurements take place on the inner surface of a hollow inclusion (Figures 6.5a, c), numerical examples using the computer programs mentioned previously have shown that those numbers still apply.

As far as measurements of change in strain are concerned, it is possible to obtain at least six independent measurements by orienting the strain rosettes in different directions. For instance, in the configuration of Figure 6.5b, there are at most seven independent measurements since three of the nine gages measure the strain parallel to the hole axis.

b) Number of Boreholes

The minimum number of boreholes required to calculate the complete state of stress depends primarily on the type of measurement used.

The complete state of stress can be determined within a single borehole by measuring changes in strain in six non parallel strain gages or by combining diametral and oblique measurements: for instance 5 oblique, 1 diametral; 4 oblique, 2 diametral; 3 oblique, 3 diametral (Figure 6.5c). This is true regardless of the isotropic, anisotropic character of the rock and the orientation of the hole with respect to the planes and axes of elastic symmetry of the rock.

When changes in diameter are measured only, it would seem to be possible to use two non parallel boreholes since there are at most three independent measurements per hole as mentioned previously. Several numerical examples using the three computer programs have shown that the requirement for a third non parallel borehole depends upon the following parameters:

- the angle between the two boreholes,
- the isotropic, anisotropic character of the rock,
- the orientation of both boreholes with respect to the

planes and axes of elastic symmetry of the rock.

In those numerical examples, the rock was assumed to be either isotropic or transversely isotropic or orthotropic. The different cases are summarized in Table 6.1. It appears in this table that, in comparison to the isotropic solution, the anisotropic character of the rock can reduce the number of boreholes to two. This applies when both boreholes are not perpendicular to any plane of elastic symmetry of the rock or when only one of the boreholes is perpendicular to a plane of elastic symmetry and is not at right angle to the other one. If we add more symmetry to the problem by forcing each borehole to be perpendicular to a plane of elastic symmetry, or by forcing one borehole to be perpendicular to a plane of elastic symmetry and the other one to be parallel to that plane, a third borehole is required. If the two boreholes axes are perpendicular to an axis of radial elastic symmetry by drilling them for instance within a plane of transverse isotropy at any angle to each other, a third borehole is also required. A similar conclusion applies when the rock is isotropic since all planes and axes within the rock are now planes and axes of elastic symmetry. Figures 6.6 illustrate the different cases of Table 6.1.

The possibility of reduction of the number of boreholes from three in the isotropic solution to two in the anisotropic one is closely related to the existence of planes of elastic symmetry perpendicular to the boreholes. If there is a plane of elastic symmetry perpendicular to a hole axis, the coeffi-

Angle between the two boreholes	Isotropic Anisotropic Character of the rock	Orientation of both boreholes with respect to the planes and axes of elastic symmetry of the rock	Requirement for a third borehole
any	Anisotropic	Each hole is not perpendicular to a plane of elastic symmetry Figure 6.6f	No
Non Perpendicular	Anisotropic	Only one hole is perpendicular to a plane of elastic symmetry Figure 6.6e	
any	isotropic	----	Yes
90°	Anisotropic	Only one hole is perpendicular to a plane of elastic symmetry Figure 6.6d	
90°	Anisotropic	Each hole is perpendicular to a plane of elastic symmetry Figures 6.6 a,b	
any	Anisotropic	Both holes are within a plane of transverse isotropy Figure 6.6c	

Table 6.1. Number of Boreholes Required for the Determination of the Complete State of Stress from Diametral Measurements Only.

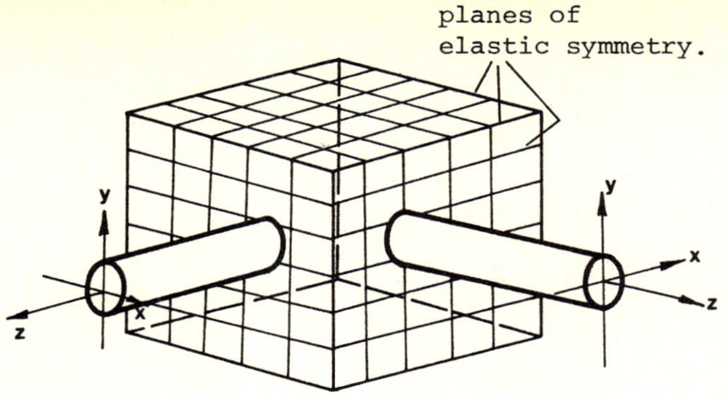

(a) Orthotropic symmetry. Boreholes are perpendicular.

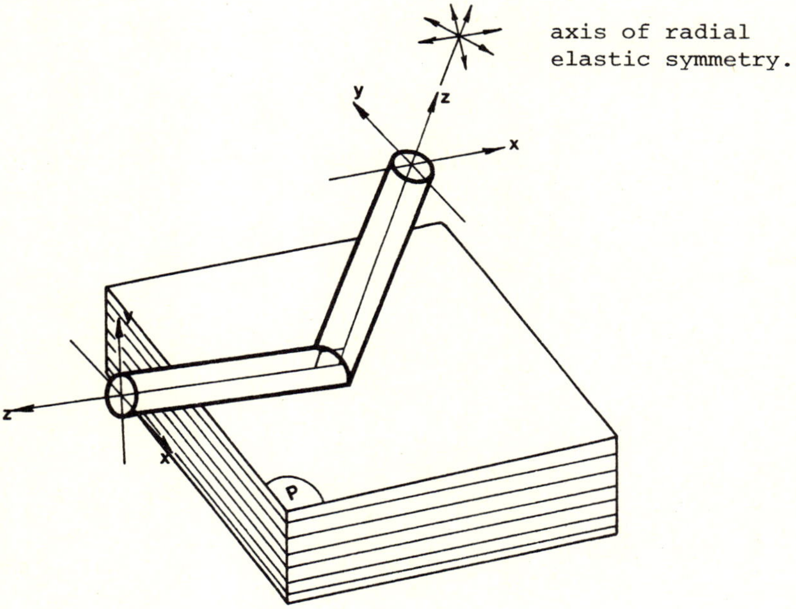

(b) Transverse isotropy. Boreholes are perpendicular to each other, parallel and perpendicular to the plane of transverse isotropy.

Figure 6.6. Illustrative examples for Table 6.1.

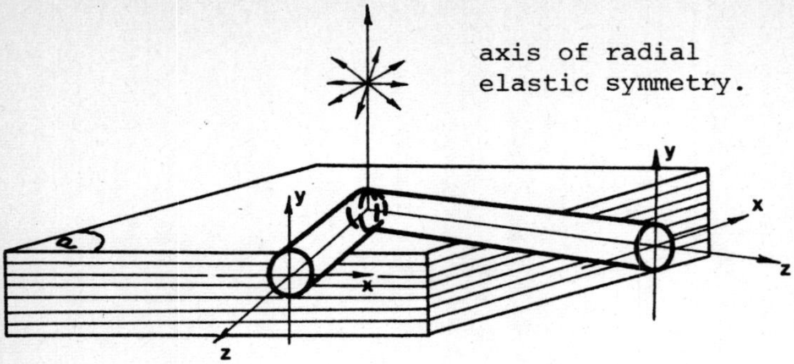

(c) Transverse isotropy. Boreholes are within the plane of transverse isotropy and do not need to be at right angle to each other.

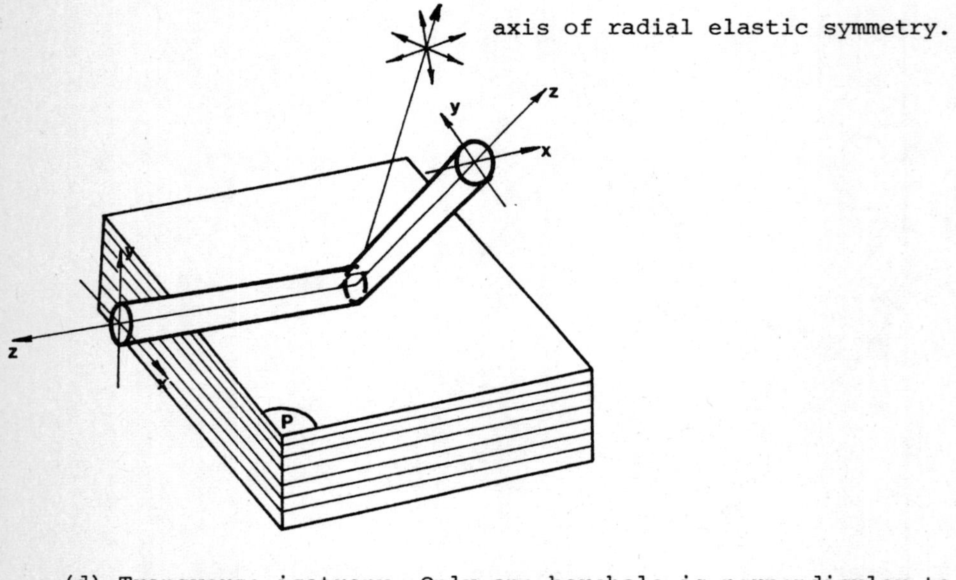

(d) Transverse isotropy. Only one borehole is perpendicular to a plane of elastic symmetry. Boreholes are perpendicular to each other.

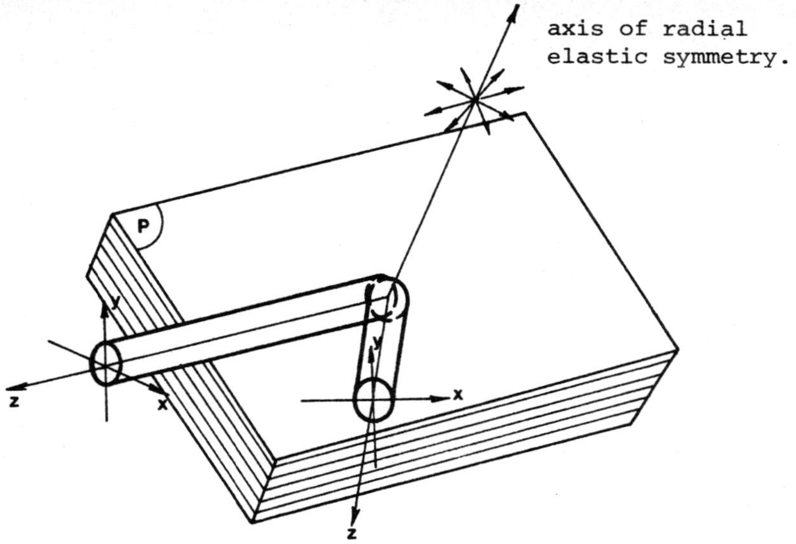

(e) Transverse isotropy. One borehole is perpendicular to a plane of elastic symmetry. Boreholes are not at right angle to each other.

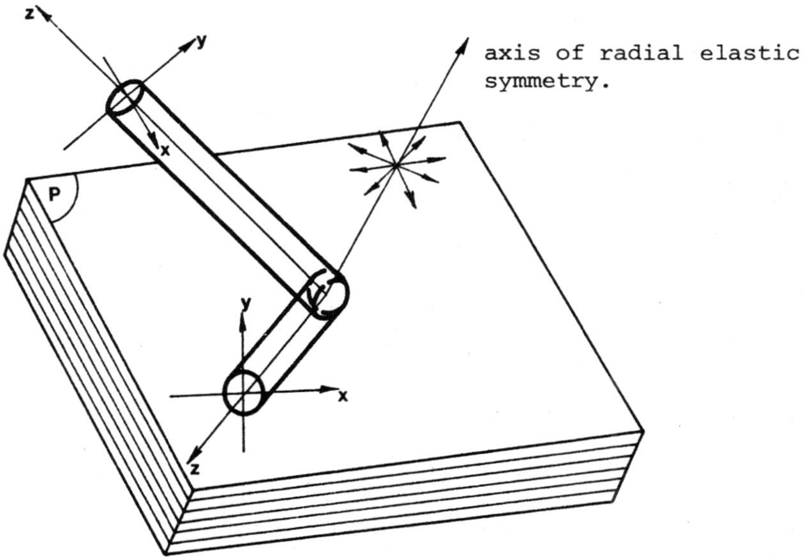

(f) Transverse isotropy. Boreholes are at any angle to each other. None of them is perpendicular to a plane of elastic symmetry.

cients of the stress components $\tau_{xz,o}$ and $\tau_{yz,o}$ in eqs (4.85) and (6.4) vanish*. Let two boreholes be each perpendicular to a plane of elastic symmetry or one be perpendicular to a plane of elastic symmetry and the other one be contained within that plane. In writing the (6×6) matrix that relates the diametral measurements to the six components of the in situ stress field expressed in a global coordinate system, because of the previous remark, some of the coefficients of that matrix vanish or are linearly dependent. The matrix becomes singular and a third borehole is required.

Let us consider a case where the rock is anisotropic and the orientation of the boreholes with respect to the planes of elastic symmetry is such that two boreholes are sufficient to calculate the complete state of stress from six diametral measurements. The question that arises is the following: "how strongly anisotropic should the rock be in order to make use of two boreholes instead of three"? This question arises naturally since when the rock is isotropic the determinant Δ of the (6×6) matrix that relates the diametral measurements to the in situ stress field components becomes zero (or very small for the computer). Theoretically, there must be cases where the rock is <u>weakly anisotropic</u> (almost isotropic) for which the determinant mentioned previously is small and for which the use of two boreholes may produce erroneous

---

* Eq (4.85) is reduced to eq (4.87)

answers. In order to investigate this problem, let consider the following numerical example.

Three boreholes BH_1, BH_2 and BH_3 are drilled within a transversely isotropic rock. The orientation of the anisotropy and of the boreholes with respect to a global coordinate system X,Y,Z is shown in Figure 6.7. In each borehole, three diametral measurements are recorded as in Figure 6.5a. The elastic properties and geometry of the hollow inclusion in each borehole are such that $a/b = 2.$; $E_{inc} = 2.\ 10^3$ MPa ; $\nu_{inc} = 0.3$. The following changes in diameter are measured*

	1 ($\theta = 0°$)	2 ($\theta = 45°$)	3 ($\theta = 90°$)
BH1	7.65 10^{-4}	8.16 10^{-4}	1.10 10^{-4}
BH_2	6.13 10^{-4}	6.35 10^{-4}	1.32 10^{-4}
BH_3	0.67 10^{-4}	2.39 10^{-4}	7.23 10^{-4}

The least square estimates of the principal components of the in situ stress field $\sigma_1, \sigma_2, \sigma_3$ and their orientation with respect to X,Y,Z are calculated from these nine measurements by assuming that the rock is isotropic (Case 5) or transversely isotropic (Case 1)**.

Let consider boreholes BH_1 and BH_2 only with their associated diametral measurements. According to Table 6.1,

---

* The changes in diameter are expressed in terms of $\Delta d_\theta / d_\theta =$ $u_r/b = (u_r/a)(a/b) = 2\,(u_r/a)$

** The elastic properties of the rock are defined in a 1,2,3 coordinate system that corresponds to the x',y',z' coordinate system in Figure 1 of Appendix 4.1.

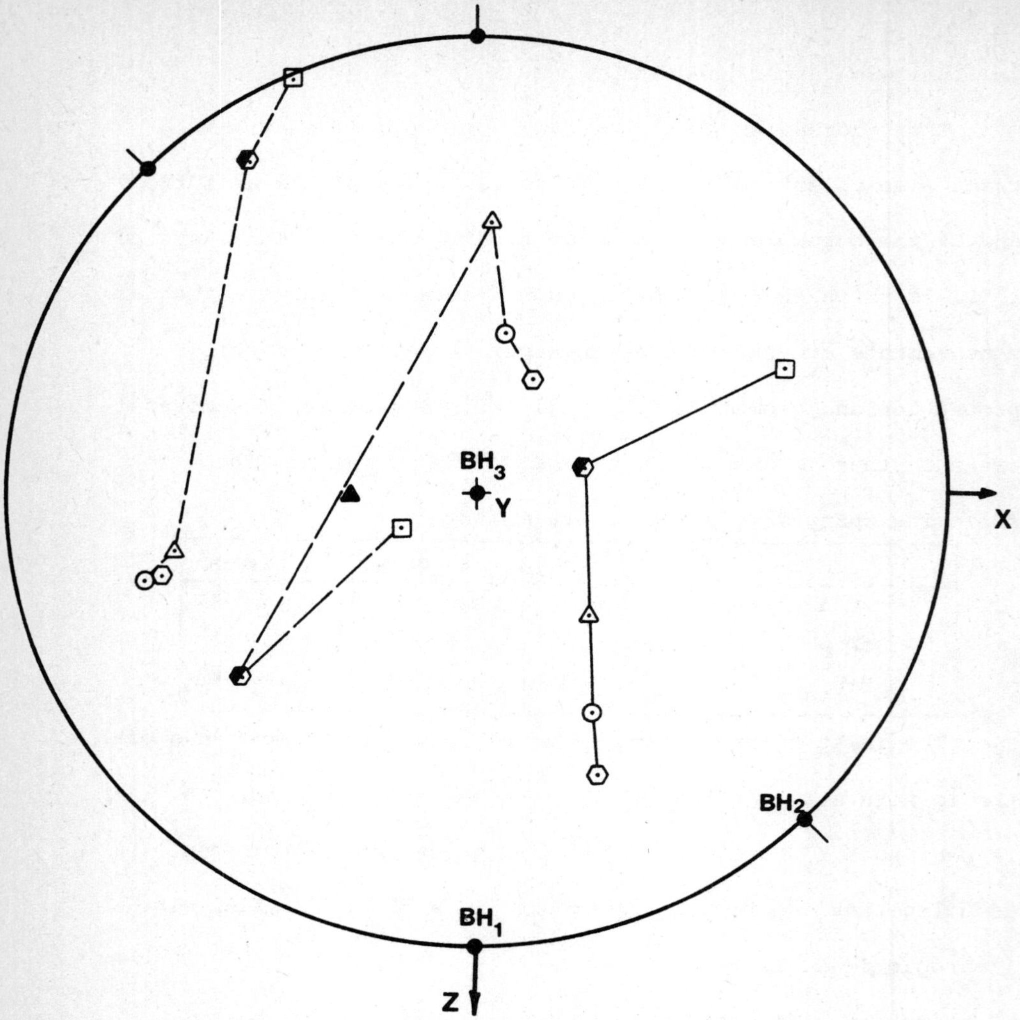

▲ Pole of plane of transverse isotropy.

——— σ_3

— — σ_2

- - - - σ_1

●— Borehole orientation.

	Case 1 ⊙	Case 2 ⬡	Case 3 ⬢	Case 4 ⊡	Case 5 △ 3 boreholes
E_1	$2.\ 10^4$	$3.\ 10^4$	$3.5\ 10^4$	$3.9\ 10^4$	$4.\ 10^4$
E_2, E_3	$4.\ 10^4$	$4.\ 10^4$	$4.\ 10^4$	$4.\ 10^4$	$4.\ 10^4$
G_{12}, G_{13}	$4.\ 10^3$	$8.\ 10^3$	$1.4\ 10^4$	$1.55\ 10^4$	$1.6\ 10^4$
G_{23}	$1.6\ 10^4$	$1.6\ 10^4$	$1.6\ 10^4$	$1.6\ 10^4$	$1.6\ 10^4$
ν_{21}, ν_{31}	0.4	0.27	0.23	0.24	0.25
ν_{23}	0.25	0.25	0.25	0.25	0.25
σ_3	0.29	-0.30	1.23	-1.63	0.69
σ_2	1.29	1.37	6.04	5.41	2.13
σ_1	3.71	5.64	10.08	62.98	8.09
Δ	Δ_o	$6.10^{-2}\Delta_o$	$2.10^{-3}\Delta_o$	$1.10^{-4}\Delta_o$	----

Figure 6.7. Orientation and magnitude of the principal in situ stress field components for different cases of transverse isotropy using two or three boreholes.

those two boreholes are sufficient to calculate the complete state of stress when the rock is transversely isotropic. Four cases of transverse isotropy are analyzed. Cases 1 and 2 correspond respectively to strongly and moderately anisotropic rocks. Cases 3 and 4 are associated with weakly anisotropic rocks. For each case, the principal components of the in situ stress field, their orientation in the X,Y,Z coordinate system, and the value of the determinant Δ are calculated from the same set of six diametral measurements. The results are summarized in Figure 6.7. The three changes in diameter in borehole BH_3 were chosen in order to make the computed principal stress field components and their orientation identical when using boreholes BH_1, BH_2, BH_3 or boreholes BH_1, BH_2 and when case 1 of anisotropy is considered. Therefore, case 1 appears as a reference for which the determinant Δ is equal to a reference value Δ_0. For the example presented in Figure 6.7, it appears that the value of the determinant Δ decreases appreciably from its initial value Δ_0 as the rock becomes less anisotropic. For a weakly anisotropic rock (cases 3 and 4) the use of two boreholes may produce erroneous answers in comparison to the isotropic solution as far as the magnitude and the orientation of the principal in situ stress field $\sigma_1, \sigma_2, \sigma_3$ is concerned.

In summary, whenever the rock is anisotropic and the orientation of the boreholes with respect to the planes of elastic symmetry of the rock is such that two boreholes are sufficient to calculate the complete state of stress, attention must be paid to the anisotropic character of the rock. The use of two boreholes is limited to moderately or strongly anisotropic rocks. For weakly anisotropic (almost isotropic) rocks a third borehole is required. In any case, it is worthwhile to evaluate the value of the determinant Δ . One should be especially careful when this quantity becomes small.

c) Influence of Rock Anisotropy on the Determination of the In Situ Stress Field

The following two questions are occasioned by the previous analysis:

- For a given set of measurements, how do anisotropy type and orientation influence the determination of the in situ stress field?
- How large an error is involved by neglecting rock anisotropy?

Since it is difficult to draw general conclusions for all types of anisotropy, the present analysis is restricted to the transversely isotropic case only. A similar analysis can be carried out for more general types of anisotropy (with up to 21 coefficients of elasticity) since the theory developed herein is general. One of the several reasons for using the transversely isotropic type is that it can be visualized as a plane and can be easily represented geometrically.

In order to answer the two questions stated previously, let us consider the instrumented hollow inclusion of Figure

6.5b. The geometric and elastic properties of the inclusion are such that $a/b = 2$, $E_{inc} = 2\ 10^3$ MPa and $\nu_{inc} = 0.3$. Three strain rosettes are embedded halfway within the inclusion $(r/a = 0.75)$ and located at $\theta = 0°, 90°$ and $225°$. Each rosette is made of three strain gages with $\psi_i = 225°$. Let us consider the following set of strain measurements

	$10^6 \varepsilon_A$	$10^6 \varepsilon_B$	$10^6 \varepsilon_C$
1 $(\theta = 0°)$	-79	-86	44
2 $(\theta = 90°)$	-430	-85	-187
3 $(\theta = 225°)$	-52	-87	-210

The hollow inclusion is located within a borehole whose axes x,y,z are parallel to the axes X,Y,Z of a global coordinate system. The orientation of the anisotropy with respect to X,Y,Z is defined by the strike β and the dip ψ of the plane of transverse isotropy as shown in Figure 6.8.

For the given set of strain measurements, the least square estimates of the principal components of the in situ stress field $\sigma_1, \sigma_2, \sigma_3$ and their orientation with respect to X,Y,Z are calculated for each of the following conditions:

(1) the rock is isotropic with elastic properties

$$E = 4\ 10^4 \text{ MPa}; \ \nu = 0.25 \Rightarrow G = 1.6\ 10^4 \text{ MPa}$$

(2) the rock has a plane of transverse isotropy dipping at at an angle ψ of 30 or 90 degrees and striking at different angles β from 0 to 90 degrees. The elastic properties of the rock are as follows: $E_1 = 2\ 10^4$ MPa; $E_2 = E_3 = 4\ 10^4$ MPa;

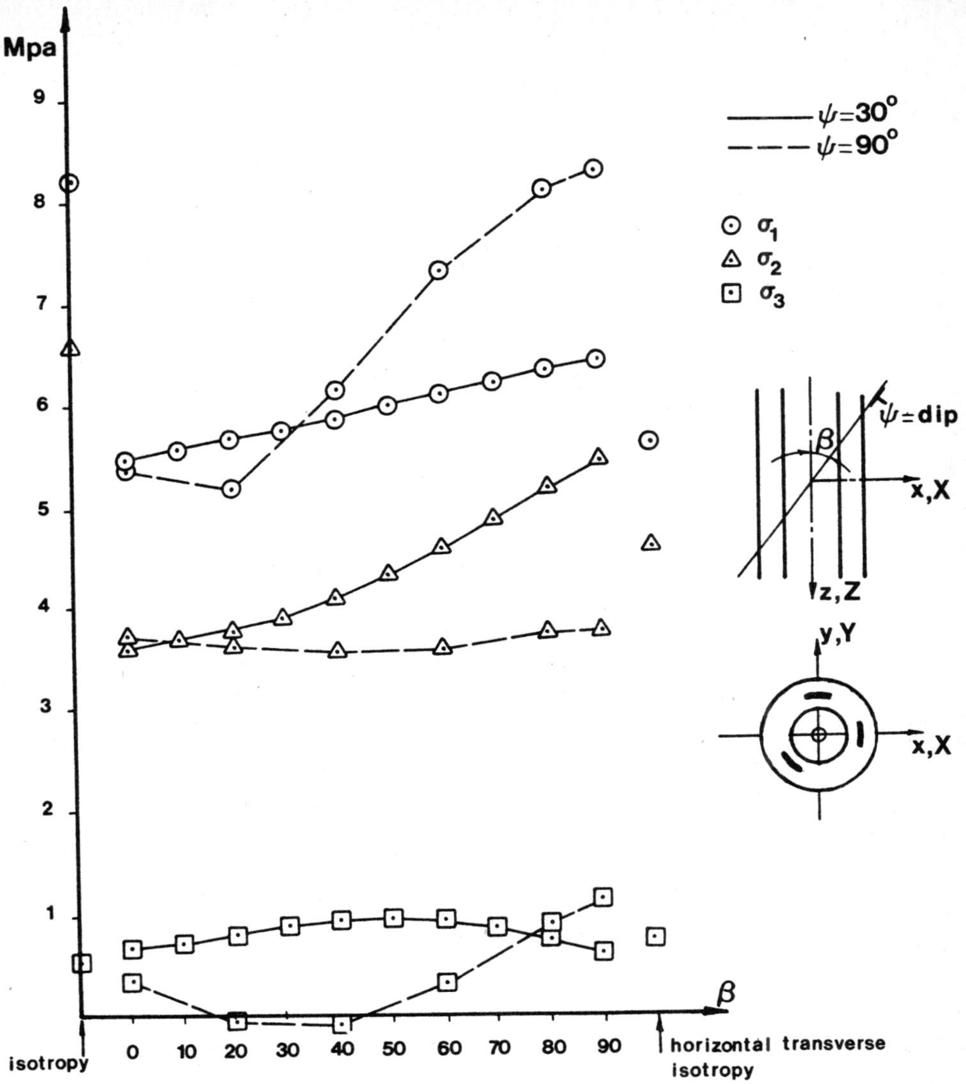

Figure 6.8. Magnitude of the principal in situ stress field components for different orientations of a plane of transverse isotropy.

$$G_{12} = G_{13} = 4\ 10^3\ \text{MPa}\ ;\ G_{23} = 1.6\ 10^4\ \text{MPa}$$

$$\nu_{21} = \nu_{31} = 0.4\ ;\ \nu_{23} = 0.25^*$$

(3) the borehole containing the inclusion is within the plane of transverse isotropy. This case is termed "horizontal transverse isotropy" for sake of clarity**.

The results are summarized in Figure 6.8 as concerns the magnitude of $\sigma_1, \sigma_2, \sigma_3$ and in Figures 6.9 and 6.10 for their orientation when ψ is respectively equal to 30 and 90 degrees.

It appears through this numerical example, that the influence of the rock anisotropy on the determination of the in situ stress field is non negligible especially for low values of the strike angle β. This influence is reduced as the plane of transverse isotropy turns and becomes perpendicular to the hole axis. This seems to apply regardless of the value of the dip angle ψ (except for $\psi = 0$). When the plane of transverse isotropy is horizontal it also has a significant influence. The stereographic projections in Figures 6.9 and 6.10 show that the directions of σ_1 and σ_2 follow a path close to a great circle as β varies between 0 and 90 degrees since the direction of σ_3 seems to be little affected by the value of the strike angle.

---

* see footnote ** page 229.

** The plane of transverse isotropy is parallel to the plane XOZ of the global coordinate system

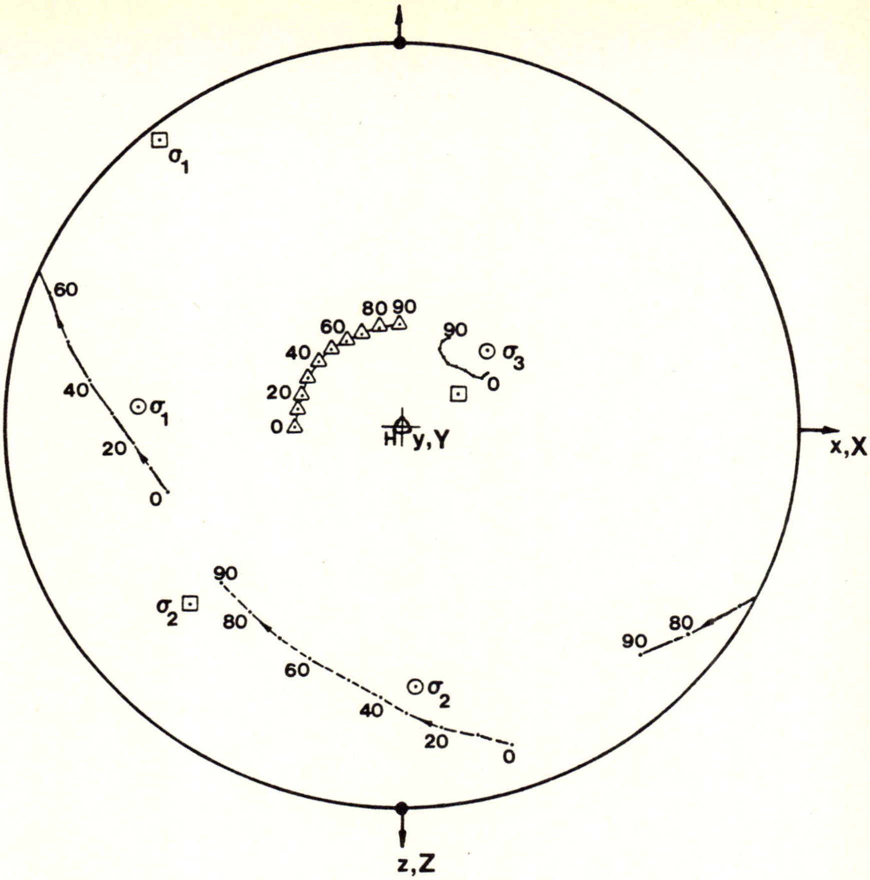

⊙ isotropic solution.

——— σ_3

— — σ_2

- - - - σ_1

0, 20, ..., 90 ; strike values β.

⊡ horizontal transverse isotropy.

△ H poles of planes of transverse isotropy.

●— borehole orientation.

Figure 6.9. Orientation of the principal in situ stress field components for different strike angles β of a plane of transverse isotropy ($\psi = 30°$) and for a horizontal transverse isotropy. Lower hemisphere stereographic projection.

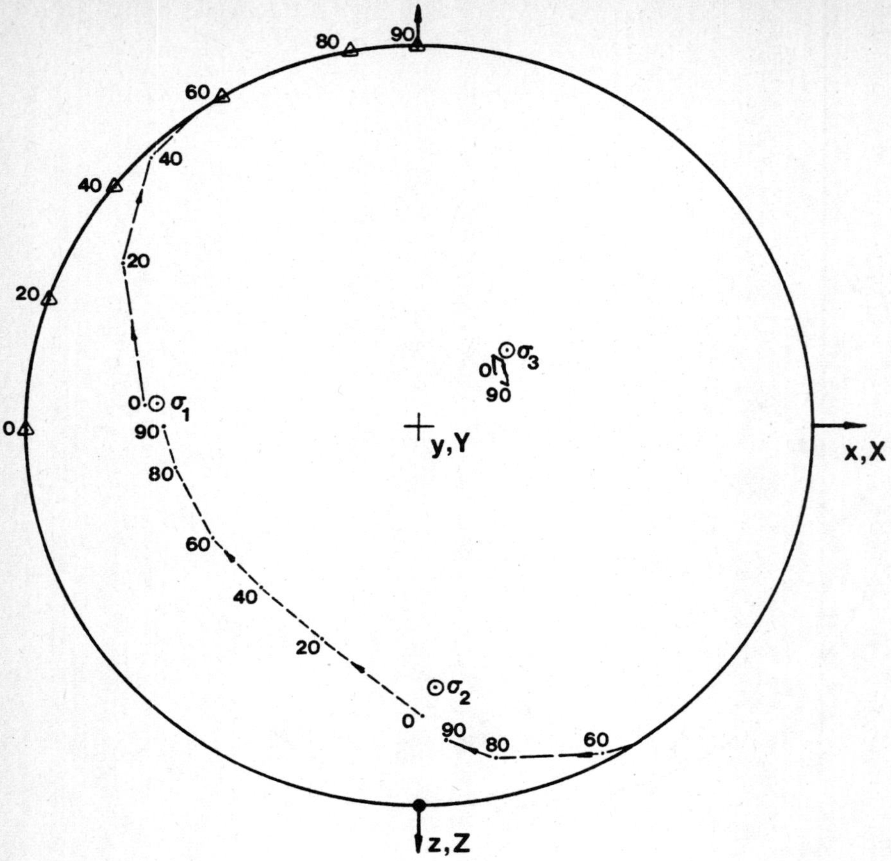

⊙ isotropic solution.

——— σ_3

— — σ_2

- - - - σ_1

0,20,.....,90; strike angles β .

△ poles of planes of transverse isotropy.

●— borehole orientation.

Figure 6.10. Orientation of the principal in situ stress field components for different strike angles β of a plane of transverse isotropy ($\psi = 90°$). Lower hemisphere stereographic projection.

In Figure 6.11, the orientation of the anisotropy is held constant ($\beta = 0°$; $\psi = 30°$) and four cases of transverse isotropy are considered; the rock becoming less anisotropic from case 1 to case 4. As a reference, case 5 is the isotropic solution. For weakly anisotropic rocks (cases 3 and 4) the influence of the rock anisotropy on the determination of the in situ stress field is negligible. For moderately and especially strongly anisotropic ones, cases 2 and 1 respectively, the rock anisotropy influences both the orientation and the magnitude of the principal in situ stress field.

In conclusion to this example, it appears that neglecting anisotropy by assuming that the rock is isotropic can create large errors as far as both the magnitude and the orientation of the principal in situ stress field are concerned. This is particularly true for moderately and especially strongly anisotropic rocks and for low values of the strike angle β. The errors can be as large as 50 to 80% and the orientation of σ_1, σ_2 can be up to 90 or 100 degrees off from the isotropic solution. This is largely reduced as the rock becomes weakly anisotropic or the plane of transverse isotropy strikes at a large angle to the hole axis. The case of a horizontal plane of transverse isotropy stands between those two extremes.

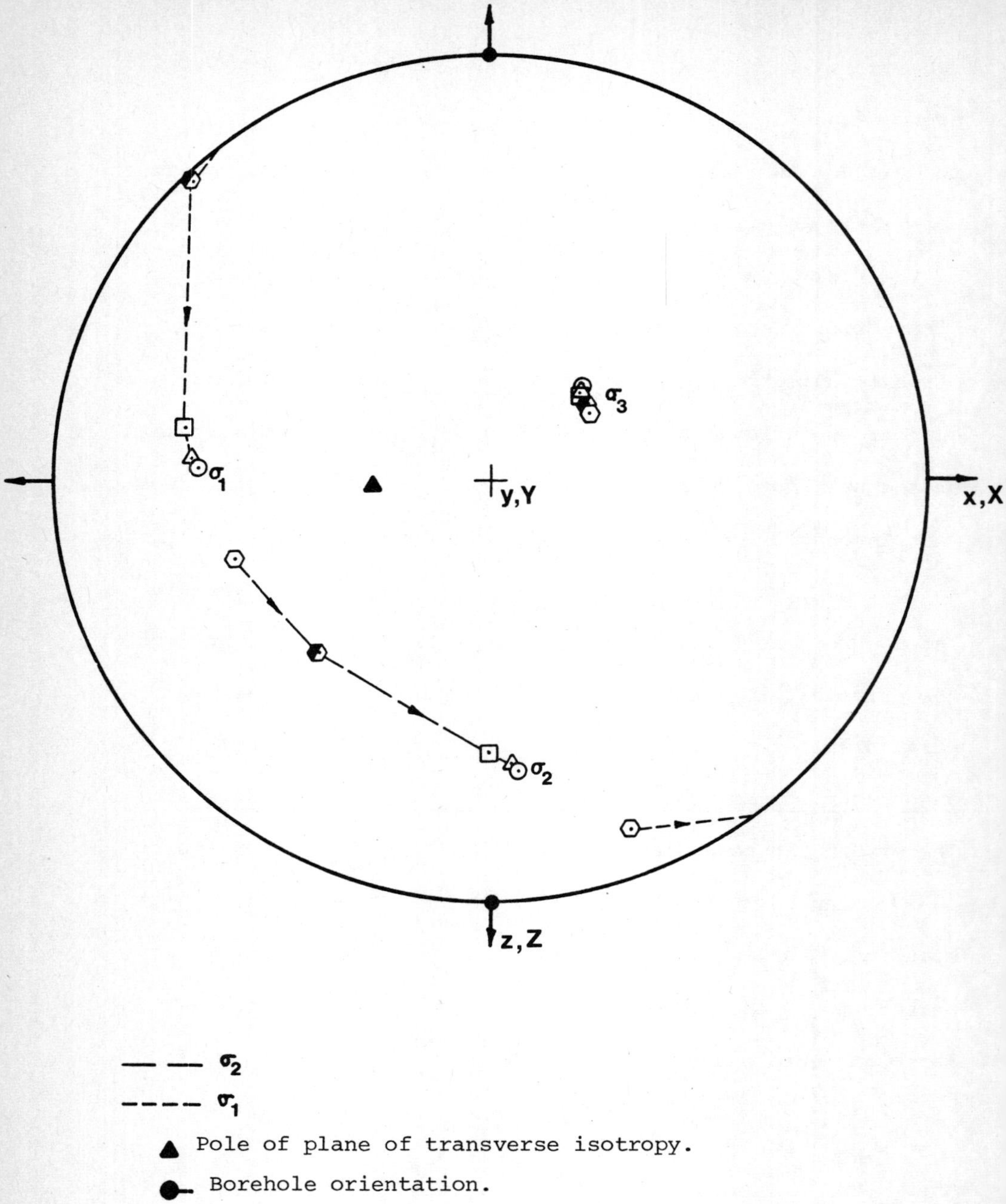

▲ Pole of plane of transverse isotropy.

●– Borehole orientation.

	Case 1	Case 2	Case 3	Case 4	Case 5 isotropic
E_1	$2\ 10^4$	$3\ 10^4$	$3.5\ 10^4$	$3.9\ 10^4$	$4\ 10^4$
$E_2\ E_3$	$4\ 10^4$	$4\ 10^4$	$4\ 10^4$	$4\ 10^4$	$4\ 10^4$
G_{12}, G_{13}	$4\ 10^3$	$8\ 10^3$	$1.4\ 10^4$	$1.55\ 10^4$	$1.6\ 10^4$
G_{23}	$1.6\ 10^4$	$1.6\ 10^4$	$1.6\ 10^4$	$1.6\ 10^4$	$1.6\ 10^4$
ν_{21}, ν_{31}	0.4	0.27	0.23	0.24	0.25
ν_{23}	0.25	0.25	0.25	0.25	0.25
σ_3	0.67	0.59	0.42	0.48	0.52
σ_2	3.62	5.29	6.16	6.47	6.58
σ_1	5.46	6.02	7.31	7.99	8.20

Figure 6.11. Determination of the principal in situ stress field for different cases of transverse isotropy ($\beta = 0^\circ$; $\Psi = 30^\circ$).

Chapter 7: Summary and Conclusions

In order to assess the influence of rock anisotropy on in situ stress measurements from a quantitative point of view, knowledge of the directional character of the rock deformability is required. In this study, the latter is described using the constitutive relations of the theory of linear elasticity for anisotropic media. Directions of elastic symmetry are assumed to coincide with the apparent directions of rock anisotropy. The extent to which this assumption is applicable must always be investigated by testing rock specimens in the laboratory and in situ. Procedures are proposed in Chapter 2 to calculate the five and nine elastic constants of rocks that can be described respectively as transversely isotropic or orthotropic materials using unconfined, triaxial or multiaxial compression laboratory tests. If more than nine elastic constants are needed, the procedure becomes more complex but may be investigated as a possible model to describe the deformability of anisotropic rocks that present directions of anisotropy inclined with respect to each other.

The elastic properties measured using laboratory tests may not be representative of the deformability of the rock in place since scale effects are important in rock mechanics. In situ tests reported in the literature for the determination of the elastic constants of anisotropic rock masses are not numerous. Additional assumptions are often stated in order to reduce the number of elastic constants that must be known. A procedure is proposed here to calculate the elastic properties of anisotropic rock masses that can be described as orthotropic or transversely isotropic by means of a dilatometer test. Since there are at most two independent measurements of diameter change obtainable from the expansion of a circular hole, general determination of the five of nine elastic constants is impossible from measurements in a single borehole. Therefore several non parallel boreholes must be drilled within the region of interest or additional assumptions are required regarding the elastic constants. However, when the anisotropy is derived from a regular joint set striking parallel to a borehole and the intact rock is isotropic, the two changes in diameter are sufficient to determine the normal and shear stiffnesses of the joint set as long as its spacing is known and the stiffnesses are assumed constant. In theory, the elastic modulus of the intact rock can be determined by pressurizing and unpressurizing the hole if the joint deformations are assumed to be non recoverable and the rock to be elastic. This procedure can be applied also if the rock is cut by three orthogonal joint sets having identical properties. This leads to a possible method to evaluate the modulus of deformation of a regularly jointed rock mass.

A prerequisite for the applicability of stress measurement techniques is that the strength of the rock is never exceeded anywhere around the hole where measurements take place. This is particularly important if the stresses to be measured are expected to be high. From a practical point of view, rock failure can occur as indicated by core discing or flaking of borehole walls. Obviously, measurements obtained from a borehole where those phenomena take place should not be interpreted using the theory of linear elasticity. Chapter 3 deals with the strength of anisotropic rocks but not exhaustively. This chapter simply reviews the experimental evidence for the directional character of the strength of anisotropic rocks and the analytical models available to account for this property. It is suggested that an appropriate failure criterion for anisotropic rocks must be sufficiently general to include the influence of the intermediate principal stress and to account for the type and orientation of the anisotropy. The general form of the failure criterion is patterned after that for the deformability. It depends on the results of laboratory tests and the apparent symmetries of the rock material.

The tensile strength of anisotropic rocks can be calculated using direct or indirect methods (e.g. the "Brazilian" test). Interpretation of test results must include the anisotropic character of the rock. In particular, if a disc of rock is loaded with two line loads, the stress distributions within the disc calculated from the theory of linear elasticity must include the anisotropic character of the rock, its symmetry if any and the orientation of the anisotropy with respect to the direction of loading. It is surprising that common practice uses the isotropic solution to calculate the tensile strength of anisotropic rocks even though an anisotropic solution is available (Okubo, 1952).

General closed form solutions are proposed in Chapter 4 for the distributions of stresses, strains and displacements at any point within an infinite linear elastic, anisotropic, continuous and homogeneous body bounded internally by a cylindrical surface of arbitrary cross section. The body is subjected to body forces and boundary stresses acting along its internal contour. The work of Lekhnitskii (1963) used as a guide line, is extended to include the influence of a boundary stress component parallel to the hole axis. The anisotropy is general and described by 21 independent elastic constants in the coordinate system attached to the hole.

From a practical point of view, these closed form solutions are complicated and simplifying assumptions are often required. The different formulations of plane strain and plane stress available in the literature are reviewed and their applicability to anisotropic media discussed. In particular, it is found that the assumptions associated with the usual plane strain and plane stress formulations imply undue restrictions on the orientation of the hole with respect to the anisotropy, the mode of loading and the mode of deformation of the anisotropic body. Instead, the generalized plane strain formulation introduced by Lekhnitskii (1963) and Milne-Thomson (1962) can be used for any mode of loading and orientation of the anisotropy. The only assumption is that stresses, strains and displacements induced by the application of the boundary stresses be the same in any cross section perpendicular to the longitudinal axis of the hole. The approximate character of the existing plane stress formulations is also emphasized. Furthermore, all of them require that the hole be perpendicular to a plane of elastic symmetry.

As a special case of the general theory, a closed form solution is derived between the components of stress, strain and displacement at any point within an infinite anisotropic body bounded internally by an infinite cylinder with a circular cross section and the components of a 3D stress field applied at infinity. The hole axis, the anisotropy and the 3D applied stress field are inclined with respect to each other. The model allows for a constant longitudinal strain.

This strain is induced by the application of the original 3D stress field on the anisotropic medium lacking a hole even though the process of drilling the hole induces zero longitudinal strain. Stresses, strains and displacements induced by the hole drilling can be calculated using the generalized plane strain formulation. The latter reduces to the classical plane strain formulation if the hole is perpendicular to a plane of elastic symmetry and is drilled in a principal direction of the applied stress field.

The general theory can be used also for other cases of boundary stresses such as the application of an internal pressure uniformly distributed along the contour of a circular hole or more complex boundary stress distributions such as those associated with borehole jacks. This should allow anisotropy to be included in the analytical formulations for the Goodman jack in association with the measurement of deformability or strength of anisotropic rock masses.

Chapter 5 analyzes the elastic equilibrium of an infinite, continuous, homogeneous and anisotropic body bounded internally by an isotropic hollow or solid inclusion of circular cross section. The inclusion is assumed to be infinitely long and perfectly bonded to the anisotropic body. Closed form solutions are developed, in matrix form, between the total components of stress, strain and displacement at any point within the inclusion located far from its ends and the components of a 3D stress field applied at infinity. The model allows for a constant longitudinal strain within the anisotropic body and the inclusion. A general expression is also derived to relate the change in length between two points located in different cross sections of the inclusion to the components of the 3D applied stress field.

Another point of interest in Chapter 5 is the influence of the value of the elastic modulus of the inclusion and its geometry upon the amount of disturbance of the stress field associated with the presence of the inclusion within the anisotropic body. Numerical examples are presented for isotropic and transversely isotropic media containing a high or a low modulus inclusion. For these examples, an inclusion is arbitrarily defined as having a high modulus if the ratios of the Young's modulus of the inclusion to the Young's moduli of the anisotropic body are all larger than 0.5. It is defined as a low modulus inclusion if these ratios are all less than 0.1. The following conclusions can be drawn:

a) at any point along the external contour or within a hollow inclusion, the principal stress field becomes more uniform as the inclusion becomes thicker. It is completely uniform for a solid inclusion as predicted by Eshelby (1957), its magnitude being mostly affected by the high or low modulus character of the inclusion. Its orientation seems to be little influenced by the orientation of the anisotropy and is close to the principal applied stress field.

b) the distribution and magnitude of the stress components σ_r, $\tau_{r\theta}$ and τ_{rz} induced by the inclusion within the anisotropic body seem to depend upon the high or low modulus character of the inclusion, its geometry and the orientation of the anisotropy. In particular, the radial stress σ_r can be as large as the applied stress field along the external contour of a thick walled and high modulus inclusion. It decays rapidly within the anisotropic body but is still non negligible at a distance of five times the external radius of the inclusion unless the ratio of the outer to inner radii of the inclusion is small (less than 2) and/or a low modulus inclusion is considered. For a transversely isotropic body, the influence of the anisotropy is found to be the largest when the plane of transverse isotropy strikes parallel to the inclusion and to decrease appreciably when it strikes perpendicular to the inclusion.

The following parameters influence the stress distributions within and around an isotropic inclusion in an infinite anisotropic medium and must be investigated further:

• type and degree of anisotropy,

• orientation of the inclusion with respect to the principal applied stress field.

In Chapter 6, the closed form solutions derived in the previous chapters are used to solve the inverse problem, that is to calculate the principal components of a 3D in situ stress field and their orientation with respect to a fixed arbitrary global coordinate system, from measurements of strains and displacements associated with overcoring in one or several boreholes. No restrictions are made on the type or the orientation of the anisotropy or the orientation of the principal in situ stress field with respect to the different boreholes. The proposed theory is general and is applicable to the following types of overcoring measurements in anisotropic rocks:

1) changes in diameter of a pilot hole and changes in strain recorded on the walls of that hole,

2) changes in length between two points located in different cross sections at the inner surface of a hollow inclusion or within

a solid inclusion. The inclusion is perfectly bonded to the walls of a pilot hole. These changes in length reduce to changes in diameter when the two points are located within the same cross section. Case 1) is then a limiting case when there is no inclusion,

3) strain measurements using strain rosettes embedded at any point and in any direction within a hollow or solid inclusion perfectly bonded to the walls of a pilot hole,

4) combination of 2) and 3).

The advantages of the types of measurements in 2) and 3) with respect to those in 1) are such that the measurements are less affected by rock heterogeneities, discontinuities or wall irregularities. Measurements can also take place in weathered or jointed rocks and no surface preparation of the walls of the pilot hole is required. However, based on the results of Chapter 5, it is found that the only limitation is that low modulus thin walled hollow inclusions or very low modulus solid inclusions must be used for two reasons:

- to reduce the effect of ignoring the size of the overcoring diameter,
- to reduce the tensile stresses at the contact between the rock and the inclusion, induced by the inclusion during overcoring.

The proposed theory is also applicable for the determination of the components of any change of the in situ state of stress using the types of measurements in 1) to 4). In this case, the limitation previously mentioned does not apply.

Six independent measurements are required to calculate the six components of the in situ stress field. The following answers to questions (i) and (iv) of Chapter 1 are proposed:

a) in a single borehole there are at most three independent measurements of diameter and five independent measurements of change in length of oblique distances recorded at the inner surface of a hollow inclusion, or at the walls of a pilot hole. It is possible to obtain six independent measurements of strain by orienting strain rosettes in different directions.

b) the minimum number of boreholes required to calculate the complete state of stress depends primarily on the type of measurement used. Additional boreholes can always be used to increase the accuracy of the calculated stress field components. If strain measurements are used, one borehole may be enough. One borehole is also enough if measurements of change in diameter and change in length of oblique distances are combined. When changes in diameter are measured only, at least two non parallel boreholes are needed. The requirement for a third non parallel borehole depends upon the angle between the two boreholes, the isotropic, anisotropic character of the rock and the orientation of both boreholes with respect to the plane and axes of elastic symmetry of the rock. Whenever those conditions are such that two boreholes can be used, it is found that attention must be paid to the degree of anisotropy of the rock. The use of two boreholes is limited to moderately or strongly anisotropic rocks. For weakly anisotropic rocks (almost isotropic) the use of two boreholes produces erroneous answers and a third borehole is required.

c) since it is difficult to draw general conclusions for all types of anisotropy, questions (iii) and (iv) of Chapter 1 are answered for the transversely isotropic case only. It appears that neglecting anisotropy by assuming that the rock is isotropic can create large errors in both the magnitude and the orientation of the in situ stress field. An example involving the measurements of change in strain within a single borehole has shown that for low values of the strike angle of the plane of transverse isotropy with respect to the hole axis and for moderately or strongly anisotropic rocks, the errors can be as large as 50 to 80% in the magnitude of the calculated principal stress field. Its orientation can be up to 100 degrees off from the isotropic solution. This is found to be largely reduced if the hole strikes perpendicular to the plane of transverse isotropy or if the rock is weakly anisotropic.

The influence of rock anisotropy on stress measurements is not restricted to the special case considered (transverse isotropy). In more general symmetry classes (9, 13 or 21 elastic constants), the specific influence of anisotropy can be investigated for a given type and orientation of anisotropy using the computer programs developed here. This also applies to rocks whose anisotropy is associated with regular joint sets and are described as equivalent continuous media as long as the joint set stiffnesses are constant.

The results of this study suggest that the influence of rock anisotropy must also be investigated in the interpretation of measurements obtained from other techniques such as "hydraulic fracturing".

As far as "undercoring" is concerned, general formulas for the influence of rock anisotropy are presented in Chapter 6 and their application should be investigated further.

In order to improve the reliability of the in situ stress determination, further study should address all non-idealities of rock behavior in the analytical formulation used to convert in situ data into rock stresses. These non-idealities include non linear elastic or inelastic rock behavior, heterogeneous and discontinuous characters of rocks, and time dependent behavior. However, it is not tractable to address all these simultaneously in a single closed form solution connecting the in situ stress field components and the measurements of strains and displacements. The most immediate extension of the work described here would be to incorporate either non linear elasticity or linear viscoelasticity.

REFERENCES

1. Agarwal, R. (1968). Sensitivity analysis of borehole deformation measurements of in situ determination when affected by borehole excentricity. Proc. 9th Symposium on Rock Mechanics, (AIME), pp. 79-83.

2. Akai, K., Yamamoto, K. and Arioka, M. (1970). Experimentelle forschung über anisotropische eingenschaften von kristallen schiefen (in German). Proc. 2nd Cong. ISRM (Belgrade), Vol. 2, 3-26.

3. Allirot, D. and Boehler, J.P. (1979). Evolution des proprietes mecaniques d'une roche stratifiee sous pression de confinement (in French). Proc. 4th Cong. ISRM (Montreux), Vol. 1, pp. 15-22.

4. Amadei, B. and Goodman, R.E. (1981a). A 3-D constitutive relation for fractured rock masses. Proc. International Symposium on the Mechanical Behavior of Structured Media, Ottawa. Selvadurai, A.P.S. (Editor). Part B, pp. 249-268. (Appendix 2.1).

5. Amadei, B. and Goodman, R.E. (1981b). Formulation of complete plane strain for regularly jointed rocks. Proc. 22nd Symposium on Rock Mechanics (M.I.T.), pp. 245-251. (Appendix 4.6).

6. Atkinson, R.H. and Ko, H.-Y. (1973). A fluid cushion, multiaxial cell for testing cubical rock specimens. Int. J. Rock Mech. Min. Sci., Vol. 10, No. 4, pp. 351-361.

7. Attewell, P.B. and Sandford, M.R. (1974). Intrinsic shear strength of a brittle, anisotropic rock. Part 1. Experimental and Mechanical Interpretation. Int. J. Rock Mech. Min. Sci., Vol. 11, No. 11, pp. 423-430.

8. Barla, G. (1974). Rock anisotropy. Theory and laboratory testing. Rock Mechanics, Müller, L. (Editor), (Course held at the Department of Mechanics of Solids, Udine, Italy), pp. 131-169.

9. Barla, G. and Wane, M.T. (1970). Stress relief method in anisotropic rocks by means of gauges applied to the end of a borehole. Int. J. Rock Mech. Min. Sci., Vol. 7, No. 2, pp. 171-182.

10. Barla, G. and Goffi, L. (1973). Prove di trazione diretta su roccia, Secondo Congresso Nazionale AIAS, Genova, (in Italian).

11. Barron, K. (1971a). Brittle fracture initiation in and ultimate failure of rocks. Part III. Int. J. Rock Mech. Min. Sci., Vol. 8, No. 6, pp. 565-575.

12. Barron, K. (1971b), Brittle fracture initiation in and ultimate failure of rocks. Part II. Int. J. Rock Mech. Min. Sci., Vol. 8, No. 6, pp. 553-563.

13. Becker, R.M. (1968). An anisotropic elastic solution for testing stress relief cores. U.S. Bureau of Mines, RI. 7143.

14. Becker, R.M. and Hooker, V.E. (1967). Some anisotropic consid erations in rock stress determinations. U.S. Bureau of Mines, RI. 6965.

15. Berry, D.S. (1968). The theory of stress determination by means of stress relief techniques in transversely isotropic medium. Tech. Rept. No. 5-68, Missouri River Division, Corps of Engineers, Omaha, Nebraska 68101.

16. Berry, D.S. (1970). The theory of determination of stress changes in a transversely isotropic medium using an instrumented cylindrical inclusion. Tech. Rept. No. MRD-1-70, Missouri River Division, Corps of Engineers, Omaha, Nebraska 68101.

17. Berry, D.S. and Fairhurst, C. (1966). Influence of rock anisotropy and time dependent deformation on the stress relief and high modulus inclusion techniques of in situ stress determination. Testing Techniques for Rock Mechanics, ASTM, STP.402. Am. Soc. Testing Mats., pp.190-206.

18. Bielenstein, H.U. and Barron, K. (1971). In Situ stresses. A summary of presentations and discussions given in Theme I at the Conference of Structural Geology to Rock Mechanics Problems. Department of Energy, Mines and Resources, Mines Branch, Ottawa.

19. Bieniawski, Z.T. (1978). A critical assessment ofselected in situ tests for rock mass deformability and stress measurements. Proc. 19th Symposium on Rock Mechanics (University of Nevada, Reno), Vol. 1, pp. 523-529.

20. Blackwood, R.L. (1977). An instrument to measure the complete stress field in soft rock or coal in a single operation. Proc. International Symposium on Field Measurements in Rock Mechanics, Zurich. Kovari, K. (Editor). Vol. 1, pp. 137-150.

21. Boehler, J.P. and Sawczuk, A. (1977). On yielding of oriented solids. Acta Mechanica 27, pp. 185-206.

22. Bonnechere, F.J. (1967). A comparative study of in situ rock stress measurements. M.S. thesis, University of Minnesota.

23. Bonnechere, F.J. and Cornet, F.M. (1977). In situ stress measurements with a borehole deformation cell. Proc. International Symposium on Field Measurements in Rock Mechanics, Zurich, Kovari, K. (Editor). Vol. 1, pp. 151-159.

24. Brady, B.H.G. and Bray, J.W. (1978). The boundary element method for determining stresses and displacements around long openings in a triaxial stress field. Int. J. Rock Mech. Min. Sci, Vol. 15, No. 1, pp. 21-28.

25. Bray, J.W. (1967). A study of jointed and fractured rock. Part I, fracture patterns and their failure characteristics. Rock Mechanics and Engineering Geology, Vol. V/2-3, pp 117-136.

26. Carroll, M.M. (1979). An effective stress law for anisotropic elastic deformation. Journal of Geophysical Research, Vol. 84, No. B13, pp. 7510-7512.

27. Casagrande, A. and Carrillo, N. (1944). Shear failure of anisotropic materials. Journal of the Boston Society of Civil Engineers, Vol. XXXl, No. 4, pp. 122-135.

28. Cook, N.E., Ko, H.-Y., and Gerstle, K.H. (1978). Variability and anisotropy of mechanical properties of the Pittsburgh coal seam. Rock Mechanics 11, pp. 3-18.

29. Crouch, S.L. (1969). A note on the stress concentrations at the bottom of a flat ended borehole. Jls. Afr. Inst. Min. Metall. No. 70, pp. 100-102.

30. Crouch, S.L. and Fairhurst, C. (1967). A four component borehole deformation gauge for the determination of in situ stresses in rock masses. Int. J. Rock Mech. Min. Sci., Vol. 4, No. 2, pp. 209-217.

31. Crouch, S.L. and Starfield, A.M. Boundary element methods in solid mechanics with applications in rock mechanics and geological engineering. University of Minnesota. In preparation.

32. Dayre, M. (1969). Anisotropie discontinue d'une formation calcaire (in French). Colloque sur la fissuration des roches. Paris. Numero special de la Revue de l'Industrie Minerale (Juillet).

33. Dayre, M. (1970). Lois de rupture d'un schiste ardoisier presentant une lineation dans le plan de schistosite (in French) Proc. 2nd Cong. ISRM (Belgrade), Vol. 2, 3-38.

34. Deklotz, E.J., Brown, J.W. and Stemler, O.A. (1966). Anisotropy of a schistose gneiss. Proc. 1st. Cong. ISRM (Lisbon), Vol. 1, pp. 465-470.

35. De la Cruz, R.V. (1969). The borehole deepening method of absolute stress measurement. Ph.D. dissertation. University of California, Berkeley.

36. De la Cruz, R.V. (1978). Modified borehole jack method for elastic property determination in rocks. Rock Mechanics 10, pp. 221-239.

37. Donath, F.A. (1961). Experimental study of shear failure in anisotropic rocks. Bull. Geol. Soc. Am, Vol. 72, pp. 985-990.

38. Donath, F.A. (1964). Strength variation and deformational behavior in anisotropic rock. In: State of Stress in the Earth's Crust. Judd, W.R. (Editor), pp. 281-297.

39. Donath, F.A. (1972a). Faulting across discontinuities in anisotropic rock. Proc. 13th Symposium on Rock Mechanics (ASCE), pp. 753-772.

40. Donath, F.A. (1972b). Effects of cohesion and granularity on deformational behavior of anisotropic rock. In: Studies in Mineralogy and Precambrian Geology. Doe, B.R. and Smith, D.K. (Editors). Geol. Soc. Am. Memoir 135, pp. 95-128.

41. Douglass, P.M. and Voight, B. (1969). Anisotropy of granites: a reflection of microscopic fabric. Geotechnique, 19, pp. 376-398.

42. Draper, N.R. and Smith, H. (1966). Applied Regression Analysis. John Wiley and Sons.

43. Drozd, K., Goodman, R.E., Heuze, F.E. and Van, T.K. (1970). On the problem of borehole strength testing. Proc. 2nd. Cong. ISRM (Belgrade), Vol. 2, 3-46.

44. Duncan, J.M. and Goodman, R.E. (1968). Finite element analysis of slopes in jointed rocks. Contract report S-68-3, U.S. Army Engineer Waterways Experiment Station, Corps of Engineers, Vicksburg, Mississippi.

45. Duncan-Fama, M.E. (1979). Analysis of a solid inclusion in situ stress measuring device. Proc. 4th Cong. ISRM (Montreux), Vol. 2, pp. 113-120.

46. Duncan-Fama, M.E. and Pender, M.J. (1980). Analysis of the hollow inclusion technique for measuring in situ rock stress. Int. J. Rock Mech. Min. Sci., Vol. 17, No. 3, pp. 137-146.

47. Duvall, W.I. (1965). The effect of anisotropy on the determination of dynamic elastic constants of rock. Trans. Society of Mining Engineers, December, pp. 309-316.

48. Duvall, W.I. (1974). See Hooker et al. (1974).

49. Einstein, H.M. and Hirschfeld, R.C. (1973). Model studies on mechanics of jointed rock. Journal of the Soil Mechanics and Foundations Division, ASCE, Vol. 99, No. SM3, pp. 229-248.

50. Eissa, E.S.A. (1980). Stress analysis of underground excavations in isotropic and stratified rock using the boundary element method. Ph.D. Dissertation, Imperial College, London.

51. Eshelby, J.D. (1957). The determination of the elastic field of an ellipsoidal inclusion and related problems. Proc. Roy. Soc. London, Ser. A, Math. Phys. Sciences, Vol. 241, No. 1226, pp. 376-396.

52. Fairhurst, C. (1967). Methods of determining in situ rock stresses at great depths. Tech. Rept. No. 1-68, Missouri River Division, Corps of Engineers, Omaha, Nebraska 68102.

53. Fayed, L.A. (1968). Shear Strength of some argillaceous rocks. Int. J. Rock Mech. Min. Sci., Vol. 5, No. 1, pp. 79-85.

54. Filon, L.N.G. (1903). On an approximate solution for the bending of a beam of rectangular cross section under any system of load with special reference to points of concentrated or discontinuous loading. Proc. Roy. Soc. London, Ser. A, Vol. 201, pp. 65-154.

55. Filon, L.N.G. (1930). On the relation between corresponding problems in plane stress and in generalized plane stress. Quarterly Journal of Mathematics, Oxford Ser., Vol. 1, pp. 289.

56. Filon, L.N.G. (1937). On antiplane stress in an elastic solid. Proc. Roy. Soc. London, Ser. A, Math. Phys. Sciences, Vol. 160, pp. 137-154.

57. Goodman, R.E., Van, T.K. and Heuze, F.E. (1972). Measurement of rock deformability in boreholes. Proc. 10th Symposium on Rock Mechanics (AIME), pp. 523-555.

58. Goodman, R.E., Taylor, R.L. and Brekke, T.L. (1968). A model for the mechanics of jointed rock. Journal of the Soil Mechanics and Foundations Division, ASCE, Vol. 94, No. SM3, pp. 637-659.

59. Goodman, R.E. (1976). Methods of Geological Engineering. West Publ.

60. Goodman, R.E. (1980). Introduction to Rock Mechanics. John Wiley and Sons.

61. Gray, W.M. and Toews, N.A. (1968). Analysis of accuracy in the determination of the ground stress tensor by means of borehole devices. Proc. 9th Symposium on Rock Mechanics, (AIME), pp. 45-72.

62. Gray, W.M. and Toews, N.A. (1975). Analysis of variance applied to data obtained by means of a six element borehole deformation gauge for stress determination. Proc. 15th Symposium on Rock Mechanics (ASCE), pp. 323-356.

63. Green, A.E. and Zerna, W. (1968). Theoretical Elasticity. Oxford, Claredon Press.

64. Handin, J. (1969). On the Coulomb-Mohr failure criterion. Journal of Geophysical Research, Vol. 74, No. 22, pp. 5343-5348.

65. Hast, N. (1979). Limit of stress measurements in the earth's crust. Rock Mechanics 11, pp. 143-150.

66. Hata, S., Tanimoto, Ch. and Kimura, K. (1979). Field measurement and consideration on deformability of the Izumi layers. Rock Mechanics, Suppl. 8, pp. 349-367.

67. Heuze, F.E. (1971). Source of errors in rock mechanics field measurements and related solutions. Int. J. Rock Mech. Min. Sci., Vol. 8, No. 4, pp. 297-310.

68. Heuze, F.E. (1980). Scale effects in the determination of rock mass strength and deformability. Rock Mechanics 12, pp. 167-192.

69. Heuze, F.E., Patrick, W.C., De la Cruz, R.V. and Voss, C.F. (1980). In situ geomechanics, Climax granite, Nevada Test Site. Preprint UCRL-85308, Lawrence Livermore Laboratory.

70. Hill, R. (1950). The Mathematical Theory of Plasticity. Claredon Press.

71. Hiltscher, R., Martna, J. and Strindell, L. (1979). The measurement of triaxial rock stresses in deep boreholes and the use of rock stress measurements in the design and construction of rock openings. Proc. 4th Cong. ISRM (Montreux), Vol. 2, pp. 227-234.

72. Hiltscher, R. (1976). Bemerkungen zur Technik der Gebirgsspannungsmessung. Rock Mechanics 8, (discussions), pp. 199-206.

73. Hiramatsu, Y. and Oka, Y. (1968). Determination of the stress in rock unaffected by boreholes or drifts, from measured strains or deformations. Int. J. Rock Mech. Min. Sci., Vol. 5, No. 4, pp. 337-353.

74. Hirashima, K. and Koga, A. (1977). Determination of stresses in anisotropic elastic medium unaffected by boreholes from measured strains or deformations. Proc. International Symposium on Field Measurements in Rock Mechanics, Zurich, Kovari K. (Editor). Vol. 1, pp. 173-182.

75. Hobbs, D.W. (1964). The tensile strength of rocks. Int. J. Rock Mech. Min. Sci, Vol. 1, No. 3, pp. 385-396.

76. Hocking, G. (1976). Three-dimensional elastic stress distribution around the flat end of a cylindrical cavity. Int. J. Rock Mech. Min. Sci., Vol. 13, No. 12, pp. 331-337.

77. Hoek, E. (1964). Fracture of anisotropic rock. J. S. Afr. Inst. Min. Metall., Vol. 64, No. 10, pp. 501-518.

78. Hoek, E. and Brown, E.T. (1980). Empirical strength criterion for rock masses. Journal of the Geotechnical Engineering Division, ASCE, Vol. 106, No. GT9, pp. 1013-1035.

79. Hooker, V.E. and Johnson, C.F. (1969). Near surface horizontal stresses including the effects of rock anisotropy. U.S. Bureau of Mines, RI. 7224.

80. Hooker, V.E., Aggson, J.R., Bickel, D.L. and Duvall, W. (1974). Improvements in the three component borehole deformation gage and overcoring techniques. U.S. Bureau of Mines, RI. 7894 with appendix by Duvall on the undercoring method.

81. Hirashima, K. and Koga, A. (1977). Proc. International Symposium on Field Measurements in Rock Mechanics, Zurich, Kovari, K. (Editor). Vol. 1, pp. 173-182.

82. Int. Society for Rock Mech. (1978a). Suggested methods for determining the strength of rock materials in triaxial compression. Int. J. Rock Mech. Min. Sci., Vol. 15, No. 2, pp. 47-51.

83. Int. Soc. for Rock Mech. (1978b). Suggested methods for determining the uniaxial compressive strength and deformability of rock materials. Int. J. Rock Mech. Min. Sci., Vol. 16, No. 2, pp. 135-140.

84. Jaeger, J.C. (1960). Shear failure of anisotropic rocks. Geol. Mag. Vol. 97, pp. 65-72.

85. Jaeger, J.C. (1964). Discussion to Donath (1964).

86. Kawamoto, T. (1963). On the state of stress and deformation around tunnel in orthotropic, elastic ground. Memoirs of the Faculty of Eng. Kumamoto Univ., Japan, Vol. 10, No. 1, pp. 1-30.

87. Kawamoto, T. (1966). On the calculation of the orthotropic elastic properties from the states of deformation around a circular hole subjected to internal pressure in orthotropic, elastic medium. Proc. 1st. Cong. ISRM (Lisbon), Vol. I, pp. 269-272.

88. Ko, H.-Y. and Gerstle, K.H. (1972). Constitutive relations of coal. Proc. 14th. Symposium on Rock Mechanics (ASCE) pp. 157-188.

89. Ko, H.-Y. and Gerstle, K.H. (1976). Elastic properties of two coals. Int. J. Rock Mech. Min. Sci., Vol. 13, pp. 81-90.

90. Kobayashi, S. (1970). Fracture criteria for anisotropic rocks. Memoirs of the Faculty of Eng., Kyoto Univ., Japan, Vol. 32, Part 3, pp. 307-333.

91. Kreyszig, E., (1972). Advanced Engineering Mathematics. John Wiley and Sons.

92. Kulhawy, F.M. (1978). Geomechanical model for rock foundation settlement. Journal of the Geotechnical Engineering Division, ASCE, Vol. 104, No. GT2, pp. 211-227.

93. Ladanyi B. and Archambault, G. (1972). Evaluation de la resistance au cisaillement d'un massif rocheux fragmente (in French), Proc. 24th Int. Geol. Congress, Montreal, Sect. 13D, pp. 249-260.

94. Ladanyi, B. and Archambault, G. (1980). Direct and indirect determination of shear strength in rock mass. Proc. AIME Annual Meeting, Paper 80-25.

95. Leeman, E.R., (1964). The measurement of stress in rock. See Appendix 4 in Fairhurst (1967).

96. Leeman, E.R. (1967). The borehole deformation type of rock stress measuring instrument. Int. J. Rock Mech. Min. Sci., Vol. 4, No. 1, pp. 23-44.

97. Leeman, E.R. (1971). The CSIR "doorstopper" and triaxial rock stress measuring instruments. Rock mechanics 3, pp. 25-50.

98. Leeman, E.R. and Hayes, D.J. (1966). A technique for determining the complete state of stress in rock using a single borehole. Proc. 1st. Cong. ISRM (Lisbon), Vol. II, pp. 17-24.

99. Lekhnitskii, S.G. (1957). Анизотропные пластинки. English translation by Tsai, S.W. (Anisotropic plates), 1968, Gordon and Breach.

100. Lekhnitskii, S.G. (1963). Theory of elasticity of an anisotropic elastic body. Holden Day, Inc. San Francisco.

101. Lerau, J., Saint-Leu, C. and Sirieys, P. (1981). Anisotropie de la dilatance des roches schisteuses. (In French). Rock Mechanics 13, pp. 185-196.

102. Liakhovitski, F.M. and Nevski, M.V. (1970). Theoretical analysis of dynamic Poisson's ratio in transversally - isotropic medium. Proc. 2nd. Cong. ISRM (Belgrade), Vol. 1, 1-19.

103. Little, R.W. (1973). Elasticity. Prentice-Hall.

104. Love, A.E.H. (1920). A Treatise on the Mathematical Theory of Elasticity. Cambridge, University Press.

105. Martino, D. and Ribacchi, R. (1972). Osservazioni su alcuni metodi di masura delle carateristiche elastiche di rocce o ammassi rocciosi, con particolare riferimento al problema dell' anisotropia. L'Industria Mineraria, pp. 193-203. (In Italian).

106. Masure, P. (1970). Comportement mecanique des roches à anisotropie planaire discontinue. (in French). Proc. 2nd. Cong. ISRM (Belgrade), Vol. 1, 1-27.

107. McGarr, A. and Gay, N.C. (1978). State of stress in the earth's crust. Ann. Rev. Earth Planet. Sci, 6, pp. 405-436.

108. McLamore, R. and Gray, K.E. (1967). The mechanical behavior of anisotropic sedimentary rocks. Journal of Engineering for Industry, Trans. of the ASME, Vol. 89, pp. 62-73.

109. McLamore, R. and Gray, K.E. (1967). A strength criterion for anisotropic rocks based upon experimental observations. Proc. 96th AIME Annual Meeting, Paper SPE 1721.

110. McClintock, F.A. and Walsh, J.B. (1962). Friction on Griffith cracks in rocks under pressure. Proc. 4th U.S. Natl. Cong. of Applied Mechanics, (ASME), Vol.II, pp. 1015-1021.

111. Merrill, R.H. (1967). Three component borehole deformation gage for determining the stress in rock. U.S. Bureau of Mines, RI. 7015.

112. Merrill, R.H. and Peterson, J.R. (1961). Deformation of a borehole in rock. U.S. Bureau of Mines, RI. 5881.

113. Michell, J.H. (1899). On the direct determination of stress in an elastic solid with application to the theory of plates. Proc. of the London Math. Soc., Vol. 31, pp. 100-124.

114. Milne-Thomson, L.M. (1962). Antiplane Elastic Systems. Springer Verlag, Berlin.

115. Morland, L.W. (1976). Elastic anisotropy of regularly jointed media. Rock Mechanics 8, pp. 35-48.

116. Muskhelishvili, N.I. (1953). Some basic problems of the mathematical theory of elasticity. Translated from the Russian by J.R.M Radok. Noordhoff, Groningen.

117. Nelson, R.A. and Hirschfeld, R.C. (1968). Modeling a jointed rock mass. M.I.T. Report R 68-70 to the U.S. Dept. of Transportation, Massachusetts Institute of Technology.

118. Nichols, T.C., Savage, W.Z. (1976). Rock strain recovery - Factor in foundation design. Rock engineering for engineering for foundations and slopes (ASCE), Vol. 1, pp. 34-54.

119. Niwa, Y., Kobayashi, S. and Hirashima, K. (1970). Stresses and deformations around a tunnel with a circular cross section in anisotropic elastic body (in Japanese). Trans. Japan Soc. Civil Eng., 178, pp. 7-17.

120. Niwa, Y. and Hirashima, K. (1970). Stress distribution around a tunnel with an arbitrary cross section excavated in anisotropic elastic ground. Memoirs of the Faculty of Eng., Kyoto Univ., Japan, Vol. 32, pp. 175-193.

121. Niwa, Y. and Hirashima, K. (1971). The theory of the determination of stress in an anisotropic elastic medium using an instrumented cylindrical inclusion. Memoirs of the Faculty of Eng., Kyoto Univ., Japan, Vol. 33, pp. 221-232.

122. Nova, R. (1980). The failure of transversely isotropic rocks in triaxial compression. Int. J. Rock Mech. Min. Sci., Vol. 17, No. 6, pp. 325-332.

123. Nur, A. and Byerlee, J.D. (1971). An exact effective stress law for elastic deformation of rocks with fluids. Journal of Geophysical Research, Vol. 76, No. 26, pp. 6414-6419.

124. Obert, L. (1966). Determination of the stress in rock. A state of the art report. See Appendix 5 in Fairhurst (1967).

125. Obert, L., Merrill, R.H. and Morgan, T.A. (1962). Borehole deformation gage for determining the stress in mine rock. U.S. Bureau of Mines, RI. 5978.

126. Obert, L. and Duvall, W.I. (1967). Rock Mechanics and the Design of Structures in Rock. John Wiley and Sons.

127. Oberti, G., Carabelli, E., Goffi, L. and Rossi, P.P. (1979). Study of an orthotropic rock mass: experimental techniques, comparative analysis of results. Proc. 4th. Cong. ISRM (Montreux), Vol. 2, pp. 485-491.

128. Okubo, H. (1952). The stress distribution of an aeolotropic circular disc compressed diametrically. Journal of

Mathematics and Physics,Vol. 31, pp. 75-83.

129. Panek, L.A. (1966). Calculation of the average ground stress components from measurements of the diametral deformation of a drillhole. U.S. Bureau of Mines, RI. 6732.

130. Pariseau, W.G. (1972). Plasticity theory for anisotropic rocks and soils. Proc. 10th Symposium on Rock Mechanics (AIME), pp. 267-295.

131. Paulmann, H.G. (1966). Messungen der festigkeits - anisotropie tektonischen ursprungs an geisteinsproben (in German). Proc. 1st. Cong. ISRM (Lisbon), Vol. I, pp. 125-131.

132. Pender, M.J. (1977). In situ stress at the site of the Rangipo underground powerhouse. New Zealand N.W.D. Central laboratories report, No. 2-77/3.

133. Peres-Rodrigues, F. (1966). Anisotropy of granites. Modulus of elasticity and ultimate strength ellipsoids, joint systems, slope attitudes and their correlations. Proc. 1st. Cong. ISRM (Lisbon), Vol. I, pp. 721-731.

134. Peres-Rodrigues, F. (1970). Anisotropy of rocks. Most probable surfaces of the ultimate stresses and of the moduli of elasticity. Proc. 2nd. Cong. ISRM (Belgrade), Vol. 1, 1-20.

135. Peres-Rodrigues, F. (1979). The anisotropy of the moduli of elasticity and the ultimate stresses in rocks and rock masses. Proc. 4th Cong. ISRM (Montreux), Vol. 2, pp. 517-523.

136. Peres-Rodriques, F. and Aires-Barros, L. (1970). Anisotropy of endogenic rocks. Correlation between micropetrographic index, ultimate strength and modulus of elasticity ellipsoids. Proc. 2nd. Cong. ISRM (Belgrade), Vol. 1, 1-23.

137. Pickering, D.J. (1970). Anisotropic elastic parameters for soils. Geotechnique, Vol. 20, No. 3, pp. 271-276.

138. Pinto, J.L. (1966). Stresses and strains in an anisotropic-orthotropic body. Proc. 1st. Cong. ISRM (Lisbon), Vol. I, pp. 625-635.

139. Pinto, J.L. (1970). Deformability of schistous rocks. Proc. 2nd. Cong. ISRM (Belgrade), Vol. 1, 2-30.

140. Pinto, J.L. (1979). Determination of the elastic constants of anisotropic bodies by diametral compression tests. Proc. 4th Cong. ISRM (Montreux), Vol. 2, pp. 359-363.

141. Pomeroy, C.D., Hobbs D.W. and Mahmoud, A. (1971). The effect of weakness plane orientation on the fracture of Barnsley

hards by triaxial compression. Int. J. Rock Mech. Min. Sci., Vol. 8, No. 3, pp. 227-238.

142. Raphael, J.M. and Goodman R.E. (1979). Strength and deformability of highly fractured rock. Journal of the Geotechnical Engineering Division, ASCE, Vol. 105, No. GT11, pp. 1285-1300.

143. Reik, G. and Zacas, M. (1978). Strength and deformation characteristics of jointed media in true triaxial compression. Int. J. Rock Mech. Min. Sci., Vol. 15, No. 6, pp. 295-303.

144. Ribacchi, R. (1977). Rock stress measurements in anisotropic rock masses. Proc. International Symposium on Field Measurements in Rock Mechanics, Zurich, Kovari, K. (Editor), Vol. 1, pp. 183-196.

145. Rocha, M., Silveira, A., Grossmann, N. and Oliveira, E. (1966). Determination of the deformability of rock masses along boreholes. Proc. 1st. Cong. ISRM (Lisbon), Vol. I, pp. 679-704.

146. Rocha, M., Silverio, A., Pedro, J.O. and Delgado, J.S. (1974). A new development of the LNEC stress tensor gauge. Proc. 3rd. Cong. ISRM (Denver), Vol. IIA, pp. 464-467.

147. Rocha, M. and Silverio, A. (1969). A new method for the complete determination of the state of stress in rock masses. Geotechnique, Vol. 19, No. 1, pp. 116-132.

148. Royea, M.J. (1969). Rock stress measurement at the Sullivan mine. Proc. 5th Can. Rock Mechanics Symposium. pp. 59-74. Dept. of Energy, Mines and Resources, Ottawa.

149. De Saint-Venant, B. (1863). Sur la distribution des elasticites autour de chaque point d'un solide ou d'un milieu de contexture quelconque, particulierement lorsqu'il est amorphe sans etre isotrope (in French). Journal de Mathematiques Pures et Appliquees, Ser. 2, Vol. 8, pp. 353-430; 257-295.

150. Salamon, M.D.G. (1968). Elastic moduli of a stratified rock mass. Int. J. Rock Mech. Min., Sci., Vol. 5, No. 6, pp. 519-527.

151. Schmidt, C.M. and Frenk, B.W. (1972). Accuracy of various techniques used to measure strain in sub-surface operations. Int. J. Rock Mech. Min. Sci., Vol. 9, pp. 1-5.

152. Serafim, J.L. (1964). Rock mechanics considerations in the design of concrete dams. In: State of stress in the Earth's Crust. Judd, W.R. (Editor), pp. 611-645.

153. Singh, B. (1973). Continuum characterization of jointed rock masses. Int. J. Rock Mech. Min. Sci, Vol. 10, No. 4, pp. 311-335.

154. Simonson, E.R. Johnson, J.N. and Buchholt, L. (1976). Anisotropic mechanical properties of a moderate and rich kerogen content oil shale. Terratek report TR 76-72, Dec. 1976.

155. Skempton, A.W. (1960). Effective stress in soils, concrete and rocks. Proc. Conference on pore pressure and suction in soils. Butterworths, London.

156. Sokolnikoff, I.S. (1941). Mathematical Theory of Elasticity. Brown Univ., Summer Session for Advanced Instruction and Research in Mechanics.

157. Takano, M. and Shidomoto, Y. (1966). Deformation test on mudstone enclosed in foundation by means of tube deformeter. Proc. 1st. Cong. ISRM (Lisbon), Vol. I, pp. 761-764.

158. Teodorescu, P.D. (1964). One hundred years of investigations in the plane problem of the theory of elasticity. Applied Mechanics Reviews, Vol.17, No. 3, pp. 175-186.

159. Timoshenko, S.P. and Goodier, J.N. (1970). Theory of Elasticity. McGraw-Hill.

160. Titchmarsch E.C. (1950). The Theory of Functions. Oxford University Press, pp. 427-432.

161. Turner, J.T. and Weiss, L.E. (1963). Structural Analysis of Metamorphic Tectonites. McGraw-Hill, Toronto, pp. 15-16.

162. Van Heerden, W.L. (1968). The effect of end of borehole configuration and stress level on stress measurements using "doorstoppers". Rep. Coun. Scient. Ind. Res. S. Afr., Meg 626.

163. Voight, B. (1966). Interpretation of in situ stress measurements. Proc. 1st. Cong ISRM (Lisbon), Vol III, pp. 332-348.

164. von Mises, R. (1928). Z. Angew. Math. Mech., 8, pp. 161-185.

165. Walsh, J.B. and Brace, W.F. (1964). A fracture criterion for brittle anisotropic rock. Journal of Geophysical Research, Vol. 69, No. 16, pp. 3449-3456.

166. Wardle, L.J. and Gerrard, C.M. (1972). The "equivalent" anisotropic properties of layered rock and soil masses. Rock Mechanics 4, pp. 155-175.

167. Worotnicki, G. and Walton, R.J. (1976). Triaxial hollow inclusion gauges for the determination of rock stress in situ. Proc. ISRM Symp. on Investigation of Stress in Rock and Advances in Shear Measurement. Sydney, supplement pp. 1-8.

Appendix 2.1

A 3-D CONSTITUTIVE RELATION FOR FRACTURED ROCK MASSES

B. AMADEI and R. E. GOODMAN
University of California, Berkeley, California, U.S.A.

ABSTRACT

The purpose of this paper is to introduce a constitutive relation describing the non-linear behavior of a discontinuous, homogeneous and anisotropic body of rock containing up to three orthogonal joint sets.

The intact rock is assumed to behave in a linear elastic manner with up to three orthogonal planes of symmetry parallel to the joint sets. The joints are modelled to behave in a non-linear inelastic fashion in compression and decompression (tension) and in a linear or non-linear elastic fashion in shear. For each joint set, the normal stiffness k_n is expressed in terms of the normal stress acting across it and properties such as the seating pressure and maximum closure.

The jointed rock mass is described as an equivalent anisotropic continuum (Duncan and Goodman, 1968).

A closed form of the constitutive relationship is derived for two loading conditions--uniaxial and hydrostatic compression. It is shown that this constitutive relation reduces to that of an isotropic material as the spacings of the joint sets increase and/or the joints become stiffer both in shear and in compression.

In uniaxial compression, both spacings and orientations of the joint sets relative to the applied stress greatly influence the response of the fractured medium. If it is assumed that the unloading of a fractured rock specimen follows essentially the same path as that of the intact rock, then a modulus of permanent deformation can be introduced and expressed in terms of joint orientations, spacings and stiffnesses. Similarly, permanent lateral strain ratios can be defined.

In triaxial compression, the compressibility of a fractured medium can be written as the sum of the compressibility of the intact rock and a term that depends on joint set spacings and their normal stiffnesses. Similarly, an equivalent bulk modulus is introduced. Compressibility and equivalent bulk modulus have been found to depend on the applied confining pressure. Both terms approach those of an isotropic medium as the applied confining pressure and/or joint spacings increase. When the three-joint sets have the same properties and

same spacings, then the fractured rock becomes isotropic, no distortion being created by normal stresses. For any triaxial state of applied stresses, there may exist joint properties and geometries for which the fractured material may behave isotropicly.

The effective stress law of Carroll (1979) for anisotropic saturated linearly elastic porous materials is extended to study the effect of pore water pressure on the response of a fractured rock described as an equivalent anisotropic continuum. This yields an effective stress law for jointed, non-porous rocks.

INTRODUCTION

Fractured rock masses constitute the usual "stuff" of engineering foundations and excavations in rock. The system of fractures is usually ordered and predictable. Ignoring the presence of the fractures is incorrect because their role is dominant in the response of the rock to load and unload.

The basic problem to be addressed in proposing constitutive relations for fractured rocks is the mechanical behavior of a single fracture surface. This surface can produce a jump in tangential stresses since it becomes an interface between bodies behaving dissimilarly. Interface problems are important in many disciplines and the methods described here may well be applicable elsewhere. However, they were engendered by our desire to calculate the behavior of rock masses in engineering and mining applications.

Because fractured rocks are non-linear in their stress/strain relations, a variety of techniques has been necessary to account for them in engineering contexts. Empirical relations for fractured rocks are probably still the most reliable, although they lack mechanistic bases. Ultimately, statistical approaches will be adopted because the family of fractures and their properties are best accommodated through statistical distributions. At present, deterministic methods of analysis are being used for tractable problems worthy of elegant computations. These methods range from equilibrium formulations, to finite element or finite difference computations with special elements or algorithms for including interfaces within the computational grid.

The discussion here centers on the stress-deformation relations for fractured rocks and makes use of an equivalent anisotropic medium concept. There is no effort to relate these constitutive equations to a given computational method. The paper begins with a description of the behavior of a single joint, takes up the concept of equivalent anisotropic media, and considers the effect of water pressures on deformations. Yield functions are not discussed owing to space limitations.

BEHAVIOR OF A SINGLE JOINT

A number of parameters are fundamental for a proper model of joint behavior. These include peak and residual shear strengths, maximum joint closure, tensile strength, shear and normal stiffness and dilatancy properties.

<u>Behavior of joints under changing normal stress with constant shear stress</u>

There are two physical constraints on normal deformations of a joint. First, an open joint has a negligible tensile strength while a healed fracture may have a finite tensile strength. Secondly, there is a limit to the amount of possible compression, a maximum closure Vmc, which must be less than or equal to the thickness of the joint. Experiments (Goodman, 1976) have shown the non-linear character of joints under changing normal stress (Fig. 1). Combining the two constraints and the general shape of the normal pressure deformation curve, a simple curve type is the hyperbola.

$$\sigma_n = A\left(\frac{v}{Vmc - v}\right)^t \quad (1)$$

where v represents the difference in normal displacements across the joint.

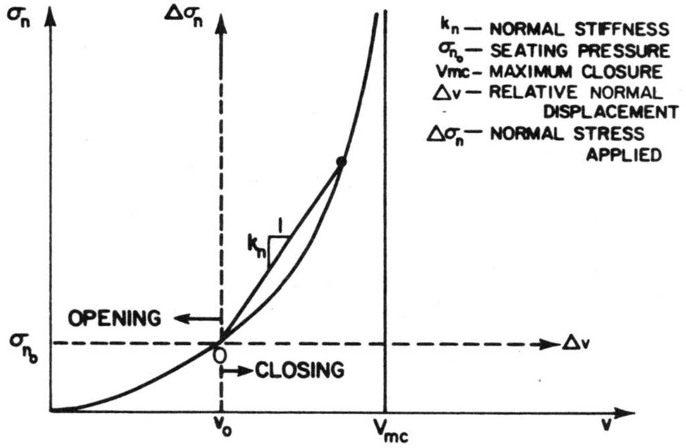

Fig. 1. Idealized normal deformation of a joint.

When the normal testing of a joint is done, there is always a seating pressure σ_{n_o} acting on a joint as the normal stress $\Delta\sigma_n$ is applied. It defines the initial conditions to measure the <u>relative normal deformation</u> Δv such that

$$\Delta\sigma_n = \sigma_n - \sigma_{n_o} \quad (2)$$

and

$$\Delta v = v - v_o \tag{3}$$

where v_o is the corresponding value of v when $\sigma_n = \sigma_{n_o}$.
Combining eqs. (1), (2), (3), and the relationship between σ_{n_o} and v_o we can define the <u>secant normal stiffness</u> k_n as follows:

$$\frac{\Delta v}{\Delta\sigma_n} = \frac{1}{k_n} = \frac{Vmc}{\Delta\sigma_n}\left[\frac{\left(\frac{\Delta\sigma_n + \sigma_{n_o}}{A}\right)^{1/t}}{1+\left(\frac{\Delta\sigma_n + \sigma_{n_o}}{A}\right)^{1/t}} - \frac{\left(\frac{\sigma_{n_o}}{A}\right)^{1/t}}{1+\left(\frac{\sigma_{n_o}}{A}\right)^{1/t}}\right] \tag{4}$$

where A and t are constants defined by curve fitting,
Vmc is the maximum closure of the joint,
Δv is the relative normal displacement (positive in compression),
and $\Delta\sigma_n$ is the normal stress applied (positive in compression).
If we choose to define k_n as the slope of the normal stress deformation curve, then

$$k_n = \frac{\partial\sigma_n}{\partial v} \tag{5}$$

In both cases, k_n is a function of the normal stress applied across the joint.

<u>Behavior of joints under changing shear stress with constant normal stress</u>

The shear deformation versus shear stress curve for a test conducted on a joint under constant normal stress can be characterized by elastic, peak and post-peak regions as depicted in Fig. 2. The peak shear stress (τ_p) is termed the <u>shear strength</u> while the minimum post-peak shear stress (τ_r) is the <u>residual strength</u>. The slope characterizing the elastic region is termed the <u>unit shear stiffness</u> k_s (Goodman, Taylor and Brekke, 1968). Peak and residual shear strengths and their relative importance as well as the shear stiffness are greatly influenced by joint type, joint properties, filling material, state of stress applied and testing procedures. The shear strength parameters are a function of the normal stress as analyzed by many authors including Patton (1966), Jaeger (1971), Ladanyi and Archambault (1970) and Barton (1974).

<u>Dilatancy of joints</u>

<u>Dilatancy</u> means volumetric change accompanying deformation. It can also connote thickening or thinning of a discontinuity undergoing shear deformation at constant normal stress. The term <u>dilation</u> is used to connote a thickening of a joint, that is, an increase in the separation of the two joint blocks, as opposed to <u>contraction</u> which connotes a closing of a joint.

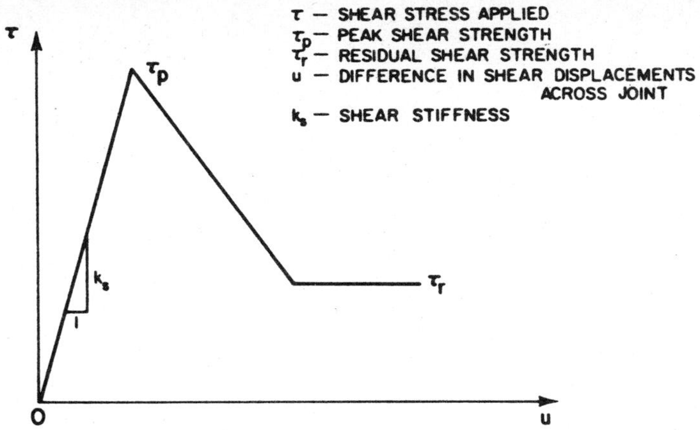

Fig. 2. Idealized shear deformation of a joint.

The mechanism of dilatancy originates mainly from surface roughness. Perfectly mating rough blocks can be forced to slide past one another only if they are free to move apart, that is, to work around the asperities. If the blocks are confined, shearing is possible only if the asperities themselves break. Those two conditions correspond respectively to joints tested in direct shear constrained to achieve constant normal stress and joints constrained to achieve constant normal deformation. The second condition can considerably strengthen a joint and has been used in rock mechanics successfully in rock bolt reinforcement.

The rate of dilatancy has been found to be influenced by surface properties, joint type, normal stress applied and shearing displacement (Goodman and Dubois, 1972; Ladanyi and Archambault, 1970). Just as normal displacements can result from shear stress, shear displacements can accompany purely normal stress. The latter are called <u>coupled shear displacements</u>.

The definitions of the normal and shear stiffnesses in previous sections are precise when there is no dilatancy. To take into account those properties, we can rewrite a more general constitutive relation for a joint as follows:

$$\begin{pmatrix} d\tau \\ d\sigma_n \end{pmatrix} = \begin{pmatrix} k_{ss} & k_{sn} \\ k_{ns} & k_{nn} \end{pmatrix} \begin{pmatrix} du \\ dv \end{pmatrix} \tag{6}$$

$$\text{where } k_{ss} = \frac{\partial \tau}{\partial u}, \; k_{sn} = \frac{\partial \tau}{\partial v}, \; k_{ns} \frac{\partial \sigma_n}{\partial u}, \; k_{nn} = \frac{\partial \sigma_n}{\partial v}. \tag{7}$$

These four stiffness components can be determined theoretically and/or experimentally (Goodman and Dubois, 1972; Heuze, 1979; Bazant and Grambarova, 1980).

CONCEPT OF AN EQUIVALENT MEDIUM

Introduction

In the present discussion, it is assumed that displacements and displacement gradients are small enough so that no distinction needs to be made between Lagrangian and Eulerian descriptions. The general constitutive equations for a linear elastic solid relates the stress and strain tensors as follows:

$$t_{ij} = C_{ijkl}\,\varepsilon_{kl} \tag{8}$$

known as the generalized Hooke's Law. It can be shown that the tensor of elasticity constants C_{ijkl} has at most 21 different coefficients based on symmetry properties of the stress and strain tensors and on the existence of a strain energy function (Lekhnitskii, 1963). Instead of eq. (8), we shall use its matrix representation or the matrix representation of its inverse relation, i.e.,

$$(T) = (C)(\varepsilon) \quad \text{or} \tag{9}$$

$$(\varepsilon) = (D)(T) \tag{10}$$

where (D) is the inverse matrix of (C).

Consider a body with three mutually perpendicular planes of elastic symmetry with normal directions n, s, t. A material that possesses this type of symmetry is called orthotropic and 9 distinct elastic constants describe the material. Then, one can write eq. (10) as follows:

$$\begin{bmatrix} \varepsilon_n \\ \varepsilon_s \\ \varepsilon_t \\ \gamma_{st} \\ \gamma_{nt} \\ \gamma_{ns} \end{bmatrix} = \begin{bmatrix} \frac{1}{E_n} & \frac{-\nu_{sn}}{E_s} & \frac{-\nu_{tn}}{E_t} & 0 & 0 & 0 \\ \frac{-\nu_{sn}}{E_s} & \frac{1}{E_s} & \frac{-\nu_{ts}}{E_t} & 0 & 0 & 0 \\ \frac{-\nu_{tn}}{E_t} & \frac{-\nu_{ts}}{E_t} & \frac{1}{E_t} & 0 & 0 & 0 \\ 0 & 0 & 0 & \frac{1}{G_{st}} & 0 & 0 \\ 0 & 0 & 0 & 0 & \frac{1}{G_{nt}} & 0 \\ 0 & 0 & 0 & 0 & 0 & \frac{1}{G_{ns}} \end{bmatrix} \begin{bmatrix} \sigma_n \\ \sigma_s \\ \sigma_t \\ \tau_{st} \\ \tau_{nt} \\ \tau_{ns} \end{bmatrix} \tag{11}$$

where E_n, E_s, E_t are the Young's moduli in n, s, t directions,

G_{st}, G_{nt}, G_{ns} are the shear moduli in the planes st, nt, and ns,

ν_{sn}, ν_{tn}, ν_{ts} are the Poisson's coefficients. Poisson's ratio ν_{ij} determines the normal strain in the symmetry direction j when a stress is added in the symmetry direction i. Poisson's ratio ν_{ij} and ν_{ji} are related by:

$$\frac{\nu_{ij}}{E_i} = \frac{\nu_{ji}}{E_j} \tag{12}$$

Now consider the same body cut by three orthogonal joint sets, each one parallel to a symmetry direction. A joint set i (i = 1, 2, 3) is characterized by its spacing S_i and its stiffnesses k_{n_i}, k_{s_i} as defined in previous sections*. Its orientation is defined with respect to a fixed coordinate system x, y, z (Fig. 3).

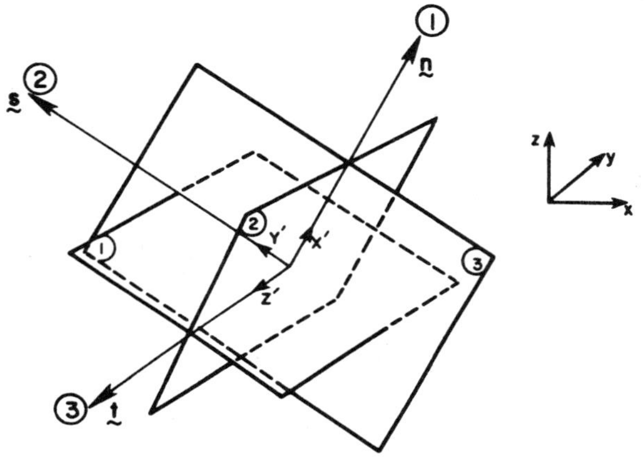

Fig. 3. Orientation of joint sets 1, 2, 3.

The concept of an equivalent anisotropic medium to represent the behavior of a fractured body will be stated first for the case of an isotropic body cut by one joint set. It will then be generalized for an orthotropic body transected by three orthogonal joint sets.

<u>Concept of an equivalent anisotropic continuum for an isotropic body transected by one joint set</u>

We assume the intact rock is linearly elastic and isotropic with two constants: a Young's modulus E and a Poisson's ratio ν. The deformations within jointed rocks subjected to both normal and shear stresses result from compressive and shear strains within the rock between the joints and from interfacial normal and shear displacements within joints. For an isotropic rock cut by one joint set, consider a basic unit consisting of a single thickness of rock and a single joint (Fig. 4a) and adopt axes (n, t) normal and parallel to the joint and therefore in the principal symmetry directions of the rock mass. The basic concept

*The influence of dilatancy and coupled shear displacements will be neglected.

of the model is to replace the jointed body by an equivalent anisotropic continuum (Duncan and Goodman, 1968). Shear and normal deformations of the equivalent continuum are equal to those of the jointed body.

In the plane (n, t), the equivalent medium is described by two moduli: (i) a shear modulus G_{nt} (Fig. 4b) such that

$$\frac{1}{G_{nt}} = \frac{1}{G} + \frac{1}{k_s S} \tag{13}$$

with $G = \frac{E}{2(1 + \nu)}$ the shear modulus of the intact rock; (ii) and a normal modulus E_n (Fig. 4c) such that

$$\frac{1}{E_n} = \frac{1}{E} + \frac{1}{k_n S} \tag{14}$$

An additional assumption is to assign the joint a negligible thickness compared with the spacing S. Therefore, the joint does not create any Poisson's effect. When dealing with an isotropic body transected by one joint set, the joint having a negligible thickness, and assuming transverse isotropy within the joint set, only four elastic constants are required to describe such a material: E, ν, $k_n S$, and $k_s S$.

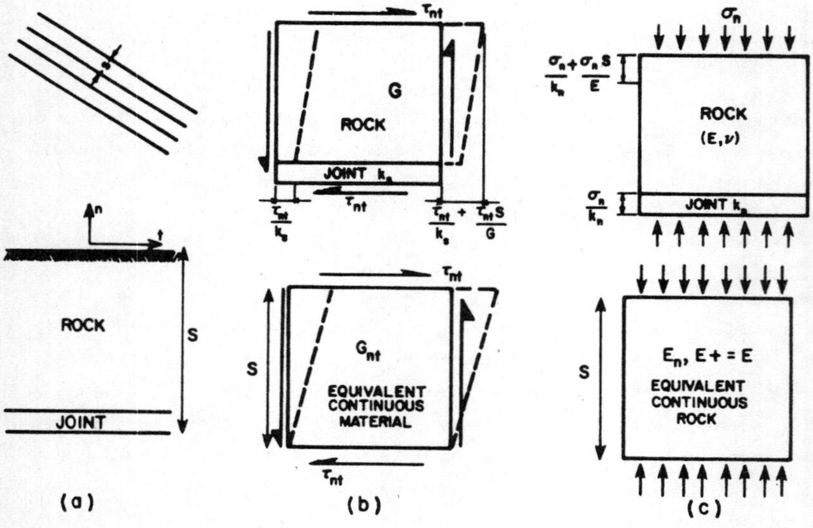

Fig. 4. Concept of the equivalent anisotropic continuum--(a) Basic unit, (b) Behavior in shear, (c) Behavior in normal compression.

The same concept can be introduced when the rock between the joints is itself anisotropic, for example, transversely anisotropic within a plane parallel to the joint set. In that case seven constants are needed to describe the material. This model can be applied to the deformability of schistose rocks.

Concept of an equivalent anisotropic continuum for an orthotropic body transected by three orthogonal joint sets

For the more general case of a body with three orthogonal joint sets, each one being parallel to a plane of elastic symmetry (Fig. 3), the jointed body can be replaced by an equivalent orthotropic continuum. Assuming again that each joint set has a negligible thickness and does not create any Poisson's effect, the constitutive relation of this equivalent orthotropic continuum is similar to eq. (11) with the diagonal moduli replaced by those of the equivalent medium. Those moduli can be defined as follows:

(i) Behavior in normal compression: in direction i (n, s, t for the intact rock or 1, 2, 3 for the joints) the equivalent Young's modulus E_i^* is defined by:

$$\frac{1}{E_i^*} = \frac{1}{E_i} + \frac{1}{k_{n_i} S_i} \tag{15}$$

(ii) Behavior in shear: in the plane ij, the equivalent shear moduli G_{ij}^* is defined by

$$\frac{1}{G_{ij}^*} = \frac{1}{G_{ij}} + \frac{1}{k_{s_i} S_i} + \frac{1}{k_{s_j} S_j} \tag{16}$$

[no summation is intended in eqs. (15) and (16)]. Then one can rewrite eq. (11):

$$(\varepsilon)_{n,s,t} = (D)_{n,s,t}(T)_{n,s,t} \quad \text{where} \tag{17}$$

$$(D)_{n,s,t} = (D_1)_{n,s,t} + (D_2)_{n,s,t} \quad \text{with} \tag{18}$$

$$(D_1)_{n,s,t} = \begin{bmatrix} \frac{1}{E_n} & \frac{-\nu_{sn}}{E_s} & \frac{-\nu_{tn}}{E_t} & 0 & 0 & 0 \\ \frac{-\nu_{sn}}{E_s} & \frac{1}{E_s} & \frac{-\nu_{ts}}{E_t} & 0 & 0 & 0 \\ \frac{-\nu_{tn}}{E_t} & \frac{-\nu_{ts}}{E_t} & \frac{1}{E_t} & 0 & 0 & 0 \\ 0 & 0 & 0 & \frac{1}{G_{st}} & 0 & 0 \\ 0 & 0 & 0 & 0 & \frac{1}{G_{nt}} & 0 \\ 0 & 0 & 0 & 0 & 0 & \frac{1}{G_{ns}} \end{bmatrix} \tag{19}$$

$$(D_2)_{n,s,t} = \begin{bmatrix} \frac{1}{k_{n1}S_1} & 0 & 0 & 0 & 0 & 0 \\ 0 & \frac{1}{k_{n2}S_2} & 0 & 0 & 0 & 0 \\ 0 & 0 & \frac{1}{k_{n3}S_3} & 0 & 0 & 0 \\ 0 & 0 & 0 & \frac{1}{k_{s2}S_2} + \frac{1}{k_{s3}S_3} & 0 & 0 \\ 0 & 0 & 0 & 0 & \frac{1}{k_{s1}S_1} + \frac{1}{k_{s3}S_3} & 0 \\ 0 & 0 & 0 & 0 & 0 & \frac{1}{k_{s1}S_1} + \frac{1}{k_{s2}S_2} \end{bmatrix} \quad (20)$$

$(D_1)_{n,s,t}$ is a constant matrix and represents the contribution of the intact rock in the deformation of the jointed rock mass. However, $(D_2)_{n,s,t}$ is not constant and represents the contribution of the joint sets. Stiffnesses k_{ni}, k_{si} $(i = 1,2,3)$ can vary with the normal stress acting on each joint set as previously described.

If we know the orientation of the joint sets with respect to a fixed system of coordinates (x, y, z), we can rewrite relation (17) in that system as

$$(\varepsilon)_{x,y,z} = (D)_{x,y,z}(T)_{x,y,z} \quad (21)$$

Closed form solutions

Closed form solutions for uniaxial and triaxial loading have been derived to demonstrate the applicability of the proposed constitutive relation. They are based on the following assumptions:

(i) the intact rock between the joints is isotropic with two constants, E and ν,

(ii) the rock has three orthogonal joint sets, two joint sets striking in the y direction and the third perpendicular to it (Fig. 5)[+],

(iii) there is no variation of the state of stress within the samples.*

Although not restricted to those assumptions, we have selected them to shorten the presentation. Furthermore, strength criteria must be introduced to verify if failure takes place in the intact rock, along the joints, or both for certain conditions of joint orientation, surface properties and applied stresses. The constraints related to joint behavior are shear and tensile strength, maximum closure and allowable amount of decompression or opening.

+The numbering of the joint sets is shown in Fig. 5.
*Compression and contraction are taken positive.

<u>Uniaxial compression (Fig. 5)</u>

If a stress σ_v is applied in the z direction and if θ is the dip angle of joint set 1, we have the following strain components:

$$\begin{cases} \varepsilon_x = \left(\dfrac{\sin^2 2\theta}{4}\left(\dfrac{1}{k_{n1}S_1} + \dfrac{1}{k_{n2}S_2} - \dfrac{1}{k_{s1}S_1} - \dfrac{1}{k_{s2}S_2}\right) - \dfrac{\nu}{E}\right)\sigma_v \\ \varepsilon_y = \dfrac{-\nu}{E}\sigma_v \\ \varepsilon_z = \left(\dfrac{\cos^4\theta}{k_{n1}S_1} + \dfrac{\sin^4\theta}{k_{n2}S_2} + \dfrac{\sin^2 2\theta}{4}\left(\dfrac{1}{k_{s1}S_1} + \dfrac{1}{k_{s2}S_2}\right) + \dfrac{1}{E}\right)\sigma_v \\ \gamma_{xz} = \left(\sin 2\theta\left(\dfrac{\cos^2\theta}{k_{n1}S_1} - \dfrac{\sin^2\theta}{k_{n2}S_2}\right) - \dfrac{\sin 4\theta}{4}\left(\dfrac{1}{k_{s1}S_1} + \dfrac{1}{k_{s2}S_2}\right)\right)\sigma_v \end{cases} \quad (22)$$

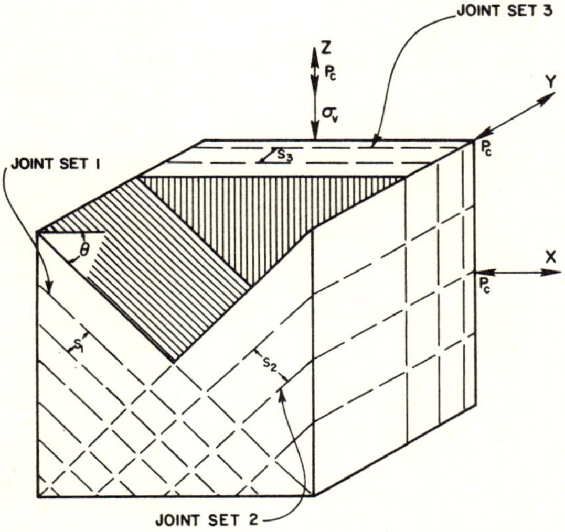

Fig. 5. Uniaxial compression (σ_V, $P_C = 0$)--Triaxial compression (P_C, $\sigma_V = 0$).

We may observe the following:

(i) For very stiff joint sets both in shear and in normal compression or tension and/or large spacings, eqs. (22) reduce to those for the isotropic intact rock.

(ii) Joint set 3 parallel to the direction of loading has no influence on strains.

(iii) When $\theta = 0$ and therefore joint set 1 is perpendicular to σ_v, we have

$$\varepsilon_x = \varepsilon_y = \frac{-\nu}{E}\sigma_v \qquad \varepsilon_z = \left(\frac{1}{E} + \frac{1}{k_{n1}S_1}\right)\sigma_v \qquad \gamma_{xz} = 0 \quad (23)$$

Joint sets 2 and 3 do not have any influence. Similarly when $\theta = 90°$, joint set 2 is perpendicular to σ_v and joint sets 1 and 3 have no influence.

(iv) When loading is not in a symmetry direction of the material, a distortion component γ_{xz} appears and thus, the principal directions of strain do not correspond with those of stress (see numerical example).

(v) It has been shown (Goodman 1976, 1980) that the unloading of a fractured rock mass follows essentially the same path as that of the intact rock. Joint compression is essentially unrecoverable. Assuming the intact rock isotropic, we can separate ε_z into two parts (Fig. 6).

$$\varepsilon_z = \frac{\sigma_v}{E_T} = \frac{\sigma_v}{E} + \frac{\sigma_v}{M} \tag{24}$$

where E_T is a "total modulus" or <u>modulus of deformation</u>,

E is the <u>elastic modulus</u> of Young's modulus defined in the unloading and reloading parts of the uniaxial stress strain curve of the fractured rock,

and M is a <u>modulus of permanent deformation</u> defined as the ratio of a stress to the permanent deformation observed upon releasing that stress to zero.

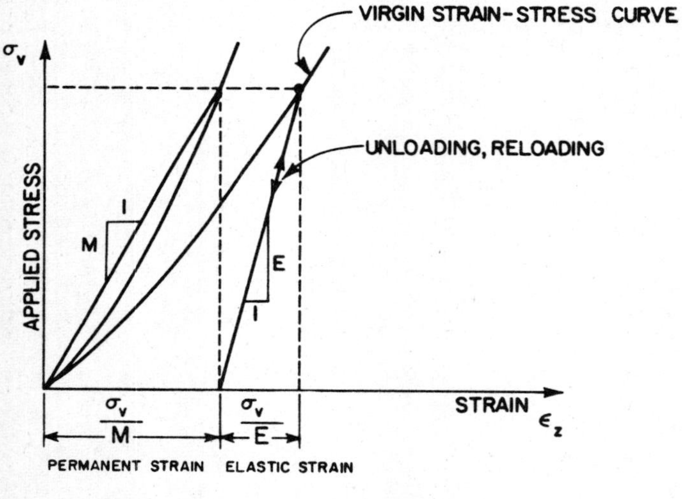

Fig. 6. Definition of the modulus of permanent deformation M.

σ_v/E and σ_v/M represent respectively the contribution of the intact rock and the joints in the vertical strain of a fractured rock sample. For the present joint configuration:

$$\frac{1}{M} = \frac{\cos^4\theta}{k_{n1}S_1} + \frac{\sin^4\theta}{k_{n2}S_2} + \frac{\sin^2 2\theta}{4}\left(\frac{1}{k_{s1}S_1} + \frac{1}{k_{s2}S_2}\right). \tag{25}$$

Due to the normal stress dependency of the normal and shear stiffnesses, the modulus M depends on applied stresses, joint orientations, spacings and stiffnesses.

(vi) In order to evaluate the importance of the lateral strains ε_x and ε_y one can define two strain ratios ν_{zx} and ν_{zy} such that

$$\nu_{zx} = \frac{-\varepsilon_x}{\varepsilon_z} \quad \text{and} \quad \nu_{zy} = \frac{-\varepsilon_y}{\varepsilon_z} \tag{26}$$

Using the same type of decomposition as in eq. (24), one can assimilate ν_{zx} and ν_{zy} as total strain ratios and define elastic strain ratios $\nu_{zx\,e}$, $\nu_{zy\,e}$ and permanent strain ratios $\nu_{zx\,p}$, $\nu_{zy\,p}$ such that

$$\nu_{zx\,e} = \frac{-\varepsilon_{x\,e}}{\varepsilon_{z\,e}}, \quad \nu_{zy\,e} = \frac{-\varepsilon_{y\,e}}{\varepsilon_{z\,e}}, \quad \nu_{zx\,p} = \frac{-\varepsilon_{x\,p}}{\varepsilon_{z\,p}}, \quad \nu_{zy\,p} = \frac{-\varepsilon_{y\,p}}{\varepsilon_{z\,p}} \tag{27}$$

where $(\varepsilon_{x\,e}, \varepsilon_{y\,e}, \varepsilon_{z\,e})$ and $(\varepsilon_{x\,p}, \varepsilon_{y\,p}, \varepsilon_{z\,p})$ are respectively the elastic and permanent components of strains ε_x, ε_y, ε_z.

If the intact rock is isotropic, those quantities are related by

$$\nu_{zx} = \frac{\nu M + \nu_{zx\,p} E}{M + E} \quad \text{and} \quad \nu_{zy} = \frac{\nu M + \nu_{zy\,p} E}{M + E} \tag{28}$$

where M and E have been defined in eq. (24).

(vii) In order to illustrate the previous discussion, the following examples have been studied:

* Three orthogonal joint sets with identical spacing, surface properties and stress dependency of the normal stiffnesses [eq. (4)] with the configuration of Fig. 5. Figures 7 and 8 show, in polar coordinates, the variations of ε_x, ε_y, ε_z and γ_{xz} with the dip angle θ of joint set 1 for a given value of joint spacing and compressive stress applied in the z direction. The strain components have been compared with those for one joint set (joint set 1). The distance AB represents the contribution of the joint set(s) on the strain components of the fractured sample. When dealing with three joint sets with the same properites, Figs. 7 and 8 are symmetric with respect to $\theta = 45°$.

* One joint set (joint set 1) with a dip angle θ. Figures 9a and 9b show respectively for given joint spacing and compressive stress applied in the z direction the variation of the total strain vatios ν_{zx}, ν_{zy} and the inclination of the principal strains when θ varies.

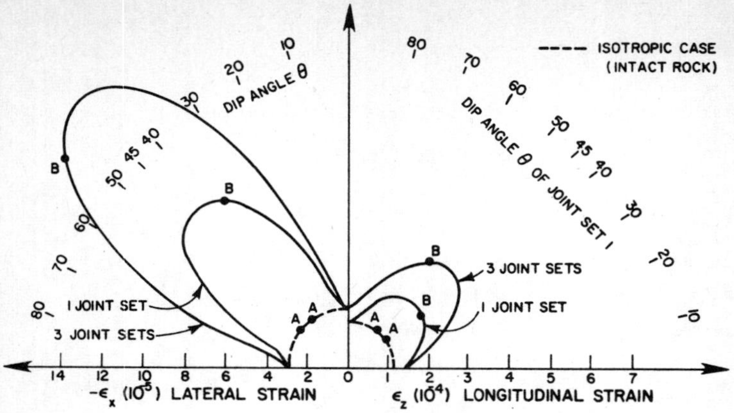

Fig. 7. Variation of strain components ε_x, ε_y, ε_z with dip angle θ of joint set 1 for an applied compressive stress σ_v = 4 MPa. Joint set properties: Spacing 1.00 m, shear stiffness 10,000 MPa/m, initial vertical stress 1 MPa, $\sigma_n = v/(0.1 - v)$ with v (mm). Intact rock properties: E = 35,000 MPa, ν = 0.25.

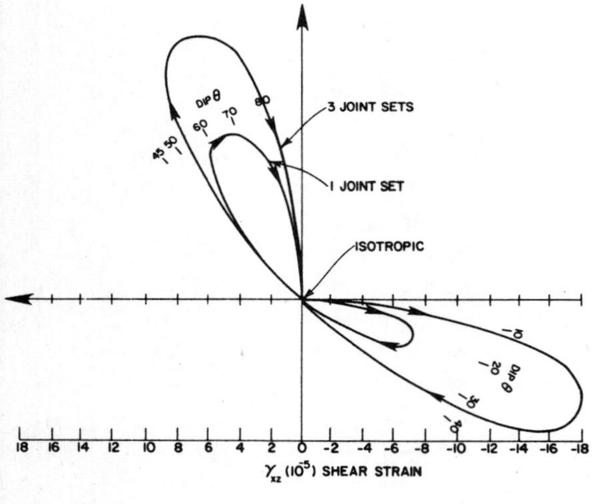

Fig. 8. Variation of strain component γ_{xz} with the dip angle θ of joint set 1 for an applied compressive stress σ_v = 4 MPa. Joint set properties: Spacing 1.00 m, shear stiffness 10,000 MPa/m, initial vertical stress 1 MPa, $\sigma_n = v/(0.1 - v)$ with v (mm). Intact rock properties: E = 35,000 MPa, ν = 0.25

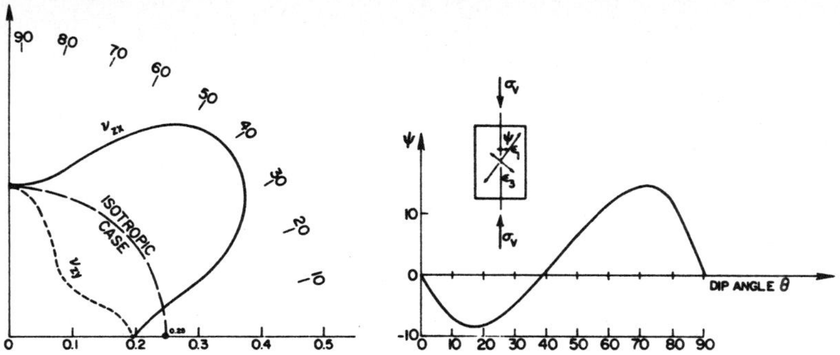

(a) (b)

Fig. 9. (a) Variation of the total strain ratios ν_{zx}, ν_{zy} with the dip angle θ. One joint set with spacing $S = 1.00$ m, applied stress $\sigma_v = 4$ MPa.
(b) Influence of the dip angle θ on the inclination of the principal strains. One joint set with spacing $S = 1.00$ m, applied stress $\sigma_v = 4.00$ MPa.

Triaxial compression (Fig. 5)

If a stress P_c is applied in x,y,z directions, for the present joint configuration, we have the following strain components:

$$\begin{cases} \varepsilon_x = \left[\left(\frac{1-2\nu}{E}\right) + \frac{\sin^4\theta}{k_{n1}S_1} + \frac{\cos^4\theta}{k_{n2}S_2} + \frac{\sin^2 2\theta}{4}\left(\frac{1}{k_{n1}S_1} + \frac{1}{k_{n2}S_2}\right)\right] P_c \\ \varepsilon_y = \left[\left(\frac{1-2\nu}{E}\right) + \frac{1}{k_{n3}S_3}\right] P_c \\ \varepsilon_z = \left[\left(\frac{1-2\nu}{E}\right) + \frac{\cos^4\theta}{k_{n1}S_1} + \frac{\sin^4\theta}{k_{n2}S_2} + \frac{\sin^2 2\theta}{4}\left(\frac{1}{k_{n1}S_1} + \frac{1}{k_{n2}S_2}\right)\right] P_c \\ \gamma_{xz} = \left[\frac{1}{k_{n1}S_1} - \frac{1}{k_{n2}S_2}\right] \sin 2\theta \; P_c \end{cases} \tag{29}$$

The volumetric strain of the rock sample is equal to:

$$\frac{\Delta V}{V} = \varepsilon_x + \varepsilon_y + \varepsilon_z = \left(3\left(\frac{1-2\nu}{E}\right) + \sum_{i=1}^{3} \frac{1}{k_{ni}S_i}\right) P_c \tag{30}$$

We define an equivalent bulk modulus K^* and the effective compressibility β_{eff} of the fractured material such that:

$$P_c = K^* \frac{\Delta V}{V} \quad \text{and} \quad \beta_{eff} = \frac{1}{K^*} \tag{31}$$

Hence,

$$\beta_{eff} = \frac{1}{K^*} = 3\left(\frac{1-2\nu}{E}\right) + \sum_{i=1}^{3} \frac{1}{k_{ni}S_i} \tag{32}$$

Therefore, the compressibility of a fractured medium can be written as the sum of the compressibility of the intact rock and a term that depends on joint spacings and normal stiffnesses. If k_{ni} (i = 1,2,3) are dependent on the value of P_c, so are K* and β_{eff}. Note that $\nu = 0.5$ is no longer a limiting value for the compressibility or the bulk modulus. Using the same decomposition as in eq. (24), the effective compressibility can be described as the sum of elastic and permanent compressibilities, the former related to the intact rock and the latter related to the joints.

According to eqs. (29), when $k_{n1}S_1 = k_{n2}S_2 = k_{n3}S_3$, the fractured material behaves isotropicly regardless of the value of the angle θ with

$$\varepsilon_x = \varepsilon_y = \varepsilon_z = \left[\left(\frac{1-2\nu}{E}\right) + \frac{1}{k_n S}\right]P_c, \quad \gamma_{xz} = 0 \tag{33}$$

In comparison with the isotropic case, the strains are increased by a constant amount due to the presence of the joints. We can generalize the last remark and state that for any triaxial state of applied stresses on a jointed rock sample, there exist joint properties and geometry for which the fractured material behaves isotropicly.

An example is presented with the same intact rock and joint properties as for uniaxial compression. The joint sets have identical spacing and surface properties. Figure 10 shows the variation of the equivalent bulk modulus defined by eq. (32)

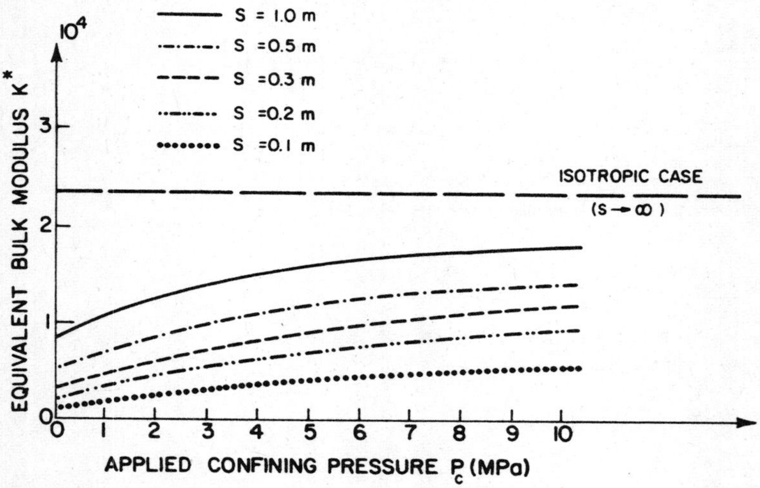

Fig. 10. Influence of the applied confining pressure P_c on the equivalent bulk modulus K* for different spacing values S.

with the confining pressure applied from an initial value of 1 MPa. One can easily see that K* approaches the values for an isotropic medium as the applied confining pressure and/or joint spacings increase.

EFFECT OF WATER--AN EFFECTIVE STRESS LAW FOR AN EQUIVALENT ANISOTROPIC CONTINUUM

The constitutive relation of the previous section has been derived for a dry fractured rock. Several effective stress laws have been proposed for linear elastic, isotropic, homogeneous and continuous materials (Skempton, 1960). Carroll (1979) derived analytically an effective stress law to describe the effect of pore fluid pressure on the linearly elastic response of saturated porous materials that exhibit anisotropy. The purpose of this section is to extend this concept to evaluate the effect of pore fluid pressure on the response of a fractured rock described as an equivalent anisotropic continuum.

Consider a representative sample of a dry fractured material large enough to contain a number of grains and fractures, the fracture being the only voids in the material. Assume that the stress strain response of that material can be expressed by a constitutive relation of the type eqs. (9) or (10). The response of a fluid-saturated fractured material may be related to that of a dry material by considering the loading

$$t_i = t_{ij}n_j \text{ on } B_o \quad \text{and} \quad t_i = P_p n_i \text{ on } B_p \tag{34}$$

where B_o and B_p are respectively the outer boundary of and the joint surface within the sample,

t_{ij} are constant applied stresses, n_j are the direction cosines of a surface.

Carroll (1979) treats this state of stress as the superposition of two loadings.

(i) a uniform hydrostatic pressure P_p along B_o and B_p. This loading gives rise to a uniform straining of the intact solid $\varepsilon_{ij}^{(1)}$, and

(ii) a constant stress, $t_{ij} - P_p\delta_{ij}$ on B_o and nothing on B_p, with induced strains $\varepsilon_{ij}^{(2)}$.

For a linear elastic material, as shown by Carroll, the total strain is equal to:

$$\varepsilon_{ij} = \varepsilon_{ij}^{(1)} + \varepsilon_{ij}^{(2)} = D_{ijkl}t_{kl} - P_p(D_{ijkk} - D_{ijkk}^{(s)}) \tag{35}$$

where $D_{ijkl}^{(s)}$ is the tensor of elastic compliances for the solid material.

The effective stress concept is that the response of a saturated material is described by the response law for the dry material with the applied stress t_{ij} replaced by the effective stress $\bar{t}_{ij}$ such that

$$\bar{t}_{ij} = C_{ijkl}\varepsilon_{kl} \quad \text{or} \quad \varepsilon_{ij} = D_{ijkl}\bar{t}_{kl} \tag{36}$$

Combining eqs. (35), (36) and the relation between the tensor D_{ijkl} and C_{ijkl}, we obtain the effective stress tensor for anisotropic deformation

$$\bar{t}_{ij} = t_{ij} - P_p(\delta_{ij} - C_{ijkl}D^{(s)}_{klmm}) \tag{37}$$

If we know the constitutive relation in a coordinate system (x,y,z) [eq. (21)], we can rewrite eqs. (35) and (37) using a matrix representation as follows:

$$(\varepsilon) = (D)(T) - P_p(H) \quad \text{where} \quad (H) = \big((D) - (D_s)\big)\begin{pmatrix}1\\1\\1\\0\\0\\0\end{pmatrix} \tag{38}$$

and

$$(\bar{T}) = (T) - P_p\ \big((I) - (C)(D_s)\big)\begin{pmatrix}1\\1\\1\\0\\0\\0\end{pmatrix} \tag{39}$$

where (D_s) is the matrix of elastic compliances for the solid material,
$(\bar{T})$ is the effective stress matrix, and
(I) is the identity matrix.

According to the constitutive relation of the previous section, the coefficients of the matrix (D) may depend on the applied stresses through joint stiffnesses. Therefore the superposition principle used to write eq. (37) is correct for two conditions:

(i) the joint stiffnesses are constant, or

(ii) the joint stiffnesses depend only on the total applied stresses.

One can see that the modification to the applied stresses (T) is not hydrostatic as it is for an isotropic material and depends on joint orientation, geometry and state of stress. Another important assumption is that the fluid pressure P_p does not change with the applied state of stress (drained conditions).

A closed form solution has been derived for the same joint configuration as in Fig. 5, the intact rock being isotropic. When a confining pressure P_c is applied on the outer boundaries of the jointed sample, it can be shown that the volumetric strain is equal to:

$$\frac{\Delta V}{V} = (P_c - P_p)\beta_{eff} + 3\ P_p\left(\frac{1 - 2\nu}{E}\right) \tag{40}$$

with β_{eff} has been defined as eq. (32). When P_p is constant, and as P_c approaches P_p, the volumetric strain converges to the value for the solid material.

Figure 11 shows the variation of the volumetric strain with the confining pressure P_c applied from an initial value of 1 MPa for different values of the fluid pressure P_p. Joints have the same spacing (S = 0.3 m) and the same properties as in the previous section. Normal stiffnesses depend only on the total applied stresses. The influence of the fluid pressure decreases as the applied confining pressure increases.

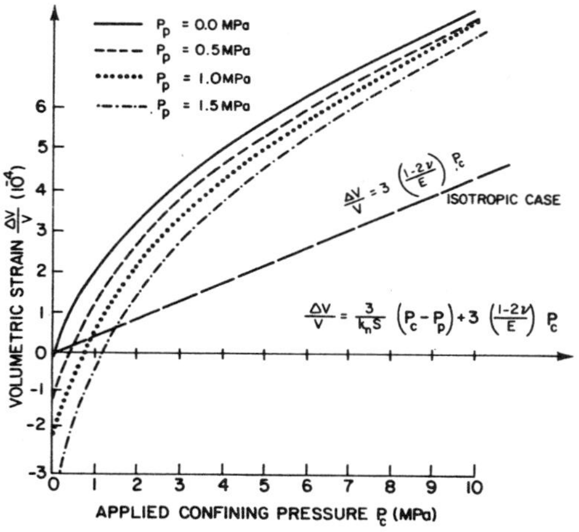

Fig. 11. Influence of the applied confining pressure P_c on the volumetric strain $\Delta V/V$ for different values of the fluid pressure P_p (theoretical example with S = 0.3 m).

CONCLUSIONS

A constitutive relation describing the non linear behavior of a discontinuous, homogeneous and anisotropic body of rock containing up to three orthogonal joint sets has been introduced. The jointed rock has been described as an equivalent anisotropic continuum. This approach allows one to study the influence of joint and intact rock properties, and state of stress on the deformability of fractured rock. We have not considered the influence of those properties on the strength of fractured rock. This will be required to complete the constitutive relationship.

The two loading conditions studied in this paper suggest that the full description of rock deformability should include not only the elastic coefficients but also the permanent deformation coefficients associated with any applied stress level. Those permanent deformation coefficients are presented in the form of a modulus, strain ratios and compressibility and are related to the behavior of joints in shear, normal compression and decompression.

This constitutive relation has been used recently at Berkeley to study the distribution of stresses and displacements around elliptical openings in a fractured rock in plane and generalized plane strain conditions. The closed form solution proposed by Lekhnitskii (1963) has been used.

REFERENCES

1. N. R. Barton, Estimating the shear strength of rock joints, Proc. 3rd. Cong. ISRM, Denver, 1974, V. 2A, p. 219.
2. Z. P. Bazant and P. Gambarova, Rough cracks in reinforced concrete, ASCE, Journal of the Structural Division, 1980, ST4, p. 819.
3. M. M. Carroll, An effective stress law for anisotropic elastic deformations, Journal of Geophysical Research, 1979, V. 84, No. B13, p. 7510.
4. J. M. Duncan and R. E. Goodman, Finite element analysis of slopes in jointed rocks, U.S. Army Corps of Engineers Report TR, 1968, No. 1-68.
5. R. E. Goodman, Methods of Geological Engineering, West Publ., 1976.
6. R. E. Goodman, Introduction to Rock Mechanics, John Wiley & Sons, 1980.
7. R. E. Goodman and J. Dubois, Duplication of dilatancy in analysis of jointed rocks, ASCE, Journal of the Soil Mechanics and Foundation Division, 1972, SM4, p. 399.
8. R. E. Goodman, R. L. Taylor, and T. L. Brekke, A model for the mechanics of jointed rock, ASCE Journal of the Soil Mechanics and Foundations Division, 1968, SM3, p. 637.
9. F. Heuze, Dilatant effects of rock joints, Proc. 4th Cong. ISRM, Montreux, 1979, pp. 169-175.
10. J. C. Jaeger, Friction of rocks and the stability of rock slopes, Rankine Lecture, Geotechnique, 1971, V. 21, p. 97.
11. B. Ladanyi and G. Archambault, Simulation of shear behaviour of a jointed rock mass, Proc. 11th U.S. Symp. on Rock Mechanics, (AIME), 1972, p. 105.
12. S. G. Lekhnitskii, Theory of elasticity of an anisotropic elastic body, Holden Day, Inc. 1963.
13. F. D. Patton, Multiple modes of shear failure in rock, Proc. 1st Cong. ISRM, Lisbon, 1966, V. 1, p. 509.
14. A. W. Skempton, Effective stress in soils, concrete and rock, Pore and suction in soils, Butterworths, London, 1960.

Appendix 4.1

Let us define three coordinate systems:

(i) X,Y,Z is a global coordinate system that can be chosen arbitrarily,

(ii) x',y',z' is a coordinate system attached to the rectilinear anisotropy. The orientation of that system with respect to the global one is defined by two angles β and δ as shown in Figure 1 (x' is in the direction normal to one plane of rectilinear anisotropy).

The direction cosines of the unit vectors in the x', y' and z' directions are given by the following quantities

$$l_{x'} = \cos\delta_{x'}\cos\beta_{x'} \qquad l_{y'} = \cos\delta_{y'}\cos\beta_{y'} \qquad l_{z'} = \cos\delta_{z'}\cos\beta_{z'}$$
$$m_{x'} = \sin\delta_{x'} \qquad m_{y'} = \sin\delta_{y'} \qquad m_{z'} = \sin\delta_{z'}$$
$$n_{x'} = \cos\delta_{x'}\sin\beta_{x'} \qquad n_{y'} = \cos\delta_{y'}\sin\beta_{y'} \qquad n_{z'} = \cos\delta_{z'}\sin\beta_{z'} \qquad (1)$$

where

	$\beta_{x',y',z'}$	$\delta_{x',y',z'}$
x'	β	δ
y'	$\beta+\pi$	$\frac{\pi}{2}-\delta$
z'	$\beta+\frac{\pi}{2}$	0

(iii) x,y,z is a coordinate system attached to a hole of arbitrary cross section. The orientation of that system with respect to the global one is defined by two angles β_h and δ_h as shown in Figure 2.

The direction cosines of the unit vectors in the x,y, and z directions are given by the following quantities

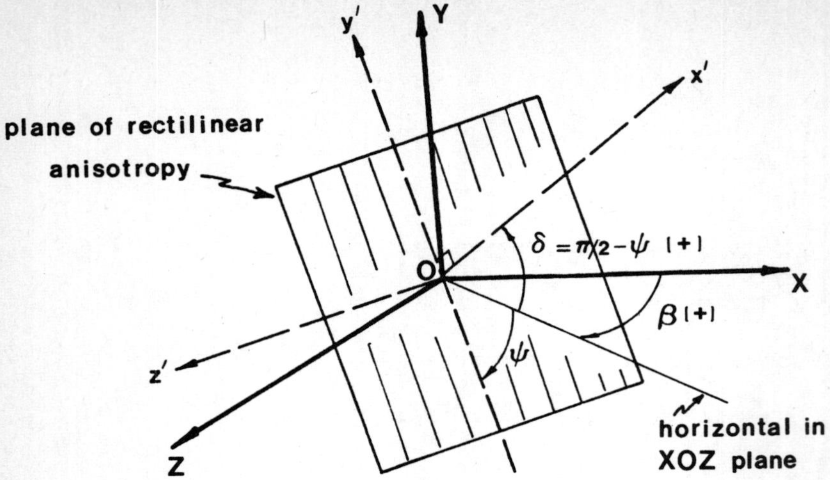

Figure 1. Orientation of the coordinate system x',y',z' with respect to the global one X,Y,Z.

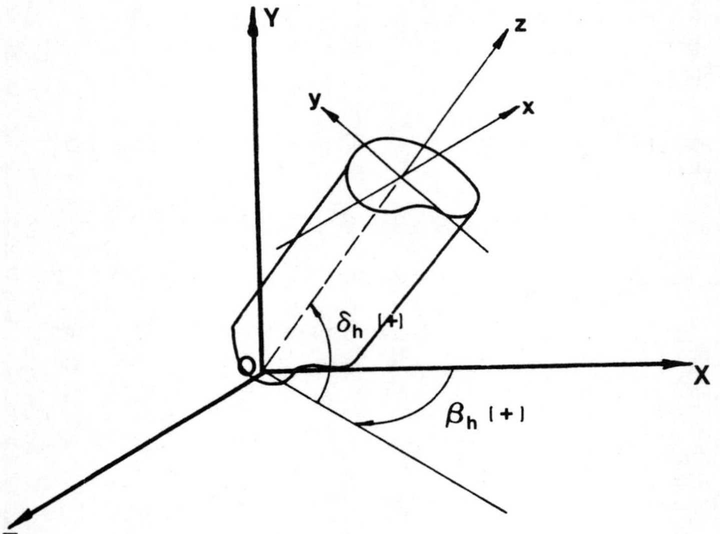

Figure 2. Orientation of the coordinate system x,y,z with respect to the global one X,Y,Z.

$$l_x = \cos\delta_x \cos\beta_x \qquad l_y = \cos\delta_y \cos\beta_y \qquad l_z = \cos\delta_z \cos\beta_z$$
$$m_x = \sin\delta_x \qquad m_y = \sin\delta_y \qquad m_z = \sin\delta_z$$
$$n_x = \cos\delta_x \sin\beta_x \qquad n_y = \cos\delta_y \sin\beta_y \qquad n_z = \cos\delta_z \sin\beta_z \tag{2}$$

where

	$\beta_{x,y,z}$	$\delta_{x,y,z}$
x	$\beta_h + \frac{3\pi}{2}$	0
y	$\beta_h + \pi$	$\frac{\pi}{2} - \delta_h$
z	β_h	δ_h

Then, let us define the following stress and strain components in matrix form

$$(\varepsilon)^t_{xyz} = (\varepsilon_x \ \varepsilon_y \ \varepsilon_z \ \gamma_{yz} \ \gamma_{xz} \ \gamma_{xy})$$
$$(\varepsilon)^t_{XYZ} = (\varepsilon_X \ \varepsilon_Y \ \varepsilon_Z \ \gamma_{YZ} \ \gamma_{XZ} \ \gamma_{XY})$$
$$(\varepsilon')^t = (\varepsilon_{x'} \ \varepsilon_{y'} \ \varepsilon_{z'} \ \gamma_{y'z'} \ \gamma_{x'z'} \ \gamma_{x'y'})$$
$$(\sigma)^t_{xyz} = (\sigma_x \ \sigma_y \ \sigma_z \ \tau_{yz} \ \tau_{xz} \ \tau_{xy})$$
$$(\sigma)^t_{XYZ} = (\sigma_X \ \sigma_Y \ \sigma_Z \ \tau_{YZ} \ \tau_{XZ} \ \tau_{XY})$$
$$(\sigma')^t = (\sigma_{x'} \ \sigma_{y'} \ \sigma_{z'} \ \tau_{y'z'} \ \tau_{x'z'} \ \tau_{x'y'}) \tag{3}$$

The constitutive relation of the anisotropic material in the x',y',z' coordinate system can be expressed as follows

$$(\varepsilon') = (H')(\sigma') \tag{4}$$

(σ') and $(\sigma)_{XYZ}$ are related by the equation

$$(\sigma') = (T'_\sigma)(\sigma)_{XYZ} \tag{5}$$

where (T_σ') is a transformation matrix defined as follows

(Goodman, 1980, Appendix 1)

$$\begin{bmatrix} l_{x'}^2 & m_{x'}^2 & n_{x'}^2 & 2m_{x'}n_{x'} & 2n_{x'}l_{x'} & 2l_{x'}m_{x'} \\ l_{y'}^2 & m_{y'}^2 & n_{y'}^2 & 2m_{y'}n_{y'} & 2n_{y'}l_{y'} & 2l_{y'}m_{y'} \\ l_{z'}^2 & m_{z'}^2 & n_{z'}^2 & 2m_{z'}n_{z'} & 2n_{z'}l_{z'} & 2l_{z'}m_{z'} \\ l_{y'}l_{z'} & m_{y'}m_{z'} & n_{y'}n_{z'} & m_{y'}n_{z'}+m_{z'}n_{y'} & n_{y'}l_{z'}+n_{z'}l_{y'} & l_{y'}m_{z'}+l_{z'}m_{y'} \\ l_{z'}l_{x'} & m_{z'}m_{x'} & n_{z'}n_{x'} & m_{x'}n_{z'}+m_{z'}n_{x'} & n_{x'}l_{z'}+n_{z'}l_{x'} & l_{x'}m_{z'}+l_{z'}m_{x'} \\ l_{x'}l_{y'} & m_{x'}m_{y'} & n_{x'}n_{y'} & m_{x'}n_{y'}+m_{y'}n_{x'} & n_{x'}l_{y'}+n_{y'}l_{x'} & l_{x'}m_{y'}+l_{y'}m_{x'} \end{bmatrix} = (T_\sigma') \tag{6}$$

Similarly (ε') and $(\varepsilon)_{xyz}$ are related by the equation

$$(\varepsilon') = (T_\varepsilon')(\varepsilon)_{xyz} \tag{7}$$

(T_ε') is a transformation matrix for strains with the properties

$$(T_\varepsilon')^{-1} = (T_\sigma')^{t} \quad ; \quad (T_\varepsilon')^{t} = (T_\sigma')^{-1} \tag{8}$$

Substituting eqs (5) and (7) into (4) and making use of eqs (8) we obtain the constitutive relation of the material in the X,Y,Z coordinate system as follows

$$(\varepsilon)_{xyz} = (T_\sigma')^{t}(H')(T_\sigma')(\sigma)_{xyz} \tag{9}$$

$(\sigma)_{XYZ}$ and $(\sigma)_{xyz}$ are related by the equation

$$(\sigma)_{xyz} = (T_\sigma)(\sigma)_{XYZ} \tag{10}$$

where (T_σ) is now defined as

$$\begin{bmatrix} l_x^2 & m_x^2 & n_x^2 & 2m_xn_x & 2n_xl_x & 2l_xm_x \\ l_y^2 & m_y^2 & n_y^2 & 2m_yn_y & 2n_yl_y & 2l_ym_y \\ l_z^2 & m_z^2 & n_z^2 & 2m_zn_z & 2n_zl_z & 2l_zm_z \\ l_yl_z & m_ym_z & n_yn_z & m_yn_z+m_zn_y & n_yl_z+n_zl_y & l_ym_z+l_zm_y \\ \vdots & & & & & \vdots \end{bmatrix} = (T_\sigma)$$

$$\begin{bmatrix} l_z l_x & m_z m_x & n_z n_x & m_x n_z + m_z n_x & n_x l_z + n_z l_x & l_x m_z + l_z m_x \\ l_x l_y & m_x m_y & n_x n_y & m_x n_y + m_y n_x & n_x l_y + n_y l_x & l_x m_y + l_y m_x \end{bmatrix} \tag{11}$$

Similarly,

$$(\varepsilon)_{xyz} = (T_\varepsilon)(\varepsilon)_{xyz} \tag{12}$$

where

$$(T_\varepsilon)^{-1} = (T_\sigma)^t \quad ; \quad (T_\varepsilon)^t = (T_\sigma)^{-1} \tag{13}$$

Substituting eqs (10) and (12) into (9) and making use of eqs (13) we obtain the constitutive relation of the material in the x,y,z coordinate system as follows

$$(\varepsilon)_{xyz} = (T_\varepsilon)(T'_\sigma)^t(H')(T'_\sigma)(T_\varepsilon)^t(\sigma)_{xyz} \tag{14}$$

or

$$(\varepsilon)_{xyz} = (A)(\sigma)_{xyz} \tag{15}$$

with

$$(A) = (T_\varepsilon)(T'_\sigma)^t(H')(T'_\sigma)(T_\varepsilon)^t \tag{16}$$

(T_ε) is defined as follows

$$\begin{bmatrix} l_x^2 & m_x^2 & n_x^2 & m_x n_x & n_x l_x & l_x m_x \\ l_y^2 & m_y^2 & n_y^2 & m_y n_y & n_y l_y & l_y m_y \\ l_z^2 & m_z^2 & n_z^2 & m_z n_z & n_z l_z & l_z m_z \\ 2 l_y l_z & 2 m_y m_z & 2 n_y n_z & m_y n_z + m_z n_y & n_y l_z + n_z l_y & l_y m_z + l_z m_y \\ 2 l_z l_x & 2 m_z m_x & 2 n_z n_x & m_x n_z + m_z n_x & n_x l_z + n_z l_x & l_x m_z + l_z m_x \\ 2 l_x l_y & 2 m_x m_y & 2 n_x n_y & m_x n_y + m_y n_x & n_x l_y + n_y l_x & l_x m_y + l_y m_x \end{bmatrix} = (T_\varepsilon) \tag{17}$$

Eq (15) takes also the form

$$\begin{bmatrix} \varepsilon_x \\ \varepsilon_y \\ \varepsilon_z \\ \gamma_{yz} \\ \gamma_{xz} \\ \gamma_{xy} \end{bmatrix} = \begin{bmatrix} a_{11} & a_{12} & a_{13} & a_{14} & a_{15} & a_{16} \\ a_{21} & a_{22} & a_{23} & a_{24} & a_{25} & a_{26} \\ a_{31} & a_{32} & a_{33} & a_{34} & a_{35} & a_{36} \\ a_{41} & a_{42} & a_{43} & a_{44} & a_{45} & a_{46} \\ a_{51} & a_{52} & a_{53} & a_{54} & a_{55} & a_{56} \\ a_{61} & a_{62} & a_{63} & a_{64} & a_{65} & a_{66} \end{bmatrix} \begin{bmatrix} \sigma_x \\ \sigma_y \\ \sigma_z \\ \tau_{yz} \\ \tau_{xz} \\ \tau_{xy} \end{bmatrix} \qquad (18)$$

Appendix 4.2

Let us consider the elastic equilibrium of an anisotropic, homogeneous and continuous body with a rectilinear anisotropy of a general form. The body is assumed to be analogous to a plate of thickness $2h$, (see Figure 4.4) the middle plane of which will be taken to coincide with the xOy plane of the coordinate system. The body is bounded internally by a cylindrical surface of arbitrary cross section, the region outside the cylindrical surface being infinite. The z axis of the coordinate system defines the longitudinal axis of the hole.

The faces of the plate are assumed to be free of force. The medium is subjected to body and surfaces forces that are assumed to be symmetrically distributed with respect to the middle plane z = 0.

Let the components of stress, strain, displacement, body and surfaces forces be functions of x,y and z. Since, by hypothesis the faces of the plate are free from force, we must have

$$\tau_{xz}(x,y,\pm h) = \tau_{yz}(x,y,\pm h) = \sigma_z(x,y,\pm h) = 0 \tag{1}$$

so that*

$$\left.\frac{\partial \tau_{xz}}{\partial x}\right|_{\pm h} = \left.\frac{\partial \tau_{xz}}{\partial y}\right|_{\pm h} = \left.\frac{\partial \tau_{yz}}{\partial x}\right|_{\pm h} = \left.\frac{\partial \tau_{yz}}{\partial y}\right|_{\pm h} = \left.\frac{\partial \sigma_{z}}{\partial x}\right|_{\pm h} = \left.\frac{\partial \sigma_{z}}{\partial y}\right|_{\pm h} = 0 \tag{2}$$

We denote the average values taken over the thickness of the plate as

. $\overline{u}, \overline{v}, \overline{w}$ for the displacements,

. $\overline{\sigma_x}, \overline{\sigma_y}, \overline{\sigma_z}, \overline{\tau_{yz}}, \overline{\tau_{xz}}, \overline{\tau_{xy}}$ for the stresses,

. $\overline{\varepsilon_x}, \overline{\varepsilon_y}, \overline{\varepsilon_z}, \overline{\gamma_{yz}}, \overline{\gamma_{xz}}, \overline{\gamma_{xy}}$ for the strains,

. $\overline{X_n}, \overline{Y_n}, \overline{Z_n}$ for the surface forces per unit area,

and $\overline{X}, \overline{Y}$ for the body forces per unit volume, so that we have for example,

$$\overline{u}\,(x,y) = \frac{1}{2h}\int_{-h}^{h} u(x,y,z)\,dz \tag{3}$$

Since the components of stress, strain, displacement, surface and body forces are functions of x,y,z, their mean values are function of x and y only. In particular,

$$\frac{\partial \overline{u}}{\partial z} = \frac{\partial \overline{v}}{\partial z} = \frac{\partial \overline{w}}{\partial z} = 0 \tag{4}$$

Let us rewrite the equations of equilibrium, equations of compatibility for strains, strain displacement relations,

---

* $f\big|_{\pm h}$ indicates the value of the function f at $z = h$ or $-h$

constitutive relations and boundary conditions in terms of the average quantities defined previously.

(i) Equations of Equilibrium

Recall the equations of equilibrium

$$\frac{\partial \sigma_x}{\partial x} + \frac{\partial \tau_{xy}}{\partial y} + \frac{\partial \tau_{xz}}{\partial z} + X = 0 \; ; \; \frac{\partial \tau_{xy}}{\partial x} + \frac{\partial \sigma_y}{\partial y} + \frac{\partial \tau_{yz}}{\partial z} + Y = 0 \; ; \; \frac{\partial \tau_{xz}}{\partial x} + \frac{\partial \tau_{yz}}{\partial y} + \frac{\partial \sigma_z}{\partial z} = 0 \tag{5}$$

(the body force component Z is assumed to be equal to zero).

If we integrate eqs (5) with respect to z between the limits $+h$ and $-h$, then divide the results by $2h$ and make use of eqs (1), we obtain

$$\frac{\partial \overline{\sigma_x}}{\partial x} + \frac{\partial \overline{\tau_{xy}}}{\partial y} + \overline{X} = 0 \; ; \; \frac{\partial \overline{\tau_{xy}}}{\partial x} + \frac{\partial \overline{\sigma_y}}{\partial y} + \overline{Y} = 0 \; ; \; \frac{\partial \overline{\tau_{xz}}}{\partial x} + \frac{\partial \overline{\tau_{yz}}}{\partial y} = 0 \tag{6}$$

To make eqs (6) similar to eqs (4.2), we introduce a potential $\overline{U}$ such that

$$\overline{X} = -\frac{\partial \overline{U}}{\partial x} \; ; \quad \overline{Y} = -\frac{\partial \overline{U}}{\partial y} \tag{7}$$

Combining eqs (2) and the last of eqs (5), it follows

$$\left. \frac{\partial \sigma_z}{\partial z} \right|_{\pm h} = 0 \tag{8}$$

(ii) Strain Displacement Relations

In terms of average quantities the strain displacement relations (eqs 4.3) take the following form:

---

* Eqs (6) and (7) can be satisfied by introducing two stress functions F and Ψ as defined in eq (4.9).

$$\bar{\varepsilon}_x = -\frac{\partial \bar{u}}{\partial x} \;;\quad \bar{\varepsilon}_y = -\frac{\partial \bar{v}}{\partial y} \;;\quad \bar{\gamma}_{xy} = -\left(\frac{\partial \bar{u}}{\partial y} + \frac{\partial \bar{v}}{\partial x}\right) ;\quad \bar{\varepsilon}_z = -\frac{1}{2h}\left(w(x,y,h) - w(x,y,-h)\right)$$

$$\bar{\gamma}_{xz} = -\frac{\partial \bar{w}}{\partial x} - \frac{1}{2h}\left(u(x,y,h) - u(x,y,-h)\right)$$

$$\bar{\gamma}_{yz} = -\frac{\partial \bar{w}}{\partial y} - \frac{1}{2h}\left(v(x,y,h) - v(x,y,-h)\right) \tag{9}$$

As the plate becomes thinner ($h \longrightarrow 0$)

$$\frac{u(x,y,h) - u(x,y,-h)}{2h} \longrightarrow \left.\frac{\partial u}{\partial z}\right|_{z=0}$$

$$\frac{v(x,y,h) - v(x,y,-h)}{2h} \longrightarrow \left.\frac{\partial v}{\partial z}\right|_{z=0}$$

$$\frac{w(x,y,h) - w(x,y,-h)}{2h} \longrightarrow \left.\frac{\partial w}{\partial z}\right|_{z=0} \tag{10}$$

(iii) Equations of Compatibility for Strains

If we integrate eqs (4.4) with respect to z between the limits $+h$ and $-h$ and divide the results by $2h$, we obtain the six following relations*

$$\frac{\partial^2 \bar{\varepsilon}_x}{\partial y^2} + \frac{\partial^2 \bar{\varepsilon}_y}{\partial x^2} = \frac{\partial^2 \bar{\gamma}_{xy}}{\partial x \partial y}$$

$$\frac{\partial^2 \bar{\varepsilon}_z}{\partial y^2} = \frac{1}{2h}\left[\frac{\partial \gamma_{yz}}{\partial y} - \frac{\partial \varepsilon_y}{\partial z}\right]_{-h}^{+h} \;;\quad \frac{\partial^2 \bar{\varepsilon}_z}{\partial x^2} = \frac{1}{2h}\left[\frac{\partial \gamma_{xz}}{\partial x} - \frac{\partial \varepsilon_x}{\partial z}\right]_{-h}^{+h}$$

$$\frac{\partial}{\partial x}\left(\frac{\partial \bar{\gamma}_{xz}}{\partial y} - \frac{\partial \bar{\gamma}_{yz}}{\partial x}\right) = \frac{1}{2h}\left[2\frac{\partial \varepsilon_x}{\partial y} - \frac{\partial \gamma_{xy}}{\partial x}\right]_{-h}^{+h}$$

$$\frac{\partial}{\partial y}\left(\frac{\partial \bar{\gamma}_{yz}}{\partial x} - \frac{\partial \bar{\gamma}_{xz}}{\partial y}\right) = \frac{1}{2h}\left[2\frac{\partial \varepsilon_y}{\partial x} - \frac{\partial \gamma_{xy}}{\partial y}\right]_{-h}^{+h}$$

$$2\frac{\partial^2 \bar{\varepsilon}_z}{\partial x \partial y} = \frac{1}{2h}\left[\frac{\partial \gamma_{yz}}{\partial x} + \frac{\partial \gamma_{xz}}{\partial y} - \frac{\partial \gamma_{xy}}{\partial z}\right]_{-h}^{+h} \tag{11}$$

---

* Where $\left[f\right]_{-h}^{+h} = f_{+h} - f_{-h}$ for any function f.

(iv) Constitutive relation

In terms of average quantities, the constitutive relation of the material (eqs 4.5) has the following form.

$$\begin{bmatrix} \overline{\varepsilon_x} \\ \overline{\varepsilon_y} \\ \overline{\varepsilon_z} \\ \overline{\gamma_{yz}} \\ \overline{\gamma_{xz}} \\ \overline{\gamma_{xy}} \end{bmatrix} = \begin{bmatrix} & & \\ & a_{ij} & \\ & i=1,6 & \\ & j=1,6 & \\ & & \end{bmatrix} \begin{bmatrix} \overline{\sigma_x} \\ \overline{\sigma_y} \\ \overline{\sigma_z} \\ \overline{\tau_{yz}} \\ \overline{\tau_{xz}} \\ \overline{\tau_{xy}} \end{bmatrix} \tag{12}$$

(v) Boundary Conditions

Eqs (4.7) can be rewritten as follows:

$$\begin{aligned} \overline{\sigma_x} \cos(n,x) + \overline{\tau_{xy}} \cos(n,y) &= \overline{X_n} \\ \overline{\tau_{xy}} \cos(n,x) + \overline{\sigma_y} \cos(n,y) &= \overline{Y_n} \\ \overline{\tau_{xz}} \cos(n,x) + \overline{\tau_{yz}} \cos(n,y) &= \overline{Z_n}^{*} \end{aligned} \tag{13}$$

Eqs (1) to (13) are valid for any plate regardless of its thickness or the orientation of the anisotropy.

Let us consider the special case when the following conditions are satisfied

$$\begin{aligned} u(x,y,+h) - u(x,y,-h) &= 0 \\ v(x,y,+h) - v(x,y,-h) &= 0 \\ w(x,y,+h) - w(x,y,-h) &= 0 \end{aligned} \tag{14}$$

---

* In the present analysis, we also assume that the surface force component Z_n varies in a parabolic manner across the plate since it has to satisfy the boundary condition at $z = \pm h$ and has to be symmetrically distributed with respect to the middle plane $z = 0$.

Then from eqs (4.3)

$$\varepsilon_x\big|_{+h} = \varepsilon_x\big|_{-h} \; ; \quad \varepsilon_y\big|_{+h} = \varepsilon_y\big|_{-h} \; ; \quad \gamma_{xy}\big|_{+h} = \gamma_{xy}\big|_{-h} \tag{15}$$

If we substitute eqs (14) into eqs (9) we obtain the following strain displacement relations

$$\bar{\varepsilon}_x = -\frac{\partial \bar{u}}{\partial x} \; ; \quad \bar{\varepsilon}_y = -\frac{\partial \bar{v}}{\partial y} \; ; \quad \bar{\varepsilon}_z = 0$$

$$\bar{\gamma}_{yz} = -\frac{\partial \bar{w}}{\partial y} \; ; \quad \bar{\gamma}_{xz} = -\frac{\partial \bar{w}}{\partial x} \; ; \quad \bar{\gamma}_{xy} = -\left(\frac{\partial \bar{u}}{\partial y} + \frac{\partial \bar{v}}{\partial x}\right) \tag{16}$$

Similarly, if we substitute eqs (15) into the fourth and fifth of eqs (11), we obtain:

$$\frac{\partial}{\partial x}\left(\frac{\partial \bar{\gamma}_{xz}}{\partial y} - \frac{\partial \bar{\gamma}_{yz}}{\partial x}\right) = 0 \; ; \quad \frac{\partial}{\partial y}\left(\frac{\partial \bar{\gamma}_{xz}}{\partial y} - \frac{\partial \bar{\gamma}_{yz}}{\partial x}\right) = 0 \tag{17}$$

From the expressions for $\bar{\gamma}_{xz}$ and $\bar{\gamma}_{yz}$ in eqs (16), eqs (17) integrate to only one equation

$$\frac{\partial \bar{\gamma}_{xz}}{\partial y} - \frac{\partial \bar{\gamma}_{yz}}{\partial x} = 0 \tag{18}$$

Since $\bar{\varepsilon}_z = 0$, the second, third and sixth of eqs (11) reduce to

$$\left[\frac{\partial \gamma_{yz}}{\partial y} - \frac{\partial \varepsilon_y}{\partial z}\right]_{-h}^{+h} = 0 \; ; \quad \left[\frac{\partial \gamma_{xz}}{\partial x} - \frac{\partial \varepsilon_x}{\partial z}\right]_{-h}^{+h} = 0$$

$$\left[\frac{\partial \gamma_{yz}}{\partial x} + \frac{\partial \gamma_{xz}}{\partial y} - \frac{\partial \gamma_{xy}}{\partial z}\right]_{-h}^{+h} = 0 \tag{19}$$

From the strain displacement relations (eqs 4.3) and the third of eqs (14), eqs (19) are <u>automatically</u> satisfied if

$$\left.\frac{\partial u}{\partial z}\right|_{+h} = \left.\frac{\partial u}{\partial z}\right|_{-h} \quad ; \quad \left.\frac{\partial v}{\partial z}\right|_{+h} = \left.\frac{\partial v}{\partial z}\right|_{-h} \tag{20}$$

Therefore, if both conditions (14) and (20) are satisfied, the six compatibility equations for strains reduce to two i.e.

$$\frac{\partial^2 \overline{\varepsilon_x}}{\partial y^2} + \frac{\partial^2 \overline{\varepsilon_y}}{\partial x^2} = \frac{\partial^2 \overline{\gamma_{xy}}}{\partial x \partial y} \quad ; \quad \frac{\partial \overline{\gamma_{xz}}}{\partial y} - \frac{\partial \overline{\gamma_{yz}}}{\partial x} = 0 \tag{21}$$

In Table 1, we compare the conditions to be satisfied for the present formulation with those to be satisfied for the case of generalized plane strain (Section 4.3.2). It can be seen that if conditions (14) and (20) are satisfied, then, equations of equilibrium, equations of compatibility for strains, strain displacement relations, constitutive relations and boundary conditions are in all respect identical between the two formulations. Stress, strain and displacement components in the generalized plane strain formulation can be replaced by their average quantities over the thickness $2h$ in the present formulation.

Furthermore, conditions (14) and (20) can be obtained by integration over the plate thickness of the following equations

$$\frac{\partial u}{\partial z} = \frac{\partial v}{\partial z} = \frac{\partial w}{\partial z} = 0 \tag{22}$$

and

$$\frac{\partial}{\partial z}\left(\frac{\partial u}{\partial z}\right) = \frac{\partial}{\partial z}\left(\frac{\partial v}{\partial z}\right) = 0 \tag{23}$$

since u, v, w are by hypothesis independent of z in the plane strain formulation.

Table 1

Generalized Plane Strain Formulation	Present Formulation (conditions (14) and (20) being satisfied)
$\cdot\ \frac{\partial \sigma_x}{\partial x} + \frac{\partial \tau_{xy}}{\partial y} - \frac{\partial \bar{U}}{\partial x} = 0$ $\frac{\partial \tau_{xy}}{\partial x} + \frac{\partial \sigma_y}{\partial y} - \frac{\partial \bar{U}}{\partial y} = 0$ $\frac{\partial \tau_{xz}}{\partial x} + \frac{\partial \tau_{yz}}{\partial y} = 0$	$\cdot\ \frac{\partial \bar{\sigma}_x}{\partial x} + \frac{\partial \bar{\tau}_{xy}}{\partial y} - \frac{\partial \bar{U}}{\partial x} = 0$ $\frac{\partial \bar{\tau}_{xy}}{\partial x} + \frac{\partial \bar{\sigma}_y}{\partial y} - \frac{\partial \bar{U}}{\partial y} = 0$ $\frac{\partial \bar{\tau}_{xz}}{\partial x} + \frac{\partial \bar{\tau}_{yz}}{\partial y} = 0$
$\cdot\ \varepsilon_x = -\frac{\partial u}{\partial x} \quad \varepsilon_y = -\frac{\partial v}{\partial y} \quad \varepsilon_z = -\frac{\partial w}{\partial z} = 0$ $\gamma_{yz} = -\frac{\partial w}{\partial y} \quad \gamma_{xz} = -\frac{\partial w}{\partial x} \quad \gamma_{xy} = -\left(\frac{\partial u}{\partial y} + \frac{\partial v}{\partial x}\right)$	$\cdot\ \bar{\varepsilon}_x = -\frac{\partial \bar{u}}{\partial x} \quad \bar{\varepsilon}_y = -\frac{\partial \bar{v}}{\partial y} \quad \bar{\varepsilon}_z = 0$ $\bar{\gamma}_{yz} = -\frac{\partial \bar{w}}{\partial y} \quad \bar{\gamma}_{xz} = -\frac{\partial \bar{w}}{\partial x} \quad \bar{\gamma}_{xy} = -\left(\frac{\partial \bar{u}}{\partial y} + \frac{\partial \bar{v}}{\partial x}\right)$
$\cdot\ \frac{\partial u}{\partial z} = \frac{\partial v}{\partial z} = \frac{\partial w}{\partial z} = 0$	$\cdot\ u(x,y,+h) = u(x,y,-h)$ $v(x,y,+h) = v(x,y,-h)$ $w(x,y,+h) = w(x,y,-h)$
$\cdot\ \frac{\partial^2 \varepsilon_x}{\partial y^2} + \frac{\partial^2 \varepsilon_y}{\partial x^2} = \frac{\partial^2 \gamma_{xy}}{\partial x \partial y}$ $\frac{\partial \gamma_{xz}}{\partial y} - \frac{\partial \gamma_{yz}}{\partial x} = 0$	$\cdot\ \frac{\partial^2 \bar{\varepsilon}_x}{\partial y^2} + \frac{\partial^2 \bar{\varepsilon}_y}{\partial x^2} = \frac{\partial^2 \bar{\gamma}_{xy}}{\partial x \partial y}$ $\frac{\partial \bar{\gamma}_{yz}}{\partial x} - \frac{\partial \bar{\gamma}_{xz}}{\partial y} = 0$
$\cdot\ (\varepsilon)_{xyz} = (A)(\sigma)_{xyz}$	$\cdot\ (\bar{\varepsilon})_{xyz} = (A)(\bar{\sigma})_{xyz}$

- $\sigma_x \cos(n,x) + \tau_{xy} \cos(n,y) = X_n$
  $\tau_{xy} \cos(n,x) + \sigma_y \cos(n,y) = Y_n$
  $\tau_{xz} \cos(n,x) + \tau_{yz} \cos(n,y) = Z_n$

- $\sigma_x = \dfrac{\partial^2 F}{\partial y^2} + \overline{U}$
  $\sigma_y = \dfrac{\partial^2 F}{\partial x^2} + \overline{U}$
  $\tau_{xy} = -\dfrac{\partial^2 F}{\partial x \partial y}$
  $\tau_{xz} = \dfrac{\partial \psi}{\partial y} \qquad \tau_{yz} = -\dfrac{\partial \psi}{\partial x}$

- $\dfrac{\partial}{\partial z}\left(\dfrac{\partial u}{\partial z}\right) = \dfrac{\partial}{\partial z}\left(\dfrac{\partial v}{\partial z}\right) = 0$

- $\overline{\sigma_x} \cos(n,x) + \overline{\tau_{xy}} \cos(n,y) = \overline{X_n}$
  $\overline{\tau_{xy}} \cos(n,x) + \overline{\sigma_y} \cos(n,y) = \overline{Y_n}$
  $\overline{\tau_{xz}} \cos(n,x) + \overline{\tau_{yz}} \cos(n,y) = \overline{Z_n}$

- $\overline{\sigma_x} = \dfrac{\partial^2 F}{\partial y^2} + \overline{U}$
  $\overline{\sigma_y} = \dfrac{\partial^2 F}{\partial x^2} + \overline{U}$
  $\overline{\tau_{xy}} = -\dfrac{\partial^2 F}{\partial x \partial y}$
  $\overline{\tau_{xz}} = \dfrac{\partial \psi}{\partial y} \; ; \; \overline{\tau_{yz}} = -\dfrac{\partial \psi}{\partial x}$

- $\left.\dfrac{\partial u}{\partial z}\right|_{+h} = \left.\dfrac{\partial u}{\partial z}\right|_{-h} \; ; \; \left.\dfrac{\partial v}{\partial z}\right|_{+h} = \left.\dfrac{\partial v}{\partial z}\right|_{-h}$

Appendix 4.3

Recall the general expression for the analytic functions $\phi_k(z_k)$ (eq 4.76) for $k=1,2,3$. Along the boundary of the circular hole $\zeta_k = e^{i\theta}$. Then,

$$\phi_k(z_k) = A_k\, i\theta + \sum_{m=1}^{\infty} A_{km}\, e^{-im\theta} \tag{1}$$

Replacing eqs (1) into the boundary conditions (eqs 4.72), we obtain for the second one, for instance

$$2\,\mathrm{Re}\left((A_1 + A_2 + \lambda_3 A_3)\, i\theta + \sum_{m=1}^{\infty} (A_{1m} + A_{2m} + \lambda_3 A_{3m})\, e^{-im\theta}\right) = \frac{P_y\theta}{2\pi} + 2\,\mathrm{Re}\left(\sum_{m=1}^{\infty} \overline{a}_m e^{-im\theta}\right) \tag{2}$$

where the following identity has been used

$$a_m e^{im\theta} + \overline{a}_m e^{-im\theta} = 2\,\mathrm{Re}\left(\overline{a}_m e^{-im\theta}\right) \tag{3}$$

Since,

$$2\,\mathrm{Re}\left((A_1 + A_2 + \lambda_3 A_3)\, i\theta\right) = \left((A_1 - \overline{A}_1) + (A_2 - \overline{A}_2) + (\lambda_3 A_3 - \overline{\lambda}_3 \overline{A}_3)\right) i\theta \tag{4}$$

then, equating in eq (2) the coefficinets of θ and $e^{-im\theta}$, it follows

$$A_1 + A_2 + \lambda_3 A_3 - \overline{A}_1 - \overline{A}_2 - \overline{\lambda}_3 \overline{A}_3 = \frac{-iP_y}{2\pi} \tag{5}$$

and for each value of m

$$A_{1m} + A_{2m} + \lambda_3 A_{3m} = \overline{a}_m \tag{6}$$

Similar equations can be obtained from the first and third of eqs (4.72).

A_{1m}, A_{2m}, A_{3m} satisfy the following system

$$\begin{aligned} A_{1m} + A_{2m} + \lambda_3 A_{3m} &= \bar{a}_m \\ \mu_1 A_{1m} + \mu_2 A_{2m} + \lambda_3 \mu_3 A_{3m} &= \bar{b}_m \\ \lambda_1 A_{1m} + \lambda_2 A_{2m} + A_{3m} &= \bar{c}_m \end{aligned} \tag{7}$$

whose solution is

$$\begin{aligned} A_{1m} &= \left(\bar{a}_m (\mu_2 - \lambda_2\lambda_3\mu_3) + \bar{b}_m (\lambda_2\lambda_3 - 1) + \bar{c}_m \lambda_3 (\mu_3 - \mu_2)\right)/\Delta \\ A_{2m} &= \left(\bar{a}_m (\lambda_1\lambda_3\mu_3 - \mu_1) + \bar{b}_m (1 - \lambda_1\lambda_3) + \bar{c}_m \lambda_3 (\mu_1 - \mu_3)\right)/\Delta \\ A_{3m} &= \left(\bar{a}_m (\mu_1\lambda_2 - \mu_2\lambda_1) + \bar{b}_m (\lambda_1 - \lambda_2) + \bar{c}_m (\mu_2 - \mu_1)\right)/\Delta \end{aligned} \tag{8}$$

with

$$\Delta = \mu_2 - \mu_1 + \lambda_2\lambda_3 (\mu_1 - \mu_3) + \lambda_1\lambda_3 (\mu_3 - \mu_2) \tag{9}$$

Similarly, A_1, A_2, A_3 and their conjuguates $\bar{A}_1, \bar{A}_2, \bar{A}_3$ must satisfy the following equations

$$\begin{aligned} A_1 + A_2 + \lambda_3 A_3 - \bar{A}_1 - \bar{A}_2 - \bar{\lambda}_3 \bar{A}_3 &= P_y / 2\pi i \\ \mu_1 A_1 + \mu_2 A_2 + \lambda_3 \mu_3 A_3 - \bar{\mu}_1 \bar{A}_1 - \bar{\mu}_2 \bar{A}_2 - \bar{\mu}_3 \bar{\lambda}_3 \bar{A}_3 &= - P_x / 2\pi i \\ \lambda_1 A_1 + \lambda_2 A_2 + A_3 - \bar{\lambda}_1 \bar{A}_1 - \bar{\lambda}_2 \bar{A}_2 - \bar{A}_3 &= - P_z / 2\pi i \end{aligned} \tag{10}$$

plus three equations resulting from the conditions of single-valuedness of the displacement components

$$\begin{aligned} p_1 A_1 + p_2 A_2 + p_3 A_3 - \bar{p}_1 \bar{A}_1 - \bar{p}_2 \bar{A}_2 - \bar{p}_3 \bar{A}_3 &= 0 \\ q_1 A_1 + q_2 A_2 + q_3 A_3 - \bar{q}_1 \bar{A}_1 - \bar{q}_2 \bar{A}_2 - \bar{q}_3 \bar{A}_3 &= 0 \\ r_1 A_1 + r_2 A_2 + r_3 A_3 - \bar{r}_1 \bar{A}_1 - \bar{r}_2 \bar{A}_2 - \bar{r}_3 \bar{A}_3 &= 0 \end{aligned} \tag{11}$$

As a result, we obtain the general expressions for the functions ϕ_1, ϕ_2, ϕ_3 as follows

$$\phi_1 = A_1 \ln \xi_1 + \frac{1}{\Delta}\sum_{m=1}^{\infty}\left\{\left(\bar{a}_m(\mu_2-\lambda_2\lambda_3\mu_3) + \bar{b}_m(\lambda_2\lambda_3-1) + \bar{c}_m\lambda_3(\mu_3-\mu_2)\right)\frac{1}{\xi_1^m}\right\}$$

$$\phi_2 = A_2 \ln \xi_2 + \frac{1}{\Delta}\sum_{m=1}^{\infty}\left\{\left(\bar{a}_m(\lambda_1\lambda_3\mu_3-\mu_1) + \bar{b}_m(1-\lambda_1\lambda_3) + \bar{c}_m\lambda_3(\mu_1-\mu_3)\right)\frac{1}{\xi_2^m}\right\}$$

$$\phi_3 = A_3 \ln \xi_3 + \frac{1}{\Delta}\sum_{m=1}^{\infty}\left\{\left(\bar{a}_m(\mu_1\lambda_2-\mu_2\lambda_1) + \bar{b}_m(\lambda_1-\lambda_2) + \bar{c}_m(\mu_2-\mu_1)\right)\frac{1}{\xi_3^m}\right\} \tag{12}$$

Recall eq (4.75b)

$$\xi_k = \frac{\frac{z_k}{a} + \sqrt{\left(\frac{z_k}{a}\right)^2 - 1 - \mu_k^2}}{1 - i\mu_k} \tag{13}$$

If $\xi_k'(z_k)$ denotes the derivative of $\xi_k(z_k)$ with respect to the variable z_k then, it can be shown that

$$\xi_k'(z_k) = \frac{\xi_k(z_k)}{a\sqrt{\left(\frac{z_k}{a}\right)^2 - 1 - \mu_k^2}} \tag{14}$$

Therefore,

$$(\xi_k^{-m})' = -m(\xi_k)^{-m-1}\xi_k' = \frac{-m(\xi_k)^{-m}}{a\sqrt{\left(\frac{z_k}{a}\right)^2 - 1 - \mu_k^2}} \tag{15a}$$

and

$$(\ln \xi_k)' = \frac{\xi_k'}{\xi_k} = \frac{1}{a\sqrt{\left(\frac{z_k}{a}\right)^2 - 1 - \mu_k^2}} \tag{15b}$$

The derivatives of the functions $\phi_k(z_k)$ with respect to the variable z_k are determined by the formulas

$$\phi_1' = \frac{1}{a\sqrt{\left(\frac{z_1}{a}\right)^2 - 1 - \mu_1^2}}\left\{A_1 - \frac{1}{\Delta}\sum_{m=1}^{\infty} m\left\{\bar{a}_m(\mu_2-\lambda_2\lambda_3\mu_3) + \bar{b}_m(\lambda_2\lambda_3-1) + \bar{c}_m\lambda_3(\mu_3-\mu_2)\right\}\frac{1}{\xi_1^m}\right\}$$

$$\phi_2' = \frac{1}{a\sqrt{\left(\frac{z_2}{a}\right)^2 - 1 - \mu_2^2}} \left\{ A_2 - \frac{1}{\Delta}\sum_{m=1}^{\infty} m \left\{ \bar{a}_m (\lambda_1\lambda_3\mu_3 - \mu_1) + \bar{b}_m (1 - \lambda_1\lambda_3) + \bar{c}_m \lambda_3 (\mu_1 - \mu_3) \right\} \frac{1}{\xi_2^m} \right\}$$

$$\phi_3' = \frac{1}{a\sqrt{\left(\frac{z_3}{a}\right)^2 - 1 - \mu_3^2}} \left\{ A_3 - \frac{1}{\Delta}\sum_{m=1}^{\infty} m \left\{ \bar{a}_m (\mu_1\lambda_2 - \mu_2\lambda_1) + \bar{b}_m (\lambda_1 - \lambda_2) + \bar{c}_m (\mu_2 - \mu_1) \right\} \frac{1}{\xi_3^m} \right\}$$

(16)

According to eqs (10) and (11), A_1, A_2, A_3 are equal to zero if the resultant forces P_x, P_y, P_z vanish.

Appendix 4.4

If we replace in eqs (1) (Appendix 4.1) x',y',z', respectively by r, θ, z then, for $\beta = 0$ and $\delta = \theta$.

$$\begin{array}{lll} l_{x'} = \cos\theta & m_{x'} = \sin\theta & n_{x'} = 0 \\ l_{y'} = -\sin\theta & m_{y'} = \cos\theta & n_{y'} = 0 \\ l_{z'} = 0 & m_{z'} = 0 & n_{z'} = 1 \end{array} \quad (1)$$

If we set in eqs (5), (7) (Appendix 4.1) x,y,z instead of X,Y,Z we obtain

$$\begin{bmatrix} \sigma_r \\ \sigma_\theta \\ \sigma_z \\ \tau_{\theta z} \\ \tau_{rz} \\ \tau_{r\theta} \end{bmatrix} = \begin{bmatrix} \cos^2\theta & \sin^2\theta & 0 & 0 & 0 & \sin 2\theta \\ \sin^2\theta & \cos^2\theta & 0 & 0 & 0 & -\sin 2\theta \\ 0 & 0 & 1 & 0 & 0 & 0 \\ 0 & 0 & 0 & \cos\theta & -\sin\theta & 0 \\ 0 & 0 & 0 & \sin\theta & \cos\theta & 0 \\ \frac{-\sin 2\theta}{2} & \frac{\sin 2\theta}{2} & 0 & 0 & 0 & \cos 2\theta \end{bmatrix} \begin{bmatrix} \sigma_x \\ \sigma_y \\ \sigma_z \\ \tau_{yz} \\ \tau_{xz} \\ \tau_{xy} \end{bmatrix} \quad (2)$$

$$\begin{bmatrix} \varepsilon_r \\ \varepsilon_\theta \\ \varepsilon_z \\ \gamma_{\theta z} \\ \gamma_{rz} \\ \gamma_{r\theta} \end{bmatrix} = \begin{bmatrix} \cos^2\theta & \sin^2\theta & 0 & 0 & 0 & \sin 2\theta/2 \\ \sin^2\theta & \cos^2\theta & 0 & 0 & 0 & -\sin 2\theta/2 \\ 0 & 0 & 1 & 0 & 0 & 0 \\ 0 & 0 & 0 & \cos\theta & -\sin\theta & 0 \\ 0 & 0 & 0 & \sin\theta & \cos\theta & 0 \\ -\sin 2\theta & \sin 2\theta & 0 & 0 & 0 & \cos 2\theta \end{bmatrix} \begin{bmatrix} \varepsilon_x \\ \varepsilon_y \\ \varepsilon_z \\ \gamma_{yz} \\ \gamma_{xz} \\ \gamma_{xy} \end{bmatrix} \quad (3)$$

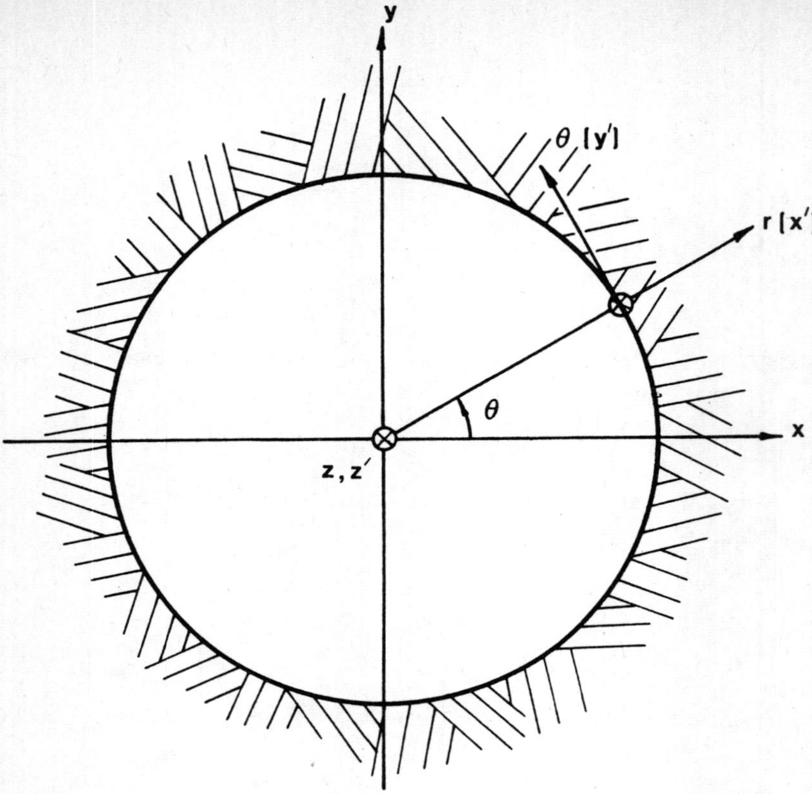

Figure 1. Definition of the coordinate system r, θ, z.

or in matrix form

$$(\sigma)_{r\theta z} = (T^*_\sigma)(\sigma)_{xyz} \tag{4}$$

and

$$(\varepsilon)_{r\theta z} = (T^*_\varepsilon)(\varepsilon)_{xyz} \tag{5}$$

Appendix 4.5

1) Let us consider the elastic equilibrium of an anisotropic homogeneous body bounded internally by an infinite cylinder with a circular cross section of radius a (Figure 1). The problem is to determine the distributions of stresses, strains and displacements around the hole drilled in the body initially subjected to a 3D stress field. The final stress distribution around the hole is the tensor sum of three stress distributions:

(i) the field stress tensor before drilling the hole with components $\sigma_{x,o}$; $\sigma_{y,o}$; $\sigma_{z,o}$; $\tau_{yz,o}$; $\tau_{xz,o}$; $\tau_{xy,o}$ in the (x,y,z) coordinate system. Let define (σ_o) as follows

$$(\sigma_o)^t = \begin{pmatrix} \sigma_{x,o} & \sigma_{y,o} & \sigma_{z,o} & \tau_{yz,o} & \tau_{xz,o} & \tau_{xy,o} \end{pmatrix} \tag{1}$$

(ii) the stress tensor induced by drilling the hole; and

(iii) the stress tensor induced by any boundary stress acting along the walls of the hole with components X_n, Y_n, Z_n. For the present closed form solution, the influence of an internal pressure q distributed uniformly along the hole is studied only.

Similarly, the strain and displacement distributions around the hole are the sum of those before drilling the hole, those induced by drilling the hole and those induced by the internal pressure q .

Problems (i), (ii), (iii) correspond respectively to

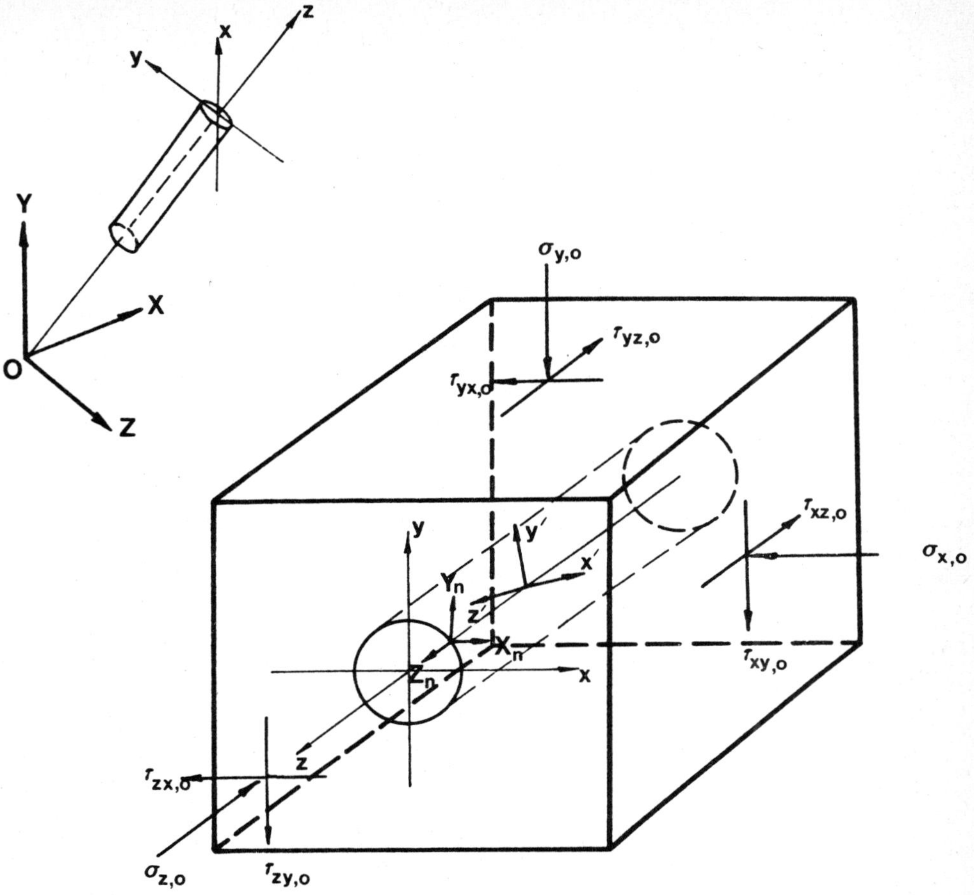

Figure 1. Schematic view of a long circular hole to be excavated in a Triaxial stress field.

problems (ii)a, (ii)b, and (i) of section 4.6.3. A generalized plane strain formulation is used to solve problems (ii) and (iii).

If the constitutive relation of the anisotropic material is known in an x',y',z' coordinate system, the general expression of the constitutive relation in the x,y,z coordinate system (see Appendix 4.1) can be written as follows

$$\begin{bmatrix} \varepsilon_x \\ \varepsilon_y \\ \varepsilon_z \\ \gamma_{yz} \\ \gamma_{xz} \\ \gamma_{xy} \end{bmatrix} = \begin{bmatrix} & & \\ & a_{ij} & \\ & i = 1,6 & \\ & j = 1,6 & \\ & & \end{bmatrix} \begin{bmatrix} \sigma_x \\ \sigma_y \\ \sigma_z \\ \tau_{yz} \\ \tau_{xz} \\ \tau_{xy} \end{bmatrix} \qquad (2a)$$

or in matrix form $$(\varepsilon)_{xyz} = (A)(\sigma)_{xyz} \qquad (2b)$$

With reference to eqs (4.31), (4.53) and (4.54) for the generalized plane formulation, the total stress components around the hole can be expressed as follows*

$$\begin{aligned}
\sigma_x &= \sigma_{x,0} + \sigma_{xh} = \sigma_{x,0} + 2\,\mathrm{Re}\,(\mu_1^2\phi_1' + \mu_2^2\phi_2' + \lambda_3\mu_3^2\phi_3') \\
\sigma_y &= \sigma_{y,0} + \sigma_{yh} = \sigma_{y,0} + 2\,\mathrm{Re}\,(\phi_1' + \phi_2' + \lambda_3\phi_3') \\
\tau_{xy} &= \tau_{xy,0} + \tau_{xy\,h} = \tau_{xy,0} - 2\,\mathrm{Re}\,(\mu_1\phi_1' + \mu_2\phi_2' + \lambda_3\mu_3\phi_3') \\
\tau_{xz} &= \tau_{xz,0} + \tau_{xz\,h} = \tau_{xz,0} + 2\,\mathrm{Re}\,(\lambda_1\mu_1\phi_1' + \lambda_2\mu_2\phi_2' + \mu_3\phi_3') \\
\tau_{yz} &= \tau_{yz,0} + \tau_{yz_h} = \tau_{yz,0} - 2\,\mathrm{Re}\,(\lambda_1\phi_1' + \lambda_2\phi_2' + \phi_3') \\
\sigma_z &= \sigma_{z,0} - \frac{1}{a_{33}}(a_{31}\sigma_{x_h} + a_{32}\sigma_{y_h} + a_{34}\tau_{yz_h} + a_{35}\tau_{xz_h} + a_{36}\tau_{xy_h})
\end{aligned} \tag{3}$$

In eq (3), the expression for σ_z makes use of the definition of the generalized plane strain formulation ($\varepsilon_{z_h}=0$).

Similarly, the total displacement components u, v, w respectively in the x,y,z directions can be expressed as

$$\begin{aligned}
u &= u_0 - 2\,\mathrm{Re}\,(p_1\phi_1 + p_2\phi_2 + p_3\phi_3) - \omega y + u_t^{(0)} \\
v &= v_0 - 2\,\mathrm{Re}\,(q_1\phi_1 + q_2\phi_2 + q_3\phi_3) + \omega x + v_t^{(0)} \\
w &= w_0 - 2\,\mathrm{Re}\,(r_1\phi_1 + r_2\phi_2 + r_3\phi_3) + w_t^{(0)}
\end{aligned} \tag{4}$$

where

u_0, v_0, w_0 are the initial displacement components before drilling the hole, $u_t^{(0)}$, $v_t^{(0)}$, $w_t^{(0)}$ and ω are rigid body translations in the x,y,z directions and a rigid body rotation around the z axis that are not accompanied by deformation .

For the influence of an internal pressure q and a three dimensional stress field applied at infinity, the coefficients

---

*$(\sigma_{x_h}, \sigma_{y_h}, \sigma_{z_h}, \tau_{yz_h}, \tau_{xz_h}, \tau_{xy_h})$ and (u_h, v_h, w_h) denote respectively the stress and displacement components induced by drilling the hole in the initial stress field and by the application of an internal pressure q.

of eqs (4.73) reduce to

$$\bar{a}_1 = -\frac{a}{2}\left(\sigma_{y,0} - i\tau_{xy,0}\right) + \frac{qa}{2} \quad ; \quad \bar{b}_1 = \frac{a}{2}\left(\tau_{xy,0} - i\sigma_{x,0}\right) + \frac{iqa}{2}$$

$$\bar{c}_1 = \frac{a}{2}\left(\tau_{yz,0} - i\tau_{xz,0}\right) \tag{5}$$

and all the other coefficients a_m, b_m, c_m $(m \geq 2)$ are zero.

The functions ϕ_i, ϕ_i' $(i=1,2,3)$ of eqs (12) and (16) in Appendix 4.3 reduce to the following

$$\begin{aligned}
\phi_1(z_1) &= \left((\mu_2-\mu_3\lambda_2\lambda_3)\bar{a}_1 + (\lambda_2\lambda_3-1)\bar{b}_1 + \lambda_3(\mu_3-\mu_2)\bar{c}_1\right) / \Delta\xi_1\\
\phi_2(z_2) &= \left((\lambda_1\lambda_3\mu_3-\mu_1)\bar{a}_1 + (1-\lambda_1\lambda_3)\bar{b}_1 + \lambda_3(\mu_1-\mu_3)\bar{c}_1\right) / \Delta\xi_2\\
\phi_3(z_3) &= \left((\mu_1\lambda_2-\mu_2\lambda_1)\bar{a}_1 + (\lambda_1-\lambda_2)\bar{b}_1 + (\mu_2-\mu_1)\bar{c}_1\right) / \Delta\xi_3
\end{aligned} \tag{6}$$

and

$$\begin{aligned}
\phi_1'(z_1) &= \frac{-1}{a\Delta\sqrt{\left(\frac{z_1}{a}\right)^2-1-\mu_1^2}}\left((\mu_2-\mu_3\lambda_2\lambda_3)\bar{a}_1 + (\lambda_2\lambda_3-1)\bar{b}_1 + \lambda_3(\mu_3-\mu_2)\bar{c}_1\right) / \xi_1\\
\phi_2'(z_2) &= \frac{-1}{a\Delta\sqrt{\left(\frac{z_2}{a}\right)^2-1-\mu_2^2}}\left((\lambda_1\lambda_3\mu_3-\mu_1)\bar{a}_1 + (1-\lambda_1\lambda_3)\bar{b}_1 + \lambda_3(\mu_1-\mu_3)\bar{c}_1\right) / \xi_2\\
\phi_3'(z_3) &= \frac{-1}{a\Delta\sqrt{\left(\frac{z_3}{a}\right)^2-1-\mu_3^2}}\left((\mu_1\lambda_2-\mu_2\lambda_1)\bar{a}_1 + (\lambda_1-\lambda_2)\bar{b}_1 + (\mu_2-\mu_1)\bar{c}_1\right) / \xi_3
\end{aligned} \tag{7}$$

In the following derivations we shall denote

$$\gamma_k = -\frac{a}{a\Delta\xi_k\sqrt{\left(\frac{z_k}{a}\right)^2-1-\mu_k^2}} \qquad (k=1,2,3) \tag{8}$$

At first, the following derivations will be carried out assuming zero internal pressure q within the hole. It will be shown later how to incorporate it

2) Combining eqs (5), (7) and (8) we obtain

$$\phi_1'(z_1) = \gamma_1\Big(-\frac{1}{2}(\mu_2-\mu_3\lambda_2\lambda_3)(\sigma_{y,0}-i\tau_{xy,0}) + \frac{1}{2}(\lambda_2\lambda_3-1)(\tau_{xy,0}-i\sigma_{x,0})$$

$$+\frac{1}{2}\lambda_3(\mu_3-\mu_2)(\tau_{yz,0}-i\tau_{xz,0})\Big)$$

$$\phi_2' = \frac{\gamma_2}{2}\Big(-(\lambda_1\lambda_3\mu_3-\mu_1)(\sigma_{y,0}-i\,\tau_{xy,0}) + (1-\lambda_1\lambda_3)(\tau_{xy,0}-i\,\sigma_{x,0})$$
$$+ \lambda_3(\mu_1-\mu_3)(\tau_{yz,0}-i\,\tau_{xz,0})\Big)$$

$$\phi_3' = \frac{\gamma_3}{2}\Big(-(\mu_1\lambda_2-\mu_2\lambda_1)(\sigma_{y,0}-i\,\tau_{xy,0}) + (\lambda_1-\lambda_2)(\tau_{xy,0}-i\,\sigma_{x,0})$$
$$+ (\mu_2-\mu_1)(\tau_{yz,0}-i\,\tau_{xz,0})\Big)$$

If we substitute the expression of $\phi_1', \phi_2', \phi_3'$ in eqs (3), the stress components take the form

$$\begin{aligned}
\sigma_x &= \sigma_{x,0} + b_1\sigma_{x,0} + c_1\sigma_{y,0} + d_1\tau_{xy,0} + e_1\tau_{yz,0} + f_1\tau_{xz,0}\\
\sigma_y &= \sigma_{y,0} + b_2\sigma_{x,0} + c_2\sigma_{y,0} + d_2\tau_{xy,0} + e_2\tau_{yz,0} + f_2\tau_{xz,0}\\
\tau_{xy} &= \tau_{xy,0} + b_3\sigma_{x,0} + c_3\sigma_{y,0} + d_3\tau_{xy,0} + e_3\tau_{yz,0} + f_3\tau_{xz,0}\\
\tau_{xz} &= \tau_{xz,0} + b_4\sigma_{x,0} + c_4\sigma_{y,0} + d_4\tau_{xy,0} + e_4\tau_{yz,0} + f_4\tau_{xz,0}\\
\tau_{yz} &= \tau_{yz,0} + b_5\sigma_{x,0} + c_5\sigma_{y,0} + d_5\tau_{xy,0} + e_5\tau_{yz,0} + f_5\tau_{xz,0}
\end{aligned} \qquad (9)$$

and the induced stress in the z direction is equal to

$$\sigma_{z_h} = -\frac{1}{a_{33}}\left(a_{31}\sigma_{x_h} + a_{32}\sigma_{y_h} + a_{34}\tau_{yz_h} + a_{35}\tau_{xz_h} + a_{36}\tau_{xy_h}\right) \qquad (10)$$

Coefficients $b_i, c_i, \ldots, f_i$ $(i=1,5)$ have the following form

$$\begin{aligned}
b_1 &= -\mathrm{Re}\left(i\gamma_1\mu_1^2(\lambda_2\lambda_3-1) + i\gamma_2\mu_2^2(1-\lambda_1\lambda_3) + \gamma_3 i\mu_3^2\lambda_3(\lambda_1-\lambda_2)\right)\\
c_1 &= -\mathrm{Re}\left(\gamma_1\mu_1^2(\mu_2-\mu_3\lambda_2\lambda_3) + \gamma_2\mu_2^2(\lambda_1\lambda_3\mu_3-\mu_1) + \gamma_3\mu_3^2\lambda_3(\mu_1\lambda_2-\mu_2\lambda_1)\right)\\
d_1 &= \mathrm{Re}\left(\gamma_1\mu_1^2(\lambda_2\lambda_3-1+i\mu_2-i\mu_3\lambda_2\lambda_3) + \gamma_2\mu_2^2(1-\lambda_1\lambda_3+i\mu_3\lambda_1\lambda_3-i\mu_1)\right.\\
&\quad \left. + \gamma_3\mu_3^2\lambda_3(\lambda_1-\lambda_2+i\mu_1\lambda_2-i\mu_2\lambda_1)\right)\\
e_1 &= \mathrm{Re}\left(\mu_1^2\gamma_1\lambda_3(\mu_3-\mu_2) + \mu_2^2\gamma_2\lambda_3(\mu_1-\mu_3) + \mu_3^2\lambda_3\gamma_3(\mu_2-\mu_1)\right)\\
f_1 &= -\mathrm{Re}\left(i\gamma_1\mu_1^2\lambda_3(\mu_3-\mu_2) + i\gamma_2\mu_2^2\lambda_3(\mu_1-\mu_3) + i\gamma_3\lambda_3\mu_3^2(\mu_2-\mu_1)\right)
\end{aligned}$$

$$\begin{aligned}
b_2 &= -\mathrm{Re}\left(i\gamma_1(\lambda_2\lambda_3-1) + i\gamma_2(1-\lambda_1\lambda_3) + i\gamma_3\lambda_3(\lambda_1-\lambda_2)\right)\\
c_2 &= -\mathrm{Re}\left(\gamma_1(\mu_2-\mu_3\lambda_2\lambda_3) + \gamma_2(\lambda_1\lambda_3\mu_3-\mu_1) + \gamma_3\lambda_3(\mu_1\lambda_2-\mu_2\lambda_1)\right)\\
d_2 &= \mathrm{Re}\left(\gamma_1(\lambda_2\lambda_3-1+i\mu_2-i\mu_3\lambda_2\lambda_3) + \gamma_2(1-\lambda_1\lambda_3+i\lambda_1\lambda_3\mu_3-i\mu_1)\right.\\
&\quad \left. + \gamma_3\lambda_3(\lambda_1-\lambda_2+i\mu_1\lambda_2-i\mu_2\lambda_1)\right)\\
e_2 &= \mathrm{Re}\left(\gamma_1\lambda_3(\mu_3-\mu_2) + \gamma_2\lambda_3(\mu_1-\mu_3) + \gamma_3\lambda_3(\mu_2-\mu_1)\right)\\
f_2 &= -\mathrm{Re}\left(i\gamma_1\lambda_3(\mu_3-\mu_2) + i\gamma_2\lambda_3(\mu_1-\mu_3) + i\gamma_3\lambda_3(\mu_2-\mu_1)\right)
\end{aligned}$$

$$\begin{aligned}
b_3 &= \mathrm{Re}\left(i\gamma_1\mu_1(\lambda_2\lambda_3-1) + i\gamma_2\mu_2(1-\lambda_1\lambda_3) + i\gamma_3\mu_3\lambda_3(\lambda_1-\lambda_2)\right)\\
c_3 &= \mathrm{Re}\left(\gamma_1\mu_1(\mu_2-\mu_3\lambda_2\lambda_3) + \gamma_2\mu_2(\lambda_1\lambda_3\mu_3-\mu_1) + \gamma_3\mu_3\lambda_3(\mu_1\lambda_2-\mu_2\lambda_1)\right)\\
d_3 &= -\mathrm{Re}\left(\gamma_1\mu_1(\lambda_2\lambda_3-1+i\mu_2-i\mu_3\lambda_2\lambda_3) + \gamma_2\mu_2(1-\lambda_1\lambda_3+i\lambda_1\lambda_3\mu_3-i\mu_1)\right.\\
&\quad \left. + \gamma_3\mu_3\lambda_3(\lambda_1-\lambda_2+i\mu_1\lambda_2-i\mu_2\lambda_1)\right)\\
e_3 &= -\mathrm{Re}\left(\mu_1\gamma_1\lambda_3(\mu_3-\mu_2) + \mu_2\gamma_2\lambda_3(\mu_1-\mu_3) + \mu_3\gamma_3\lambda_3(\mu_2-\mu_1)\right)\\
f_3 &= \mathrm{Re}\left(i\mu_1\gamma_1\lambda_3(\mu_3-\mu_2) + i\gamma_2\mu_2\lambda_3(\mu_1-\mu_3) + i\gamma_3\mu_3\lambda_3(\mu_2-\mu_1)\right)
\end{aligned}$$

$$\begin{aligned}
b_4 &= -\mathrm{Re}\left(i\gamma_1\lambda_1\mu_1(\lambda_2\lambda_3-1) + i\gamma_2\lambda_2\mu_2(1-\lambda_1\lambda_3) + i\gamma_3\mu_3(\lambda_1-\lambda_2)\right)\\
c_4 &= -\mathrm{Re}\left(\gamma_1\lambda_1\mu_1(\mu_2-\mu_3\lambda_2\lambda_3) + \gamma_2\lambda_2\mu_2(\lambda_1\lambda_3\mu_3-\mu_1) + \gamma_3\mu_3(\mu_1\lambda_2-\mu_2\lambda_1)\right)\\
d_4 &= \mathrm{Re}\left(\gamma_1\mu_1\lambda_1(\lambda_2\lambda_3-1+i\mu_2-i\mu_3\lambda_2\lambda_3) + \gamma_2\mu_2\lambda_2(1-\lambda_1\lambda_3+i\lambda_1\lambda_3\mu_3\right.\\
&\quad \left. -i\mu_1) + \gamma_3\mu_3(\lambda_1-\lambda_2+i\mu_1\lambda_2-i\mu_2\lambda_1)\right)\\
e_4 &= \mathrm{Re}\left(\gamma_1\mu_1\lambda_1\lambda_3(\mu_3-\mu_2) + \gamma_2\mu_2\lambda_2\lambda_3(\mu_1-\mu_3) + \mu_3\gamma_3(\mu_2-\mu_1)\right)\\
f_4 &= -\mathrm{Re}\left(i\gamma_1\mu_1\lambda_1\lambda_3(\mu_3-\mu_2) + i\gamma_2\mu_2\lambda_2\lambda_3(\mu_1-\mu_3) + i\gamma_3\mu_3(\mu_2-\mu_1)\right)
\end{aligned}$$

$$\begin{aligned}
b_5 &= \mathrm{Re}\left(i\gamma_1\lambda_1(\lambda_2\lambda_3-1) + i\gamma_2\lambda_2(1-\lambda_1\lambda_3) + i\gamma_3(\lambda_1-\lambda_2)\right)\\
c_5 &= \mathrm{Re}\left(\gamma_1\lambda_1(\mu_2-\mu_3\lambda_2\lambda_3) + \gamma_2\lambda_2(\lambda_1\lambda_3\mu_3-\mu_1) + \gamma_3(\mu_1\lambda_2-\mu_2\lambda_1)\right)\\
d_5 &= -\mathrm{Re}\left(\gamma_1\lambda_1(\lambda_2\lambda_3-1+i\mu_2-i\mu_3\lambda_2\lambda_3) + \gamma_2\lambda_2(1-\lambda_1\lambda_3+i\lambda_1\lambda_3\mu_3-i\mu_1)\right.\\
&\quad \left. + \gamma_3(\lambda_1-\lambda_2+i\mu_1\lambda_2-i\mu_2\lambda_1)\right)\\
e_5 &= -\mathrm{Re}\left(\lambda_1\gamma_1\lambda_3(\mu_3-\mu_2) + \lambda_2\gamma_2\lambda_3(\mu_1-\mu_3) + \gamma_3(\mu_2-\mu_1)\right)\\
f_5 &= \mathrm{Re}\left(i\gamma_1\lambda_1\lambda_3(\mu_3-\mu_2) + i\gamma_2\lambda_2\lambda_3(\mu_1-\mu_3) + i\gamma_3(\mu_2-\mu_1)\right)
\end{aligned}$$

Combining eq (10) and the contribution of the induced stress components in eqs (9), we obtain

$$\begin{aligned}
-\sigma_{zh}\, a_{33} = {} & a_{31}\left(b_1 \sigma_{x,0} + c_1 \sigma_{y,0} + d_1 \tau_{xy,0} + e_1 \tau_{yz,0} + f_1 \tau_{xz,0} \right) \\
& + a_{32}\left(b_2 \sigma_{x,0} + c_2 \sigma_{y,0} + d_2 \tau_{xy,0} + e_2 \tau_{yz,0} + f_2 \tau_{xz,0} \right) \\
& + a_{34}\left(b_5 \sigma_{x,0} + c_5 \sigma_{y,0} + d_5 \tau_{xy,0} + e_5 \tau_{yz,0} + f_5 \tau_{xz,0} \right) \\
& + a_{35}\left(b_4 \sigma_{x,0} + c_4 \sigma_{y,0} + d_4 \tau_{xy,0} + e_4 \tau_{yz,0} + f_4 \tau_{xz,0} \right) \\
& + a_{36}\left(b_3 \sigma_{x,0} + c_3 \sigma_{y,0} + d_3 \tau_{xy,0} + e_3 \tau_{yz,0} + f_3 \tau_{xz,0} \right)
\end{aligned} \tag{11}$$

or

$$\begin{aligned}
\sigma_{zh} = {} & -\sigma_{x,0}\left(a_{31} b_1 + a_{32} b_2 + a_{34} b_5 + a_{35} b_4 + a_{36} b_3 \right)/a_{33} \\
& -\sigma_{y,0}\left(a_{31} c_1 + a_{32} c_2 + a_{34} c_5 + a_{35} c_4 + a_{36} c_3 \right)/a_{33} \\
& -\tau_{xy,0}\left(a_{31} d_1 + a_{32} d_2 + a_{34} d_5 + a_{35} d_4 + a_{36} d_3 \right)/a_{33} \\
& -\tau_{yz,0}\left(a_{31} e_1 + a_{32} e_2 + a_{34} e_5 + a_{35} e_4 + a_{36} e_3 \right)/a_{33} \\
& -\tau_{xz,0}\left(a_{31} f_1 + a_{32} f_2 + a_{34} f_5 + a_{35} f_4 + a_{36} f_3 \right)/a_{33}
\end{aligned} \tag{12}$$

Thus, the total stress component in the z direction is equal to

$$\sigma_z = \sigma_{z,0} + \sigma_{zh} \tag{13}$$

If we rewrite eqs (9) and (10) in matrix form we obtain for the induced stress components

$$\begin{bmatrix} \sigma_{xh} \\ \sigma_{yh} \\ \sigma_{zh} \\ \tau_{yzh} \\ \tau_{xzh} \\ \tau_{xyh} \end{bmatrix} = \begin{bmatrix} f_{11}^{(h)} & f_{12}^{(h)} & 0 & f_{14}^{(h)} & f_{15}^{(h)} & f_{16}^{(h)} \\ f_{21}^{(h)} & f_{22}^{(h)} & 0 & f_{24}^{(h)} & f_{25}^{(h)} & f_{26}^{(h)} \\ f_{31}^{(h)} & f_{32}^{(h)} & 0 & f_{34}^{(h)} & f_{35}^{(h)} & f_{36}^{(h)} \\ f_{41}^{(h)} & f_{42}^{(h)} & 0 & f_{44}^{(h)} & f_{45}^{(h)} & f_{46}^{(h)} \\ f_{51}^{(h)} & f_{52}^{(h)} & 0 & f_{54}^{(h)} & f_{55}^{(h)} & f_{56}^{(h)} \\ f_{61}^{(h)} & f_{62}^{(h)} & 0 & f_{64}^{(h)} & f_{65}^{(h)} & f_{66}^{(h)} \end{bmatrix} \begin{bmatrix} \sigma_{x,0} \\ \sigma_{y,0} \\ \sigma_{z,0} \\ \tau_{yz,0} \\ \tau_{xz,0} \\ \tau_{xy,0} \end{bmatrix} \tag{14}$$

or

$$(\sigma)^{(h)}_{xyz} = \left(F_\sigma^{(h)}\right)(\sigma_0) \tag{15}$$

where

$$\begin{array}{lll}
f^{(h)}_{11} = b_1 & f^{(h)}_{21} = b_2 & a_{33} f^{(h)}_{31} = -(a_{31} b_1 + a_{32} b_2 + a_{34} b_5 + a_{35} b_4 + a_{36} b_3) \\
f^{(h)}_{12} = c_1 & f_{22} = c_2 & a_{33} f^{(h)}_{32} = -(a_{31} c_1 + a_{32} c_2 + a_{34} c_5 + a_{35} c_4 + a_{36} c_3) \\
f^{(h)}_{14} = e_1 & f^{(h)}_{24} = e_2 & a_{33} f^{(h)}_{34} = -(a_{31} e_1 + a_{32} e_2 + a_{34} e_5 + a_{35} e_4 + a_{36} e_3) \\
f^{(h)}_{15} = f_1 & f^{(h)}_{25} = f_2 & a_{33} f^{(h)}_{35} = -(a_{31} f_1 + a_{32} f_2 + a_{34} f_5 + a_{35} f_4 + a_{36} f_3) \\
f^{(h)}_{16} = d_1 & f^{(h)}_{26} = d_2 & a_{33} f^{(h)}_{36} = -(a_{31} d_1 + a_{32} d_2 + a_{34} d_5 + a_{35} d_4 + a_{36} d_3)
\end{array}$$

$$\begin{array}{lll}
f^{(h)}_{41} = b_5 & f^{(h)}_{51} = b_4 & f^{(h)}_{61} = b_3 \\
f^{(h)}_{42} = c_5 & f^{(h)}_{52} = c_4 & f^{(h)}_{62} = c_3 \\
f^{(h)}_{44} = e_5 & f^{(h)}_{54} = e_4 & f^{(h)}_{64} = e_3 \\
f^{(h)}_{45} = f_5 & f^{(h)}_{55} = f_4 & f^{(h)}_{65} = f_3 \\
f^{(h)}_{46} = d_5 & f^{(h)}_{56} = d_4 & f^{(h)}_{66} = d_3
\end{array}$$

If we add to eq (15) the components of initial stress field, the relationship between the total components of stress and the initial stress field can be written as follows

$$(\sigma)_{xyz} = (F_\sigma)(\sigma_0) \tag{16}$$

where

$$(F_\sigma) = \left(F_\sigma^{(h)}\right) + (I) \tag{17}$$

and (I) is the identity matrix

Combining eq (15) or (16) with eq (2b) we can then relate the induced and total strain components to the initial stress field as such

$$(\varepsilon)^{(h)}_{xyz} = (A)(F^{(h)}_{\sigma})(\sigma_o) \tag{18a}$$

and

$$(\varepsilon)_{xyz} = (A)(F_{\sigma})(\sigma_o) \tag{18b}$$

In terms of cylindrical coordinates (r, θ, z), eqs (18a) and (18b) become

$$(\varepsilon)^{(h)}_{r\theta z} = (T^*_{\varepsilon})(A)(F^{(h)}_{\sigma})(\sigma_o) \tag{19a}$$

and

$$(\varepsilon)_{r\theta z} = (T^*_{\varepsilon})(A)(F_{\sigma})(\sigma_o) \tag{19b}$$

where (T^*_{ε}) is the transformation matrix defined in Appendix 4.4.

3) The components of the induced displacements are equal to*

$$\begin{aligned} u_h &= -2\mathrm{Re}\,(p_1\phi_1 + p_2\phi_2 + p_3\phi_3) \\ v_h &= -2\mathrm{Re}\,(q_1\phi_1 + q_2\phi_2 + q_3\phi_3) \\ w_h &= -2\mathrm{Re}\,(r_1\phi_1 + r_2\phi_2 + r_3\phi_3) \end{aligned} \tag{20}$$

Let denote

$$\Delta_k = \frac{1}{\Delta \xi_k} \qquad (k = 1, 2, 3) \tag{21a}$$

Then, combining eqs (5), (6) and (21a) we obtain

---

* if we exclude temporarily rigid body rotation and translations

$$\emptyset_1 = \frac{\Delta_1 a}{2}\Big(-i\sigma_{x,0}\left(\lambda_2\lambda_3-1\right)-\sigma_{y,0}\left(\mu_2-\mu_3\lambda_2\lambda_3\right)+\tau_{xy,0}\left(\lambda_2\lambda_3-1+i\left(\mu_2-\mu_3\lambda_2\lambda_3\right)\right)$$
$$+\tau_{yz,0}\lambda_3\left(\mu_3-\mu_2\right)-i\lambda_3\tau_{xz,0}\left(\mu_3-\mu_2\right)\Big)$$

$$\emptyset_2 = \frac{\Delta_2 a}{2}\Big(-i\left(1-\lambda_1\lambda_3\right)\sigma_{x,0}-\left(\lambda_1\lambda_3\mu_3-\mu_1\right)\sigma_{y,0}+\tau_{xy,0}\left(1-\lambda_1\lambda_3+i\left(\lambda_1\lambda_3\mu_3-\mu_1\right)\right)$$
$$+\tau_{yz,0}\lambda_3\left(\mu_1-\mu_3\right)-i\lambda_3\tau_{xz,0}\left(\mu_1-\mu_3\right)\Big)$$

$$\emptyset_3 = \frac{\Delta_3 a}{2}\Big(-i\left(\lambda_1-\lambda_2\right)\sigma_{x,0}-\left(\mu_1\lambda_2-\mu_2\lambda_1\right)\sigma_{y,0}+\tau_{xy,0}\left(\lambda_1-\lambda_2+i\left(\mu_1\lambda_2-\mu_2\lambda_1\right)\right)$$
$$+\tau_{yz,0}\left(\mu_2-\mu_1\right)-i\tau_{xz,0}\left(\mu_2-\mu_1\right)\Big)$$

$$-u_h = 2\mathrm{Re}\left(p_1\emptyset_1\right)+2\mathrm{Re}\left(p_2\emptyset_2\right)+2\mathrm{Re}\left(p_3\emptyset_3\right)$$

or

$$\Rightarrow -u_h/a = -\sigma_{x,0}\,\mathrm{Re}\left(i\Delta_1 p_1\left(\lambda_2\lambda_3-1\right)+i\Delta_2 p_2\left(1-\lambda_1\lambda_3\right)+i\Delta_3 p_3\left(\lambda_1-\lambda_2\right)\right)$$
$$-\sigma_{y,0}\,\mathrm{Re}\left(\Delta_1 p_1\left(\mu_2-\mu_3\lambda_2\lambda_3\right)+\Delta_2 p_2\left(\lambda_1\lambda_3\mu_3-\mu_1\right)+\Delta_3 p_3\left(\mu_1\lambda_2-\mu_2\lambda_1\right)\right)$$
$$+\tau_{xy,0}\,\mathrm{Re}\Big(\Delta_1 p_1\left(\lambda_2\lambda_3-1+i\left(\mu_2-\mu_3\lambda_2\lambda_3\right)\right)+\Delta_2 p_2\big(1-\lambda_1\lambda_3+$$
$$i\left(\lambda_1\lambda_3\mu_3-\mu_1\right)\big)+\Delta_3 p_3\left(\lambda_1-\lambda_2+i\left(\mu_1\lambda_2-\mu_2\lambda_1\right)\right)\Big)$$
$$+\tau_{yz,0}\,\mathrm{Re}\left(\Delta_1 p_1\lambda_3\left(\mu_3-\mu_2\right)+\Delta_2 p_2\lambda_3\left(\mu_1-\mu_3\right)+\Delta_3 p_3\left(\mu_2-\mu_1\right)\right)$$
$$-\tau_{xz,0}\,\mathrm{Re}\left(i\Delta_1 p_1\lambda_3\left(\mu_3-\mu_2\right)+i\Delta_2 p_2\lambda_3\left(\mu_1-\mu_3\right)+i\Delta_3 p_3\left(\mu_2-\mu_1\right)\right)$$

(21b)

Similar expressions can be obtained for the displacement components v_h and w_h. The coefficients p_i must be replaced respectively by q_i and r_i $(i=1,2,3)$. Those expressions can be summarized as follows

$$u_h/a = b_1'\sigma_{x,0}+c_1'\sigma_{y,0}+f_1'\tau_{xy,0}+d_1'\tau_{yz,0}+e_1'\tau_{xz,0}$$

$$v_h/a = b_2'\sigma_{x,0}+c_2'\sigma_{y,0}+f_2'\tau_{xy,0}+d_2'\tau_{yz,0}+e_2'\tau_{xz,0}$$

$$w_h/a = b_3'\sigma_{x,0}+c_3'\sigma_{y,0}+f_3'\tau_{xy,0}+d_3'\tau_{yz,0}+e_3'\tau_{xz,0} \qquad (22)$$

where

$$b'_1 = Re\left(i\Delta_1 p_1 (\lambda_2\lambda_3 - 1) + i\Delta_2 p_2 (1-\lambda_1\lambda_3) + i\Delta_3 p_3 (\lambda_1 - \lambda_2)\right)$$

$$c'_1 = Re\left(\Delta_1 p_1 (\mu_2 - \mu_3\lambda_2\lambda_3) + \Delta_2 p_2 (\lambda_1\lambda_3\mu_3 - \mu_1) + \Delta_3 p_3 (\mu_1\lambda_2 - \mu_2\lambda_1)\right)$$

$$d'_1 = -Re\left(\Delta_1 p_1 \lambda_3 (\mu_3 - \mu_2) + \Delta_2 p_2 \lambda_3 (\mu_1 - \mu_3) + \Delta_3 p_3 (\mu_2 - \mu_1)\right)$$

$$e'_1 = Re\left(i\Delta_1 p_1 \lambda_3 (\mu_3 - \mu_2) + i\Delta_2 p_2 \lambda_3 (\mu_1 - \mu_3) + i\Delta_3 p_3 (\mu_2 - \mu_1)\right)$$

$$f'_1 = -Re\left(\Delta_1 p_1 (\lambda_2\lambda_3 - 1 + i(\mu_2 - \mu_3\lambda_2\lambda_3)) + \Delta_2 p_2 (1 - \lambda_1\lambda_3 + i (\lambda_1\lambda_3\mu_3 - \mu_1)) + \Delta_3 p_3 (\lambda_1 - \lambda_2 + i(\mu_1\lambda_2 - \mu_2\lambda_1))\right)$$

(22')

$(b'_2, c'_2, d'_2, e'_2, f'_2)$ and $(b'_3, c'_3, d'_3, e'_3, f'_3)$ are similar to $(b'_1, c'_1, d'_1, e'_1, f'_1)$ but the coefficients p_i are replaced respectively by q_i and r_i $(i = 1,2,3)$.

Let define the radial and tangential displacements as shown in Figure 2, such that

$$\begin{aligned} u_r &= u \cos\theta + v \sin\theta \\ v_\theta &= v \cos\theta - u \sin\theta \end{aligned}$$

(23)

Then, combining eqs (22) and (23), the relationship between the displacement components in the r, θ, z coordinate system and the initial stress field takes the following form

$$\begin{bmatrix} \dfrac{u_{rh}}{a} \\ \dfrac{v_{\theta h}}{a} \\ \dfrac{w_h}{a} \end{bmatrix} = \begin{bmatrix} \cos\theta & \sin\theta & 0 \\ -\sin\theta & \cos\theta & 0 \\ 0 & 0 & 1 \end{bmatrix} \begin{bmatrix} b'_1 & c'_1 & 0 & d'_1 & e'_1 & f'_1 \\ b'_2 & c'_2 & 0 & d'_2 & e'_2 & f'_2 \\ b'_3 & c'_3 & 0 & d'_3 & e'_3 & f'_3 \end{bmatrix} \begin{bmatrix} \sigma_{x,0} \\ \sigma_{y,0} \\ \sigma_{z,0} \\ \tau_{yz,0} \\ \tau_{xz,0} \\ \tau_{xy,0} \end{bmatrix}$$

(24)

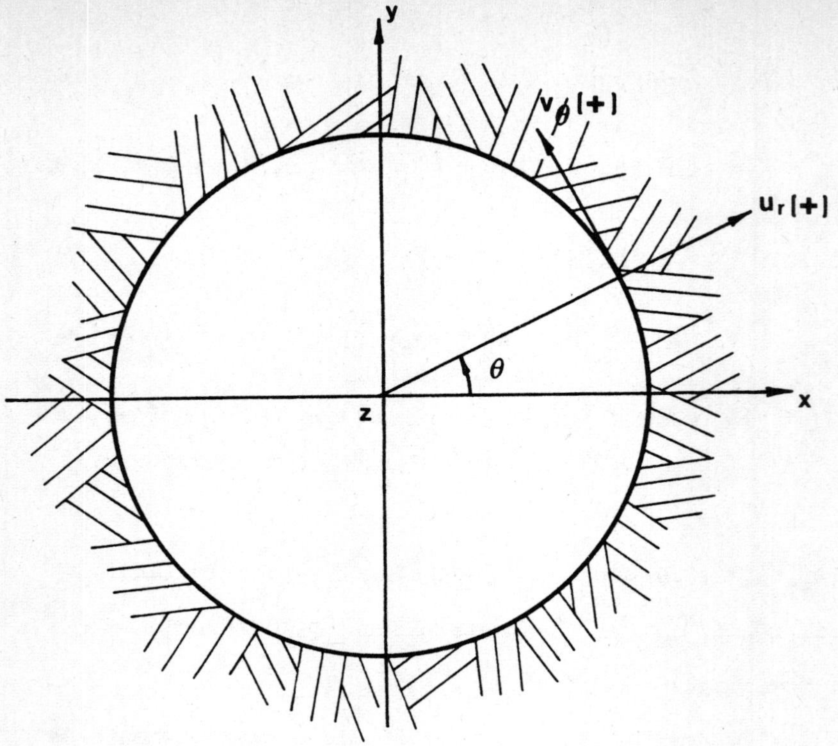

Figure 2. Definition of the radial and tangential displacement components.

If we add to eqs (24) the contribution of the rigid body translations and rotation, we can write

$$\begin{bmatrix} \frac{u_{rh}}{a} \\ \frac{v_{\theta h}}{a} \\ \frac{w_h}{a} \end{bmatrix} = \left(F_u^{(h)}\right) \begin{bmatrix} \sigma_{x,0} \\ \sigma_{y,0} \\ \sigma_{z,0} \\ \tau_{yz,0} \\ \tau_{xz,0} \\ \tau_{xy,0} \end{bmatrix} + \begin{bmatrix} \frac{u_t^{(0)}}{a}\cos\theta + \frac{v_t^{(0)}}{a}\sin\theta \\ \frac{v_t^{(0)}}{a}\cos\theta - \frac{u_t^{(0)}}{a}\sin\theta + \omega\frac{r}{a} \\ \frac{w_t^{(0)}}{a} \end{bmatrix} \tag{25}$$

where $\left(F_u^{(h)}\right)$ is matrix product ot the (3×3) and (3×6) matrices in eqs (24).

4) Let us calculate the <u>initial displacement components</u> u_0, v_0, w_0.

Due to the initial stress field there are initial displacements at each point x,y,z of the anisotropic material. Those components can be calculated by integration of the initial strain components.

According to eqs (2b), the initial strain components in the x,y,z coordinate system, are related to the initial stress field as follows

$$(\varepsilon_0)_{xyz} = (A)(\sigma_0) \tag{26}$$

or in cylindrical coordinates

$$(\varepsilon_0)_{r\theta z} = (T_\varepsilon^*)(A)(\sigma_0) \tag{27}$$

In cylindrical coordinates, initial strain and displacement components are also related through the following equations

$$\varepsilon_{r_0} = -\frac{\partial u_{r_0}}{\partial r} \qquad \varepsilon_{\theta_0} = -\frac{u_{r_0}}{r} - \frac{1}{r}\frac{\partial v_{\theta_0}}{\partial \theta} \qquad \gamma_{r\theta_0} = -\frac{1}{r}\frac{\partial u_{r_0}}{\partial \theta} - \frac{\partial v_{\theta_0}}{\partial r} + \frac{v_{\theta_0}}{r}$$

$$\gamma_{\theta z_0} = -\cancelto{0}{\frac{\partial v_{\theta_0}}{\partial z}} - \frac{1}{r}\frac{\partial w_0}{\partial \theta} \qquad \gamma_{rz_0} = -\cancelto{0}{\frac{\partial u_{r_0}}{\partial z}} - \frac{\partial w_0}{\partial r} \qquad \varepsilon_{z_0} = -\frac{\partial w_0}{\partial z} \tag{28}$$

where we assumed that u_{r_0}, v_{θ_0} are independent of z and functions of r, θ only. w_0 is assumed to be a function of r, θ and z.

Integrating the first and second of eqs (28) we obtain

$$u_{r_0} = -r\,\varepsilon_{r_0} - g_1(\theta) \tag{29}$$

and

$$v_{\theta_0} = -r\left(\frac{\sin 2\theta}{2}\left(\varepsilon_{y_0} - \varepsilon_{x_0}\right) + \frac{\cos 2\theta}{2}\gamma_{xy_0}\right) + \int_0^{\theta} g_1(\theta)\,d\theta - g_2(r)$$

g_1 and g_2 are arbitrary functions of θ and r respectively and appear as a result of integration of eqs (28).

Substituting eqs (29) into the third of eqs (28) and expressing $\gamma_{r\theta_0}$ in terms of $\varepsilon_{x,0}$, $\varepsilon_{y,0}$ and $\gamma_{xy,0}$, it can be shown that g_1 and g_2 must satisfy the following equation

$$\left(\frac{dg_1(\theta)}{d\theta} + \int_0^{\theta} g_1(\theta)\,d\theta\right) + \left(r\frac{dg_2(r)}{dr} - g_2(r)\right) = 0 \tag{30}$$

As $g_1(\theta)$ is a function of θ only and $g_2(r)$ is a function of r only, each part of eq (30) can be set equal to an arbitrary constant k. If we set $k = 0$ and integrate both parts of eq (30) we obtain

$$g_1(\theta) = -u^{(o)}\cos\theta - v^{(o)}\sin\theta$$

and

$$g_2(r) = -\omega_{xy}^{(o)}\,r \tag{31}$$

where $u_1^{(o)}, v_1^{(o)}, \omega_{xy}^{(o)}$ are constants of integration. Substitute eqs (31) into eqs (29), the radial and tangential displacements take the form

$$u_{r_o} = -r\,\varepsilon_{r_o} + u^{(o)}\cos\theta + v^{(o)}\sin\theta$$
$$v_{\theta_o} = -r\left(\frac{\sin 2\theta}{2}\left(\varepsilon_{y_o} - \varepsilon_{x_o}\right) + \frac{\cos 2\theta}{2}\,\gamma_{xy_o}\right) + v^{(o)}\cos\theta - u^{(o)}\sin\theta + \omega_{xy}^{(o)}\,r \tag{32a}$$

Integrating the three last of eqs (28), the displacement component in the z direction is then equal to

$$w_o = -z\,\varepsilon_{z_o} - r\left(\gamma_{xz_o}\cos\theta + \gamma_{yz_o}\sin\theta\right) + w^{(o)} \tag{32b}$$

$u^{(o)}, v^{(o)}, w^{(o)}$ are rigid body translations and $\omega_{xy}^{(o)}$ is a rigid body rotation associated with the initial displacement components.

Combining eqs (26), (32a) and (32b) we can write at any point (r, θ, z) the initial displacement components as follows

$$-\frac{u_{r_o}}{r} = \sigma_{x,o}\left(a_{11}\cos^2\theta + a_{21}\sin^2\theta + a_{61}\frac{\sin 2\theta}{2}\right) + \sigma_{y,o}\left(a_{12}\cos^2\theta + a_{22}\sin^2\theta + a_{62}\frac{\sin 2\theta}{2}\right) + \tau_{xy,o}\left(a_{16}\cos^2\theta + a_{26}\sin^2\theta + a_{66}\frac{\sin 2\theta}{2}\right) + \sigma_{z,o}\left(a_{13}\cos^2\theta + a_{23}\sin^2\theta + a_{63}\frac{\sin 2\theta}{2}\right) + \tau_{yz,o}\left(a_{14}\cos^2\theta + a_{24}\sin^2\theta + a_{64}\frac{\sin 2\theta}{2}\right) + \tau_{xz,o}\left(a_{15}\cos^2\theta + a_{25}\sin^2\theta + a_{65}\frac{\sin 2\theta}{2}\right) - \frac{u^{(o)}}{r}\cos\theta - \frac{v^{(o)}}{r}\sin\theta$$

$$
\begin{aligned}
-\frac{V_{\theta_0}}{r} = {} & \sigma_{x,0}\left(\frac{\sin 2\theta}{2}\left(a_{21}-a_{11}\right)+a_{61}\frac{\cos 2\theta}{2}\right)+\sigma_{y,0}\left(\frac{\sin 2\theta}{2}\left(a_{22}-a_{12}\right)\right. \\
& \left.+a_{62}\frac{\cos 2\theta}{2}\right)+\tau_{xy,0}\left(\frac{\sin 2\theta}{2}\left(a_{26}-a_{16}\right)+a_{66}\frac{\cos 2\theta}{2}\right) \\
& +\sigma_{z,0}\left(\frac{\sin 2\theta}{2}\left(a_{23}-a_{13}\right)+a_{63}\frac{\cos 2\theta}{2}\right) \\
& +\tau_{yz,0}\left(\frac{\sin 2\theta}{2}\left(a_{24}-a_{14}\right)+a_{64}\frac{\cos 2\theta}{2}\right) \\
& +\tau_{xz,0}\left(\frac{\sin 2\theta}{2}\left(a_{25}-a_{15}\right)+a_{65}\frac{\cos 2\theta}{2}\right) \\
& -\frac{v^{(0)}}{r}\cos\theta+\frac{u^{(0)}}{r}\sin\theta-\omega_{xy}^{(0)}\,.
\end{aligned}
$$

$$
\begin{aligned}
-\frac{W_0}{r} = {} & \sigma_{x,0}\left(a_{41}\sin\theta+a_{51}\cos\theta+\frac{z}{r}a_{31}\right)+ \\
& \sigma_{y,0}\left(a_{42}\sin\theta+a_{52}\cos\theta+\frac{z}{r}a_{32}\right)+ \\
& \tau_{xy_0}\left(a_{46}\sin\theta+a_{56}\cos\theta+\frac{z}{r}a_{36}\right)+ \\
& \sigma_{z,0}\left(a_{43}\sin\theta+a_{53}\cos\theta+\frac{z}{r}a_{33}\right)+ \\
& \tau_{yz,0}\left(a_{44}\sin\theta+a_{45}\cos\theta+\frac{z}{r}a_{34}\right)+ \\
& \tau_{xz,0}\left(a_{45}\sin\theta+a_{55}\cos\theta+\frac{z}{r}a_{35}\right) \\
& -\frac{w^{(0)}}{r}
\end{aligned}
\qquad (33)
$$

Since $\cos^2\theta = (1+\cos 2\theta)/2$ and $\sin^2\theta = (1-\cos 2\theta)/2$, one can rewrite $-\frac{u_{ro}}{r}$ in eqs (33) in terms of $\cos 2\theta$ and $\sin 2\theta$ as follows

$$
\begin{aligned}
-\frac{u_{ro}}{r} = {} & \sigma_{x,o}\left(\left(\frac{a_{11}+a_{21}}{2}\right) + \frac{\cos 2\theta}{2}(a_{11}-a_{21}) + a_{61}\frac{\sin 2\theta}{2}\right) \\
& + \sigma_{y,o}\left(\left(\frac{a_{12}+a_{22}}{2}\right) + \frac{\cos 2\theta}{2}(a_{12}-a_{22}) + a_{62}\frac{\sin 2\theta}{2}\right) \\
& + \tau_{xy,o}\left(\left(\frac{a_{16}+a_{26}}{2}\right) + \frac{\cos 2\theta}{2}(a_{16}-a_{26}) + a_{66}\frac{\sin 2\theta}{2}\right) \\
& + \sigma_{z,o}\left(\left(\frac{a_{13}+a_{23}}{2}\right) + \frac{\cos 2\theta}{2}(a_{13}-a_{23}) + a_{63}\frac{\sin 2\theta}{2}\right) \\
& + \tau_{yz,o}\left(\left(\frac{a_{14}+a_{24}}{2}\right) + \frac{\cos 2\theta}{2}(a_{14}-a_{24}) + a_{64}\frac{\sin 2\theta}{2}\right) \\
& + \tau_{xz,o}\left(\left(\frac{a_{15}+a_{25}}{2}\right) + \frac{\cos 2\theta}{2}(a_{15}-a_{25}) + a_{65}\frac{\sin 2\theta}{2}\right) \\
& - \frac{u^{(o)}}{r}\cos\theta - \frac{v^{(o)}}{r}\sin\theta .
\end{aligned}
\tag{34}
$$

In matrix form, eqs (33) take the following form

$$
\begin{bmatrix} \frac{u_{ro}}{a} \\ \frac{v_{\theta o}}{a} \\ \frac{w_o}{a} \end{bmatrix}
= \left(F_u^{(o)}\right)
\begin{bmatrix} \sigma_{x,o} \\ \sigma_{y,o} \\ \sigma_{z,o} \\ \tau_{yz,o} \\ \tau_{xz,o} \\ \tau_{xy,o} \end{bmatrix}
+
\begin{bmatrix} \frac{u^{(o)}}{a}\cos\theta + \frac{v^{(o)}}{a}\sin\theta \\ \frac{v^{(o)}}{a}\cos\theta - \frac{u^{(o)}}{a}\sin\theta + \omega_{xy}^{(o)}\frac{r}{a} \\ \frac{w^{(o)}}{a} \end{bmatrix}
\tag{35a}
$$

where

$$
\begin{aligned}
f_{1i_u}^{(o)} &= -\left(\frac{r}{a}\right)\left(\left(\frac{a_{1i}+a_{2i}}{2}\right) + \frac{\cos 2\theta}{2}(a_{1i}-a_{2i}) + \frac{\sin 2\theta}{2}a_{6i}\right) \\
f_{2i_u}^{(o)} &= -\left(\frac{r}{a}\right)\left(\frac{\sin 2\theta}{2}(a_{2i}-a_{1i}) + \frac{\cos 2\theta}{2}a_{6i}\right) \\
f_{3i_u}^{(o)} &= -\frac{z}{a}a_{3i} - \frac{r}{a}(a_{4i}\sin\theta + a_{5i}\cos\theta) \qquad (i = 1,6)
\end{aligned}
\tag{35b}
$$

5) Adding eqs (25) and (35a) we obtain the total displacement components u_r, v_θ and w such that

$$\begin{bmatrix} \frac{u_r}{a} \\ \frac{v_\theta}{a} \\ \frac{w}{a} \end{bmatrix} = (F_u) \begin{bmatrix} \sigma_{x,0} \\ \sigma_{y,0} \\ \sigma_{z,0} \\ \tau_{yz,0} \\ \tau_{xz,0} \\ \tau_{xy,0} \end{bmatrix} + \begin{bmatrix} \left(\frac{u_t^{(0)} + u^{(0)}}{a}\right)\cos\theta + \left(\frac{v_t^{(0)} + v^{(0)}}{a}\right)\sin\theta \\ \left(\frac{v_t^{(0)} + v^{(0)}}{a}\right)\cos\theta - \left(\frac{u_t^{(0)} + u^{(0)}}{a}\right)\sin\theta + (\omega + \omega_{xy}^{(0)})\frac{r}{a} \\ \left(\frac{w_t^{(0)} + w^{(0)}}{a}\right) \end{bmatrix} \tag{35c}$$

with

$$(F_u) = (F_u^{(h)}) + (F_u^{(0)})$$

In the rest of this analysis we assume zero rigid body rotation and translations. Therefore, the second term on the right hand side of eqs (35c) vanishes.

6) Among the three displacement components, the radial component u_r is by far the most important when dealing with stress measurements (Chapter 6). Let us calculate the general expression of u_r/a along the contour of the circular hole.

- Replacing $r = a$ into eq (34) we obtain

$$\frac{u_{r0}}{a} = M_{u_1}^{(0)} + M_{u_2}^{(0)}\cos 2\theta + M_{u_3}^{(0)}\sin 2\theta \tag{36}$$

where

$$M_{u_1}^{(0)} = -\frac{1}{2}\Big(\sigma_{x,0}(a_{11}+a_{21}) + \sigma_{y,0}(a_{12}+a_{22}) + \sigma_{z,0}(a_{13}+a_{23}) + \tau_{yz,0}(a_{14}+a_{24}) + \tau_{xz,0}(a_{15}+a_{25}) + \tau_{xy,0}(a_{16}+a_{26})\Big)$$

$$M_{u_2}^{(0)} = -\frac{1}{2}\Big(\sigma_{x,0}(a_{11}-a_{21}) + \sigma_{y,0}(a_{12}-a_{22}) + \sigma_{z,0}(a_{13}-a_{23}) + \tau_{yz,0}(a_{14}-a_{24}) + \tau_{xz,0}(a_{15}-a_{25}) + \tau_{xy,0}(a_{16}-a_{26})\Big)$$

$$M_{u_3}^{(0)} = -\frac{1}{2}\left(a_{61}\sigma_{x,0} + a_{62}\sigma_{y,0} + a_{63}\sigma_{z,0} + a_{64}\tau_{yz,0} + a_{65}\tau_{xz,0} + a_{66}\tau_{xy,0}\right)$$

- For any point along the contour of the hole ($r = a$), the coefficients $\mathcal{E}_k$ appearing in eq (21a) are equal to $e^{i\theta} = \cos\theta + i\sin\theta$. Therefore,

$$\Delta_1 = \Delta_2 = \Delta_3 = e^{-i\theta}/\Delta = (\cos\theta - i\sin\theta)/\Delta \tag{37}$$

Substituting eq (37) into the expressions of u_h and v_h (eq 21b) yields

$$\begin{aligned}
-\frac{u_h}{a} = & -\sigma_{x,0}\cos\theta\,\mathrm{Re}\left(\frac{ip_1}{\Delta}(\lambda_2\lambda_3-1)+\frac{ip_2}{\Delta}(1-\lambda_1\lambda_3)+\frac{ip_3}{\Delta}(\lambda_1-\lambda_2)\right)\\
& +\sigma_{x,0}\sin\theta\,\mathrm{Re}\left(\frac{i^2p_1}{\Delta}(\lambda_2\lambda_3-1)+\frac{i^2p_2}{\Delta}(1-\lambda_1\lambda_3)+\frac{i^2p_3}{\Delta}(\lambda_1-\lambda_2)\right)\\
& -\sigma_{y,0}\cos\theta\,\mathrm{Re}\left(\frac{p_1}{\Delta}(\mu_2-\mu_3\lambda_2\lambda_3)+\frac{p_2}{\Delta}(\lambda_1\lambda_3\mu_3-\mu_1)+\frac{p_3}{\Delta}(\mu_1\lambda_2-\mu_2\lambda_1)\right)\\
& +\sigma_{y,0}\sin\theta\,\mathrm{Re}\left(\frac{ip_1}{\Delta}(\mu_2-\mu_3\lambda_2\lambda_3)+\frac{ip_2}{\Delta}(\lambda_1\lambda_3\mu_3-\mu_1)+\frac{ip_3}{\Delta}(\mu_1\lambda_2-\mu_2\lambda_1)\right)\\
& +\tau_{xy,0}\cos\theta\,\mathrm{Re}\left(\frac{p_1}{\Delta}(\lambda_2\lambda_3-1+i(\mu_2-\mu_3\lambda_2\lambda_3))+\frac{p_2}{\Delta}(1-\lambda_1\lambda_3+\right.\\
& \qquad\left. i(\lambda_1\lambda_3\mu_3-\mu_1))+\frac{p_3}{\Delta}(\lambda_1-\lambda_2+i(\mu_1\lambda_2-\mu_2\lambda_1))\right)\\
& -\tau_{xy,0}\sin\theta\,\mathrm{Re}\left(\frac{ip_1}{\Delta}(\lambda_2\lambda_3-1+i(\mu_2-\mu_3\lambda_2\lambda_3))+\frac{ip_2}{\Delta}(1-\lambda_1\lambda_3+\right.\\
& \qquad\left. i(\lambda_1\lambda_3\mu_3-\mu_1))+\frac{ip_3}{\Delta}(\lambda_1-\lambda_2+i(\mu_1\lambda_2-\mu_2\lambda_1))\right)\\
& +\tau_{yz,0}\cos\theta\,\mathrm{Re}\left(\frac{p_1}{\Delta}\lambda_3(\mu_3-\mu_2)+\frac{p_2\lambda_3}{\Delta}(\mu_1-\mu_3)+\frac{p_3}{\Delta}(\mu_2-\mu_1)\right)\\
& -\tau_{yz,0}\sin\theta\,\mathrm{Re}\left(\frac{ip_1}{\Delta}\lambda_3(\mu_3-\mu_2)+\frac{ip_2\lambda_3}{\Delta}(\mu_1-\mu_3)+\frac{ip_3}{\Delta}(\mu_2-\mu_1)\right)\\
& -\tau_{xz,0}\cos\theta\,\mathrm{Re}\left(\frac{ip_1}{\Delta}\lambda_3(\mu_3-\mu_2)+\frac{ip_2\lambda_3}{\Delta}(\mu_1-\mu_3)+\frac{ip_3}{\Delta}(\mu_2-\mu_1)\right)\\
& +\tau_{xz,0}\sin\theta\,\mathrm{Re}\left(\frac{i^2p_1}{\Delta}\lambda_3(\mu_3-\mu_2)+\frac{i^2p_2\lambda_3}{\Delta}(\mu_1-\mu_3)+\frac{i^2p_3}{\Delta}(\mu_2-\mu_1)\right)
\end{aligned} \tag{38}$$

or in a more general form

$$\frac{u_h}{a} = m_{u_h} \cos\theta + n_{u_h} \sin\theta \qquad (39)$$

A similar expression can be obtained for v_h/a by replacing all the coefficients p_i in eq (38) by q_i $(i=1,2,3)$. Then,

$$\frac{v_h}{a} = m_{v_h} \cos\theta + n_{v_h} \sin\theta \qquad (40)$$

Combining eqs (23), (39) and (40), we can write

$$\frac{u_{rh}}{a} = M_{u_1}^{(h)} + M_{u_2}^{(h)} \cos 2\theta + M_{u_3}^{(h)} \sin 2\theta \qquad (41)$$

where

$$M_{u_1}^{(h)} = \frac{m_{u_h} + n_{v_h}}{2} \; ; \quad M_{u_2}^{(h)} = \frac{m_{u_h} - n_{v_h}}{2} \; ; \quad M_{u_3}^{(h)} = \frac{n_{u_h} + m_{v_h}}{2}$$

Adding eqs (36) and (41) we obtain the general expression of the ratio u_r/a along the contour

$$\frac{u_r}{a} = M_1 + M_2 \cos 2\theta + M_3 \sin 2\theta \qquad (42)$$

where

$$M_1 = M_{u_1}^{(o)} + M_{u_1}^{(h)} \; ; \quad M_2 = M_{u_2}^{(o)} + M_{u_2}^{(h)} \; ; \quad M_3 = M_{u_3}^{(o)} + M_{u_3}^{(h)}$$

Coefficients M_1, M_2, M_3 are functions of the six stress field components, the material properties and the orientation of the hole with respect to the directions of anisotropy.

M_1 is an invariant quantity. Indeed,

Let u_{r_1} be the radial displacement at $(r=a,\ \theta=\theta_1)$ and

u_{r_2} be the radial displacement at $(\theta_2 = \theta_1 + \frac{\pi}{2})$ for the same material properties and loading conditions, then according to eq (42)

$$\frac{u_{r_1}}{a} = M_1 + M_2 \cos 2\theta_1 + M_3 \sin 2\theta_1$$

and

$$\frac{u_{r_2}}{a} = M_1 - M_2 \cos 2\theta_1 - M_3 \sin 2\theta_1$$

Then,

$$\frac{u_{r_1} + u_{r_2}}{a} = 2 M_1$$

Eq (42) is one form of the ratio u_r/a. An other form can be obtained directly from eq (35c) i.e.

$$\frac{u_r}{a} = f_{11_u} \sigma_{x,0} + f_{12_u} \sigma_{y,0} + f_{13_u} \sigma_{z,0} + f_{14_u} \tau_{yz_0} + f_{15_u} \tau_{xz_0} + f_{16_u} \tau_{xy_0} \tag{43}$$

In eq (43) the coefficients $f_{1i_u} (i = 1, 6)$ are now functions of the angle θ, the material properties and the orientation of the hole with respect to the directions of anisotropy.

7) Let us consider the special case when there is a plane of elastic symmetry perpendicular to the hole axis.

Recall eqs (4.35), (4.36) and (4.38). If we substitute the conditions $\lambda_1 = \lambda_2 = \lambda_3 = 0$ into eqs (6) and (7) we obtain

$$\phi_1(z_1) = (\mu_2 \bar{a}_1 - \bar{b}_1) / \Delta \xi_1$$

$$\phi_2(z_2) = (\bar{b}_1 - \mu_1 \bar{a}_1) / \Delta \xi_2$$

$$\phi_3(z_3) = \bar{c}_1 / \xi_3 \tag{44}$$

with

$$\Delta = \mu_2 - \mu_1 .$$

and

$$\phi_1'(z_1) = -\frac{1}{a\Delta\sqrt{\left(\frac{z_1}{a}\right)^2 - 1 - \mu_1^2}}\left(\mu_2\bar{a}_1 - \bar{b}_1\right)\frac{1}{\varepsilon_1}$$

$$\phi_2'(z_2) = -\frac{1}{a\Delta\sqrt{\left(\frac{z_2}{a}\right)^2 - 1 - \mu_2^2}}\left(\bar{b}_1 - \mu_1\bar{a}_1\right)\frac{1}{\varepsilon_2}$$

$$\phi_3'(z_3) = -\frac{\bar{c}_1\left(\mu_2 - \mu_1\right)}{a\Delta\sqrt{\left(\frac{z_3}{a}\right)^2 - 1 - \mu_3^2}}\frac{1}{\varepsilon_3} \qquad (45)$$

As shown in section 4.4, the generalized plane strain problem can be considered as the sum of two problems

• <u>a plane problem</u> involving stresses and displacements in the xOy plane only, whose components are

$$\begin{aligned}
\sigma_x &= \sigma_{x,0} + 2\,\mathrm{Re}\left(\mu_1^2\phi_1' + \mu_2^2\phi_2'\right)\\
\sigma_y &= \sigma_{y,0} + 2\,\mathrm{Re}\left(\phi_1' + \phi_2'\right)\\
\tau_{xy} &= \tau_{xy,0} - 2\,\mathrm{Re}\left(\mu_1\phi_1' + \mu_2\phi_2'\right)\\
\sigma_z &= \sigma_{z,0} - \frac{1}{a_{33}}\left(a_{31}\sigma_{x_h} + a_{32}\sigma_{y_h} + a_{36}\tau_{xy_h}\right)
\end{aligned} \qquad (46a)$$

and

$$\begin{aligned}
u &= u_0 - 2\,\mathrm{Re}\left(p_1\phi_1 + p_2\phi_2\right)\\
v &= v_0 - 2\,\mathrm{Re}\left(q_1\phi_1 + q_2\phi_2\right)
\end{aligned} \qquad (46b)$$

• <u>an antiplane problem</u> with components

$$\begin{aligned}
\tau_{xz} &= \tau_{xz,0} + 2\,\mathrm{Re}\left(\mu_3\phi_3'\right)\\
\tau_{yz} &= \tau_{yz,0} - 2\,\mathrm{Re}\left(\phi_3'\right)
\end{aligned} \qquad (47a)$$

and

$$w = w_0 - 2\,\mathrm{Re}\left(r_3\phi_3\right) \qquad (47b)$$

Combining the thirds of eqs (5) and (44), we can see that for any value of x,y and therefore for any value of

$\varepsilon_3,\ \phi_3(z_3) = 0$ <u>if and only</u> if $\tau_{xz,0}$ and $\tau_{yz,0}$ are zero. If those conditions are satisfied, then according to eqs (47a), the induced and total stress components τ_{xz}, τ_{yz} vanish and the induced component of longitudinal displacement is also zero. According to eq (32b), unless ε_{z_0} is zero, the total component of displacement w is equal to $-z\,\varepsilon_{z_0}$.

If conditions (4.35), (4.36) and (4.38) are satisfied then in eq (22)

$$d_1' = e_1' = d_2' = e_2' = b_3' = c_3' = f_3' = 0 \tag{48}$$

Thus, in eqs (25)

$$\begin{bmatrix} \frac{u_{rh}}{a} \\ \frac{v_{\theta h}}{a} \\ \frac{w_h}{a} \end{bmatrix} = \begin{bmatrix} f_{11_u}^{(h)} & f_{12_u}^{(h)} & 0 & 0 & 0 & f_{16_u}^{(h)} \\ f_{21_u}^{(h)} & f_{22_u}^{(h)} & 0 & 0 & 0 & f_{26_u}^{(h)} \\ 0 & 0 & 0 & f_{34_u}^{(h)} & f_{35_u}^{(h)} & 0 \end{bmatrix} \begin{bmatrix} \sigma_{x,0} \\ \sigma_{y,0} \\ \sigma_{z,0} \\ \tau_{yz,0} \\ \tau_{xz,0} \\ \tau_{xy,0} \end{bmatrix} \tag{49}$$

and in eqs (35a)

$$\begin{bmatrix} \frac{u_{r0}}{a} \\ \frac{v_{\theta 0}}{a} \\ \frac{w_0}{a} \end{bmatrix} = \begin{bmatrix} f_{11_u}^{(0)} & f_{12_u}^{(0)} & f_{13_u}^{(0)} & 0 & 0 & f_{16_u}^{(0)} \\ f_{21_u}^{(0)} & f_{22_u}^{(0)} & f_{23_u}^{(0)} & 0 & 0 & f_{26_u}^{(0)} \\ f_{31_u}^{(0)} & f_{32_u}^{(0)} & f_{33_u}^{(0)} & f_{34_u}^{(0)} & f_{35_u}^{(0)} & f_{36_u}^{(0)} \end{bmatrix} \begin{bmatrix} \sigma_{x,0} \\ \sigma_{y,0} \\ \sigma_{z,0} \\ \tau_{yz,0} \\ \tau_{xz,0} \\ \tau_{xy,0} \end{bmatrix} \tag{50}$$

adding those equations, we obtain for eqs (35c)

$$\begin{bmatrix} \frac{u_r}{a} \\ \frac{v_\theta}{a} \\ \frac{w}{a} \end{bmatrix} = \begin{bmatrix} f_{11_u} & f_{12_u} & f_{13_u} & 0 & 0 & f_{16_u} \\ f_{21_u} & f_{22_u} & f_{23_u} & 0 & 0 & f_{26_u} \\ f_{31_u} & f_{32_u} & f_{33_u} & f_{34_u} & f_{35_u} & f_{36_u} \end{bmatrix} \begin{bmatrix} \sigma_{x,0} \\ \sigma_{y,0} \\ \sigma_{z,0} \\ \tau_{yz,0} \\ \tau_{xz,0} \\ \tau_{xy,0} \end{bmatrix} \tag{51}$$

Then, from eqs (51), eq (43) is replaced by the following

$$\frac{u_r}{a} = f_{11_u}\sigma_{x,0} + f_{12_u}\sigma_{y,0} + f_{13_u}\sigma_{z,0} + f_{16_u}\tau_{xy,0} \tag{52}$$

8) Let us consider the special case of an <u>isotropic material</u>

with two constants: a Young modulus E and a Poisson's ratio ν . Since the material is isotropic there is at least one plane of elastic symmetry perpendicular to the hole axis.

For an isotropic material the coefficients in eq (2a) are such that

$$a_{11} = a_{22} = a_{33} = \frac{1}{E} \; ; \; a_{12} = a_{13} = a_{23} = -\frac{\nu}{E} \; ; \; a_{44} = a_{55} = a_{66} = \frac{2(1+\nu)}{E}$$

and all the other coefficients a_{ij} are zero. Then, from eq (4.14)

$$\beta_{11} = \beta_{22} = \frac{1-\nu^2}{E} \; ; \; \beta_{12} = -\frac{\nu}{E}(1+\nu) \; ; \; \beta_{44} = \beta_{55} = \beta_{66} = \frac{2(1+\nu)}{E}$$

and all the other coefficients β_{ij} are zero.

All the equations developed for the previous special case are still valid where

- μ_1, μ_2 are solutions of the equation

$$l_4(\mu) = \beta_{11}\mu^4 + (2\beta_{12} + \beta_{66})\mu^2 + \beta_{22} = \beta_{11}(\mu^4 + 2\mu^2 + 1) = 0$$

Then $\mu_1 = \mu_2 = i$ for any $\beta_{11} \neq 0$.

- μ_3 is solution of the equation

$$l_2(\mu) = \beta_{44}(\mu^2 + 1) = 0$$

Then $\mu_3 = i$ for any $\beta_{44} \neq 0$.

- p_1, p_2, q_1, q_2, r_3 are such that

$$p_i = \beta_{11}\mu_i^2 + \beta_{12} \; ; \; q_i = \beta_{12}\mu_i + \frac{\beta_{22}}{\mu_i} \quad (i = 1,2)$$

and

$$r_3 = -\beta_{44}/\mu_3$$

If we calculate the induced and total displacement components along the contour of the hole using eqs (46b) and (47b), we obtain the general expressions of the total radial displacement as follows

$$\frac{u_r}{a} = f_{11_u}\sigma_{x,0} + f_{12_u}\sigma_{y,0} + f_{13_u}\sigma_{z,0} + f_{16_u}\tau_{xy,0} \tag{53}$$

where

$$-f_{11_u} = \frac{(1-\nu^2)}{E}(1+2\cos 2\theta) + \frac{\nu^2}{E}$$

$$-f_{12_u} = \frac{(1-\nu^2)}{E}(1-2\cos 2\theta) + \frac{\nu^2}{E}$$

$$-f_{13_u} = -\frac{\nu}{E} \;;\quad -f_{16_u} = 4\sin 2\theta\,\frac{(1-\nu^2)}{E} \tag{53a}$$

or

$$\frac{u_r}{a} = M_1 + M_2\cos 2\theta + M_3\sin 2\theta \tag{54}$$

where*

$$M_1 = -\frac{1}{E}\left(\sigma_{x,0}+\sigma_{y,0}-\nu\sigma_{z,0}\right) \;;\quad M_2 = -\frac{2}{E}(1-\nu^2)\left(\sigma_{x,0}-\sigma_{y,0}\right)$$

$$M_3 = -\frac{4}{E}(1-\nu^2)\tau_{xy,0} \tag{54a}$$

Similarly, the total displacement component w along the contour of the hole takes the form

$$-\frac{w}{a} = \frac{z}{aE}\left(\sigma_{z,0}-\nu\left(\sigma_{x,0}+\sigma_{y,0}\right)\right) + \frac{4(1+\nu)}{E}\left(\tau_{yz,0}\sin\theta + \tau_{xz,0}\cos\theta\right) \tag{55}$$

---

* if $\tau_{xy,0}$ is zero then $M_3 = 0$.

9) Let us now introduce an internal pressure q distributed uniformly along the hole.

Substituting eqs (5) into eqs (6) or (7) and making use of eqs (8) and (21a) we obtain

$$\begin{aligned} \phi_1 &= \phi_{1q=0} + q\Delta_1 a\left((\mu_2-\mu_3\lambda_2\lambda_3)+i(\lambda_2\lambda_3-1)\right)/2 \\ \phi_2 &= \phi_{2q=0} + q\Delta_2 a\left((\lambda_1\lambda_3\mu_3-\mu_1)+i(1-\lambda_1\lambda_3)\right)/2 \\ \phi_3 &= \phi_{3q=0} + q\Delta_3 a\left((\mu_1\lambda_2-\mu_2\lambda_1)+i(\lambda_1-\lambda_2)\right)/2 \end{aligned} \tag{56}$$

and

$$\begin{aligned} \phi'_1 &= \phi'_{1q=0} + q\gamma_1\left((\mu_2-\mu_3\lambda_2\lambda_3)+i(\lambda_2\lambda_3-1)\right)/2 \\ \phi'_2 &= \phi'_{2q=0} + q\gamma_2\left((\lambda_1\lambda_3\mu_3-\mu_1)+i(1-\lambda_1\lambda_3)\right)/2 \\ \phi'_3 &= \phi'_{3q=0} + q\gamma_3\left((\mu_1\lambda_2-\mu_2\lambda_1)+i(\lambda_1-\lambda_2)\right)/2 \end{aligned} \tag{57}$$

If we substitute eqs (57) into eqs (3), eq (16) is replaced by the following

$$(\sigma)_{xyz} = (F_\sigma)(\sigma_0) + q\,(F^*) \tag{58}$$

where

$$f^*_{11} = \ell_1\ ;\quad f^*_{21} = \ell_2\ ;\quad f^*_{41} = \ell_5\ ;\quad f^*_{51} = \ell_4\ ;\quad f^*_{61} = \ell_3$$

$$f^*_{31} = -\left(a_{31}\ell_1 + a_{32}\ell_2 + a_{34}\ell_5 + a_{35}\ell_4 + a_{36}\ell_3\right)/a_{33}$$

with

$$\ell_1 = \mathrm{Re}\left(\gamma_1\mu_1^2\left((\mu_2-\mu_3\lambda_2\lambda_3)+i(\lambda_2\lambda_3-1)\right) + \gamma_2\mu_2^2\left((\lambda_1\lambda_3\mu_3-\mu_1) + i(1-\lambda_1\lambda_3)\right) + \lambda_3\gamma_3\mu_3^2\left((\mu_1\lambda_2-\mu_2\lambda_1)+i(\lambda_1-\lambda_2)\right)\right)$$

$$\ell_2 = \mathrm{Re}\left(\gamma_1\left((\mu_2-\mu_3\lambda_2\lambda_3)+i(\lambda_2\lambda_3-1)\right) + \gamma_2\left((\lambda_1\lambda_3\mu_3-\mu_1) + i(1-\lambda_1\lambda_3)\right) + \lambda_3\gamma_3\left((\mu_1\lambda_2-\mu_2\lambda_1)+i(\lambda_1-\lambda_2)\right)\right)$$

$$\ell_3 = -\mathrm{Re}\left(\gamma_1\mu_1\left((\mu_2-\mu_3\lambda_2\lambda_3)+i(\lambda_2\lambda_3-1)\right) + \gamma_2\mu_2\left((\lambda_1\lambda_3\mu_3-\mu_1) + i(1-\lambda_1\lambda_3)\right) + \lambda_3\gamma_3\mu_3\left((\mu_1\lambda_2-\mu_2\lambda_1)+i(\lambda_1-\lambda_2)\right)\right)$$

$$\ell_4 = Re\left(\lambda_1\gamma_1\mu_1\left((\mu_2-\mu_3\lambda_2\lambda_3)+i(\lambda_2\lambda_3-1)\right)+\lambda_2\gamma_2\mu_2\left((\lambda_1\lambda_3\mu_3-\mu_1)+ i(1-\lambda_1\lambda_3)\right)+\gamma_3\mu_3\left((\mu_1\lambda_2-\mu_2\lambda_1)+i(\lambda_1-\lambda_2)\right)\right)$$

$$\ell_5 = -Re\left(\gamma_1\lambda_1\left((\mu_2-\mu_3\lambda_2\lambda_3)+i(\lambda_2\lambda_3-1)\right)+\gamma_2\lambda_2\left((\lambda_1\lambda_3\mu_3-\mu_1)+ i(1-\lambda_1\lambda_3)\right)+\gamma_3\left((\mu_1\lambda_2-\mu_2\lambda_1)+i(\lambda_1-\lambda_2)\right)\right)$$

Similarly, eqs (18b) and (19b) take the following forms

$$(\varepsilon)_{xyz} = (A)(F_\sigma)(\sigma_0) + q(A)(F^*) \tag{59}$$

and

$$(\varepsilon)_{r\theta z} = (T_\varepsilon^*)(A)(F_\sigma)(\sigma_0) + q\,(T_\varepsilon^*)(A)(F^*) \tag{60}$$

If we substitute eqs (56) into eqs (4), the induced displacement components in the r,θ,z coordinate system take the form

$$\begin{bmatrix} \frac{u_{rh}}{a} \\ \frac{v_{\theta h}}{a} \\ \frac{w_h}{a} \end{bmatrix} = \begin{bmatrix} \frac{u_{rh}}{a} \\ \frac{v_{\theta h}}{a} \\ \frac{w_h}{a} \end{bmatrix}_{q=0} + q \begin{bmatrix} \cos\theta & \sin\theta & 0 \\ -\sin\theta & \cos\theta & 0 \\ 0 & 0 & 1 \end{bmatrix} \begin{bmatrix} U_1 \\ U_2 \\ U_3 \end{bmatrix} \tag{61}$$

where

$$U_1 = -Re\left(\Delta_1 p_1\left((\mu_2-\mu_3\lambda_2\lambda_3)+i(\lambda_2\lambda_3-1)\right)+\Delta_2 p_2\left((\lambda_1\lambda_3\mu_3-\mu_1)+ i(1-\lambda_1\lambda_3)\right)+\Delta_3 p_3\left((\mu_1\lambda_2-\mu_2\lambda_1)+i(\lambda_1-\lambda_2)\right)\right)$$

$$U_2 = -Re\left(\Delta_1 q_1\left((\mu_2-\mu_3\lambda_2\lambda_3)+i(\lambda_2\lambda_3-1)\right)+\Delta_2 q_2\left((\lambda_1\lambda_3\mu_3-\mu_1)+ i(1-\lambda_1\lambda_3)\right)+\Delta_3 q_3\left((\mu_1\lambda_2-\mu_2\lambda_1)+i(\lambda_1-\lambda_2)\right)\right)$$

$$U_3 = -Re\left(\Delta_1 r_1\left((\mu_2-\mu_3\lambda_2\lambda_3)+i(\lambda_2\lambda_3-1)\right)+\Delta_2 r_2\left((\lambda_1\lambda_3\mu_3-\mu_1)+ i(1-\lambda_1\lambda_3)\right)+\Delta_3 r_3\left((\mu_1\lambda_2-\mu_2\lambda_1)+i(\lambda_1-\lambda_2)\right)\right)$$

Appendix 4.6

FORMULATION OF COMPLETE PLANE STRAIN PROBLEMS FOR REGULARLY JOINTED ROCKS

B. Amadei and R. E. Goodman

Department of Civil Engineering
University of California, Berkeley
Berkeley, California

1. INTRODUCTION

In engineering practice, rock is usually assumed to be linearly elastic, homogeneous and isotropic. This paper focuses on a procedure for evaluating the behavior of rock that is "anisotropic," i.e., not isotropic. This is the usual characteristic of schists, slates, and other metamorphic rocks with fabrics having parallel arrangements of long or flat minerals. Banded rocks like gneiss or bedded rock like Flysch deposits that include interlayered mixtures of different components also display anisotropy. Even isotropic rocks like some granites, sandstones, and limestones, will behave anisotropically if cut by regular sets of joints.

In the case of anisotropy derived from regular discontinuities, the theory of linear elasticity is invalid. However, numerical techniques, like the finite element method, can be used to evaluate deformations if the jointed rock mass is discretized in terms of linear elastic elements for "intact" material and joint elements for discontinuity planes. Another procedure is to replace the jointed rock by a homogeneous, anisotropic and continuous medium, the behavior of which is equivalent to the behavior of the jointed rock (Duncan and Goodman, 1968; Morland, 1976). This procedure can be considered as a special case of those proposed for stratified rock masses (Pinto, 1966; Salomon, 1968).

When an infinitely long hole is drilled in an initially stressed medium, a plane strain condition is usually assumed. However, there is no reason that the hole must follow a principal stress direction or that the material must be isotropic. When one or both of those conditions are not satisfied, the usual plane strain definition no longer applies.

The paper begins with a more general formulation of plane strain to be applied when a hole is drilled in an orthotropic or transversely isotropic initially stressed medium. The hole may be inclined with the principal stress directions and the planes and/or axes of elastic symmetry of the medium may be generally directed with respect to the geometrical symmetry of the problem.

The formulation is then applied to calculate the displacement and stress distributions around a circular hole drilled in a regularly jointed rock described as an equivalent anisotropic continuum.

2. FORMULATION OF COMPLETE PLANE STRAIN FOR ANISOTROPIC ROCK MASSES

2.1 Geometry and definition of the problem

Consider a circular hole in an anisotropic elastic homogeneous body (Fig. 1). The body possesses rectilinear anisotropy. Only orthotropic and transversely isotropic materials will be considered in this analysis. Let x, y, z be a system of cartesian coordinates with the z axis defining the geometrical axial symmetry of the hole. The orientation of the hole is defined with respect to a fixed global coordinate system X, Y, Z. The hole is supposed to be infinitely long and to remain constant in length.

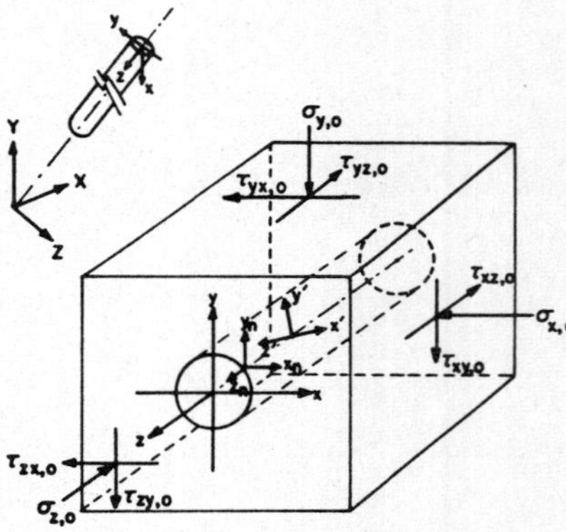

FIGURE 1. Schematic view of a long circular hole to be excavated in a triaxial stress field.

The body has planes and/or axes of elastic symmetry (x', y', z') with respect to directions independent of the x,y,z directions while the principal stresses at infinity (field stresses) will in general have directions unrelated to either of these coordinate bases.

The problem is to determine the distributions of stresses, strains and displacements around the hole drilled in the body initially subjected to field stresses. The final stress distribution around the hole is the tensor sum of three stress distributions (Fig. 1):

(i) the field stress tensor, before drilling the hole, with components $\sigma_{x,0}$; $\sigma_{y,0}$; $\sigma_{z,0}$; $\tau_{yz,0}$; $\tau_{xz,0}$; $\tau_{xy,0}$;
(ii) the stress tensor induced by drilling the hole; and
(iii) the stress tensor induced by any boundary stresses acting along the walls of the hole, with components X_n, Y_n, Z_n.

Similarly the strain and displacement distributions

around the hole are the sum of those before drilling the hole, those induced by drilling the hole and those induced by the boundary stresses.

The strain components are related to the stress components through the constitutive relation of the anisotropic body. In the system of coordinates x,y,z, the constitutive relation can be written as follows:

$$\begin{bmatrix}\varepsilon_x\\ \varepsilon_y\\ \varepsilon_z\\ \gamma_{yz}\\ \gamma_{xz}\\ \gamma_{xy}\end{bmatrix} = \begin{bmatrix} a_{11} & a_{12} & a_{13} & a_{14} & a_{15} & a_{16}\\ a_{21} & a_{22} & a_{23} & a_{24} & a_{25} & a_{26}\\ a_{31} & a_{32} & a_{33} & a_{34} & a_{35} & a_{36}\\ a_{41} & a_{42} & a_{43} & a_{44} & a_{45} & a_{46}\\ a_{51} & a_{52} & a_{53} & a_{54} & a_{55} & a_{56}\\ a_{61} & a_{62} & a_{63} & a_{64} & a_{65} & a_{66}\end{bmatrix} \begin{bmatrix}\sigma_x\\ \sigma_y\\ \sigma_z\\ \tau_{yz}\\ \tau_{xz}\\ \tau_{xy}\end{bmatrix} \quad (1)$$

or

$$(\varepsilon) = (A)(\sigma) \quad (2)$$

Infinitesimal strains and displacements are related through the following equations:

$$\varepsilon_x = \frac{-\partial u}{\partial x}\ ;\ \varepsilon_y = \frac{-\partial v}{\partial y}\ ;\ \varepsilon_z = \frac{-\partial w}{\partial z}\ ;\ \gamma_{yz} = -\left(\frac{\partial w}{\partial y} + \frac{\partial v}{\partial z}\right);$$

$$\gamma_{xz} = -\left(\frac{\partial u}{\partial z} + \frac{\partial w}{\partial x}\right);\ \gamma_{xy} = -\left(\frac{\partial u}{\partial y} + \frac{\partial v}{\partial x}\right) \quad (3)$$

In the following analysis, we adopt the convention of compressive stresses and contractile strains taken as positive with the sense of positive normal stresses defining the sense of positive shear stresses. The displacement components u,v,w respectively in the x,y, z directions are taken positive if directed in the positive directions of the coordinate axes.

Although the hole is infinitely long and remains constant in length, the independence of the various axes of symmetry mentioned at the beginning of this section implies that the deformation of the medium will not (in general) be in plane strain, in "sensu stricto."

2.2 Complete plane strain conditions

Before going to complete-plane-strain, it is helpful to review the main assumptions in the usual definition of plane strain. If the hole is infinitely long and remains constant in length, it is commonly assumed that at all points in the body the induced displacement components u,v,w are independent of z and that w is also independent of x and y. Therefore, according to eq. (3):

$$\varepsilon_z = \gamma_{yz} = \gamma_{xz} = 0 \quad (4)$$

Then, among the six components of strains in eq. (3) only ε_x, ε_y, and γ_{xy} are non zero. Eq. (4) implies that at all points in the medium the induced stresses τ_{xz} and τ_{yz} are zero and according to eq. (1) a plane of elastic symmetry must be perpendicular to the hole axis ($a_{4i} = a_{5i} = a_{46} = a_{56} = 0$ for i = 1,2, 3). This result imposes undue restrictions on the applicability of plane strain methods: the long axis of the hole must coincide with a principal stress direction and one plane of elastic symmetry of the anisotropic material must coincide with the plane xOy.

The essential notion in the plane strain concept defined is that induced displacements, induced and resultant stresses, strains and boundary conditions are to be identical in all planes perpendicular to the hole axis, i.e.,

$$\frac{\partial u}{\partial z} = \frac{\partial v}{\partial z} = \frac{\partial w}{\partial z} = 0 \quad (5)$$

According to eq. (5), the process of drilling the hole in the stressed medium and the process of applying boundary stresses along the hole after drilling must induce zero longitudinal strain. The induced displacements u,v,w are functions of x and y only. The five other strain components and the six induced stresses components may be non-zero at all points in the medium. This plane strain concept is called generalized plane strain (Lekhitskii, 1963; Milne Thomson, 1962) or complete plane strain (Brady and Bray, 1978). This approach can be used when the planes of elastic symmetry of the medium do not correspond with the geometrical symmetry of the hole and/or the hole is not in a direction of a principal field stress.

The stress components being independent of z, and assuming zero body force, the equations of equilibrium have the following form:

$$\frac{\partial \sigma_x}{\partial x} + \frac{\partial \tau_{xy}}{\partial y} = 0 \qquad \frac{\partial \tau_{xy}}{\partial x} + \frac{\partial \sigma_y}{\partial y} = 0 \qquad \frac{\partial \tau_{xz}}{\partial x} + \frac{\partial \tau_{yz}}{\partial y} = 0 \quad (6)$$

In general, there are six compatibility conditions for strains. From the assumptions of complete plane strain they reduce to two, i.e.,

$$\frac{\partial^2 \varepsilon_x}{\partial y^2} + \frac{\partial^2 \varepsilon_y}{\partial x^2} = \frac{\partial^2 \gamma_{xy}}{\partial x \partial y} \quad \text{and} \quad \frac{\partial \gamma_{xz}}{\partial y} - \frac{\partial \gamma_{yz}}{\partial x} = 0 \quad (7)$$

The third equation of (6) and the second equation of (7) result from the assumptions of complete plane strain. They vanish for the usual plane strain conditions.

2.3 Formulation of the problem

The object is to determine the distributions of stresses, strains and displacements induced by drilling a circular hole and by the boundary stresses (if any) acting along the hole after drilling. The hole is located in an elastic anisotropic homogeneous body initially subjected to field stresses. A complete plane strain formulation is assumed. For any problem of elastostatics, stresses, strains and displacements must satisfy the following equations:

(i) equations of equilibrium (eqs. 6),
(ii) equations of compatibility for strains (eqs. 7),
(iii) strain-displacement relations (eqs. 3),
(iv) constitutive relations (eqs. 1), and
(v) boundary conditions.

Lekhnitskii (1963) showed that the equations of equilibrium are satisifed if two stress functions F(x,y) and ψ(x,y) are introduced such that the stress components are defined by:

$$\sigma_x = \frac{\partial^2 F}{\partial y^2} \quad \sigma_y = \frac{\partial^2 F}{\partial x^2} \quad \tau_{xy} = \frac{-\partial^2 F}{\partial x \partial y} \quad \tau_{xz} = \frac{\partial \psi}{\partial y}$$

$$\tau_{yz} = \frac{-\partial \psi}{\partial x} \quad (8)$$

If the process of drilling the hole and the process of applying the boundary stresses induce zero longitudinal strain, according to eq. (1), for each process the induced stress in the z direction is equal to:

$$\sigma_z = \frac{-1}{a_{33}}(a_{31}\sigma_x + a_{32}\sigma_y + a_{34}\tau_{yz} + a_{35}\tau_{xz} + a_{36}\tau_{xy}) \tag{9}$$

where σ_x, σ_y, τ_{xz}, τ_{yz}, and τ_{xy} are the stresses induced in the x,y,z coordinate system. Combining eqs. (7), (8), and making use of eq. (9), the two stress functions satisfy the following system of differential equations:

$$\begin{cases} L_3F + L_2\psi = 0 \\ L_4F + L_3\psi = 0 \end{cases} \tag{10}$$

where L_3, L_3, L_4 are three differential operators defined as follows:

$$\begin{cases} L_2 = \beta_{44}\frac{\partial^2}{\partial x^2} - 2\beta_{45}\frac{\partial^2}{\partial x\partial y} + \beta_{55}\frac{\partial^2}{\partial y^2} \\ L_3 = -\beta_{24}\frac{\partial^3}{\partial x^3} + (\beta_{25} + \beta_{46})\frac{\partial^3}{\partial x^2\partial y} - (\beta_{14} + \beta_{56})\frac{\partial^3}{\partial x\partial y^2} + \beta_{15}\frac{\partial^3}{\partial y^3} \\ L_4 = \beta_{22}\frac{\partial^4}{\partial x^4} - 2\beta_{26}\frac{\partial^4}{\partial x^3\partial y} + (2\beta_{12} + \beta_{66})\frac{\partial^4}{\partial x^2\partial y^2} - 2\beta_{16}\frac{\partial^4}{\partial x\partial y^3} + \beta_{11}\frac{\partial^4}{\partial y^4} \end{cases} \tag{11}$$

$$\text{where } \beta_{ij} = a_{ij} - \frac{a_{i3}a_{j3}}{a_{33}} \quad (i,j = 1,2,3,4,5,6) \tag{12}$$

The two functions F and ψ must also satisfy the boundary conditions. No body forces have been assumed. The gravity stress has been incorporated into one of the field stresses. Solving eqs. (10) in terms of F we are left with a sixth order differential equation:

$$(L_4L_2 - L_3^2)F = 0 \tag{13}$$

Making use of the solutions of eqs. (10) and the relations between stress components and the two stress functions [eqs. (8)], the distribution of the six components of induced stresses and the three components of induced displacements around a circular hole drilled in an elastic, anisotropic body initially subject to a 3-D field stress can be calculated for different loading and boundary conditions (Lekhnitskii, 1963; Milne-Thomson, 1962). However, there exist directions of anisotropy for which the stresses τ_{xz} and τ_{yz} cannot be determined using this method even though the field stresses $\tau_{xz,0}$ and $\tau_{yz,0}$ are non zero. Those cases correspond to the condition $L_3 = 0$. According to eqs. (11) and (13), this is satisfied when $a_{4i} = a_{5i} = a_{46} = a_{56} = 0$ (for i = 1,2,3) or equivalently when a plane of elastic symmetry is perpendicular to the hole axis. The following conditions of anisotropy correspond to the condition $L_3 = 0$

(i) orthotropic material with one plane of elastic symmetry perpendicular to the hole axis and the two other planes parallel to the hole axis,
(ii) transverse isotropy in a plane striking parallel to the hole axis,
(iii) transverse isotropy in a plane perpendicular to the hole axis, and
(iv) isotropy.

Cases (ii), (iii), and (iv) can be derived from case (i).

If L_3 is equal to zero, according to eq. (10) F and ψ are solutions of two independent differential equations. For the anisotropy condition (i), eqs. (10) can be written as follows:

$$\begin{cases} \beta_{44}\frac{\partial^2\psi}{\partial x^2} + \beta_{55}\frac{\partial^2\psi}{\partial y^2} - 2\beta_{45}\frac{\partial^2\psi}{\partial x\partial y} = 0 \\ \beta_{22}\frac{\partial^4 F}{\partial x^4} - 2\beta_{26}\frac{\partial^4 F}{\partial x^3\partial y} + (2\beta_{12} + \beta_{66})\frac{\partial^4 F}{\partial x^2\partial y^2} - 2\beta_{16}\frac{\partial^4 F}{\partial x\partial y^3} + \beta_{11}\frac{\partial^4 F}{\partial y^4} = 0 \end{cases} \tag{14}$$

If the two planes of elastic symmetry of the orthotropic material which are parallel to the hole axis are parallel to planes x0z and y0z, then eqs. (14) reduce to:

$$\begin{cases} \frac{\partial^2\psi}{\partial x^2} + \frac{\partial^2\psi}{\partial Y^2} = 0 \quad \text{with} \quad Y = y\sqrt{\frac{\beta_{44}}{\beta_{55}}} \\ \beta_{22}\frac{\partial^4 F}{\partial x^4} + (2\beta_{12} + \beta_{66})\frac{\partial^4 F}{\partial x^2\partial y^2} + \beta_{11}\frac{\partial^4 F}{\partial y^4} = 0 \end{cases} \tag{15}$$

For a material that possesses a transverse isotropy in a plane perpendicular to the hole axis or for an isotropic material the first differential equation becomes a Laplacian equation and the second a Biharmonic equation.

Using eqs. (8), it can be seen that eqs. $(14)_2$ and $(15)_2$ connect only the stress components acting in the plane perpendicular to the hole axis. Eqs. $(14)_1$ and and $(15)_1$ connect only the stress components τ_{xz} and τ_{yz}.

If the field stresses are such that $\tau_{xz,0}$ and $\tau_{yz,0}$ are zero, eqs. $(14)_1$ or $(15)_1$ vanish and we are left with either eqs. $(14)_2$ or $(15)_2$. This is a plane strain condition, in the usual sense, and the method proposed by Lekhnikskii can be used to calculate the induced stress and displacement distributions around the hole in the x0y plane.

If $\tau_{xz,0}$ and $\tau_{yz,0}$ do not vanish, we have a complete plane strain condition, which can be considered as the sum of two problems:

(i) a plane strain problem involving stresses and displacements in the plane x0y which can be solved using the method proposed by Lekhnikskii (1963),
(ii) a longitudinal problem or antiplane problem (Filon, 1937) involving τ_{xz} and τ_{yz} which can be solved using the method proposed by Fairhurst (1968) and Berry (1968).

Therefore the choice between plane and complete plane strain conditions depends upon two parameters: the orientation of the directions of anisotropy with respect to the geometrical symmetry of the problem and the values of the field stresses (see Appendix 1).

3. A 3-D CONSTITUTIVE RELATION FOR FRACTURED ROCK MASSES

Consider a body with three mutually perpendicular

planes of elastic symmetry with normal directions n,s, t (orthotropic material). Let the body be cut by three orthogonal joint sets, each one parallel to a symmetry direction. A joint set i (i = 1,2,3) has a spacing S_i and its orientation defined with respect to a fixed coordinate system x,y,z.

A constitutive relation describing the non-linear behavior of a discontinuous, homogeneous and anisotropic body of rock containing up to three orthogonal joint sets has been proposed in the literature (Duncan and Goodman, 1968; Amadei and Goodman, 1981).

Let the intact rock between the joints be linearly elastic and isotropic. As shown in the two papers mentioned above, the constitutive relation for the jointed rock described as an <u>equivalent anisotropic continuum</u> relates the stress and strain components in n,s,t coordinate system as follows: $(\varepsilon) = (A)(\sigma)$, with (A) =

$$\begin{bmatrix} \left(\frac{1}{E}+\frac{1}{kn_1S_1}\right) & \frac{-\nu}{E} & \frac{-\nu}{E} & 0 & 0 & 0 \\ \frac{-\nu}{E} & \left(\frac{1}{E}+\frac{1}{kn_2S_2}\right) & \frac{-\nu}{E} & 0 & 0 & 0 \\ \frac{-\nu}{E} & \frac{-\nu}{E} & \left(\frac{1}{E}+\frac{1}{kn_3S_3}\right) & 0 & 0 & 0 \\ 0 & 0 & 0 & \left(\frac{1}{G}+\frac{1}{ks_2S_2}+\frac{1}{ks_3S_3}\right) & 0 & 0 \\ 0 & 0 & 0 & 0 & \left(\frac{1}{G}+\frac{1}{ks_1S_1}+\frac{1}{ks_3S_3}\right) & 0 \\ 0 & 0 & 0 & 0 & 0 & \left(\frac{1}{G}+\frac{1}{ks_1S_1}+\frac{1}{ks_2S_2}\right) \end{bmatrix} \quad (16)$$

where k_{ni}, k_{si}, S_i are respectively the normal and shear stiffnesses and the spacing of joint set i (i = 1,2,3)
and E, ν, G are respectively the Young's modulus, the Poisson's ratio, and the shear modulus of the intact rock.

This constitutive relation reduces to that of an isotropic material as the spacings of the joint sets increase and/or the joints become stiffer both in shear and in normal compression.

4. FORMULATION OF COMPLETE PLANE STRAIN FOR REGULARLY JOINTED ROCKS

Consider a body cut by three orthogonal joint sets, the intact rock being isotropic. The constitutive relation of the jointed rock [eq. (16)] can be rewritten in the coordinate system x,y,z if the orientation of joint set 1 with respect to that system is known. The orientation of that joint set is defined by its dip angle and its strike <u>with respect to the hole axis</u>. Using the transformation laws for the stress and strain tensors, the constitutive relation (16) can be rewritten in the x,y,z coordinate system as relation (1).

In the previous formulation of complete plane strain we assumed the coefficients a_{ij} of eq. (1) to be constant. Therefore if we want to apply the same procedure for regularly jointed rocks, the stiffness coefficients k_{ni}, k_{si} (i = 1,2,3) must also be constant.

Using the same procedure as in section 2, we can calculate the distribution of displacements and stresses around a hole drilled in a body cut by one, two, or three orthogonal joint sets described as an equivalent anisotropic material. The purpose of the present analysis is to study the influences of joint set orientations and properties on the distribution of stresses and displacements around the hole. Joint properties can always be expressed in terms of two sets of products $k_{ni}S_i$ and $k_{si}S_i$ (i = 1,2,3) as shown in eq. (16).

As a first example, consider the complete plane strain formulation for regularly jointed rocks for the special case where the hole is drilled in a principal stress direction. Further, let the material directions differ from the geometrical symmetry directions of the problem. An example is shown in Fig. 2a. An infinitely long hole of radius a = 0.1 m is cut by a joint set that strikes at different angles from the hole axis and dips at an angle of 90 degrees. The joint properties are expressed in terms of two products: k_nS = 6000 MPa and k_sS = 3000 MPa.

The field stress components relative to the hole local axes are given by $\sigma_{x,0}$ = 4 MPa; $\sigma_{y,0}$ = 2 MPa; $\sigma_{z,0}$ = 5 MPa and $\tau_{xy,0} = \tau_{xz,0} = \tau_{yz,0} = 0$. The intact rock properties are a Young's Modulus E = 20,000 MPa and a Poisson's ratio ν = 0.25. Figs. 2b, 2c, and 2d show respectively the polar distribution of total tangential stresses, induced radial and tangential displacements along the wall of the hole (θ varying from 0 to 90°) for different strike angles. When the joint set is perpendicular to the hole axis (strike equals 90°), stress and displacement distributions in the xOy plane are those for an isotropic medium. For the given joint properties, the tangential stress distribution is not highly affected by the orientation of the joint set except for θ less than 20° where the tangential stress decreases as the strike decreases. However, for the given joint properties the radial and the tangential displacement distributions are highly affected by the value of the strike angle. The worst condition occurs when the joint set is parallel to the hole axis. For this joint set orientation, Fig. 2e shows the distribution of the total tangential stresses along the hole for different values of the products k_nS and k_sS ($k_nS = 2 \cdot k_sS$). An increase in those products would correspond for example to an increase in joint spacing for constant shear and normal stiffnesses. It can be seen that decreasing the value of those products enhances the influence of the strike angle on the tangential stress distribution at low θ values. As far as the displacements are concerned, for a given strike angle, a decrease in the values of k_nS and k_sS enhances radial and tangential displacements.

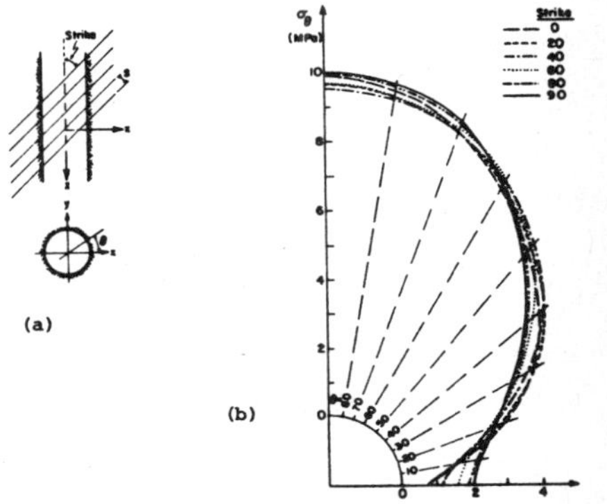

FIGURE 2. (a) Geometry of the example. (b) Distribution of total tangential stresses along the wall of the hole.

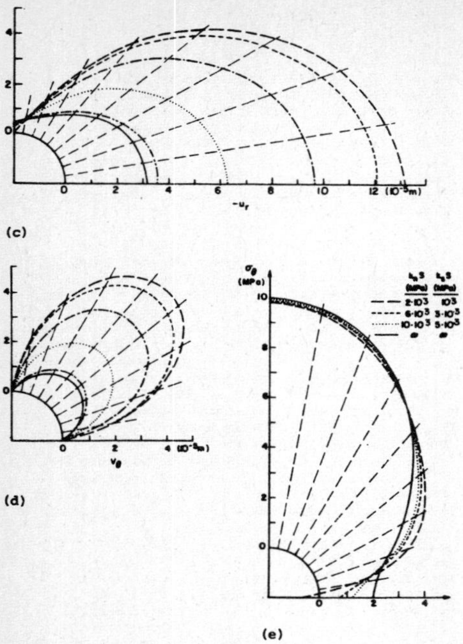

FIGURE 2. (c) Distribution of induced radial displacements.
(d) Distribution of induced tangential displacements.
(e) Influence of the values of k_nS, k_sS on the total tangential stress distribution (Strike angle = 0).

Joint inclination with respect to the hole axis also influences the orientation of the induced principal stresses around the hole as shown in the following example. Consider the same problem as in Fig. 2a except that the hole is now pressurized over its whole length with an internal pressure of 10 MPa. To simplify the explanation assume the hole is horizontal. It is cut by a joint set with a dip angle of 90 degrees. The intact rock has the same properties as for the previous example. The joint has the following properties: $k_nS = 2 \cdot 10^4$ MPa and $k_sS = 10^3$ MPa. In Fig. 3 the orientation of the induced principal stresses acting along a quarter of the hole wall has been represented using a stereographic projection (lower hemisphere) for different joint strike values. The principal stresses are those induced by the hole pressure only. The great circles shown in Fig. 3 are a family of planes tangent to the wall of the hole. It can be seen that except for a joint set parallel or perpendicular to the hole axis, the principal stress orientations are greatly affected by joint inclination.

Finally, consider an infinitely long hole, with radius a, cut by a joint set striking parallel to the

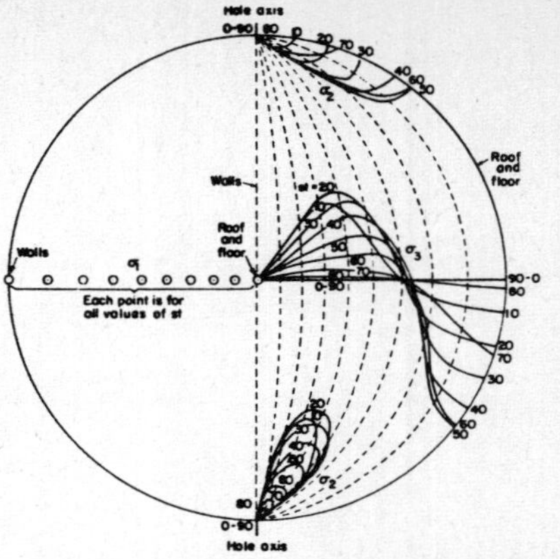

FIGURE 3. Orientation of induced principal stresses acting along a quarter of the surface of a circular pressurized hole cut by one joint set striking at different angles st with a dip angle of 90 degrees. Great circles are a family of planes tangent to the hole (lower hemisphere projection).

hole axis and perpendicular to the x axis (Fig. 4). The joint set has a spacing S, and normal and shear stiffnesses respectively k_n and k_s. Let the hole be pressurized over its entire length with an internal pressure q. The purpose of this example is to study the influence of the products k_nS and k_sS on the distribution of the displacements along the walls of the hole induced by q only.

Using the method proposed by Lekhnitskii to solve this problem, the radial and tangential displacements at a point along the wall of the hole located at an angle θ from the x axis can be computed. Assuming that the change in hole diameter is due to the radial component of displacement only, one can calculate the changes of diameter ΔAA and ΔBB respectively for θ = 0 and θ = 90°.

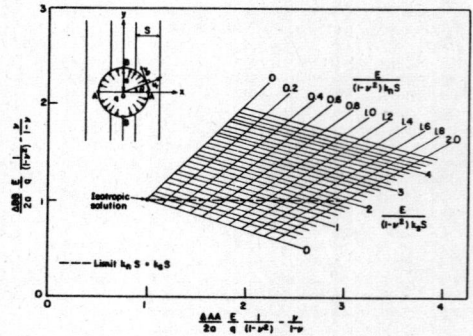

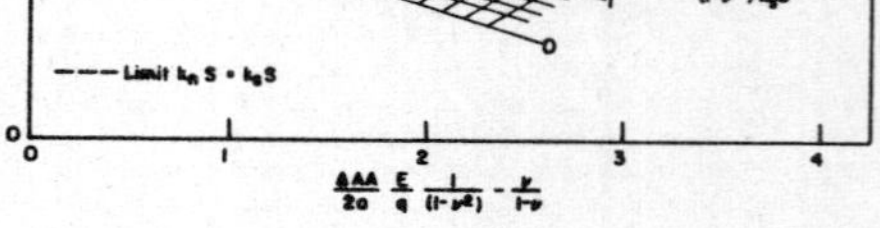

FIGURE 4.

Fig. 4 shows the variation of two dimensionless quantities containing ΔAA and ΔBB with respect to k_nS and k_sS. When k_nS and k_sS approach infinity, the solution converges to that for the isotropic case and

$$\frac{\Delta AA}{2a}\frac{E}{q}\frac{1}{(1-\nu^2)} - \frac{\nu}{1-\nu} = \frac{\Delta BB}{2a}\frac{E}{q}\frac{1}{(1-\nu^2)} - \frac{\nu}{1-\nu} = 1 \qquad (17)$$

Theoretically, by measuring the two changes of diameter ΔAA and ΔBB it would be possible to calculate the products k_nS and k_sS and the stiffnesses k_n and k_s if the spacing is known. However, this procedure implies a knowledge of the two intact rock properties E and ν. If we assume that the unloading of a fractured rock mass follows essentially the same path as that of the intact rock (Goodman 1976, 1980), it would be possible to back-calculate the Young modulus of the intact rock by pressurizing and unpressurizing the hole and by measuring the recoverable part of the radial displacement, i.e.,

$$\Delta AA = \Delta BB = \frac{2qa(1+\nu)}{E} \qquad (18)$$

Knowing q, a, ΔAA, ΔBB and assuming a value for ν, it would then be possible to calculate E.

Note that if the hole is not pressurized over its entire length, a generalized plane stress condition may be assumed (Leknitskii, pp. 139-141) and a similar procedure can be carried out. Fig. 4 is still applicable if the quantities $E/(1-\nu^2)$ and $\nu/(1-\nu)$ appearing in the dimensionless quantities are modified to E and ν respectively.

5. DISCUSSION AND CONCLUSIONS

The determination of stress and displacement distributions around a long hole drilled in an initially stressed anisotropic medium can be achieved by using the concept of generalized or complete plane strain. This approach can be used when the planes of elastic symmetry of the medium do not correspond with the geometrical symmetry of the hole and/or the hole is not in a direction of principal field stress. Stresses, induced displacements and boundary conditions are identical on all planes perpendicular to the axis of the hole. This approach allows the longitudinal displacements to be non zero.

When the field stresses and/or the boundary conditions against the wall of the hole are acting on a narrow section of the hole, the concepts of generalized plane stress (Leknitskii, 1963) or complete plane stress (Brady and Bray, 1978) can be introduced.

The concept of an equivalent medium has been used to model the mechanical behavior or regularly jointed rocks. The applicability of this concept depends on the ratio between the joint set spacings and the problem to be dealt with. For example, when dealing with a borehole the appropriate spacings may be of the order of a few centimeters. However, for large cavities they may be as large as a few decimeters.

It was also assumed that the joint normal and shear stiffnesses are constant. However, it is known (Goodman, 1976) that those quantities can be normal stress dependent. A numerical procedure can be constructed to incorporate that dependency as well as an eventual dilatancy component. Joint properties always appeared in terms of sets of products $k_{ni}Si$ and $k_{si}Si$ ($i = 1,2,3$). It would be interesting to study the influence of the ratios between these products and the intact rock properties on the displacement and stress distributions.

A failure criterion must be superimposed on the present analysis to investigate the zones of potential failure around the hole. This failure criterion may be of the empirical type (Hoek and Brown, 1980) or based on mechanistic or physical models (Ladanyi and Archambault, 1980).

The complete plane strain formulation can be incorporated into numerical models. Eissa (1980) proposed an indirect formulation of the boundary element method to calculate the distribution of stresses and displacements around cavities of any shape in isotropic or anisotropic rocks. However, his model does not apply when the directions of anisotropy do not correspond with the geometrical symmetry of the cavity.

A few examples limited to the case of one joint set, have been presented here. They have shown the influence of the joint set orientation and properties on the stress and displacement distributions and on the orientation of the induced principal stresses around a hole drilled in a principal stress direction. A similar approach can be used when dealing with two or three orthogonal joint sets, when the hole is not drilled in a principal stress direction or when the anisotropy is due to parallel arrangements of flat minerals (foliated rocks).

For in-situ stress techniques that require the measurement of displacements in or around a hole, it would be misleading to neglect rock anisotropy and its orientation when calculating the field stress components. (Procedures for including the anisotropy are being developed presently.)

In the last example, a procedure is proposed to calculate the properties of a joint set parallel to a borehole dilatometer. The deformations of two orthogonal diameters have to be recorded, as is possible, for example, with the "tube deformeter" developed by Takano and Shidomoto (1966). Similarly, the Young's modulus of the intact rock can be estimated by pressurizing and unpressurizing the hole. If the joint spacing is known, its normal and shear stiffnesses (assumed constant) can be calculated. Therefore, knowing the orientation and the properties of a joint set in a rock mass, it would be possible to evaluate the modulus of deformation and the modulus of permanent deformation of the fractured rock mass (Amadei and Goodman, 1981) for different directions of loading and joint set spacings. This leads to a method similar to those proposed by Heuze (1971), Heuze et al (1980), and by Raphael and Goodman (1979). A figure similar to Fig. 4 can be proposed for other joint set orientations.

6. REFERENCES

Amadei, B. and Goodman, R. E., 1981, "A 3-D Constitutive Relation for Fractured Rock Masses," Proc. Int. Symp. on the Mechanical Behavior of Structured Media, Ottawa, May.

Berry, D. S., 1968, "The Theory of Stress Determination by Means of Stress Relief Techniques in a Transversely Isotropic Medium," Tech. Rept. No. 5-68, Missouri River Division, Corps of Engineers, Omaha, Nebraska 68101, Oct.

Brady, B. H. G. and Bray, J. W., 1978, "The Boundary Element Method for Determining Stresses and Displacements Around Long Openings in a Triaxial

Stress Field," Int. J. Rock Mech. Min. Sci., Vol. 15, pp. 21-28.

Duncan, J. M. and Goodman, R. E., 1968, "Finite Element Analysis of Slopes in Jointed Rocks," Contract Rept. No. 568-3, U.S. Army Engineer Waterways Experiment Station, Corps of Engineers, Vicksburg, Mississippi, Feb.

Eissa, E. S. A., "Stress Analysis of Underground Excavations in Isotropic and Stratified Rock Using the Boundary Element Method," Thesis presented to the Univ. of London, Imperial College, England in 1979, in partial fulfillment of the requirements for the degree of Doctor of Philosophy.

Fairhurst, C., 1968, "Methods of Determining In-Situ Rock Stresses at Great Depths," Tech. Rept. No. 1-68, Missouri River Division, Corps of Engineers, Omaha, Nebraska 68102, Feb.

Filon, L. N. G., 1937, "On Antiplane Stress in an Elastic Solid," Proc. Roy. Soc. Lond., A 160, pp. 137-154.

Goodman, R. E., 1976, Methods of Geological Engineering, West Publ.

Goodman, R. E., 1980, Introduction to Rock Mechanics, J. Wiley & Sons.

Heuze, F. E., 1971, "Source of Errors in Rock Mechanics Field Measurements, and Related Solution," Int. J. Rock Mech. Min. Sci., Vol. 8, pp. 297-310.

Heuze, F. E., Patrick, W. E., and de la Cruz, R. V., 1980, "In Situ Geomechanics, Climax Granite, Nevada Test Site," Report UCRL 85308, Dec.

Hoek, E. and Brown, E. T. 1980, "Empirical Strength Criterion for Rock Masses," Journal of the Geotechnical Eng. Division, ASCE, Vol. 106, No. GT9, pp. 1013-1035, Sept.

Ladanyi, B. and Archambault, G., 1980, "Direct and Indirect Determination of Shear Strength in Rock Mass," Proc. AIME Annual Meeting, Paper 80-25, Feb.

Lekhnitskii, S. G., 1963, Theory of Elasticity of an Anisotropic Elastic Body, Holden Day, Inc.

Milne Thomson, L. M., 1962, Antiplane Elastic Systems, Springer Verlag.

Morland, L. W., 1976, "Elastic Anisotropy of Regularly Jointed Media," Rock Mechanics 8, pp. 35-48.

Pinto, J. L., 1966, "Stresses and Strains in an Anisotropic Orthotropic Body," Proc. 1st Cong. ISRM, Vol. 1, pp. 625-635.

Raphael, J. M. and Goodman, R. E., 1979, "Strength and Deformability of Highly Fractured Rock," Journal of the Geotechnical Eng. Division, ASCE, Vol. 105, No. GT11, pp. 1285-1300, Nov.

Salomon, M. D. G., 1968, "Elastic Moduli of a Stratified Rock Mass," Int. J. Rock Mech. Min. Sci., Vol. 5, No. 6, pp. 519-527.

Takano, M. and Shidomoto, Y., 1966, "Deformation Test on Mudstone Enclosed in a Foundation by Means of Tube Deformeter," Proc. 1st Cong. ISRM, Vol. 1, pp. 761-764.

APPENDIX 1. Types of Plane Strain Problems

	Orientation of the directions of anisotropy with respect to the geometrical symmetry of the problem	Field stresses applied at infinity	Type of plane strain condition
$L_3 \neq 0$	Any orientation with $L_3 \neq 0$	$\sigma_{x,0}$; $\sigma_{y,0}$; $\tau_{yz,0}$; $\tau_{xz,0}$; $\tau_{xy,0}$	Complete plane strain*
$L_3 = 0$	• Orthotropy with one plane of symmetry perpendicular to the hole axis • Transverse isotropy parallel or perpendicular to the hole axis • Isotropy	$\tau_{xz,0} = \tau_{yz,0} = 0$	Plane strain
		$\tau_{xz,0} \neq 0$ $\tau_{yz,0} \neq 0$	Complete plane strain = plane strain + antiplane strain

*τ_{xz}, τ_{yz}, γ_{xz}, γ_{yz} can be induced even though $\tau_{xz,0}$ and/or $\tau_{yz,0}$ are zero.

Appendix 4.7

1) Let us consider the elastic equilibrium of an anisotropic, homogeneous body bounded internally by a cylinder of circular cross section of radius a. The anisotropic body is now analogous to a thin plate, with one plane of elastic symmetry perpendicular to its middle plane. The problem is to determine the distributions of average components of stress, strain, and displacement (across the plate thickness) induced

• by drilling the hole in a 2D stress field parallel to the plane of the plate with average components $\overline{\sigma_{x,0}}, \overline{\sigma_{y,0}}, \overline{\tau_{xy,0}}$ and,

• by the application of an internal pressure (if any) q distributed uniformly along the contour of the hole*.

A generalized plane stress formulation is used to solve this problem. As shown in section 4.5.3, the average values of stress and displacement induced by $\overline{\sigma_{x,0}}, \overline{\sigma_{y,0}}, \overline{\tau_{xy,0}}$ and q are equal to**

$$\begin{aligned} \overline{\sigma_{xh}} &= 2\mathrm{Re}\left(\mu_1^2\phi_1' + \mu_2^2\phi_2'\right) & \overline{u_h} &= -2\mathrm{Re}\left(p_1\phi_1 + p_2\phi_2\right) \\ \overline{\sigma_{yh}} &= 2\mathrm{Re}\left(\phi_1' + \phi_2'\right) & & \\ \overline{\tau_{xyh}} &= -2\mathrm{Re}\left(\mu_1\phi_1' + \mu_2\phi_2'\right) & \overline{v_h} &= -2\mathrm{Re}\left(q_1\phi_1 + q_2\phi_2\right) \end{aligned} \tag{1}$$

where μ_1, μ_2 are the roots of eq (4.66) and $\phi_1, \phi_2, \phi_1', \phi_2'$ are equal to the two first of eqs (44) and (45) in Appendix 4.5. However, $\overline{a_1}$ and $\overline{b_1}$ are now equal to

---

* q is also assumed to be an average quantity across the plate thickness.

** rigid body rotation and translations are set equal to zero.

$$\overline{a}_1 = -\frac{a}{2}\left(\overline{\sigma}_{y,0} - i\,\overline{\tau}_{xy,0}\right) + \frac{qa}{2}$$

$$\overline{b}_1 = \frac{a}{2}\left(\overline{\tau}_{xy,0} - i\,\overline{\sigma}_{x,0}\right) + i\frac{qa}{2}$$

(2)

Note that in eqs (2), a bar over a_1 and b_1 means the conjugate value of those quantities. A bar over $\sigma_{x,0}, \sigma_{y,0}$ or $\tau_{xy,0}$ indicates an average quantity

p_1, p_2, q_1, q_2 take the form

$$p_k = a_{11}\mu_k^2 + a_{12} - a_{16}\mu_k$$

$$q_k = a_{12}\mu_k + \frac{a_{22}}{\mu_k} - a_{26} \qquad (3)$$

with $k = 1, 2$.

At first, the following derivations will be carried out assuming zero internal pressure q within the hole. It will be shown later how to incorporate it.

2) Combining the expressions for $\phi_1'(z_1), \phi_2'(z_2)$ (Eqs (45), Appendix 4.5) with eqs (2) and the quantities γ_k $(k=1,2)$ defined in eq (8) (Appendix 4.5), the average induced stress components take the form

$$\begin{bmatrix} \overline{\sigma}_{xh} \\ \overline{\sigma}_{yh} \\ \overline{\tau}_{xyh} \end{bmatrix} = \begin{bmatrix} b_1 & c_1 & d_1 \\ b_2 & c_2 & d_2 \\ b_3 & c_3 & d_3 \end{bmatrix} \begin{bmatrix} \overline{\sigma}_{x,0} \\ \overline{\sigma}_{y,0} \\ \overline{\tau}_{xy,0} \end{bmatrix} \qquad (4)$$

or

$$(\overline{\sigma})_{xy}^{(h)} = \left(F_\sigma^{(h)}\right)\left(\overline{\sigma}_0\right)_{xy} \qquad (5)$$

where

$$b_1 = -\,\mathrm{Re}\left(i\gamma_2\mu_2^2 - i\gamma_1\mu_1^2\right)$$
$$c_1 = -\,\mathrm{Re}\left(\gamma_1\mu_1^2\mu_2 - \gamma_2\mu_1\mu_2^2\right)$$
$$d_1 = \mathrm{Re}\left(\gamma_1\mu_1^2\left(i\mu_2-1\right) + \gamma_2\mu_2^2\left(1-i\mu_1\right)\right)$$
$$b_2 = -\,\mathrm{Re}\left(i\gamma_2 - i\gamma_1\right)$$
$$c_2 = -\,\mathrm{Re}\left(\gamma_1\mu_2 - \mu_1\gamma_2\right)$$
$$d_2 = \mathrm{Re}\left(\gamma_1\left(i\mu_2-1\right) + \gamma_2\left(1-i\mu_1\right)\right)$$
$$b_3 = \mathrm{Re}\left(i\gamma_2\mu_2 - i\gamma_1\mu_1\right)$$
$$c_3 = \mathrm{Re}\left(\gamma_1\mu_1\mu_2 - \gamma_2\mu_1\mu_2\right)$$
$$d_3 = -\,\mathrm{Re}\left(\gamma_1\mu_1\left(i\mu_2-1\right) + \gamma_2\mu_2\left(1-i\mu_1\right)\right)$$

Since

$$\begin{bmatrix}\bar{\varepsilon}_{x_h}\\ \bar{\varepsilon}_{y_h}\\ \bar{\gamma}_{xy_h}\end{bmatrix} = \begin{bmatrix}a_{11} & a_{12} & a_{16}\\ a_{21} & a_{22} & a_{26}\\ a_{61} & a_{62} & a_{66}\end{bmatrix}\begin{bmatrix}\bar{\sigma}_{x_h}\\ \bar{\sigma}_{y_h}\\ \bar{\tau}_{xy_h}\end{bmatrix} \tag{6}$$

or

$$\left(\bar{\varepsilon}\right)_{xy}^{(h)} = \left(A\right)\left(\bar{\sigma}\right)_{xy}^{(h)} \tag{7}$$

Then, combining eqs (5) and (7) we obtain

$$\left(\bar{\varepsilon}\right)_{xy}^{(h)} = \left(A\right)\left(F_\sigma^{(h)}\right)\left(\bar{\sigma}_0\right)_{xy} \tag{8}$$

From eq (3) (<u>Appendix 4.4</u>)

$$\begin{bmatrix}\bar{\varepsilon}_{r_h}\\ \bar{\varepsilon}_{\theta_h}\\ \bar{\gamma}_{r\theta_h}\end{bmatrix} = \begin{bmatrix}\cos^2\theta & \sin^2\theta & \frac{\sin 2\theta}{2}\\ \sin^2\theta & \cos^2\theta & -\frac{\sin 2\theta}{2}\\ -\sin 2\theta & \sin 2\theta & \cos 2\theta\end{bmatrix}\begin{bmatrix}\bar{\varepsilon}_{x_h}\\ \bar{\varepsilon}_{y_h}\\ \bar{\gamma}_{xy_h}\end{bmatrix} \tag{9}$$

or

$$\left(\bar{\varepsilon}\right)_{r\theta}^{(h)} = \left(T_\varepsilon^*\right)\left(\bar{\varepsilon}\right)_{xy}^{(h)} \tag{10}$$

Combining eqs (8) and (10) we obtain

$$(\overline{\varepsilon})^{(h)}_{r\theta} = (T^*_\varepsilon)(A)(F^{(h)}_\sigma)(\overline{\sigma_o})_{xy} \tag{11}$$

3) Combining the expressions for $\phi_1(z_1), \phi_2(z_2)$ (Eqs (44), Appendix 4.5) with eqs (2) and the quantities Δ_k $(k=1,2)$ defined in eq (21a) (Appendix 4.5) the induced displacement components $\overline{u}_h, \overline{v}_h$ take the form

$$\begin{bmatrix} \dfrac{\overline{u}_h}{a} \\ \dfrac{\overline{v}_h}{a} \end{bmatrix} = \begin{bmatrix} b_1' & c_1' & f_1' \\ b_2' & c_2' & f_2' \end{bmatrix} \begin{bmatrix} \overline{\sigma}_{x,o} \\ \overline{\sigma}_{y,o} \\ \overline{\tau}_{xy,o} \end{bmatrix} \tag{12}$$

where

$$\begin{aligned}
b_1' &= \mathrm{Re}\,(i\Delta_2 p_2 - i\Delta_1 p_1) \\
c_1' &= \mathrm{Re}\,(p_1\mu_2\Delta_1 - p_2\mu_1\Delta_2) \\
f_1' &= -\mathrm{Re}\,(\Delta_1 p_1 (i\mu_2 - 1) + \Delta_2 p_2 (1 - i\mu_1)) \\
b_2' &= \mathrm{Re}\,(i\Delta_2 q_2 - i\Delta_1 q_1) \\
c_2' &= \mathrm{Re}\,(q_1\mu_2\Delta_1 - q_2\mu_1\Delta_2) \\
f_2' &= -\mathrm{Re}\,(\Delta_1 q_1 (i\mu_2 - 1) + \Delta_2 q_2 (1 - i\mu_1))
\end{aligned}$$

Then, rewrite eqs (12) in terms of average radial and tangential displacements

$$\begin{bmatrix} \dfrac{\overline{u}_{rh}}{a} \\ \dfrac{\overline{v}_{\theta h}}{a} \end{bmatrix} = \begin{bmatrix} \cos\theta & \sin\theta \\ -\sin\theta & \cos\theta \end{bmatrix} \begin{bmatrix} b_1' & c_1' & f_1' \\ b_2' & c_2' & f_2' \end{bmatrix} \begin{bmatrix} \overline{\sigma}_{x,o} \\ \overline{\sigma}_{y,o} \\ \overline{\tau}_{xy,o} \end{bmatrix} \tag{13}$$

or

$$\begin{bmatrix} \dfrac{\overline{u}_{rh}}{a} \\ \dfrac{\overline{v}_{\theta h}}{a} \end{bmatrix} = \begin{bmatrix} f_{11}^{(h)} & f_{12}^{(h)} & f_{13}^{(h)} \\ f_{21}^{(h)} & f_{22}^{(h)} & f_{23}^{(h)} \end{bmatrix} \begin{bmatrix} \overline{\sigma}_{x,o} \\ \overline{\sigma}_{y,o} \\ \overline{\tau}_{xy,o} \end{bmatrix} \tag{14}$$

From (14) we can see that

$$\frac{\overline{u}_{rh}}{a} = f_{11}^{(h)}\overline{\sigma}_{x,0} + f_{12}^{(h)}\overline{\sigma}_{y,0} + f_{13}^{(h)}\overline{\tau}_{xy,0} \tag{15}$$

For any point along the contour ($r = a$), the coefficients $\Delta_1, \Delta_2, \Delta_3$ can be expressed in terms of $\cos\theta$ and $\sin\theta$ (eq (37), Appendix 4.5), and substitute into the expressions for b_1', c_1', f_1' and b_2', c_2', f_2'. Using the same procedure than in Appendix 4.5, it can be shown that $\overline{u}_{rh}/a$ takes the form

$$\frac{\overline{u}_{rh}}{a} = M_1 + M_2\cos 2\theta + M_3 \sin 2\theta \tag{16}$$

Coefficients M_1, M_2, M_3 are now functions of $\overline{\sigma}_{x,0}, \overline{\sigma}_{y,0}, \overline{\tau}_{xy,0}$ and the material properties.

It can be shown that for an isotropic material with a Young modulus E and a Poisson's ratio ν and for a point located along the contour ($r = a$), the coefficients $f_{1i}^{(h)}$ ($i = 1,2,3$) are equal to

$$\begin{aligned} f_{11}^{(h)} &= -\left((1+\nu) + (3-\nu)\cos 2\theta\right)/2E \\ f_{12}^{(h)} &= -\left((1+\nu) - (3-\nu)\cos 2\theta\right)/2E \\ f_{13}^{(h)} &= -(3-\nu)\sin 2\theta / E \end{aligned} \tag{17a}$$

and

$$\begin{aligned} M_1 &= -(1+\nu)(\overline{\sigma}_{x,0} + \overline{\sigma}_{y,0})/2E \\ M_2 &= -(3-\nu)(\overline{\sigma}_{x,0} - \overline{\sigma}_{y,0})/2E \\ M_3 &= -(3-\nu)\overline{\tau}_{xy,0}/E \end{aligned} \tag{17b}$$

4) Let us introduce an internal pressure q distributed

uniformly along the hole. Combining the expressions for $\phi_1, \phi_2, \phi'_1, \phi'_2$ (Eqs (44) and (45) Appendix 4.5) with eqs (2) we obtain

$$\begin{aligned}
\phi_1 &= \phi_{1\,q=0} + q\,\Delta_1 a\,(\mu_2 - i)/2 \\
\phi_2 &= \phi_{2\,q=0} + q\,\Delta_2 a\,(i - \mu_1)/2 \\
\phi'_1 &= \phi'_{1\,q=0} + q\,\gamma_1\,(\mu_2 - i)/2 \\
\phi'_2 &= \phi'_{2\,q=0} + q\,\gamma_2\,(i - \mu_1)/2
\end{aligned} \tag{17}$$

The induced stress components are equal to

$$(\bar{\sigma})^{(h)}_{xy} = (F^{(h)}_{\sigma})(\bar{\sigma}_o)_{xy} + q\,(F^*) \tag{18}$$

where

$$\begin{aligned}
f^*_{11} &= \mathrm{Re}\left(\gamma_1 \mu_1^2 (\mu_2 - i) + \gamma_2 \mu_2^2 (i - \mu_1)\right) \\
f^*_{12} &= \mathrm{Re}\left(\gamma_1 (\mu_2 - i) + \gamma_2 (i - \mu_1)\right) \\
f^*_{13} &= -\mathrm{Re}\left(\gamma_1 \mu_1 (\mu_2 - i) + \gamma_2 \mu_2 (i - \mu_1)\right)
\end{aligned}$$

Combining eqs (7) and (18) and making use of eq (10)

$$(\bar{\varepsilon})^{(h)}_{xy} = (A)(F^{(h)}_{\sigma})(\bar{\sigma}_o)_{xy} + q\,(A)(F^*) \tag{19}$$

$$(\bar{\varepsilon})^{(h)}_{r\theta} = (T^*_{\varepsilon})(A)(F^{(h)}_{\sigma})(\bar{\sigma}_o)_{xy} + q\,(T^*_{\varepsilon})(A)(F^*) \tag{20}$$

The induced displacement components in the (r, θ) coordinate system are equal to

$$\begin{bmatrix} \dfrac{\bar{u}_{rh}}{a} \\ \dfrac{\bar{v}_{\theta h}}{a} \end{bmatrix} = \begin{bmatrix} \dfrac{\bar{u}_{rh}}{a} \\ \dfrac{\bar{v}_{\theta h}}{a} \end{bmatrix}_{q=0} + q \begin{bmatrix} \cos\theta & \sin\theta \\ -\sin\theta & \cos\theta \end{bmatrix} \begin{bmatrix} U_1 \\ U_2 \end{bmatrix} \tag{21}$$

with

$$\begin{aligned}
U_1 &= -\mathrm{Re}\left(\Delta_1 p_1 (\mu_2 - i) + \Delta_2 p_2 (i - \mu_1)\right) \\
U_2 &= -\mathrm{Re}\left(\Delta_1 q_1 (\mu_2 - i) + \Delta_2 q_2 (i - \mu_1)\right)
\end{aligned}$$

Appendix 4.8

1) Let us consider the elastic equilibrium of an anisotropic, homogeneous body bounded internally by a cylinder of circular cross section of radius a . The body has three orthogonal planes of elastic symmetry, each one perpendicular to a coordinate axis x,y or z. The hole is pressurized by an internal pressure q uniformly distributed along its contour. The problem is to determine the distribution of the displacement components induced by the internal pressure q. Two formulations are proposed

- Classical plane strain formulation when the hole is very long.
- Generalized plane stress formulation when the hole is pressurized over a short length.

2) Consider first the classical plane strain formulation when the hole is infinitely long and pressurized over its whole length.

Recall eqs (44), (46b), (5) and (21a) in Appendix 4.5. The induced displacement components are equal to*

$$u_h = -2\,\mathrm{Re}\left(p_1 \phi_1(z_1) + p_2 \phi_2(z_2)\right)$$
$$v_h = -2\,\mathrm{Re}\left(q_1 \phi_1(z_1) + q_2 \phi_2(z_2)\right) \tag{1}$$

where

$$\phi_1 = \frac{qa\Delta_1}{2}(\mu_2 - i) \quad ; \quad \phi_2 = \frac{qa\Delta_2}{2}(i - \mu_1) \tag{2}$$

---

* if we neglect rigid body rotation and translations.

Substituting eqs (2) into (1) we obtain

$$\frac{uh}{a} = -q\,\mathrm{Re}\left(\Delta_1 p_1 (\mu_2 - i) + \Delta_2 p_2 (i - \mu_1)\right)$$

$$\frac{vh}{a} = -q\,\mathrm{Re}\left(\Delta_1 q_1 (\mu_2 - i) + \Delta_2 q_2 (i - \mu_1)\right) \tag{3}$$

μ_1, μ_2 are solutions of the equation

$$l_4(\mu) = \beta_{11}\mu^4 - 2\beta_{16}\mu^3 + (2\beta_{12} + \beta_{66})\mu^2 - 2\beta_{26}\mu + \beta_{22} = 0 \tag{4}$$

Since β_{16}, β_{26} must be equal to zero for the present orientation of the anisotropy with respect to the x,y,z coordinate system, eq (4) reduces to the following

$$\beta_{11}\mu^4 + (2\beta_{12} + \beta_{66})\mu^2 + \beta_{22} = 0 \tag{5}$$

It can be shown that it has four roots $\mu_1, \mu_2, \bar{\mu}_1, \bar{\mu}_2$ such that

$$\mu_1 = i\beta_1\ ;\ \mu_2 = i\beta_2\ ;\ \bar{\mu}_1 = -i\beta_1\ ;\ \bar{\mu}_2 = -i\beta_2 \tag{6}$$

β_1, β_2 are positive quantities and are defined as follows

$$\begin{matrix}\beta_1\\ \beta_2\end{matrix} = \sqrt{\frac{(2\beta_{12} + \beta_{66}) \mp \sqrt{(2\beta_{12} + \beta_{66})^2 - 4\beta_{11}\beta_{22}}}{2\beta_{11}}} \tag{7}$$

Furthermore,

$$\beta_1\beta_2 = \sqrt{\frac{\beta_{22}}{\beta_{11}}}\ ;\qquad \beta_1^2 + \beta_2^2 = \frac{2\beta_{12} + \beta_{66}}{\beta_{11}} \tag{8}$$

For the present anisotropy

$$p_k = \beta_{11}\mu_k^2 + \beta_{12}\ ;\ q_k = \beta_{12}\mu_k + \frac{\beta_{22}}{\mu_k} \quad (k = 1,2) \tag{9}$$

Let us calculate u_h/a and v_h/a at any point along the contour of the hole ($r = a$). Then, (eq (37), Appendix 4.5)

$$\Delta_1 = \Delta_2 = (\cos\theta - i\sin\theta)/\Delta \quad ; \quad \Delta = \mu_2 - \mu_1 \tag{10}$$

Substituting eqs (9), (10), (6) into eqs (3) we obtain

$$\frac{u_h}{a} = -q\cos\theta\left(\beta_{12} + \beta_{11}\beta_1\beta_2 - \beta_{11}(\beta_1+\beta_2)\right)$$

$$\frac{v_h}{a} = -q\sin\theta\left(\beta_{12} + \frac{\beta_{22}}{\beta_1\beta_2} - \beta_{22}\frac{(\beta_1+\beta_2)}{\beta_1\beta_2}\right) \tag{11}$$

Using eqs (23) (Appendix 4.5), the radial and tangential components of displacement are equal to

$$\frac{u_{rh}}{a} = q\left(\beta_{11}(\beta_1+\beta_2-\beta_1\beta_2)\cos^2\theta + \frac{\beta_{22}}{\beta_1\beta_2}(\beta_1+\beta_2-1)\sin^2\theta - \beta_{12}\right)$$

$$\frac{v_{\theta h}}{a} = q\sin\theta\cos\theta\left(\frac{\beta_{22}}{\beta_1\beta_2}(\beta_1+\beta_2-1) - \beta_{11}(\beta_1+\beta_2-\beta_1\beta_2)\right) \tag{12}$$

If the rock material is isotropic with a Young modulus E and a Poisson's ratio ν, then

$$\beta_1 = \beta_2 = 1 \quad ; \quad \beta_{11} = \beta_{22} = \frac{1-\nu^2}{E} \quad ; \quad \beta_{12} = -\nu\frac{(1+\nu)}{E} \tag{13}$$

Combining eqs (13) and (12) we obtain

$$\frac{u_{rh}}{a} = \frac{q(1+\nu)}{E} \quad ; \quad \frac{v_{\theta h}}{a} = 0 \tag{14}$$

Recall eq (12). Since $\cos^2\theta$ and $\sin^2\theta$ can be expressed in terms of $\cos 2\theta$ only, this implies that there are at most two independent values of radial displacement u_r. Furthermore, the sum of two radial displacements 90° apart is an invariant along the contour. Indeed, let u_{r_1} be the radial displacement at $r=a, \theta=\theta_1$, and u_{r_2} be the radial displacement at $r=a, \theta_2=\theta_1+\frac{\pi}{2}$ for the same material properties and value of q, then according to eq (12)

$$\frac{u_{r_1h}}{a} = q\left(\beta_{11}(\beta_1+\beta_2-\beta_1\beta_2)\cos^2\theta_1 + \frac{\beta_{22}}{\beta_1\beta_2}(\beta_1+\beta_2-1)\sin^2\theta_1 - \beta_{12}\right)$$

$$\frac{u_{r_2h}}{a} = q\left(\beta_{11}(\beta_1+\beta_2-\beta_1\beta_2)\sin^2\theta_1 + \frac{\beta_{22}}{\beta_1\beta_2}(\beta_1+\beta_2-1)\cos^2\theta_1 - \beta_{12}\right)$$

Then,

$$\frac{u_{r_1h}+u_{r_2h}}{qa} = \beta_{11}(\beta_1+\beta_2-\beta_1\beta_2) + \frac{\beta_{22}}{\beta_1\beta_2}(\beta_1+\beta_2-1) - 2\beta_{12} \qquad (15)$$

3) Then, consider the <u>generalized plane stress formulation</u> when the hole is pressurized over a short length.

Eqs (1), (2) and (3) are still valid. In eqs (5), (7), (8) and (9), all the coefficients β_{ij}' must be replaced by coefficients a_{ij} $(i=1,2)$. Then, eqs (11) and (12) become

$$\frac{\overline{u}_h}{a} = -q\cos\theta\left(a_{12}+a_{11}\beta_1\beta_2 - a_{11}(\beta_1+\beta_2)\right)$$

$$\frac{\overline{v}_h}{a} = -q\sin\theta\left(a_{12}+\frac{a_{22}}{\beta_1\beta_2} - a_{22}\frac{(\beta_1+\beta_2)}{\beta_1\beta_2}\right) \qquad (16)$$

and

$$\frac{\overline{u}_{rh}}{a} = q\left(a_{11}(\beta_1+\beta_2-\beta_1\beta_2)\cos^2\theta + \frac{a_{22}}{\beta_1\beta_2}(\beta_1+\beta_2-1)\sin^2\theta - a_{12}\right)$$

$$\frac{\overline{v}_{\theta h}}{a} = q\sin\theta\cos\theta\left(\frac{a_{22}}{\beta_1\beta_2}(\beta_1+\beta_2-1) - a_{11}(\beta_1+\beta_2-\beta_1\beta_2)\right) \qquad (17)$$

If the rock material is isotropic with a Young modulus E and a Poisson's ratio ν, then

$$\beta_1=\beta_2=1 \;;\quad a_{11}=a_{22}=\frac{1}{E} \;;\quad a_{12}=\frac{-\nu}{E}$$

and

$$\frac{\overline{u}_{rh}}{a} = q\,\frac{(1+\nu)}{E} \quad ; \quad \frac{\overline{v}_{\theta h}}{a} = 0 \tag{18}$$

Condition (15) is again satisfied and can be written as follows

$$\frac{\overline{u}_{r_1h} + \overline{u}_{r_2h}}{qa} = a_{11}\left(\beta_1+\beta_2-\beta_1\beta_2\right) + \frac{a_{22}}{\beta_1\beta_2}\left(\beta_1+\beta_2-1\right) - 2a_{12} \tag{19}$$

Appendix 5.1

1) Recall the general expressions for the stress function $\phi(r,\theta)$ (eq 5.15) for a hollow inclusion:

$$\phi(r,\theta) = a_o + b_o \ln r + c_o r^2 + A_o^* \theta + \left(a_1 r + \frac{c_1}{r} + d_1 r^3\right)_{\cos\theta}^{\sin\theta} + \sum_{n=2}^{\infty} \left(a_n r^n + b_n r^{2+n} + c_n r^{-n} + d_n r^{2-n}\right)_{\cos n\theta}^{\sin n\theta} \quad (1)$$

Substituting eq (5.15) into eqs (5.8) we obtain the general expressions for the stress components σ_r, σ_θ and $\tau_{r\theta}$ as follows:

$$\rightarrow \sigma_r = A_{o_r} + A_{1_r}\cos\theta + B_{1_r}\sin\theta + \sum_{n=2}^{\infty}\left(A_{n_r}\cos n\theta + B_{n_r}\sin n\theta\right) \quad (2)$$

with

$$A_{o_r} = \frac{b_o}{r^2} + 2c_o$$

$$A_{1_r} = -\frac{2c_1'}{r^3} + 2d_1' r \quad ; \quad B_{1_r} = -\frac{2c_1}{r^3} + 2d_1 r$$

$$A_{n_r} = -\left(n a_n' r^{n-2}(n-1) + b_n' r^n (n+1)(n-2) + n(n+1)c_n' r^{-(n+2)} + d_n' r^{-n}(n-1)(n+2)\right)$$

$$B_{n_r} = -\left(n a_n r^{n-2}(n-1) + b_n r^n (n+1)(n-2) + n(n+1)c_n r^{-(n+2)} + d_n r^{-n}(n-1)(n+2)\right)$$

$$\rightarrow \sigma_\theta = A_{o_\theta} + A_{1_\theta}\cos\theta + B_{1_\theta}\sin\theta + \sum_{n=2}^{\infty}\left(A_{n_\theta}\cos n\theta + B_{n_\theta}\sin n\theta\right) \quad (3)$$

with

$$A_{o_\theta} = -\frac{b_o}{r^2} + 2c_o$$

$$A_{1_\theta} = \frac{2c_1'}{r^3} + 6d_1' r \quad ; \quad B_{1_\theta} = \frac{2c_1}{r^3} + 6d_1 r$$

$$A_{n_\theta} = a_n' n(n-1) r^{n-2} + b_n'(n+1)(n+2) r^n + c_n' n(n+1) r^{-(n+2)} + (2-n)(1-n) d_n' r^{-n}$$

$$B_{n_\theta} = a_n n(n-1) r^{n-2} + b_n (n+1)(n+2) r^n + c_n n(n+1) r^{-(n+2)} + (2-n)(1-n) d_n r^{-n}$$

$$\longrightarrow \tau_{r\theta} = A_{0_{r\theta}} + A_{1_{r\theta}} \cos\theta + B_{1_{r\theta}} \sin\theta + \sum_{n=2}^{\infty} \left(A_{n_{r\theta}} \cos n\theta + B_{n_{r\theta}} \sin n\theta \right) \quad (4)$$

with

$$A_{0_{r\theta}} = \frac{A_0^*}{r^2}$$

$$A_{1_{r\theta}} = \frac{2c_1}{r^3} - 2d_1 r \quad ; \quad B_{1_{r\theta}} = -\frac{2c_1'}{r^3} + 2d_1' r$$

$$A_{n_{r\theta}} = -\left(n a_n r^{n-2}(n-1) + n b_n r^n (n+1) - n c_n r^{-(n+2)}(n+1) - n d_n r^{-n}(n-1) \right)$$

$$B_{n_{r\theta}} = \left(n a_n' r^{n-2}(n-1) + n b_n' r^n (n+1) - n c_n' r^{-(n+2)}(n+1) - n d_n' r^{-n}(n-1) \right)$$

The stress component in the z direction is defined by

$$\longrightarrow \sigma_z = E\,\varepsilon_{z_0} + \nu \left(\sigma_r + \sigma_\theta \right) \quad (5)$$

The coefficients entering into eqs (2), (3), and (4) can be calculated from the boundary conditions for stresses at $r = a$ and at $r = b$.

(i) <u>at $r = a$</u> the expressions for σ_r and $\tau_{r\theta}$ in eqs (2) and (4) must be equal to the two first of eqs (5.11) with conditions (5.12) and (5.13) being satisfied. Since the summations appearing in eqs (5.11) take place between $n = 2$ and $n = N + 1$, the coefficients to be calculated are reduced to $b_0, c_0, c_1, d_1, c_1', d_1', A_0^*, a_n, b_n, c_n, d_n, a_n', b_n', c_n', d_n'$ for n varying between 2 and $N + 1$.

Eqs (2) and (4) are equal to the two first of eqs (5.11) for any value of θ if and only if*

---

* See uniqueness theorem in Titchmarsch (1950).

$$\frac{b_0}{a^2} + 2c_0 = A_0$$

$$\frac{-2c'_1}{a^3} + 2d'_1 a = A_1$$

$$\frac{-2c_1}{a^3} + 2d_1 a = B_1$$

$$A_0^* = 0$$

$$-\left(n a'_n a^{n-2}(n-1) + b'_n a^n (n+1)(n-2) + n(n+1) c'_n a^{-(n+2)} + d'_n a^{-n}(n-1)(n+2)\right) = A_n$$

$$-\left(n a_n a^{n-2}(n-1) + b_n a^n (n+1)(n-2) + n(n+1) c_n a^{-(n+2)} + d_n a^{-n}(n-1)(n+2)\right) = B_n$$

$$-\left(n a_n a^{n-2}(n-1) + n b_n a^n (n+1) - n c_n a^{-(n+2)}(n+1) - n d_n a^{-n}(n-1)\right) = C_n$$

$$\left(n a'_n a^{n-2}(n-1) + n b'_n a^n (n+1) - n c'_n a^{-(n+2)}(n+1) - n d'_n a^{-n}(n-1)\right) = D_n$$

(for $n = 2, N+1$) (6)

(ii) at $\underline{r = b}$, σ_r and $\tau_{r\theta}$ must vanish for any value of θ i.e.

$$\frac{b_0}{b^2} + 2c_0 = 0$$

$$\frac{-2c'_1}{b^3} + 2d'_1 b = 0 \quad ; \quad -\frac{2c_1}{b^3} + 2d_1 b = 0$$

$$n a'_n b^{n-2}(n-1) + b'_n b^n (n+1)(n-2) + n(n+1) c'_n b^{-(n+2)} + d'_n b^{-n}(n-1)(n+2) = 0$$

$$n a_n b^{n-2}(n-1) + b_n b^n (n+1)(n-2) + n(n+1) c_n b^{-(n+2)} + d_n b^{-n}(n-1)(n+2) = 0$$

$$n a_n b^{n-2}(n-1) + n b_n b^n (n+1) - n c_n b^{-(n+2)}(n+1) - n d_n b^{-n}(n-1) = 0$$

$$n a'_n b^{n-2}(n-1) + n b'_n b^n (n+1) - n c'_n b^{-(n+2)}(n+1) - n d'_n b^{-n}(n-1) = 0$$

(for $n = 2, N+1$) (7)

The solution of the system made of eqs (6) and (7) can be written as follows

$$A_0^* = 0 \quad ; \quad b_0 = \alpha_0 A_0 \quad ; \quad c_0 = \beta_0 A_0$$
$$c_1 = \beta_1 B_1 \quad ; \quad d_1 = \delta_1 B_1$$
$$c_1' = \beta_1 A_1 \quad ; \quad d_1' = \delta_1 A_1 \tag{8a}$$

where

$$\frac{\alpha_0}{a^2} = \frac{1}{1-m^2} \quad ; \quad \beta_0 = \frac{-m^2}{2(1-m^2)} \quad ; \quad \frac{\beta_1}{a^3} = \frac{1}{2(m^4-1)}$$
$$\delta_1 a = \frac{m^4}{2(m^4-1)} \tag{8b}$$

with

$$m = a/b \tag{8c}$$

Furthermore, for $n = 2, N+1$

$$\begin{aligned}
a_n &= U_{11}^{(n)} B_n + U_{12}^{(n)} C_n & a_n' &= U_{11}^{(n)\prime} A_n + U_{12}^{(n)\prime} D_n \\
b_n &= U_{21}^{(n)} B_n + U_{22}^{(n)} C_n & b_n' &= U_{21}^{(n)\prime} A_n + U_{22}^{(n)\prime} D_n \\
c_n &= U_{31}^{(n)} B_n + U_{32}^{(n)} C_n & c_n' &= U_{31}^{(n)\prime} A_n + U_{32}^{(n)\prime} D_n \\
d_n &= U_{41}^{(n)} B_n + U_{42}^{(n)} C_n & d_n' &= U_{41}^{(n)\prime} A_n + U_{42}^{(n)\prime} D_n
\end{aligned} \tag{9}$$

with

$$a^{n-2} U_{11}^{(n)} = \frac{m^{n-2}}{n(n-1)} \frac{T_{11}^{(n)}}{T^{(n)}} \quad ; \quad a^{n-2} U_{12}^{(n)} = \frac{-m^{n-2}}{n(n-1)} \frac{T_{12}^{(n)}}{T^{(n)}}$$

$$a^{n} U_{21}^{(n)} = \frac{-m^{n}}{(n+1)} \frac{T_{21}^{(n)}}{T^{(n)}} \quad ; \quad a^{n} U_{22}^{(n)} = \frac{m^{n}}{n+1} \frac{T_{22}^{(n)}}{T^{(n)}}$$

$$a^{-(n+2)} U_{31}^{(n)} = \frac{m^{-(n+2)}}{n(n+1)} \frac{T_{31}^{(n)}}{T^{(n)}} \quad ; \quad a^{-(n+2)} U_{32}^{(n)} = \frac{-m^{-(n+2)}}{n(n+1)} \frac{T_{32}^{(n)}}{T^{(n)}}$$

$$a^{-n} U_{41}^{(n)} = \frac{-m^{-n}}{(n-1)} \frac{T_{41}^{(n)}}{T^{(n)}} \quad ; \quad a^{-n} U_{42}^{(n)} = \frac{m^{-n}}{n-1} \frac{T_{42}^{(n)}}{T^{(n)}} \tag{10}$$

$$U_{i1}^{(n)'} = U_{i1}^{(n)} \quad ; \quad U_{i2}^{(n)'} = -U_{i2}^{(n)} \qquad (i=1,2,3,4) \tag{10}$$

and

$$T_{11}^{(n)} = -2n m^{n} + 2n m^{-n}(1-n) + 2n^{2} m^{-(n+2)}$$

$$T_{12}^{(n)} = -2(n-2)m^{n} - 2(n+2)(1-n)m^{-n} - 2n^{2}m^{-(n+2)}$$

$$T_{21}^{(n)} = -2m^{n-2} + 2m^{-(n+2)}(n+1) - 2n m^{-n}$$

$$T_{22}^{(n)} = -2m^{n-2} - 2m^{-(n+2)}(n+1) + 2(n+2)m^{-n}$$

$$T_{31}^{(n)} = 2n^{2}m^{n-2} - 2(n+1)n m^{n} + 2n m^{-n}$$

$$T_{32}^{(n)} = 2n^{2}m^{n-2} - 2(n-2)(n+1)m^{n} - 2(n+2)m^{-n}$$

$$T_{41}^{(n)} = -2m^{n-2}(1-n) - 2n m^{n} + 2m^{-(n+2)}$$

$$T_{42}^{(n)} = -2m^{n-2}(1-n) - 2(n-2)m^{n} - 2m^{-(n+2)}$$

$$T^{(n)} = 4m^{2n-2} - 4n^{2}m^{-4} - 4n^{2} + 4m^{-2n-2} + m^{-2}(8n^{2}-8) \tag{11}$$

The general expressions for the strain and displacement components $\varepsilon_r, \varepsilon_\theta, \gamma_{r\theta}, u_r, v_\theta$ may be obtained from eqs (5.4) and (5.5). Substituting eqs (2) and (3) into the first of eqs (5.5), then integrating with respect to r, yields

$$-u_r = \frac{(1-\nu^2)}{E}\Bigg(-\frac{b_0}{r}\frac{1}{(1-\nu)} + 2c_0 r\frac{(1-2\nu)}{(1-\nu)} + \left(\frac{c_1}{r^2}\frac{1}{(1-\nu)} + d_1 r^2\frac{(1-4\nu)}{(1-\nu)}\right)_{\cos\theta}^{\sin\theta} -$$

$$\sum_{n=2}^{N+1}\Bigg(\frac{a_n n r^{n-1}}{(1-\nu)} + b_n r^{n+1}\left(\frac{n}{(1-\nu)} + \frac{(4\nu-2)}{(1-\nu)}\right) - \frac{c_n n r^{-(n+1)}}{(1-\nu)}$$

$$- d_n r^{1-n}\left(\frac{n}{(1-\nu)} + \frac{(2-4\nu)}{(1-\nu)}\right)\Bigg)_{\cos n\theta}^{\sin n\theta} + g(\theta)\Bigg) - \nu r \varepsilon_{z0} \tag{12}$$

where $g(\theta)$ is an arbitrary function of integration. Combining the seconds of eqs (5.4) and (5.5), using eq (12) and integrating with respect to θ we obtain

$$-v_\theta = \frac{(1-\nu^2)}{E}\Bigg(\Big(\frac{c_1}{r^2}\frac{1}{(1-\nu)} + \frac{d_1 r^2}{(1-\nu)}(5-4\nu)\Big)^{-\cos\theta}_{\sin\theta} + \sum_{n=2}^{N+1}\Big(\frac{a_n n r^{n-1}}{(1-\nu)} + b_n r^{n+1}\frac{(n+4-4\nu)}{(1-\nu)} + \frac{c_n n r^{-(n+1)}}{(1-\nu)} + \frac{d_n r^{-(n-1)}}{(1-\nu)}(n-4+4\nu)\Big)^{-\cos n\theta}_{\sin n\theta} - \int g(\theta)\,d\theta + f(r)\Bigg) \tag{13}$$

where $f(r)$ is another arbitrary function of integration. Combining eq (4), the lasts of eqs (5.4) and (5.5), eqs (12) and (13), $f(r)$ and $g(\theta)$ must satisfy the following equation

$$\left(\frac{\partial g(\theta)}{\partial\theta} + \int g(\theta)\,d\theta\right) + \left(r\frac{\partial f(r)}{\partial r} - f(r)\right) = 0 \tag{14}$$

As shown for eq (30) in Appendix 4.5, the solution of eq (14) takes the form

$$g(\theta) = S_1\cos\theta + S_2\sin\theta \quad ; \quad f(r) = \varpi_1^* r \tag{15}$$

where S_1, S_2, ϖ_1^* are arbitrary functions of integration.

Substituting eqs (15) into eqs (12) and (13) yields

$$-u_r = \frac{1}{2G}\Bigg[-\frac{b_0}{r} + 2c_0 r(1-2\nu) + \Big(\frac{c_1}{r^2} + d_1 r^2(1-4\nu)\Big)\sin\theta + \Big(\frac{c_1'}{r^2} + d_1' r^2(1-4\nu)\Big)\cos\theta - \sum_{n=2}^{N+1}\Big(n a_n r^{n-1} + b_n r^{n+1}(n+4\nu-2) - c_n n r^{-(n+1)} - d_n r^{1-n}(n-4\nu+2)\Big)\sin n\theta - \sum_{n=2}^{N+1}\Big(n a_n' r^{n-1} + b_n' r^{n+1}(n+4\nu-2) - n c_n' r^{-(n+1)} - d_n' r^{1-n}(n-4\nu+2)\Big)\cos n\theta\Bigg] - \nu r \varepsilon_{z_0} - s_1^+\cos\theta - s_2^+\sin\theta. \tag{16}$$

and

$$-v_\theta = \frac{1}{2G}\Big[-\Big(\frac{c_1}{r^2} + d_1 r^2(5-4\nu)\Big)\cos\theta + \Big(\frac{c_1'}{r^2} + d_1' r^2(5-4\nu)\Big)\sin\theta - \sum_{n=2}^{N+1}\Big(a_n n r^{n-1} + b_n r^{n+1}(n+4-4\nu) + n c_n r^{-(n+1)} + d_n r^{-(n-1)}(n-4+4\nu)\Big)\cos n\theta + \sum_{n=2}^{N+1}\Big(a_n' n r^{n-1} + b_n' r^{n+1}(n+4-4\nu) + n c_n' r^{-(n+1)} + d_n' r^{-(n-1)}(n-4+4\nu)\Big)\sin n\theta\Big] + s_1^+ \sin\theta - s_2^+\cos\theta - \varpi_1^+ r \tag{17}$$

where s_1^+, s_2^+ are rigid body translations in the x,y, directions and ϖ_1^+ is a rigid body rotation around the z axis such that

$$s_1^+ = -\frac{(1-\nu^2)}{E} S_1 \quad ; \quad s_2^+ = -\frac{(1-\nu^2)}{E} S_2 \quad ; \quad \varpi_1^+ = -\frac{(1-\nu^2)}{E}\varpi_1^* \tag{18}$$

If we substitute eqs (8a), (8b), (9) and (10) into eqs (2), (3), (4), (16) and (17), then, stress and displacement components take the form

$$\bullet\sigma_r = \Big(\Big(\frac{a}{r}\Big)^2\frac{1}{1-m^2} - \frac{m^2}{1-m^2}\Big)A_0 + \Big(\frac{m^4}{m^4-1}\Big(\frac{r}{a}\Big) - \Big(\frac{a}{r}\Big)^3\frac{1}{m^4-1}\Big)(A_1\cos\theta + B_1\sin\theta)$$
$$-\sum_{n=2}^{N+1}\Big(\Big(\frac{r}{a}\Big)^{n-2} m^{n-2}\frac{T_{11}^{(n)}}{T^{(n)}} - \Big(\frac{r}{a}\Big)^n (n-2) m^n \frac{T_{21}^{(n)}}{T^{(n)}} + \Big(\frac{r}{a}\Big)^{-(n+2)} m^{-(n+2)}\frac{T_{31}^{(n)}}{T^{(n)}} - \Big(\frac{r}{a}\Big)^{-n}(n+2)m^{-n}\frac{T_{41}^{(n)}}{T^{(n)}}\Big) B_n \sin n\theta$$
$$-\sum_{n=2}^{N+1}\Big(-\Big(\frac{r}{a}\Big)^{n-2} m^{n-2}\frac{T_{12}^{(n)}}{T^{(n)}} + \Big(\frac{r}{a}\Big)^n m^n (n-2)\frac{T_{22}^{(n)}}{T^{(n)}} - \Big(\frac{r}{a}\Big)^{-(n+2)} m^{-(n+2)}\frac{T_{32}^{(n)}}{T^{(n)}} + \Big(\frac{r}{a}\Big)^{-n}(n+2)m^{-n}\frac{T_{42}^{(n)}}{T^{(n)}}\Big) C_n \sin n\theta$$
$$-\sum_{n=2}^{N+1}\Big(\Big(\frac{r}{a}\Big)^{n-2} m^{n-2}\frac{T_{11}^{(n)}}{T^{(n)}} - \Big(\frac{r}{a}\Big)^n (n-2) m^n \frac{T_{21}^{(n)}}{T^{(n)}} + \Big(\frac{r}{a}\Big)^{-(n+2)} m^{-(n+2)}\frac{T_{31}^{(n)}}{T^{(n)}} - \Big(\frac{r}{a}\Big)^{-n}(n+2)m^{-n}\frac{T_{41}^{(n)}}{T^{(n)}}\Big) A_n \cos n\theta \tag{19}$$

$$-\sum_{n=2}^{N+1}\left(\left(\frac{r}{a}\right)^{n-2} m^{n-2}\frac{T_{12}^{(n)}}{T^{(n)}} - \left(\frac{r}{a}\right)^{n} m^{n}(n-2)\frac{T_{22}^{(n)}}{T^{(n)}} + \left(\frac{r}{a}\right)^{-(n+2)} m^{-(n+2)}\frac{T_{32}^{(n)}}{T^{(n)}}\right.$$

$$\left. - \left(\frac{r}{a}\right)^{-n}(n+2)\, m^{-n}\frac{T_{42}^{(n)}}{T^{(n)}}\right) D_n \cos n\theta .$$

(19)

$$\cdot\sigma_\theta = -\left(\left(\frac{a}{r}\right)^2\frac{1}{1-m^2} + \frac{m^2}{1-m^2}\right)A_0 + \left(\left(\frac{a}{r}\right)^3\frac{1}{m^4-1} + 3\left(\frac{r}{a}\right)\frac{m^4}{m^4-1}\right)\left(B_1 \sin\theta +\right.$$

$$\left. A_1 \cos\theta\right)$$

$$+\sum_{n=2}^{N+1}\left(\left(\frac{r}{a}\right)^{n-2} m^{n-2}\frac{T_{11}^{(n)}}{T^{(n)}} - \left(\frac{r}{a}\right)^{n}(n+2)\, m^{n}\frac{T_{21}^{(n)}}{T^{(n)}} + \left(\frac{r}{a}\right)^{-(n+2)} m^{-(n+2)}\frac{T_{31}^{(n)}}{T^{(n)}}\right.$$

$$\left. - \left(\frac{r}{a}\right)^{-n}(n-2)\, m^{-n}\frac{T_{41}^{(n)}}{T^{(n)}}\right) B_n \sin n\theta$$

$$+\sum_{n=2}^{N+1}\left(-\left(\frac{r}{a}\right)^{n-2} m^{n-2}\frac{T_{12}^{(n)}}{T^{(n)}} + \left(\frac{r}{a}\right)^{n}(n+2)\, m^{n}\frac{T_{22}^{(n)}}{T^{(n)}} - \left(\frac{r}{a}\right)^{-(n+2)} m^{-(n+2)}\frac{T_{32}^{(n)}}{T^{(n)}}\right.$$

$$\left. + \left(\frac{r}{a}\right)^{-n}(n-2)\, m^{-n}\frac{T_{42}^{(n)}}{T^{(n)}}\right) C_n \sin n\theta$$

$$+\sum_{n=2}^{N+1}\left(\left(\frac{r}{a}\right)^{n-2} m^{n-2}\frac{T_{11}^{(n)}}{T^{(n)}} - \left(\frac{r}{a}\right)^{n}(n+2)\, m^{n}\frac{T_{21}^{(n)}}{T^{(n)}} + \left(\frac{r}{a}\right)^{-(n+2)} m^{-(n+2)}\frac{T_{31}^{(n)}}{T^{(n)}}\right.$$

$$\left. - \left(\frac{r}{a}\right)^{-n}(n-2)\, m^{-n}\frac{T_{41}^{(n)}}{T^{(n)}}\right) A_n \cos n\theta$$

$$+\sum_{n=2}^{N+1}\left(\left(\frac{r}{a}\right)^{n-2} m^{n-2}\frac{T_{12}^{(n)}}{T^{(n)}} - \left(\frac{r}{a}\right)^{n} m^{n}(n+2)\frac{T_{22}^{(n)}}{T^{(n)}} + \left(\frac{r}{a}\right)^{-(n+2)} m^{-(n+2)}\frac{T_{32}^{(n)}}{T^{(n)}}\right.$$

$$\left. - \left(\frac{r}{a}\right)^{-n}(n-2)\, m^{-n}\frac{T_{42}^{(n)}}{T^{(n)}}\right) D_n \cos n\theta \qquad (20)$$

$$\cdot\tau_{r\theta} = \left(\frac{m^4}{m^4-1}\left(\frac{r}{a}\right) - \left(\frac{a}{r}\right)^3\frac{1}{m^4-1}\right)\left(A_1 \sin\theta - B_1 \cos\theta\right)$$

$$-\sum_{n=2}^{N+1}\left(\left(\frac{r}{a}\right)^{n-2} m^{n-2}\frac{T_{11}^{(n)}}{T^{(n)}} - \left(\frac{r}{a}\right)^{n} n\, m^{n}\frac{T_{21}^{(n)}}{T^{(n)}} - \left(\frac{r}{a}\right)^{-(n+2)} m^{-(n+2)}\frac{T_{31}^{(n)}}{T^{(n)}}\right.$$

$$\left. + \left(\frac{r}{a}\right)^{-n} n\, m^{-n}\frac{T_{41}^{(n)}}{T^{(n)}}\right) B_n \cos n\theta$$

$$-\sum_{n=2}^{N+1}\left(-\left(\frac{r}{a}\right)^{n-2} m^{n-2}\frac{T_{12}^{(n)}}{T^{(n)}} + \left(\frac{r}{a}\right)^{n} n\, m^{n}\frac{T_{22}^{(n)}}{T^{(n)}} + \left(\frac{r}{a}\right)^{-(n+2)} m^{-(n+2)}\frac{T_{32}^{(n)}}{T^{(n)}}\right.$$

$$\left. - \left(\frac{r}{a}\right)^{-n} n\, m^{-n}\frac{T_{42}^{(n)}}{T^{(n)}}\right) C_n \cos n\theta$$

(21)

$$+\sum_{n=2}^{N+1}\left(\left(\frac{r}{a}\right)^{n-2} m^{n-2}\frac{T_{11}^{(n)}}{T^{(n)}} - \left(\frac{r}{a}\right)^{n} n m^{n}\frac{T_{21}^{(n)}}{T^{(n)}} - \left(\frac{r}{a}\right)^{-(n+2)} m^{-(n+2)}\frac{T_{31}^{(n)}}{T^{(n)}}\right.$$

$$\left. + \left(\frac{r}{a}\right)^{-n} n m^{-n}\frac{T_{41}^{(n)}}{T^{(n)}}\right) A_n \sin n\theta$$

$$+\sum_{n=2}^{N+1}\left(\left(\frac{r}{a}\right)^{n-2} m^{n-2}\frac{T_{12}^{(n)}}{T^{(n)}} - \left(\frac{r}{a}\right)^{n} n m^{n}\frac{T_{22}^{(n)}}{T^{(n)}} - \left(\frac{r}{a}\right)^{-(n+2)} m^{-(n+2)}\frac{T_{32}^{(n)}}{T^{(n)}}\right.$$

$$\left. + \left(\frac{r}{a}\right)^{-n} n m^{-n}\frac{T_{42}^{(n)}}{T^{(n)}}\right) D_n \sin n\theta. \qquad (21)$$

$$\frac{u_r}{a} = -\frac{1}{2G}\left\{ A_0\left(-\left(\frac{a}{r}\right)\frac{1}{1-m^2} - \left(\frac{r}{a}\right)\frac{m^2(1-2\nu)}{(1-m^2)}\right) + \left(\left(\frac{a}{r}\right)^2\frac{1}{2(m^4-1)} + \frac{m^4(1-4\nu)}{2(m^4-1)}\left(\frac{r}{a}\right)^2\right)\right.$$

$$\left(B_1\sin\theta + A_1\cos\theta\right)$$

$$-\sum_{n=2}^{N+1}\left(\left(\frac{r}{a}\right)^{n-1}\frac{m^{n-2}}{(n-1)}\frac{T_{11}^{(n)}}{T^{(n)}} - (n+4\nu-2)\left(\frac{r}{a}\right)^{n+1}\frac{m^{n}}{(n+1)}\frac{T_{21}^{(n)}}{T^{(n)}} - \left(\frac{r}{a}\right)^{-(n+1)}\frac{m^{-(n+2)}}{(n+1)}\frac{T_{31}^{(n)}}{T^{(n)}}\right.$$

$$\left. + (n-4\nu+2)\left(\frac{r}{a}\right)^{1-n}\frac{m^{-n}}{n-1}\frac{T_{41}^{(n)}}{T^{(n)}}\right) B_n \sin n\theta$$

$$-\sum_{n=2}^{N+1}\left(-\left(\frac{r}{a}\right)^{n-1}\frac{m^{n-2}}{(n-1)}\frac{T_{12}^{(n)}}{T^{(n)}} + \frac{(n+4\nu-2)}{(n+1)}\left(\frac{r}{a}\right)^{n+1} m^{n}\frac{T_{22}^{(n)}}{T^{(n)}} + \left(\frac{r}{a}\right)^{-(n+1)}\frac{m^{-(n+2)}}{(n+1)}\frac{T_{32}^{(n)}}{T^{(n)}}\right.$$

$$\left. - (n-4\nu+2)\left(\frac{r}{a}\right)^{1-n}\frac{m^{-n}}{(n-1)}\frac{T_{42}^{(n)}}{T^{(n)}}\right) C_n \sin n\theta$$

$$-\sum_{n=2}^{N+1}\left(\left(\frac{r}{a}\right)^{n-1}\frac{m^{n-2}}{(n-1)}\frac{T_{11}^{(n)}}{T^{(n)}} - \frac{(n+4\nu-2)}{(n+1)}\left(\frac{r}{a}\right)^{n+1} m^{n}\frac{T_{21}^{(n)}}{T^{(n)}} - \left(\frac{r}{a}\right)^{-(n+1)}\frac{m^{-(n+2)}}{(n+1)}\frac{T_{31}^{(n)}}{T^{(n)}}\right.$$

$$\left. + (n-4\nu+2)\left(\frac{r}{a}\right)^{1-n}\frac{m^{-n}}{(n-1)}\frac{T_{41}^{(n)}}{T^{(n)}}\right) A_n \cos n\theta$$

$$+\sum_{n=2}^{N+1}\left(-\left(\frac{r}{a}\right)^{n-1}\frac{m^{n-2}}{(n-1)}\frac{T_{12}^{(n)}}{T^{(n)}} + \frac{(n+4\nu-2)}{(n+1)}\left(\frac{r}{a}\right)^{n+1} m^{n}\frac{T_{22}^{(n)}}{T^{(n)}} + \left(\frac{r}{a}\right)^{-(n+1)}\frac{m^{-(n+2)}}{(n+1)}\frac{T_{32}^{(n)}}{T^{(n)}}\right.$$

$$\left.\left. - (n-4\nu+2)\left(\frac{r}{a}\right)^{1-n}\frac{m^{-n}}{(n-1)}\frac{T_{42}^{(n)}}{T^{(n)}}\right) D_n \cos n\theta \right\}$$

$$+ \nu\left(\frac{r}{a}\right)\varepsilon_{z_0} + \frac{\Delta_1^+}{a}\cos\theta + \frac{\Delta_2^+}{a}\sin\theta$$

$$(22)$$

$$\frac{v_\theta}{a} = -\frac{1}{2G}\left\{ (A_1 \sin\theta - B_1\cos\theta)\left(\left(\frac{a}{r}\right)^2 \frac{1}{2(m^4-1)} + \frac{m^4(5-4\nu)}{2(m^4-1)}\left(\frac{r}{a}\right)^2\right)\right.$$

$$- \sum_{n=2}^{N+1}\left(\left(\left(\frac{r}{a}\right)^{n-1}\frac{m^{n-2}}{(n-1)}\frac{T_{11}^{(n)}}{T^{(n)}} - (n+4-4\nu)\left(\frac{r}{a}\right)^{n+1}\frac{m^n}{(n+1)}\frac{T_{21}^{(n)}}{T^{(n)}} + \left(\frac{r}{a}\right)^{-(n+1)}\frac{m^{-(n+2)}}{(n+1)}\frac{T_{31}^{(n)}}{T^{(n)}}\right.\right.$$

$$\left.\left. - (n-4+4\nu)\left(\frac{r}{a}\right)^{1-n}\frac{m^{-n}}{(n-1)}\frac{T_{41}^{(n)}}{T^{(n)}}\right) B_n \cos n\theta\right)$$

$$- \sum_{n=2}^{N+1}\left(-\left(\frac{r}{a}\right)^{n-1}\frac{m^{n-2}}{(n-1)}\frac{T_{12}^{(n)}}{T^{(n)}} + \frac{(n+4-4\nu)}{(n+1)}\left(\frac{r}{a}\right)^{n+1} m^n \frac{T_{22}^{(n)}}{T^{(n)}} - \left(\frac{r}{a}\right)^{-(n+1)}\frac{m^{-(n+2)}}{(n+1)}\frac{T_{32}^{(n)}}{T^{(n)}}\right.$$

$$\left. + (n-4+4\nu)\left(\frac{r}{a}\right)^{1-n}\frac{m^{-n}}{n-1}\frac{T_{42}^{(n)}}{T^{(n)}}\right) C_n \cos n\theta$$

$$+ \sum_{n=2}^{N+1}\left(\left(\left(\frac{r}{a}\right)^{n-1}\frac{m^{n-2}}{(n-1)}\frac{T_{11}^{(n)}}{T^{(n)}} - (n+4-4\nu)\left(\frac{r}{a}\right)^{n+1}\frac{m^n}{(n+1)}\frac{T_{21}^{(n)}}{T^{(n)}} + \left(\frac{r}{a}\right)^{-(n+1)}\frac{m^{-(n+2)}}{(n+1)}\frac{T_{31}^{(n)}}{T^{(n)}}\right.\right.$$

$$\left.\left. - (n-4+4\nu)\left(\frac{r}{a}\right)^{1-n}\frac{m^{-n}}{(n-1)}\frac{T_{41}^{(n)}}{T^{(n)}}\right) A_n \sin n\theta\right)$$

$$- \sum_{n=2}^{N+1}\left(-\left(\frac{r}{a}\right)^{n-1}\frac{m^{n-2}}{(n-1)}\frac{T_{12}^{(n)}}{T^{(n)}} + \frac{(n+4-4\nu)}{(n+1)}\left(\frac{r}{a}\right)^{n+1} m^n \frac{T_{22}^{(n)}}{T^{(n)}} - \left(\frac{r}{a}\right)^{-(n+1)}\frac{m^{-(n+2)}}{(n+1)}\frac{T_{32}^{(n)}}{T^{(n)}}\right.$$

$$\left.\left. + (n-4+4\nu)\left(\frac{r}{a}\right)^{1-n}\frac{m^{-n}}{(n-1)}\frac{T_{42}^{(n)}}{T^{(n)}}\right) D_n \sin n\theta \right\}$$

$$+ \frac{s_2^+}{a}\cos\theta - \frac{s_1^+}{a}\sin\theta + \varpi_1^+\left(\frac{r}{a}\right) \tag{23}$$

In particular, when $r = a$ eqs (22) and (23) become

$$\frac{u_r}{a} = (g_0 A_0 + \nu\varepsilon_{z0}) + g_1 B_1 \sin\theta + g_1 A_1 \cos\theta + \sum_{n=2}^{N+1}(h_n B_n + k_n C_n)\sin n\theta$$
$$+ \sum_{n=2}^{N+1}(h_n A_n - k_n D_n)\cos n\theta + \frac{s_1^+}{a}\cos\theta + \frac{s_2^+}{a}\sin\theta \tag{24a}$$

where

$$g_0 = -\frac{1}{2G}\left(-\frac{\alpha_0}{a^2} + 2\beta_0(1-2\nu)\right)$$

$$g_1 = \frac{-1}{2G}\left(\frac{\beta_1}{a^3} + \delta_1 a\,(1-4\nu)\right)$$

$$h_n = \frac{1}{2G}\left(n\,a^{n-2}U_{11}^{(n)} + a^n(n+4\nu-2)\,U_{21}^{(n)} - n\,a^{-(n+2)}U_{31}^{(n)} - a^{-n}(n-4\nu+2)\,U_{41}^{(n)}\right)$$

$$k_n = \frac{1}{2G}\left(n\,a^{n-2}U_{12}^{(n)} + a^n(n+4\nu-2)\,U_{22}^{(n)} - n\,a^{-(n+2)}U_{32}^{(n)} - a^{-n}(n-4\nu+2)\,U_{42}^{(n)}\right)$$

(24b)

and

$$\frac{v_\theta}{a} = r_1 B_1\cos\theta - r_1 A_1 \sin\theta + \sum_{n=2}^{N+1}(s_n B_n + t_n C_n)\cos n\theta + \sum_{n=2}^{N+1}(-s_n A_n + t_n D_n)\sin n\theta + \frac{s_2^+}{a}\cos\theta - \frac{s_1^+}{a}\sin\theta + \sigma_1^+$$

(25a)

where

$$r_1 = \frac{-1}{2G}\left(-\left(\frac{\beta_1}{a^3} + \delta_1 a\,(5-4\nu)\right)\right)$$

$$s_n = \frac{1}{2G}\left(n a^{n-2}U_{11}^{(n)} + a^n(n+4-4\nu)\,U_{21}^{(n)} + n a^{-(n+2)}U_{31}^{(n)} + a^{-n}(n-4+4\nu)\,U_{41}^{(n)}\right)$$

$$t_n = \frac{1}{2G}\left(n a^{n-2}U_{12}^{(n)} + a^n(n+4-4\nu)\,U_{22}^{(n)} + n a^{-(n+2)}U_{32}^{(n)} + a^{-n}(n-4+4\nu)\,U_{42}^{(n)}\right)$$

(25b)

2) Recall the general expression for the stress function $\phi(r,\theta)$ (eq 5.16) for a <u>full inclusion</u>

$$\phi(r,\theta) = c_0 r^2 + (d_1 r^3)_{\cos\theta}^{\sin\theta} + \sum_{n=2}^{\infty}(a_n r^n + b_n r^{2+n})_{\cos n\theta}^{\sin n\theta} \qquad (26)$$

The general expressions for the stress and displacement components $\sigma_r, \sigma_\theta, \tau_{r\theta}, u_r$ and v_θ can be obtained by setting $b_0, A_0^*, c_1, c_1', c_n, d_n, c_n', d_n'$ to zero into eqs (2), (3), (4), (16) and (17). The non zero terms in eqs (8b) and (10) reduce to

$$\beta_0 = 1/2 \quad ; \quad \delta_1 a = 1/2$$

$$a^{n-2}U_{11}^{(n)} = -1/2(n-1) \; ; \quad a^{n-2}U_{12}^{(n)} = (n-2)/2(n-1)n$$

$$a^n U_{21}^{(n)} = 1/2(n+1) \; ; \quad a^n U_{22}^{(n)} = -1/2(n+1)$$

(27)

Those quantities can be substituted into the general expressions for the stress and displacement components to obtain equations similar to (19), (20), (21), (22), and (23).*
In particular, when $r = a$ eqs (24a) and (25a) are still valid but eqs (24b) and (25b) are now replaced by the followings

$$g_0 = -(1-2\nu)/2G$$

$$g_1 = -(1-4\nu)/4G \quad ; \quad r_1 = (5-4\nu)/4G$$

$$h_n = \left(\frac{n+4\nu-2}{n+1} - \frac{n}{n-1}\right)\frac{1}{4G} \quad ; \quad k_n = \left(\frac{n-2}{n-1} - \frac{n+4\nu-2}{n+1}\right)\frac{1}{4G}$$

$$s_n = \left(\frac{n+4-4\nu}{n+1} - \frac{n}{n-1}\right)\frac{1}{4G} \quad ; \quad t_n = \left(\frac{n-2}{n-1} - \frac{n+4-4\nu}{n+1}\right)\frac{1}{4G} \tag{28}$$

3) Remarks:

If we substitute eqs (10) and (11) into eqs (24b) and (25b) it can be shown that for any value of n

$$k_n = s_n \tag{29}$$

For a full inclusion only, the following relation is also satisfied in addition to eq (29)

$$h_n = t_n \tag{30}$$

---

* Those equations can also be obtained directly by setting $m \to \infty$ into eqs (19) to (23).

Appendix 5.2

1) As far as ψ is concerned, recall eq (5.9)

$$\nabla^2\psi = \frac{\partial^2\psi}{\partial r^2} + \frac{1}{r}\frac{\partial\psi}{\partial r} + \frac{1}{r^2}\frac{\partial^2\psi}{\partial\theta^2} = 0 \tag{1}$$

If we use the method of separation of variables, then a solution of eq (1) takes the form

$$\psi(r,\theta) = X(r)Y(\theta) \tag{2}$$

Substituting this expression and its derivatives into eq (1) we obtain

$$\frac{r^2}{X}\left(X'' + \frac{1}{r}X'\right) = -\frac{Y''}{Y} \tag{3}$$

The expression on the left depends only on r while the right side depends only on θ. Thus, both expressions must be equal to a constant, say, n^2. This yields two linear differential equations

$$\frac{d^2Y}{d\theta^2} + n^2Y = 0 \tag{4}$$

and

$$r^2\frac{d^2X}{dr^2} + r\frac{dX}{dr} - n^2X = 0 \tag{5}$$

Eq (5) is also known as Cauchy Equation*

Two cases to be considered:

(i) n is different from zero: Eqs (4) and (5) have the following solutions

---

* see solution, for instance, in Kreyszig (1972) pp.69-70.

$$Y(\theta) = Y_n(\theta) = C_1 \cos n\theta + C_2 \sin n\theta$$

$$X(r) = C_1'' r^n + \frac{C_2''}{r^n}$$

Hence, the functions $\Psi(r,\theta) = \Psi_n(r,\theta) = X(r)\,Y_n(\theta)$ written out

$$\Psi_n(r,\theta) = \left(a_n^* r^n + \frac{b_n^*}{r^n}\right)\cos n\theta + \left(c_n^* r^n + \frac{d_n^*}{r^n}\right)\sin n\theta \tag{6}$$

are solutions of eq (1) for any non zero values of n. Therefore, a general solution of eq (1) can be written as follows

$$\Psi(r,\theta) = \sum_{n=1}^{\infty} \Psi_n(r,\theta) \tag{7}$$

since the Laplace equation is linear and homogeneous.

(ii) <u>n is equal to zero</u>: Eqs (4) and (5) have the following solutions

$$Y(\theta) = a_1 + b_1\theta \quad ; \qquad X(r) = A \ln r + B$$

Hence, $\Psi(r,\theta)$, written out

$$\Psi(r,\theta) = a_0^* + b_0^* \ln r + c_0^* \theta \ln r + d_0^* \theta \tag{8}$$

is solution of eq (1) when n is equal to zero.

Since n can take any value, the general solution of eq (1) is obtained by adding eqs (7) and (8) together i.e.

$$\Psi(r,\theta) = a_0^* + b_0^* \ln r + c_0^* \theta \ln r + d_0^* \theta + \sum_{n=1}^{\infty}\left(\left(a_n^* r^n + \frac{b_n^*}{r^n}\right)\cos n\theta + \left(c_n^* r^n + \frac{d_n^*}{r^n}\right)\sin n\theta\right) \tag{9}$$

Substituting eq (9) into eqs (5.8), we obtain the general expressions for the stress components τ_{rz} and $\tau_{\theta z}$ as follows

$$\tau_{rz} = c_0^* \frac{\ln r}{r} + \frac{d_0^*}{r} + \sum_{n=1}^{\infty} n\left(-\left(a_n^* r^{n-1} + \frac{b_n^*}{r^{n+1}}\right)\sin n\theta + \left(c_n^* r^{n-1} + \frac{d_n^*}{r^{n+1}}\right)\cos n\theta\right)$$

$$\tau_{\theta z} = -\frac{b_0^*}{r} - \frac{c_0^* \theta}{r} - \sum_{n=1}^{\infty} n\left(\left(a_n^* r^{n-1} - \frac{b_n^*}{r^{n+1}}\right)\cos n\theta + \left(c_n^* r^{n-1} - \frac{d_n^*}{r^{n+1}}\right)\sin n\theta\right) \tag{10}$$

Coefficients $c_0^*, d_0^*, a_n^*, b_n^*, c_n^*, d_n^*$ can be calculated from the boundary conditions for stresses at $r = a$ and at $r = b$.

(i) <u>at $r = a$</u> the expressions for τ_{rz} in eqs (10) must be equal to the last of eqs (5.11) i.e.

$$c_0^* \frac{\ln a}{a} + \frac{d_0^*}{a} = 0$$

$$n\left(c_n^* a^{n-1} + \frac{d_n^*}{a^{n+1}}\right) = E_n \; ; \; -n\left(a_n^* a^{n-1} + \frac{b_n^*}{a^{n+1}}\right) = F_n \quad (n = 1, N) \tag{11}$$

(ii) <u>at $r = b$</u> τ_{rz} must vanish for any value of θ i.e.

$$c_0^* \frac{\ln b}{b} + \frac{d_0^*}{b} = 0$$

$$c_n^* b^{n-1} + \frac{d_n^*}{b^{n+1}} = 0 \quad ; \; a_n^* b^{n-1} + \frac{b_n^*}{b^{n+1}} = 0 \qquad (n = 1, N) \tag{12}$$

The solution of the system made of eqs (11) and (12) can be written as follows

$$c_0^* = d_0^* = 0$$

$$a_n^* = \psi_1^{(n)} F_n \; ; \; b_n^* = \psi_2^{(n)} F_n \; ; \; c_n^* = \psi_3^{(n)} E_n \; ; \; d_n^* = \psi_4^{(n)} E_n \tag{13}$$

where for $n = 1, N$

$$a^{n-1}\psi_1^{(n)} = \frac{-m^{2n}}{n(m^{2n}-1)} \quad ; \quad \frac{\psi_2^{(n)}}{a^{n+1}} = \frac{1}{n(m^{2n}-1)}$$

$$a^{n-1}\psi_3^{(n)} = \frac{m^{2n}}{n(m^{2n}-1)} \quad ; \quad \frac{\psi_4^{(n)}}{a^{n+1}} = \frac{-1}{n(m^{2n}-1)} \tag{14}$$

with m defined in eq (8c) of <u>Appendix 5.1</u>.

The general expressions for the strain and displacement components $\gamma_{\theta z}$, γ_{rz} and w may be obtained from eqs (5.4) and (5.5). w can be calculated with a procedure similar to the one used for u_r and v_θ in <u>Appendix 5.1</u>. This yields

$$W = -\frac{1}{G}\left(\sum_{n=1}^{N}\left(\left(c_n^* r^n - \frac{d_n^*}{r^n}\right)\cos n\theta - \left(a_n^* r^n - \frac{b_n^*}{r^n}\right)\sin n\theta\right) - b_0^*\theta\right)$$
$$- z\varepsilon_{z_0} + \Delta_3^+ \tag{15}$$

where Δ_3^+ is a rigid body translation parallel to the z axis.

Since W must be a single valued function of the angle θ, b_0^* must vanish in eq (15).

If we substitute eqs (13) and (14) into eqs (10) and (15), then, stress and displacement components take the form

$$\tau_{rz} = \sum_{n=1}^{N}\left(\frac{m^{2n}}{m^{2n}-1}\left(\frac{r}{a}\right)^{n-1} - \left(\frac{a}{r}\right)^{n+1}\frac{1}{m^{2n}-1}\right)\left(E_n\cos n\theta + F_n\sin n\theta\right)$$
$$\tau_{\theta z} = \sum_{n=1}^{N}\left(\frac{m^{2n}}{m^{2n}-1}\left(\frac{r}{a}\right)^{n-1} + \left(\frac{a}{r}\right)^{n+1}\frac{1}{m^{2n}-1}\right)\left(F_n\cos n\theta - E_n\sin n\theta\right)$$
$$\frac{W}{a} = -\frac{z}{a}\varepsilon_{z_0} - \frac{1}{G}\sum_{n=1}^{N}\left(\frac{m^{2n}}{n(m^{2n}-1)}\left(\frac{r}{a}\right)^{n} + \left(\frac{a}{r}\right)^{n}\frac{1}{n(m^{2n}-1)}\right)\left(E_n\cos n\theta + F_n\sin n\theta\right)$$
$$+ \frac{\Delta_3^+}{a} \tag{16}$$

In particular when $r = a$ the displacement component W is equal to

$$\frac{W}{a} = -\frac{z}{a}\varepsilon_{z_0} + \sum_{n=1}^{N}\left(u_n E_n\cos n\theta + u_n F_n\sin n\theta\right) + \frac{\Delta_3^+}{a} \tag{17}$$

where

$$u_n = \frac{1}{G}\left(\frac{1+m^{2n}}{n(1-m^{2n})}\right) \tag{18}$$

2) For a <u>full inclusion</u> the general expressions for the stress and displacement components can be obtained directly

by setting $m \rightarrow \infty$ into eqs (16), (17) and (18). In particular,

$$u_n = -\frac{1}{nG} \qquad (19)$$

Appendix 5.3

1) Recall the general expressions for the initial displacement components u_{r_0}/a, v_{θ_0}/a, w_0/a, the induced displacement components u_{r_h}/a, v_{θ_h}/a, w_h/a and the total displacement components u_r/a, v_θ/a, w/a in Appendix 4.5 (eqs 35a, 25, 35c).
Let us calculate the values of those displacement components along the contour of the circular hole ($r = a$) for sub-problem (B_1).

With a procedure similar to the one used for the radial displacement u_r/a in eqs (36) to (42) of Appendix 4.5, it can be shown that along the contour of the hole

$$\frac{u_{r_0}}{a} = M_{u_1}^{(0)} + M_{u_2}^{(0)}\cos 2\theta + M_{u_3}^{(0)}\sin 2\theta + \frac{u^{(0)}}{a}\cos\theta + \frac{v^{(0)}}{a}\sin\theta$$

$$\frac{v_{\theta_0}}{a} = M_{v_2}^{(0)}\cos 2\theta + M_{v_3}^{(0)}\sin 2\theta + \omega_{xy}^{(0)} + \frac{v^{(0)}}{a}\cos\theta - \frac{u^{(0)}}{a}\sin\theta$$

$$\frac{w_0}{a} = M_{w_1}^{(0)} + M_{w_2}^{(0)}\cos\theta + M_{w_3}^{(0)}\sin\theta + \frac{w^{(0)}}{a} \tag{1}$$

where

$$M_{u_1}^{(0)} = -\frac{1}{2}\Big[\sigma_{x,0}(a_{11}+a_{21}) + \sigma_{y,0}(a_{12}+a_{22}) + \sigma_{z,0}(a_{13}+a_{23}) + \tau_{yz,0}(a_{14}+a_{24}) + \tau_{xz,0}(a_{15}+a_{25}) + \tau_{xy,0}(a_{16}+a_{26})\Big]$$

$$M_{u_2}^{(0)} = -\frac{1}{2}\Big[\sigma_{x,0}(a_{11}-a_{21}) + \sigma_{y,0}(a_{12}-a_{22}) + \sigma_{z,0}(a_{13}-a_{23}) + \tau_{yz,0}(a_{14}-a_{24}) + \tau_{xz,0}(a_{15}-a_{25}) + \tau_{xy,0}(a_{16}-a_{26})\Big]$$

$$M_{u_3}^{(0)} = -\frac{1}{2}\Big[a_{61}\sigma_{x,0} + a_{62}\sigma_{y,0} + a_{63}\sigma_{z,0} + a_{64}\tau_{yz,0} + a_{65}\tau_{xz,0} + a_{66}\tau_{xy,0}\Big]$$

$$M_{v_2}^{(0)} = -\frac{1}{2}\Big[a_{61}\sigma_{x,0} + a_{62}\sigma_{y,0} + a_{63}\sigma_{z,0} + a_{64}\tau_{yz,0} + a_{65}\tau_{xz,0} + a_{66}\tau_{xy,0}\Big]$$

$$M_{v_3}^{(0)} = -\frac{1}{2}\Big[\sigma_{x,0}(a_{21}-a_{11}) + \sigma_{y,0}(a_{22}-a_{12}) + \sigma_{z,0}(a_{23}-a_{13}) + \tau_{yz,0}(a_{24}-a_{14}) + \tau_{xz,0}(a_{25}-a_{15}) + \tau_{xy,0}(a_{26}-a_{16})\Big]$$

$$M_{w_1}^{(0)} = -\frac{z}{a}\varepsilon_{z_0}$$

$$M_{w_2}^{(0)} = -\left(a_{51}\sigma_{x,0} + a_{52}\sigma_{y,0} + a_{53}\sigma_{z,0} + a_{54}\tau_{yz,0} + a_{55}\tau_{xz,0} + a_{56}\tau_{xy,0}\right)$$

$$M_{w_3}^{(0)} = -\left(a_{41}\sigma_{x,0} + a_{42}\sigma_{y,0} + a_{43}\sigma_{z,0} + a_{44}\tau_{yz,0} + a_{45}\tau_{xz,0} + a_{46}\tau_{xy,0}\right) \tag{2}$$

Similarly,

$$\frac{u_{rh}}{a} = M_{u_1}^{(h)} + M_{u_2}^{(h)}\cos 2\theta + M_{u_3}^{(h)}\sin 2\theta + \frac{u_{t_1}^{(o)}}{a}\cos\theta + \frac{v_{t_1}^{(o)}}{a}\sin\theta$$

$$\frac{v_{\theta h}}{a} = M_{v_1}^{(h)} + M_{v_2}^{(h)}\cos 2\theta + M_{v_3}^{(h)}\sin 2\theta + \omega_1 + \frac{v_{t_1}^{(o)}}{a}\cos\theta - \frac{u_{t_1}^{(o)}}{a}\sin\theta$$

$$\frac{w_h}{a} = M_{w_2}^{(h)}\cos\theta + M_{w_3}^{(h)}\sin\theta + \frac{w_{t_1}^{(o)}}{a} \tag{3}$$

where

$$M_{u_1}^{(h)} = \frac{1}{2}\Big[\sigma_{x,0}(S_{11}-T'_{11}) + \sigma_{y,0}(S_{22}-T'_{22}) - \tau_{xy,0}(S_{12}-T'_{12}) - \tau_{yz,0}(S_{23}-T'_{13}) + \tau_{xz,0}(S_{13}-T'_{13})\Big]$$

$$M_{u_2}^{(h)} = \frac{1}{2}\Big[\sigma_{x,0}(S_{11}+T'_{11}) + \sigma_{y,0}(S_{22}+T'_{22}) - \tau_{xy,0}(S_{12}+T'_{12}) - \tau_{yz,0}(S_{23}+T'_{13}) + \tau_{xz,0}(S_{13}+T'_{13})\Big]$$

$$M_{u_3}^{(h)} = \frac{1}{2}\Big[-\sigma_{x,0}(T_{11}-S'_{11}) - \sigma_{y,0}(T_{22}-S'_{22}) + \tau_{xy,0}(T_{12}-S'_{12}) + \tau_{yz,0}(T_{23}-S'_{23}) - \tau_{xz,0}(T_{13}-S'_{13})\Big]$$

$$M_{v_1}^{(h)} = \frac{1}{2}\Big[\sigma_{x,0}(S'_{11}+T_{11}) + \sigma_{y,0}(S'_{22}+T_{22}) - \tau_{xy,0}(S'_{12}+T_{12}) - \tau_{yz,0}(S'_{23}+T_{23}) + \tau_{xz,0}(S'_{13}+T_{13})\Big]$$

$$M_{v_2}^{(h)} = \frac{1}{2}\Big[\sigma_{x,0}(S'_{11}-T_{11}) + \sigma_{y,0}(S'_{22}-T_{22}) - \tau_{xy,0}(S'_{12}-T_{12}) - \tau_{yz,0}(S'_{23}-T_{23}) + \tau_{xz,0}(S'_{13}-T_{13})\Big]$$

$$M_{v_3}^{(h)} = \frac{1}{2}\Big[-\sigma_{x,0}(T'_{11}+S_{11}) - \sigma_{y,0}(T'_{22}+S_{22}) + \tau_{xy,0}(T'_{12}+S_{12}) + \tau_{yz,0}(T'_{23}+S_{23}) - \tau_{xz,0}(T'_{13}+S_{13})\Big]$$

$$M_{w_2}^{(h)} = \left(\sigma_{x,0}S''_{11} + \sigma_{y,0}S''_{22} - \tau_{xy,0}S''_{12} - \tau_{yz,0}S''_{23} + \tau_{xz,0}S''_{13}\right)$$

$$M_{w_3}^{(h)} = \left(-\sigma_{x,0}T''_{11} - \sigma_{y,0}T''_{22} + \tau_{xy,0}T''_{12} + \tau_{yz,0}T''_{23} - \tau_{xz,0}T''_{13}\right) \tag{4}$$

with

$$S_{11} = \mathrm{Re}\left(\frac{ip_1}{\Delta}(\lambda_2\lambda_3-1) + \frac{ip_2}{\Delta}(1-\lambda_1\lambda_3) + \frac{ip_3}{\Delta}(\lambda_1-\lambda_2)\right)$$

$$S_{22} = \mathrm{Re}\left(\frac{p_1}{\Delta}(\mu_2-\lambda_2\lambda_3\mu_3) + \frac{p_2}{\Delta}(\lambda_1\lambda_3\mu_3-\mu_1) + \frac{p_3}{\Delta}(\mu_1\lambda_2-\mu_2\lambda_1)\right)$$

$$S_{12} = \mathrm{Re}\Big(\frac{p_1}{\Delta}\left(\lambda_2\lambda_3-1+i(\mu_2-\mu_3\lambda_2\lambda_3)\right) + \frac{p_2}{\Delta}\left(1-\lambda_1\lambda_3+i(\lambda_1\lambda_3\mu_3-\mu_1)\right) + \frac{p_3}{\Delta}\left(\lambda_1-\lambda_2+i(\mu_1\lambda_2-\mu_2\lambda_1)\right)\Big)$$

$$S_{23} = \mathrm{Re}\left(\frac{p_1\lambda_3}{\Delta}(\mu_3-\mu_2) + \frac{p_2\lambda_3}{\Delta}(\mu_1-\mu_3) + \frac{p_3}{\Delta}(\mu_2-\mu_1)\right)$$

$$S_{13} = \mathrm{Re}\left(\frac{ip_1\lambda_3}{\Delta}(\mu_3-\mu_2) + \frac{ip_2\lambda_3}{\Delta}(\mu_1-\mu_3) + \frac{ip_3}{\Delta}(\mu_2-\mu_1)\right)$$

$$T_{11} = \mathrm{Re}\left(\frac{i^2p_1}{\Delta}(\lambda_2\lambda_3-1) + \frac{i^2p_2}{\Delta}(1-\lambda_1\lambda_3) + \frac{i^2p_3}{\Delta}(\lambda_1-\lambda_2)\right)$$

$$T_{22} = \mathrm{Re}\left(\frac{ip_1}{\Delta}(\mu_2-\lambda_2\lambda_3\mu_3) + \frac{ip_2}{\Delta}(\lambda_1\lambda_3\mu_3-\mu_1) + \frac{ip_3}{\Delta}(\mu_1\lambda_2-\mu_2\lambda_1)\right)$$

$$T_{12} = \mathrm{Re}\Big(\frac{ip_1}{\Delta}\left(\lambda_2\lambda_3-1+i(\mu_2-\mu_3\lambda_2\lambda_3)\right) + \frac{ip_2}{\Delta}\left(1-\lambda_1\lambda_3+i(\lambda_1\lambda_3\mu_3-\mu_1)\right) + \frac{ip_3}{\Delta}\left(\lambda_1-\lambda_2+i(\mu_1\lambda_2-\mu_2\lambda_1)\right)\Big)$$

$$T_{23} = \mathrm{Re}\left(\frac{ip_1}{\Delta}\lambda_3(\mu_3-\mu_2) + \frac{ip_2\lambda_3}{\Delta}(\mu_1-\mu_3) + \frac{ip_3}{\Delta}(\mu_2-\mu_1)\right)$$

$$T_{13} = \mathrm{Re}\left(\frac{i^2p_1}{\Delta}\lambda_3(\mu_3-\mu_2) + \frac{i^2p_2\lambda_3}{\Delta}(\mu_1-\mu_3) + \frac{i^2p_3}{\Delta}(\mu_2-\mu_1)\right) \qquad (5)$$

$(S'_{ij}, T'_{ij}), (S''_{ij}, T''_{ij})$ (for $i,j = 1,2,3$) are similar to (S_{ij}, T_{ij}) but the coefficients p_k $(k=1,2,3)$ must be replaced respectively by q_k and r_k.

2) Subproblem (B_2) corresponds to the elastic equilibrium of a body loaded along its internal circular contour by boundary stresses with components X_n, Y_n, Z_n. They are obtained by substituting eqs (5.10) into eq (5.2). Furthermore, a_{ox}, a_{oy}, a_{oz} must vanish and condition (5.13) must be satisfied. This yields,

$$X_n = \sum_{m=1}^{N} (a_{m_x} \cos m\theta + b_{m_x} \sin m\theta)$$

$$Y_n = \sum_{m=1}^{N} (a_{m_y} \cos m\theta + b_{m_y} \sin m\theta)$$

$$Z_n = \sum_{m=1}^{N} (a_{mz} \cos m\theta + b_{mz} \sin m\theta) \qquad (6)$$

with

$$a_{1_y} = b_{1_x}$$

Since, as shown in Figure 5.2, the deformation of the body must take place under zero axial strain, a generalized plane strain formulation can be used to calculate the distributions of stresses, strains and displacements induced by X_n, Y_n, Z_n .

The stress and displacement components at any point around the circular contour can be calculated using eqs (4.31) , (4.54) and (4.53). The general expressions for the functions $\phi_k(z_k)$ and $\phi'_k(z_k)$ $(k=1,2,3)$ are proposed in Appendix 4.3 in terms of the coefficients $\bar{a}_m, \bar{b}_m, \bar{c}_m$ $(m=1,N)$ defined in eqs (4.73).

Let us calculate first the values of the displacement components along the circular hole ($r = a$). Substituting into eqs (12) (Appendix 4.3) the coefficients $\bar{a}_m, \bar{b}_m, \bar{c}_m$ defined in eqs (4.73) with $A_1 = A_2 = A_3 = 0$ and $\xi_1 = \xi_2 = \xi_3 = e^{i\theta}$ then combining the expressions for ϕ_1, ϕ_2, ϕ_3 with eqs (4.53), the induced displacements expressed in terms of radial, tangential and longitudinal components take the form

$$\frac{u^*_{rh}}{a} = \sum_{m=1}^{N} \cos(m+1)\theta \left(\frac{X_m^{(1)} + Y_m^{(2)}}{2}\right) + \sin(m+1)\theta \left(\frac{X_m^{(2)} - Y_m^{(1)}}{2}\right)$$
$$+ \sum_{m=1}^{N} \cos(m-1)\theta \left(\frac{X_m^{(1)} - Y_m^{(2)}}{2}\right) + \sin(m-1)\theta \left(\frac{-Y_m^{(1)} - X_m^{(2)}}{2}\right)$$
$$+ \frac{u_{t_2}^{(o)}}{a} \cos\theta + \frac{v_{t_2}^{(o)}}{a} \sin\theta$$

$$\frac{v_{\theta h}^{*}}{a} = \sum_{m=1}^{N} \cos(m+1)\theta \left(\frac{X_m^{(2)} - Y_m^{(1)}}{2}\right) + \sin(m+1)\theta \left(\frac{-X_m^{(1)} - Y_m^{(2)}}{2}\right)$$
$$+ \sum_{m=1}^{N} \cos(m-1)\theta \left(\frac{X_m^{(2)} + Y_m^{(1)}}{2}\right) + \sin(m-1)\theta \left(\frac{X_m^{(1)} - Y_m^{(2)}}{2}\right)$$
$$+ \omega_z + \frac{v_{t_z}^{(0)}}{a} \cos\theta - \frac{u_{t_z}^{(0)}}{a} \sin\theta .$$
$$\frac{w_h^{*}}{a} = \sum_{m=1}^{N} \cos m\theta \, X_m^{(3)} - \sin m\theta \, Y_m^{(3)} + \frac{w_{t_z}^{(0)}}{a}$$

(7)

where

$u_{t_z}^{(0)}, v_{t_z}^{(0)}, w_{t_z}^{(0)}$ are rigid body translations in the x,y,z directions and ω_z is a rigid body rotation around the z axis,

$X_m^{(1)}, Y_m^{(1)}, X_m^{(2)}, Y_m^{(2)}, X_m^{(3)} Y_m^{(3)} (m = 1, N)$ are such that

$$\frac{X_m^{(1)} + Y_m^{(2)}}{2} = \frac{1}{2m}\Big(-b_{my}(f_{u_1} + f_{v_2}) + a_{my}(f_{u_2} - f_{v_1}) + b_{mx}(f_{u_3} + f_{v_4}) + a_{mx}(f_{v_3} - f_{u_4}) + b_{mz}(f_{u_5} + f_{v_6}) + a_{mz}(f_{v_5} - f_{u_6})\Big)$$

$$\frac{X_m^{(2)} - Y_m^{(1)}}{2} = \frac{1}{2m}\Big(b_{my}(f_{u_2} - f_{v_1}) + a_{my}(f_{v_2} + f_{u_1}) + b_{mx}(f_{v_3} - f_{u_4}) - a_{mx}(f_{v_4} + f_{u_3}) + b_{mz}(f_{v_5} - f_{u_6}) - a_{mz}(f_{v_6} + f_{u_5})\Big)$$

$$\frac{X_m^{(1)} - Y_m^{(2)}}{2} = \frac{1}{2m}\Big(b_{my}(f_{v_2} - f_{u_1}) + a_{my}(f_{u_2} + f_{v_1}) + b_{mx}(f_{u_3} - f_{v_4}) - a_{mx}(f_{u_4} + f_{v_3}) + b_{mz}(f_{u_5} - f_{v_6}) - a_{mz}(f_{u_6} + f_{v_5})\Big)$$

$$\frac{Y_m^{(1)} + X_m^{(2)}}{2} = \frac{1}{2m}\Big(-b_{my}(f_{v_1} + f_{u_2}) + a_{my}(f_{v_2} - f_{u_1}) + b_{mx}(f_{v_3} + f_{u_4}) + a_{mx}(f_{u_3} - f_{v_4}) + b_{mz}(f_{v_5} + f_{u_6}) + a_{mz}(f_{u_5} - f_{v_6})\Big)$$

$$X_m^{(3)} = \frac{1}{m}\left(-b_{my} f_{w_1} + a_{my} f_{w_2} + b_{mx} f_{w_3} - a_{mx} f_{w_4} + b_{mz} f_{w_5} - a_{mz} f_{w_6}\right)$$

$$Y_m^{(3)} = \frac{1}{m}\left(-b_{my} f_{w_2} - a_{my} f_{w_1} + b_{mx} f_{w_4} + a_{mx} f_{w_3} + b_{mz} f_{w_6} + a_{mz} f_{w_5}\right)$$

(8)

with

$$f_{u_1} = -Re\left(\frac{\Gamma_a}{\Delta}\right) \quad ; f_{v_1} = -Re\left(\frac{\Gamma'_a}{\Delta}\right) \quad ; f_{w_1} = -Re\left(\frac{\Gamma''_a}{\Delta}\right)$$

$$f_{u_2} = -Re\left(\frac{i\Gamma_a}{\Delta}\right) \quad ; f_{v_2} = -Re\left(\frac{i\Gamma'_a}{\Delta}\right) \quad ; f_{w_2} = -Re\left(\frac{i\Gamma''_a}{\Delta}\right)$$

$$f_{u_3} = -Re\left(\frac{\Gamma_b}{\Delta}\right) \quad ; f_{v_3} = -Re\left(\frac{\Gamma'_b}{\Delta}\right) \quad ; f_{w_3} = -Re\left(\frac{\Gamma''_b}{\Delta}\right)$$

$$f_{u_4} = -Re\left(\frac{i\Gamma_b}{\Delta}\right) \quad ; f_{v_4} = -Re\left(\frac{i\Gamma'_b}{\Delta}\right) \quad ; f_{w_4} = -Re\left(\frac{i\Gamma''_b}{\Delta}\right)$$

$$f_{u_5} = -Re\left(\frac{\Gamma_c}{\Delta}\right) \quad ; f_{v_5} = -Re\left(\frac{\Gamma'_c}{\Delta}\right) \quad ; f_{w_5} = -Re\left(\frac{\Gamma''_c}{\Delta}\right)$$

$$f_{u_6} = -Re\left(\frac{i\Gamma_c}{\Delta}\right) ; f_{v_6} = -Re\left(\frac{i\Gamma'_c}{\Delta}\right) ; f_{w_6} = -Re\left(\frac{i\Gamma''_c}{\Delta}\right) \qquad (9)$$

and

$$\Gamma_a = p_1(\mu_2 - \mu_3\lambda_2\lambda_3) + p_2(\lambda_1\lambda_3\mu_3 - \mu_1) + p_3(\mu_1\lambda_2 - \mu_2\lambda_1)$$

$$\Gamma_b = p_1(\lambda_2\lambda_3 - 1) + p_2(1 - \lambda_1\lambda_3) + p_3(\lambda_1 - \lambda_2)$$

$$\Gamma_c = p_1\lambda_3(\mu_3 - \mu_2) + p_2\lambda_3(\mu_1 - \mu_3) + p_3(\mu_2 - \mu_1) \qquad (10)$$

$(\Gamma'_a, \Gamma'_b, \Gamma'_c)$ and $(\Gamma''_a, \Gamma''_b, \Gamma''_c)$ are similar to $(\Gamma_a, \Gamma_b, \Gamma_c)$ but the coefficients p_k $(k = 1,2,3)$ must be replaced respectively by q_k and r_k.

As far as the stress components at any point around the circular contour are concerned, they are calculated by substituting $\bar{a}_m, \bar{b}_m, \bar{c}_m$ of eqs (4.73) into eqs (16) of Appendix 4.3. Then, the expressions for ϕ'_1, ϕ'_2 and ϕ'_3 can be combined with eqs (4.31) and (4.54).

3) The total displacement components along the contour of the hole associated to problem (B) are obtained by adding

eqs (1) and (3) to eqs (7). This yields eqs (5.20) with the following additional relations

$$
\begin{aligned}
u_t^{(o)} &= u_{t_1}^{(o)} + u_{t_2}^{(o)} \\
v_t^{(o)} &= v_{t_1}^{(o)} + v_{t_2}^{(o)} \\
w_t^{(o)} &= w_{t_1}^{(o)} + w_{t_2}^{(o)} \\
\omega &= \omega_1 + \omega_2
\end{aligned}
\qquad (11)
$$

Appendix 5.4

Let us consider a point P_1 whose components in the x,y,z coordinate system are defined by (Figure 1)

$$\overrightarrow{OP_1} = (r\cos\theta,\ r\sin\theta,\ z) \tag{1}$$

After deformation that takes place with components u_r, v_θ and w, point P_1 occupies a new position P''_1 such that

$$\overrightarrow{OP''_1} = \left((r+u_r)\cos\theta - v_\theta\sin\theta,\ (r+u_r)\sin\theta + v_\theta\cos\theta,\ z+w\right) \tag{2}$$

Consider two points P_1 and P_2 located on the inner surface of a hollow inclusion as shown in Figure 2. They are such that

$$\overrightarrow{OP_1} = (b\cos\theta, b\sin\theta, L) \ ; \quad \overrightarrow{OP_2} = (-b\cos\theta, -b\sin\theta, -L) \tag{3}$$

Let $s_\theta = |\overrightarrow{P_2P_1}|$ be the initial distance between the two points given by

$$s_\theta = 2\sqrt{b^2 + L^2} \tag{4}$$

After deformation points P_1 and P_2 occupy new positions P''_1, P''_2 such that

$$\overrightarrow{OP''_1} = \left((b+u_r)\cos\theta - v_\theta\sin\theta,\ (b+u_r)\sin\theta + v_\theta\cos\theta,\ L+w\right)$$

$$\overrightarrow{OP''_2} = \left(-(b+u_r)\cos\theta + v_\theta\sin\theta,\ -(b+u_r)\sin\theta - v_\theta\cos\theta,\ -L-w\right) \tag{5}$$

u_r, v_θ, w are the displacement components at point P_1. In writing eqs (5) we used the fact that u_r and v_θ are functions of $\sin n\theta$ and $\cos n\theta$ with even values of n and w

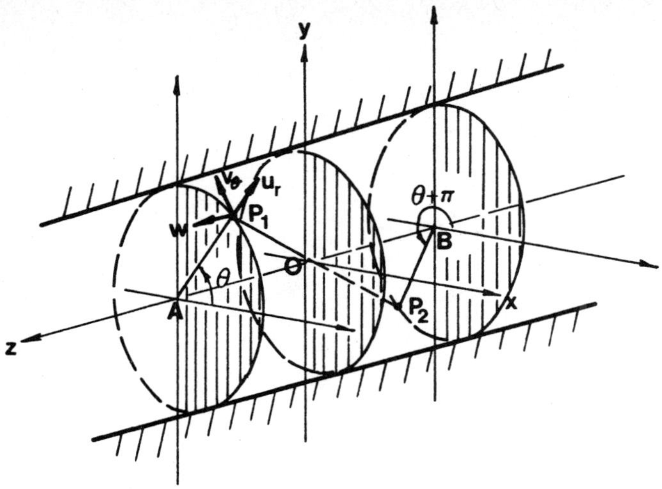

Figure 1. Definition of point P_1 and displacement components u_r, v_θ, w.

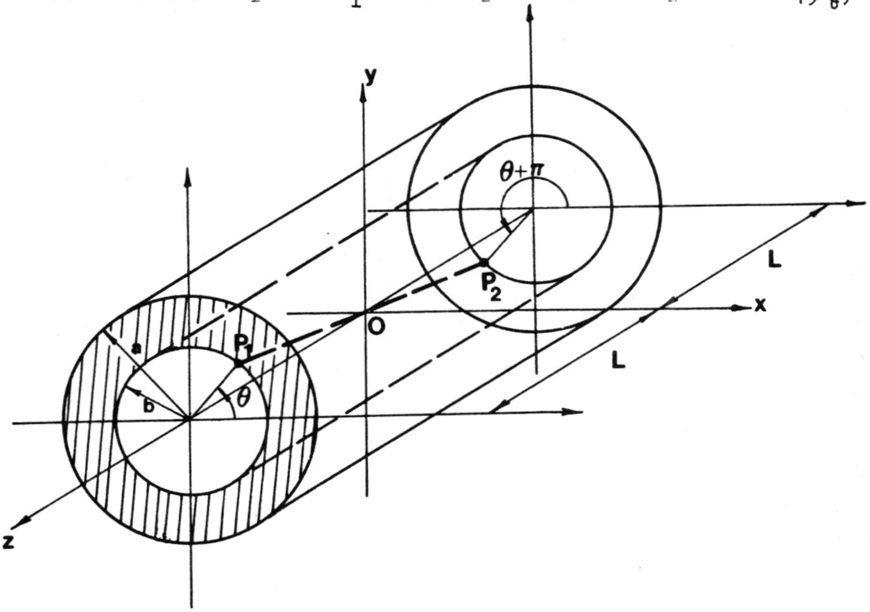

Figure 2. Variation in length of the oblique distance P_1P_2 .
Geometry of the problem.

is function of $\sin n\vartheta$ and $\cos n\vartheta$ with odd values of n.

Let $s''_\theta = \left|\overrightarrow{P''_2 P''_1}\right|$ be the distance between points P''_1 and P''_2. From eqs (5) it is equal to

$$s''_\theta = 2\sqrt{(b+u_r)^2 + v_\theta^2 + (L+w)^2} \tag{6}$$

If we assume that u_r, v_θ and w are small quantities with respect to L and b then

$$s''_\theta \simeq 2\sqrt{b^2 + L^2 + 2(bu_r + Lw)} \tag{7}$$

Therefore, from eqs (4) and (7) the change in length Δs_θ between points P_1 and P_2 is defined, after rearrangement, by

$$\Delta s_\theta \simeq 2\sqrt{b^2+L^2}\left(\sqrt{1 + \frac{2(bu_r + Lw)}{b^2+L^2}} - 1\right) \tag{8}$$

Since $\frac{2(bu_r + Lw)}{b^2+L^2}$ is an infinitesimal small quantity, eq (8) is also equal to

$$\Delta s_\theta \simeq \frac{2(bu_r + Lw)}{\sqrt{b^2+L^2}} \tag{9}$$

Combining eqs (4) and (9) and using dimensionless quantities we obtain

$$\left(\frac{\Delta s_\theta}{2a}\right)\left(\frac{s_\theta}{2a}\right) \simeq \left(\frac{b}{a}\right)\left(\frac{u_r}{a}\right) + \left(\frac{L}{a}\right)\left(\frac{w}{a}\right) \tag{10}$$

This leads to eq (5.44) using a matrix form of the right hand side of eq (10).

Appendix 6.1

Suppose we have a model under consideration that can be written in the form

$$(Y) = (X)(\beta) + (\varepsilon) \tag{1}$$

where

(Y) is a $(n, 1)$ matrix of observations,

(X) is a (n, p) matrix of known form,

(β) is a $(p, 1)$ matrix of parameters,

(ε) is a $(n, 1)$ matrix of errors.

The components of matrix (ε) are assumed to be uncorrelated random variables with mean zero and variance σ^2 i.e. in matrix form

$$E((\varepsilon)) = (0) \quad ; \quad V((\varepsilon)) = (I)\sigma^2 \tag{2}$$

where (0) is the zero column matrix of the same length as (ε),

(I) is the identity matrix,

$V((\varepsilon))$ is the variance covariance matrix of (ε) whose diagonal elements $V(\varepsilon_i)$ for $i = 1, 2, \ldots n$ are all σ^2 and whose off diagonal elements Covariance $(\varepsilon_i, \varepsilon_j)$ for $i \neq j = 1, 2, \ldots n$ are all zero.

It can be shown that the least square estimate of (β) is the matrix (b) that minimizes the quantity $(\varepsilon)^t(\varepsilon)$ and is such that

$$(X)^t(X)(b) = (X)^t(Y) \tag{3}$$

Eq (3) can be solved for (b) if $(X)^{+}(X)$ is non singular. The components of matrix (Y) that are predicted from the regression are obtained from

$$(\hat{y}) = (X)(b) \tag{4}$$

and the matrix of residuals is given by

$$(e) = (y) - (\hat{y}) \tag{5}$$

A measure of the usefulness of the regression is defined by the following quantity

$$R^2 = \frac{\sum_{i=1}^{n} (\hat{y}_i - \bar{y})^2}{\sum_{i=1}^{n} (y_i - \bar{y})^2} = \frac{(b)^{+}(x)^{+}(y) - n\bar{y}^2}{(y)^{+}(y) - n\bar{y}^2} \tag{6}$$

where $\bar{y} = \left(\sum_{i=1}^{n} y_i\right) / n$

R^2 is called the coefficient of multiple determination or the square of the multiple correlation coefficient R. If $\hat{y}_i = y_i$ for $i = 1, \ldots, n$ then $R^2 = 1$.

Let s^2 be the mean square about the regression and equal to

$$s^2 = \frac{\sum_{i=1}^{n} (y_i - \hat{y}_i)^2}{n-p} = \frac{(y)^{+}(y) - (b)^{+}(x)^{+}(y)}{n-p} \tag{7}$$

If the coefficients of matrix (X) are accurately known, s^2 is determined solely by the errors involved in making the observations (Y) and provides an estimate based on $n-p$ degrees of freedom of the variance about the regression which is also equal to σ^2 .*

s is therefore an estimate of the standard deviation about the regression.

The variance, covariance matrix of matrix (b) is equal to

$$V((b)) = ((X)^+(X))^{-1}\sigma^2 \tag{8}$$

The diagonal terms of the variance covariance matrix and their square root provide respectively the variances and standard errors of the estimates b_i ($i = 1,2 \ldots n$). The off diagonal terms provide the covariances of b_i, b_j ($i \neq j = 1,2,\ldots n$).

If $\sigma^2 = s^2$, the estimated variance covariance matrix of (b) is equal to

$$V((b)) = ((X)^+(X))^{-1}s^2 \tag{9}$$

The square roots of the diagonal terms provide the estimated standard errors of b_i denoted s_{b_i} ($i=1,2 \ldots n$) and the off diagonal terms the estimated covariances of b_i, b_j ($i \neq j = 1,2,\ldots n$).

If we also assume that (ε) follows a n dimensional multivariate normal distribution such that eqs (2) are satisfied, we can assign 100 $(1-\alpha)$% confidence limits for each b_i separately by calculating for $i = 1,2,\ldots n$

$$b_i \pm t\left(n-p, 1-\frac{\alpha}{2}\right) \times s_{b_i} \tag{10}$$

---

* Another way to obtain an estimate of σ^2 is to make repeat measurements of (Y) at the same value of (X). This estimate is said to represent 'pure error' only.

where $t(n-p, 1-\frac{\alpha}{2})$ is the $(1-\frac{\alpha}{2})$ percentage point of a Student's t distribution with $n-p$ degrees of freedom. In other words, the true value of β_i is such that

$$b_i - t(n-p, 1-\frac{\alpha}{2}) \times s_{b_i} < \beta_i < b_i + t(n-p, 1-\frac{\alpha}{2}) \times s_{b_i} \tag{11}$$

and this statement is made with 100 $(1-\alpha)\%$ of confidence.

As far as stress measurements by overcoring techniques are concerned (β) is equal to $(\sigma_o)_{xyz}$ or $(\sigma_o)_{XYZ}$ and $p = 6$. For the undercoring techniques $p = 3$ and (β) is replaced by $(\sigma_o)_{xy}$ defined in eq (5) of Appendix 4.7. For both cases n is the number of measurements. Using eqs (3) and (10) we can determine the least square estimate of the in situ stress field components and the confidence limits for each of them separately.

Appendix 6.2: Program Berni 1

1) Berni 1 is a two option program:

(i) It calculates the distributions of stresses, strains and displacements within an isotropic circular inclusion (hollow or solid) perfectly bonded to an infinite anisotropic body. The body is assumed to be either isotropic or transversely isotropic or orthotropic and to be subjected to a 3D stress field applied at infinity. It can also contain up to three orthogonal joint sets, each one being parallel to a plane of elastic symmetry of the intact material. The geometry of the problem is defined in Figure 5.1. The orientation of the anisotropy and the hole where the inclusion is located with respect to a fixed arbitrary global coordinate system X,Y,Z is defined in Figures 1 and 2 of Appendix 4.1. The program calculates also the distribution of stresses induced by the inclusion within the anisotropic body. This option is called by setting the input variable INDEX to zero,

(ii) It solves the inverse problem by calculating the least square estimates of the principal components of a 3D in situ stress field and their orientation with respect to a fixed arbitrary global coordinate system X,Y,Z from measurements of change in strain recorded within a solid or a hollow inclusion. Strain rosettes are oriented as shown in Figure 6.4. The program has been written for a maximum number of 6 rosettes (18 strain measurements). Each rosette can be embedded at different depths within the inclusion. This option is called by setting the input variable INDEX to a non zero positive value.

2) The flow chart of the program can be written as follows:

Input data (see text)

↓

Hollow ⟶ **Calculate** u_k, h_n, k_n, s_n, t_n $(n=2, N+1;\ k=1, N)$ g_0, g_2, r_2

Eqs (24b), (25b), (28) Appendix 5.1

Eqs (18), (19) Appendix 5.2

↓

Orien 2 ⟶ **Calculate the components** a_{ij} **of the Stress Strain Matrix** (A) **in the x,y,z coordinate system attached to the hole (eq (16), Appendix 4.1)**

Orienh

Transf

↓

Calculate the components β_{ij} **(Eq (4.14))**

↓

Zpolr ⟶ **Calculate the roots** μ_1, μ_2, μ_3 **of eq (4.25) and their conjugates. When there is one plane of elastic symmetry perpendicular to the hole axis check that** $\ell_4(\mu_1) = \ell_4(\mu_2) = \ell_2(\mu_3) = 0$

↓

Calculate $\lambda_1, \lambda_2, \lambda_3$ **(eqs (4.27))**

Calculate p_k, q_k, r_k **(k = 1,2,3) (Eqs 4.33a,b)**

↓

Calculate the coefficients of matrices (A_x), (C_x), (D_x), (F_x), (G_x)

↓

Multiply the coefficients of (A_x) **and** (C_x) **by** 10^6 **. This is used to verify if the determinant of** (A_x) **does not become small.**

↓

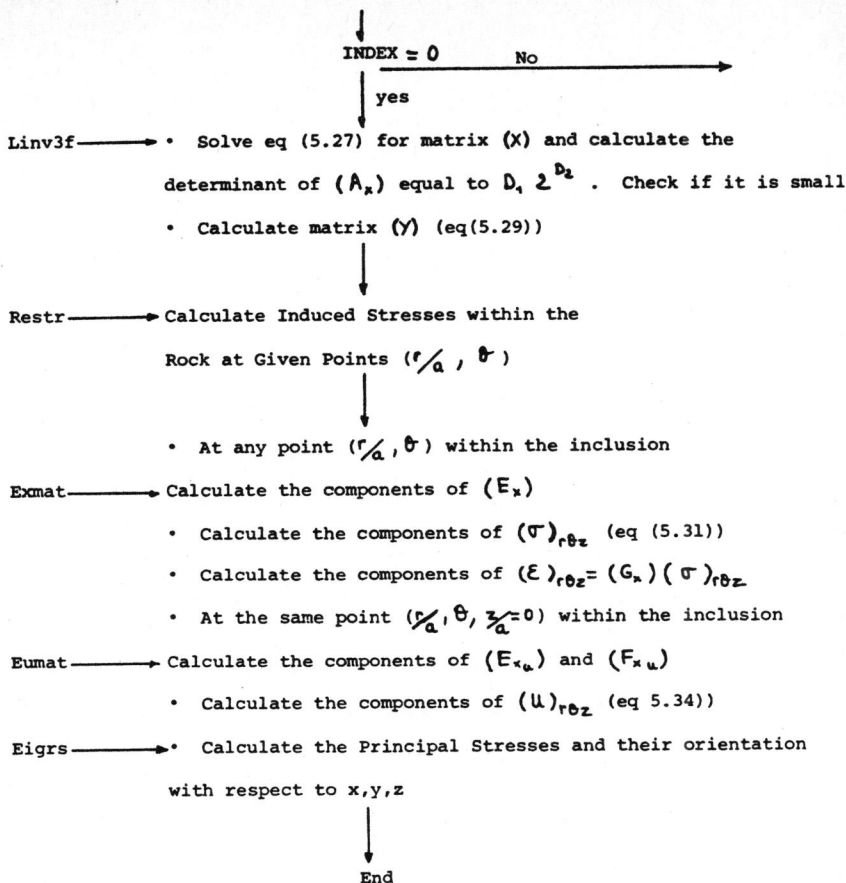

INDEX = 0 —— No ——→

↓ yes

Linv3f ——→ • Solve eq (5.27) for matrix (X) and calculate the determinant of (A_x) equal to $D_1 2^{D_2}$. Check if it is small

• Calculate matrix (Y) (eq(5.29))

↓

Restr ——→ Calculate Induced Stresses within the Rock at Given Points $(r/a, \theta)$

↓

• At any point $(r/a, \theta)$ within the inclusion

Exmat ——→ Calculate the components of (E_x)

• Calculate the components of $(\sigma)_{r\theta z}$ (eq (5.31))

• Calculate the components of $(\varepsilon)_{r\theta z} = (G_x)(\sigma)_{r\theta z}$

• At the same point $(r/a, \theta, z/a = 0)$ within the inclusion

Eumat ——→ Calculate the components of (E_{xu}) and (F_{xu})

• Calculate the components of $(u)_{r\theta z}$ (eq 5.34))

Eigrs ——→ • Calculate the Principal Stresses and their orientation with respect to x,y,z

↓

End

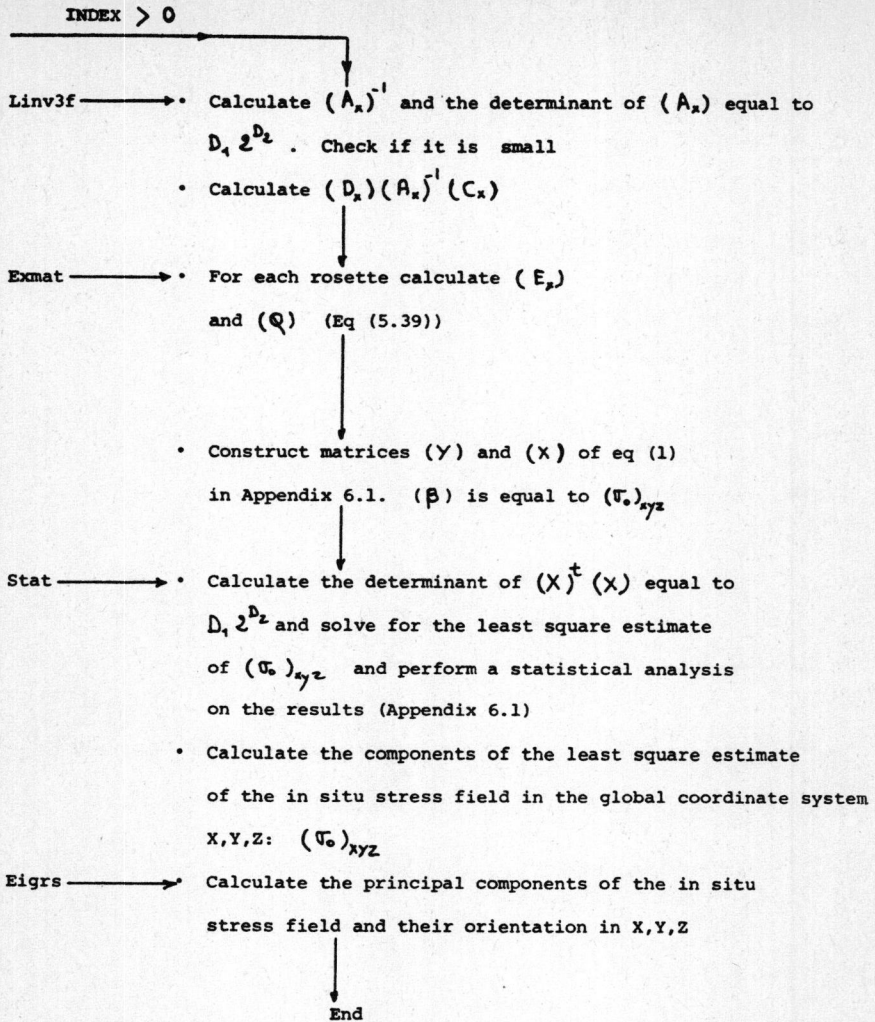

INDEX > 0

↓

Linv3f ⟶ • Calculate $(A_x)^{-1}$ and the determinant of (A_x) equal to $D_1 2^{D_2}$. Check if it is small

• Calculate $(D_x)(A_x)^{-1}(C_x)$

↓

Exmat ⟶ • For each rosette calculate (E_x) and (Q) (Eq (5.39))

↓

• Construct matrices (Y) and (X) of eq (1) in Appendix 6.1. (β) is equal to $(\sigma_o)_{xyz}$

↓

Stat ⟶ • Calculate the determinant of $(X)^t(X)$ equal to $D_1 2^{D_2}$ and solve for the least square estimate of $(\sigma_o)_{xyz}$ and perform a statistical analysis on the results (Appendix 6.1)

• Calculate the components of the least square estimate of the in situ stress field in the global coordinate system X,Y,Z: $(\sigma_o)_{XYZ}$

Eigrs ⟶ • Calculate the principal components of the in situ stress field and their orientation in X,Y,Z

↓

End

3) Program Berni 1 consists of one main program and several subroutines that can be described as follows:

(i) Hollow:

For given geometric and elastic properties of the inclusion, this subroutine calculates $u_k, h_n, k_n, \Delta_n, t_n$ $(n=2, N+1; k=1, N)$ g_o, g_1 and r_1 as defined in eqs (24b) (25b) (28) of Appendix 5.1, and in eqs (18) and (19) of Appendix 5.2,

(ii) Trans: Subroutine to calculate the transpose of a matrix,

(iii) Promat: Subroutine to calculate the product of two matrices,

(iv) Orien 2: Subroutine to calculate the direction cosines l, m, n of the unit vectors of a local coordinate system x',y',z' attached to a plane with respect to a global coordinate system X,Y,Z (Figure 1, Appendix 4.1). The orientation of the plane is defined by three numbers:

STR: strike of the plane defined from -Z axis in the clockwise direction and in the east quadrant

DIP: angle of dip

DIPDIR: parameter related to the direction of the normal x'

= 1 if directed in -Z, X(-Z) direction

= 2 if directed in X, X(Z) direction

= 3 if directed in Z, Z(-X) direction

= 4 if directed in -X, (-X)(-Z) direction

= 5 if horizontal (perpendicular to Y direction),

(v) Orienh: Subroutine to calculate the directions cosines

of the unit vectors of a local coordinate system x,y,z attached to a hole with respect to a global coordinate system X,Y,Z (eq (2) Appendix 4.1). The orientation of the hole is defined by two numbers β_h and δ_h (Figure 2, Appendix 4.1),

(vi) Transf: Subroutine to calculate the coefficients of matrices (T_σ') or (T_σ) respectively in eqs (6) and (11) of Appendix 4.1,

(vii) Exmat: Subroutine to calculate the coefficients of matrix (E_x) at any point $(r/a, \theta)$ within an inclusion,

(viii) Eumat: Subroutine to calculate the coefficients of matrices (E_{xu}) and (F_{xu}) at any point $(r/a, z/a, \theta)$ within an inclusion,

(ix) Restr: Subroutine to calculate at any point $(r/a, \theta)$ within the anisotropic body the state of stress induced by the inclusion only,

(x) Stat: Subroutine to calculate the least square solution of the system $(Y) = (X)\ (\beta) + (\varepsilon)$.

Three additional library subroutines are used;

(xi) Zpolr: Subroutine to calculate the zeros of a polynomial with real coefficients using Laguerre's method,

(xii) Eigrs: Subroutine to calculate the eigenvalues and eigenvectors of a real symmetrix matrix in symmetric storage mode,

(xiii) Linv3f: Subroutine to calculate the inverse and the determinant of a matrix with a full storage mode. It solves also the system $(A)\ (X) = (B)$. The determinant of the matrix is calculated from two output numbers D_1 and D_2 such that

Determinant = $D_1 * (2 ** D_2)$.

4) Input data:

<u>1st Card</u>: (4F10.4, I5)

M: ratio outer/inner radius of the inclusion

E: Young's Modulus of the inclusion,

MUI: Poisson's ratio of the inclusion,

B: = 0 for a hollow inclusion,

= 1 for a solid inclusion,

N: variable appearing in the summation term of eqs (5.10) (see additional remarks in section 5)).

<u>2nd Card</u> (2F10.4, I5, 2F10.4, I5)

STR, DIP, DIPDIR: Parameters to describe the orientation of an <u>arbitrary</u> plane of elastic symmetry, defined as a <u>plane of reference</u>, within the anisotropic body. x',y',z' is a coordinate system attached to that plane with x' in the direction of its normal (see description of Subroutine Orien 2 in section 3)(iv))

BETAH, DELTH: Angles β_h and δ_h to describe the orientation of the hole with respect to X,Y,Z (Figure 2, Appendix 4.1)

INDEX = 0 A 3D stress field is applied at infinity and the program calculates the distribution of stresses, strains and displacements within the inclusion, >0 Inverse problem.

3rd Card (9F8.2)

E1 : Young's modulus in x' direction,

E2 : " y' " ,

E3 : " z' " ,

G12 : Shear Modulus in x'y' plane,

G13 : " x'z' " ,

G23 : " y'z' " ,

MU21 : Poisson's ratio $\nu_{y'x'}$,

MU31 : " $\nu_{z'x'}$,

MU32 : " $\nu_{z'y'}$,

4th Card (6E10.4, 3F6.2)

KN (1) : normal stiffness of joint set 1 perpendicular to x' direction

KN (2) : normal stiffness of joint set 2 perpendicular to y' direction

KN (3) : normal stiffness of joint set 3 perpendicular to z' direction

KS (1) : shear stiffness of joint set 1,

KS (2) : shear stiffness of joint set 2,

KS (3) : shear stiffness of joint set 3,

SJ (1) : spacing of joint set 1,

SJ (2) : spacing of joint set 2,

SJ (3) : spacing of joint set 3,

5th Card (INDEX = 0) (6F10.4, I5, F10.4)

SFO (1, 1) : $\sigma_{x,0}$
SFO (2, 1) : $\sigma_{y,0}$
SFO (3, 1) : $\sigma_{z,0}$
SFO (4, 1) : $\tau_{yz,0}$
SFO (5, 1) : $\tau_{xz,0}$
SFO (6, 1) : $\tau_{xy,0}$

Components in the x,y,z coordinate system attached to the hole of the 3D stress field applied at infinity

NRAINC
TETINC

When INDEX = 0, the program generates a grid in the inclusion with $1/M \leq r/a \leq 1$ and $0° \leq \theta \leq 90°$
The incremental values of θ and r/a are equal to

$$\theta_{inc} = \text{TETINC}$$

$$(r/a)_{inc} = (1. - 1/M) / \text{NRAINC}$$

Stresses, strains and displacements are calculated at each point (r/a, θ).

6th Card (INDEX = 0)(F10.4, I5, F10.4)

RAMAX
NRAROC
TETROC

When INDEX = 0, the program generates a grid in the body with $1 \leq r/a \leq \text{RAMAX}$ and $0° \leq \theta \leq 180°$
The incremental values of θ and r/a are equal to

$$\theta_{inc} = \text{TETROC}$$

$$(r/a)_{inc} = (\text{RAMAX} - 1.) / \text{NRAROC}$$

The stresses induced by the inclusion within the body are calculated at each point (r/a, θ).

<u>5th Card</u> (INDEX $>$ 0) (I5, F10.4)

NROS : number of strain rosettes,

TSTU : value of the $(1-\frac{\alpha}{2})$ percentage point of a Student's t distribution with n - 6 degrees of freedom and 100 $(1-\alpha)$ percents confidence limits ($n = 3 \times NROS$).

<u>6th Card</u> (INDEX $>$ 0) (3E10.4, 3F10.4)

One card for each rosette i ($i = 1, NROS$) with the following informations

ROS (i,1) : <u>measured</u> strain ε_{A_i}		
ROS (i,2) : <u>measured</u> strain ε_{B_i}		
ROS (i,3) : <u>measured</u> strain ε_{C_i}	see Figure 6.4.	
ROS (i,4) : ψ_i angle		
ROS (i,5) : r_i/a		
ROS (i,6) : θ_i		

5) Remarks:

(i) N can be set equal to any value between 3 and 10,

(ii) For a solid inclusion and for INDEX = 0, the value of **M** is used to generate a grid within the inclusion where stresses, strains and displacements are calculated. If INDEX $\neq$ 0, the value of M can be left blank,

(iii) If there is no inclusion and the strain rosettes are glued directly on the anisotropic body set M to a value close to 1 and **E** to a small value. For instance $M = 1.0001$ and $E = 10^{-5}$ MPa.

(iv) If there is no joint sets parallel to the planes of

elastic symmetry of the anisotropic intact body, set KN(I), KS(I) and SJ(I) for I = 1,3 to large values.

(v) If the anisotropic character of the body is defined by 13 elastic coefficients (one plane of elastic symmetry) or 21 in the x',y',z' coordinate system attached to the plane of reference, the present program can be used by modifying the components of matrix (H1) (see program listing).

Appendix 6.3 : Program Berni 2

1) Program Berni 2 is a two option program:

(i) It calculates the distributions of total stresses, strains and displacements around a circular hole drilled in an infinite anisotropic body. The body is assumed to be either isotropic or transversely isotropic or orthotropic and to be subjected to a 3D stress field applied at infinity. It can also contain up to three orthogonal joints sets, each one being parallel to a plane of elastic symmetry of the intact material. The program calculates also the distributions of stresses, strains and displacements induced by the hole drilling and/or by the application of an internal pressure uniformly distributed along the hole. The geometry of the problem is defined in Figure 1 of Appendix 4.5. The orientation of the hole, and of the anisotropy with respect to a fixed arbitrary global coordinate system X,Y,Z is defined in Figures 1 and 2 of Appendix 4.1.

(ii) It solves the inverse problem by calculating the least square estimates of the principal components of a 3D in situ stress field and their orientation with respect to a fixed arbitrary coordinate system X,Y,Z from measurements of change in strain recorded at the surface of the hole. This option was written in order to check program Berni 1 when the inclusion vanishes. Strain rosettes are oriented as shown in Figure 6.4 with $a/b = 1$ and $r_i/a = 1$. The program has been written for a maximum number of 6 rosettes (18 measurements).

Options (i) and (ii) are called by setting the input variable INDEX respectively to zero and to a non zero positive value. Both options make use of the theory developed in chapters 4 and 6 and of the statistical analysis of Appendix 6.1. The flow chart of the program can be written as follows:

Input Data (See text)

↓

Orien 2 → Calculate the components a_{ij} of the stress strain matrix (A) in the x,y,z coordinate system attached to the hole (eq (6), Appendix 4.1)

Orienh

Transf

↓

Calculate the components β_{ij} (Eq (4.14))

↓

Zpolr → Calculate the roots μ_1, μ_2, μ_3 of eq (4.25) and their conjugates. When there is one plane of elastic symmetry perpendicular to the hole axis check that $\ell_4(\mu_1) = \ell_4(\mu_2) = \ell_2(\mu_3) = 0$

↓

Calculate $\lambda_1, \lambda_2, \lambda_3$ (Eqs (4.27))

Calculate p_k, q_k, r_k (k = 1,2,3) (Eqs 4.33a,b)

↓

INDEX = 0 — No →

↓ Yes

Fxmat → • At any point $(r/a, \theta)$ calculate the components of (F_σ) and (F^*)

• Calculate the components of $(\sigma)_{xyz}$ (eq (58) Appendix 4.5)

Tetss → • Calculate the components of $(\sigma)_{r\theta z}$

• Calculate the components of $(\varepsilon)_{xyz}$ (eq (59) Appendix 4.5)

Tetss → • Calculate the components of $(\varepsilon)_{r\theta z}$

Eigrs → • Calculate the principal stresses and their orientation with respect to x,y,z

↓

Fumat ⟶ • At any point (r/a, θ, $z/a = 0$) calculate the components of matrices $(F_u^{(u)})$, $(F_u^{(*)})$ and $(F_u)_q$

• Calculate the induced displacement components $(u_h)_{r\theta z}$ (eq (61), Appendix 4.5)

• Calculate the initial displacement components $(u_o)_{r\theta z}$ (eq (35a), Appendix 4.5)

↓

End

INDEX > 0

↓

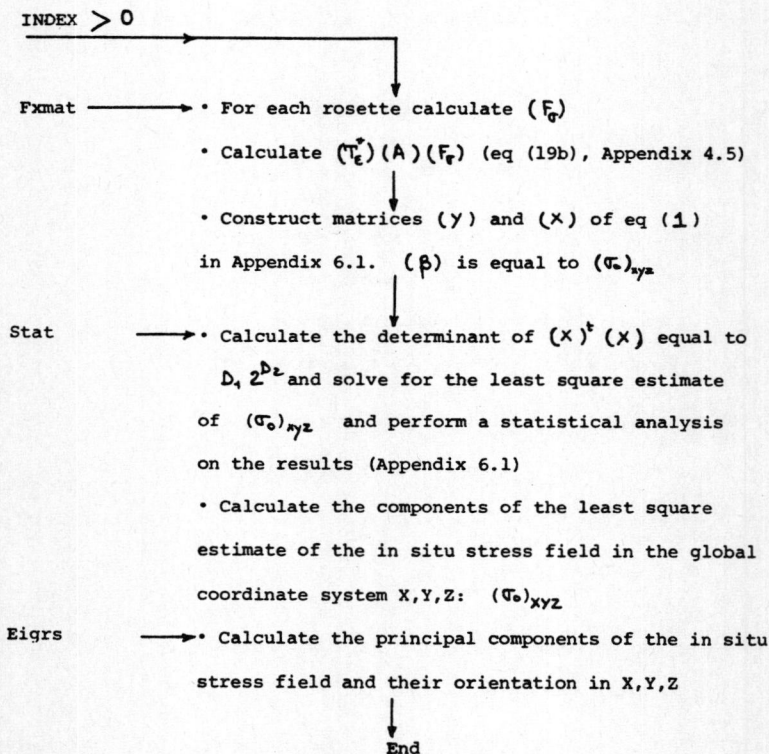

Fxmat ⟶ • For each rosette calculate (F_σ)

• Calculate $(T_\varepsilon^*)(A)(F_\sigma)$ (eq (19b), Appendix 4.5)

↓

• Construct matrices (y) and (X) of eq (1) in Appendix 6.1. (β) is equal to $(\sigma_o)_{xyz}$

↓

Stat ⟶ • Calculate the determinant of $(X)^t(X)$ equal to $D_1 2^{D_2}$ and solve for the least square estimate of $(\sigma_o)_{xyz}$ and perform a statistical analysis on the results (Appendix 6.1)

• Calculate the components of the least square estimate of the in situ stress field in the global coordinate system X,Y,Z: $(\sigma_o)_{XYZ}$

Eigrs ⟶ • Calculate the principal components of the in situ stress field and their orientation in X,Y,Z

↓

End

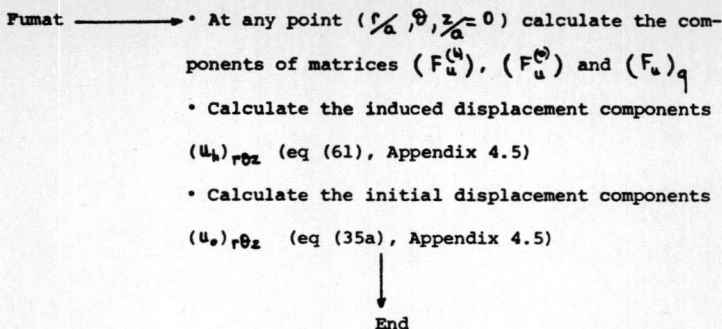

2) Program Berni 2 consists of one main program and several subroutines that can be described as follows:

Subroutines Trans, Promat, Orien 2, Orienh, Transf, Stat, Zpolr, Eigrs, Linv3f are indentical to those in Berni 1.

(i) Fxmat: Subroutine to calculate the coefficients of matrices (F_σ) and (F^*) at any point $(r/a, \theta)$

(ii) Fumat: Subroutine to calculate the coefficients of matrices $(F_u^{(h)})$, $(F_u^{(\omega)})$, and $(F_u)_q$ at any point $(r/a, \theta, z/a)$

(iii) Tetss: Subroutine to calculate the coefficients of matrices (T_σ^*) and (T_ε^*) in eqs (4) and (5) of Appendix 4.4.

3) Input Data:

   1st Card (2F10.4, I5, 2F10.4, I5)

   See second input data card of Berni 1

   2nd Card (9F8.2)

   See third input data card of Berni 1

   3rd Card (6E10.4, 3F6.2)

   See fourth input data card of Berni 1

   4th Card (7F10.4) (INDEX = 0)

   SFO (1,1) : $\sigma_{x,o}$

   SFO (2,1) : $\sigma_{y,o}$ Components in the x,y,z coordinate

   SFO (3,1) : $\sigma_{z,o}$ system attached to the hole of the

   SFO (4,1) : $\tau_{yz,o}$ 3D stress field applied at infinity

   SFO (5,1) : $\tau_{xz,o}$

   SFO (6,1) : $\tau_{xy,o}$

   QINT : internal pressure in the hole.

<u>5th Card</u> (F10.4, I5, F10.4) (INDEX = 0)

RAMAX, NRAROC, TETROC: When INDEX = 0, the program generates a grid in the body with $1 \leqslant r/a \leqslant$ RAMAX and $0^\circ \leqslant \theta \leqslant 180^\circ$. The incremental values of θ and r/a are equal to

$$\theta_{inc} = \text{TETROC}$$

$$(r/a)_{inc} = (\text{RAMAX} - 1.)/\text{NRAROC}$$

Stresses, strains and displacements are calculated at each point $(r/a, \theta)$

<u>4th Card</u> (I5, F10.4)(INDEX > 0)

See fifth input data card of Berni 1

<u>5th Card</u> (3E10.4, 3F10.4) (INDEX > 0)

See sixth input data card of Berni 1 with ROS (i,5) = 1.

4) Remarks:

Remarks 5) (iv) and (v) of Berni 1 still apply.

Appendix 6.4 : Program Berni 3

1) Program Berni 3 is a two option program:

(i) It calculates the least square estimates of the principal components of a 3D in situ stress field and their orientation with respect to a fixed arbitrary global coordinate system X,Y,Z from measurements of change in strain recorded within a solid or a hollow inclusion. Those measurements take place into one or several boreholes. If several boreholes are used, all the inclusions have the same geometric and elastic properties. Strain rosettes are oriented as shown in Figure 6.4. The program has been written for a maximum number of 4 rosettes (12 measurements) per inclusion and 3 boreholes.

(ii) It calculates the least square estimates of the principal components of a 3D in situ stress field and their orientation with respect to a fixed arbitrary global coordinate system X,Y,Z from measurements of change in length (change in diameters and change in length of oblique distances) at the inner surface of a hollow inclusion. Those measurements take place in one or several boreholes. If several boreholes are used, all the inclusion have the same geometric and elastic properties. The program has been written for a maximum number of 6 measurements of change in length per borehole and 3 boreholes.

Both options make use of the theory developed in chapters 5 and 6 and of the statistical analysis of Appendix 6.1. If the total number of measurements is equal to 6, the statistical analysis is deleted. The orientation of the anisotropy and the holes with respect to a fixed arbitrary global coordinate system X,Y,Z is defined in Figures 1 and 2 of Appendix 4.1. The anisotropic body is assumed to be either isotropic or transversely isotropic or orthotropic. It can also contain up to three orthogonal joint sets each one being perpendicular to a plane of elastic symmetry of the intact body. Options (i) and (ii) are called by setting the input variable INDEX respectively to zero and to a non zero positive value.

2) The flow chart of the program can be written as follows:

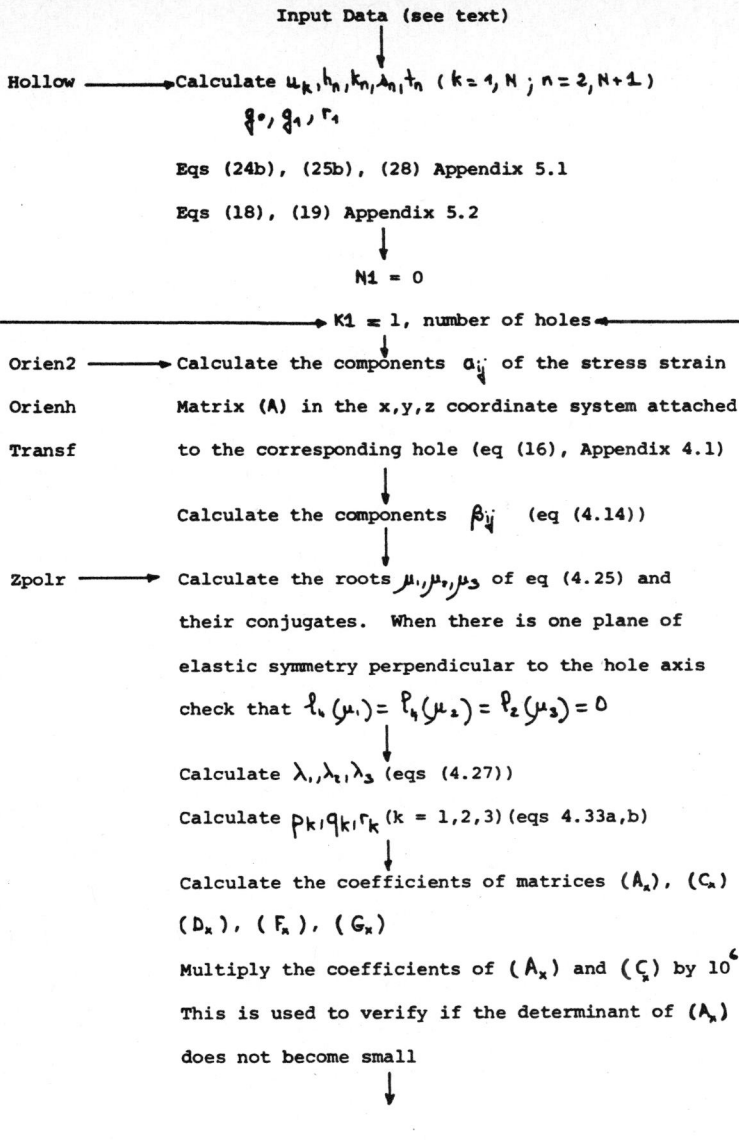

Input Data (see text)

↓

Hollow ——→ Calculate $u_k, h_n, k_n, \lambda_n, t_n$ ($k = 1, N$; $n = 2, N+1$)
g_0, g_1, r_1

Eqs (24b), (25b), (28) Appendix 5.1

Eqs (18), (19) Appendix 5.2

↓

N1 = 0

→ K1 = 1, number of holes ←

↓

Orien2 ——→ Calculate the components a_{ij} of the stress strain
Orienh — Matrix (A) in the x,y,z coordinate system attached
Transf — to the corresponding hole (eq (16), Appendix 4.1)

↓

Calculate the components β_{ij} (eq (4.14))

↓

Zpolr ——→ Calculate the roots μ_1, μ_2, μ_3 of eq (4.25) and their conjugates. When there is one plane of elastic symmetry perpendicular to the hole axis check that $l_4(\mu_1) = l_4(\mu_2) = l_2(\mu_3) = 0$

↓

Calculate $\lambda_1, \lambda_2, \lambda_3$ (eqs (4.27))

Calculate p_k, q_k, r_k (k = 1,2,3) (eqs 4.33a,b)

↓

Calculate the coefficients of matrices (A_x), (C_x) (D_x), (F_x), (G_x)

Multiply the coefficients of (A_x) and (C_x) by 10^6

This is used to verify if the determinant of (A_x) does not become small

↓

K1 = K1 + 1

N1 = N1 + 1

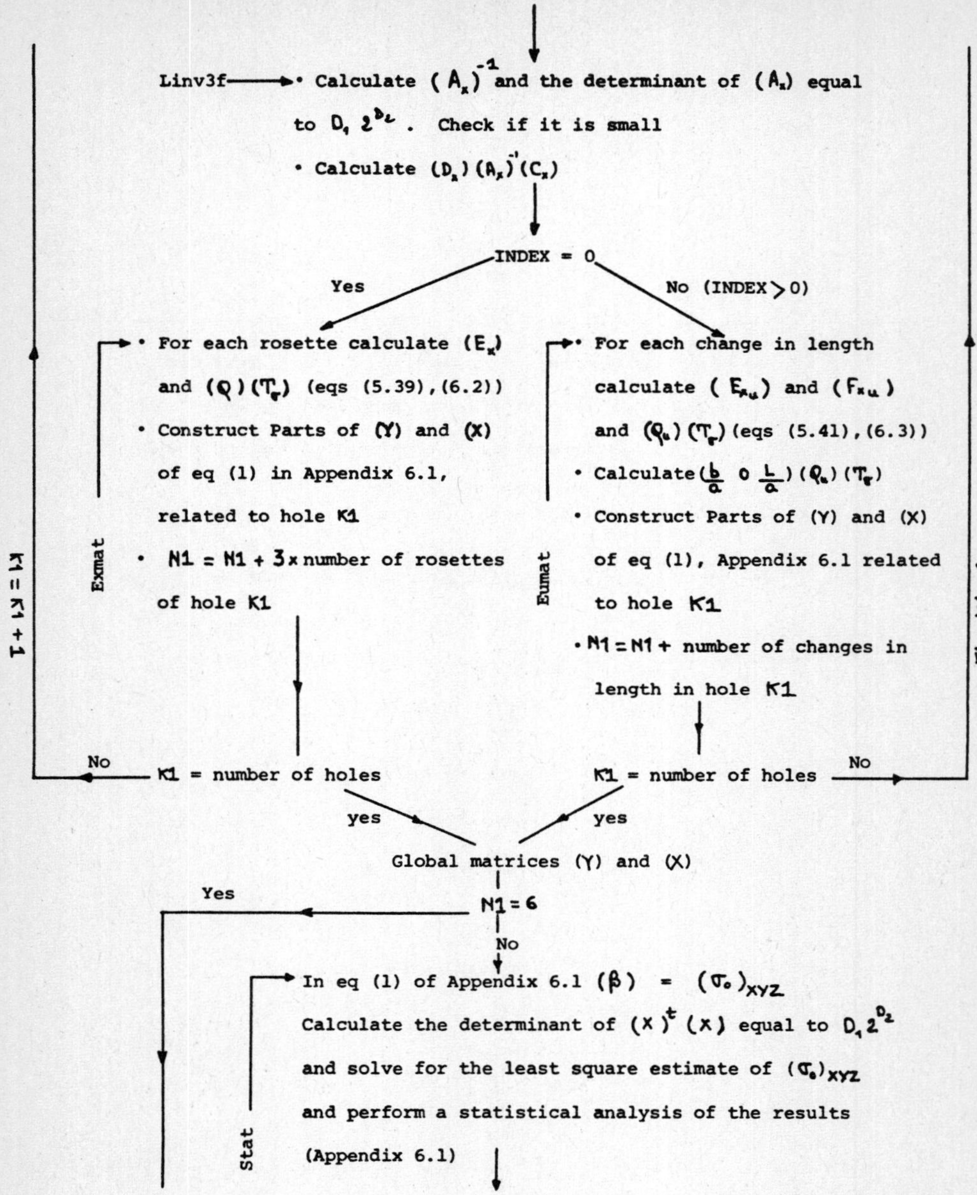
Linv3f
• Calculate $(A_x)^{-1}$ and the determinant of (A_x) equal to $D_1 2^{D_2}$. Check if it is small
• Calculate $(D_x)(A_x)^{-1}(C_x)$
INDEX = 0
Yes
No (INDEX > 0)
• For each rosette calculate (E_x) and $(Q)(T_\sigma)$ (eqs (5.39),(6.2))
• Construct Parts of (Y) and (X) of eq (1) in Appendix 6.1, related to hole K1
• N1 = N1 + 3 x number of rosettes of hole K1
Exmat
• For each change in length calculate (E_{xu}) and (F_{xu}) and $(Q_u)(T_\sigma)$ (eqs (5.41),(6.3))
• Calculate $(\frac{b}{a}\ 0\ \frac{L}{a})(Q_u)(T_\sigma)$
• Construct Parts of (Y) and (X) of eq (1), Appendix 6.1 related to hole K1
• N1 = N1 + number of changes in length in hole K1
Eumat
K1 = K1 + 1
K1 = K1 + 1
No
K1 = number of holes
K1 = number of holes
No
yes
yes
Global matrices (Y) and (X)
Yes
N1 = 6
No
In eq (1) of Appendix 6.1 $(\beta) = (\sigma_0)_{XYZ}$
Calculate the determinant of $(X)^t(X)$ equal to $D_1 2^{D_2}$ and solve for the least square estimate of $(\sigma_0)_{XYZ}$ and perform a statistical analysis of the results (Appendix 6.1)
Stat

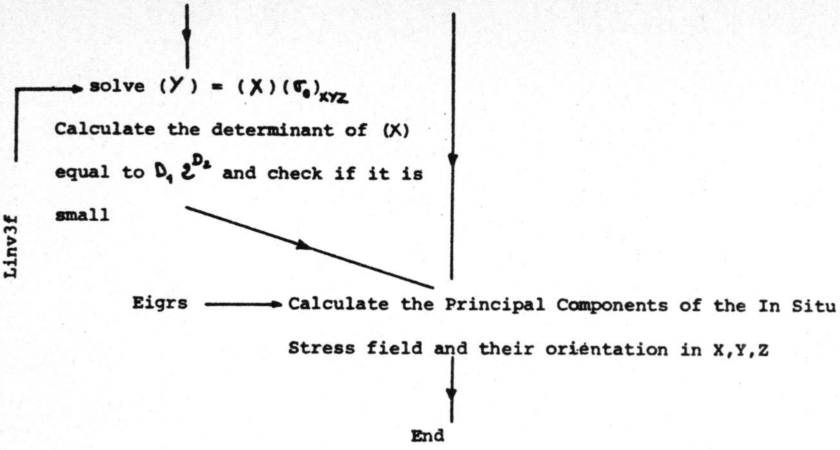
solve $(Y) = (X)(\sigma_0)_{XYZ}$
Calculate the determinant of (X)
equal to $D_1 2^{D_2}$ and check if it is
small
Linv3f
Eigrs
Calculate the Principal Components of the In Situ
Stress field and their orientation in X,Y,Z
End

3) Program Berni 3 consists of one main program and the same subroutines than in Berni 1.

4) Input Data:

<u>1st Card</u> (4F10.4, I5)

See first input data card of Berni 1

<u>2nd Card</u> (2F10.4, 2I5)

STR
DIP — see program Berni 1 (2nd input data card)
DIPDIR

INDEX = 0 Calculate the in situ stress field components from measurements of <u>change in strain</u>

> 0 Calculate the in situ stress field components from measurements of <u>change in length</u>

3rd Card (9F8.2)

See third input data card of Berni 1

4th Card (6E10.4, 3F6.2)

See fourth input data card of Berni 1

5th Card (I5, 7F10.4)

NHOLES: number of holes

TSTU: value of the $(1-\frac{\alpha}{2})$ percentage point of a Student's t distribution with $N1-6$ degrees of freedom and $(1-\alpha)$ 100 percents confidence limits. $N1$ is the total number of measurements (at most 36 when INDEX = 0 and 18 when INDEX > 0)

BETH(I), BETH(I): Angles β_h and δ_h of hole I (I = 1, NHOLES) with respect to a fixed global coordinate system X,Y,Z

6th Card (3I5)(INDEX = 0)

NROS (K) : Number of rosettes in hole K (K = 1, NHOLES)

7th Card (3E10.4, 3F10.4) (INDEX = 0)

For each hole K there is a set of N2= NROS(K) cards.

There is one card per rosette I (I = 1, N2) with the following informations

ROS (I,1,K) = measured strain ε_{A_I}

ROS (I,2,K) = measured strain ε_{B_I}

ROS (I,3,K) = measured strain ε_{C_I}

ROS (I,4,K) = ψ_I angle

ROS (I,5,K) = r_I/a

ROS (I,6,K) = θ_I

(ROS (I,1,K) to ROS (I,6,K): see Figure 6.4)

6th Card (3I5) (INDEX $>$ 0)

NGAGE (K) : number of changes in length in hole K (K=1, NHOLES)

7th Card (3E10.4, 3F10.4) (INDEX $>$ 0)

For each hole K there is a set of N2 = NGAGE (K) cards

There is one card per change in length I (I = 1,N2) with the following informations

GAGE (I,1,K) = $\Delta s_{\theta_I}/s_{\theta_I}$

GAGE (I,2,K) = $s_{\theta_I}/2a$

GAGE (I,3,K) = L_I/a

GAGE (I,4,K) = θ_I

(see Figure 5.4)

s_{θ_I} corresponds to the initial length between points P_1 and P_2 of Figure 5.4, Δs_{θ_I} is the measured change in length between those two points,

$2a$ is the diameter of the hole ,

L_I is the z coordinate of point P_1 ,

θ_I gives the position of point P_1 along the inner contour of the inclusion

5) Remarks:

(i) if several boreholes are used, all the inclusions have the same geometric and elastic properties. Similarly, all the holes have the same diameter $2a$. The variable N appearing in the summation term of eqs (5.10) is also the same and can be set equal to any value between 3 and 10.

(ii) the program was written to account for measurements of change in length at the inner surface of a hollow inclusion.

If the inclusion is solid and if those measurements take place within the solid inclusion, the present program still applies. Points P_1 and P_2 in Figure 5.4 are now located on a circle with radius b . In the first input data card M must be set equal to a/b.

(iii) If there is no inclusion and measurements take place directly on the anisotropic body, set M to a value close to 1 and E to a small value. For instance $M = 1.001$ and $E = 10^{-5}$ MPa.

(iv) Remarks (iv) and (v) in Berni 1 still apply

(v) if the total number of measurements **N1** is equal to 6, the statistical analysis is deleted and TSTU can be left blank.

(vi) in each inclusion, the strain rosettes can be embedded at different depths. The present program can be modified if the strain gages are inclined with respect to the θ, z axes or if other orientations are used.

(vii) in each inclusion, changes in length can be measured in different cross sections. Note also, in comparison to eq (6.3) that $\frac{\Delta s_\theta}{s_\theta}$ is an input instead of $\frac{\Delta s_\theta}{2a}$. This results from the identity

$$\left(\frac{\Delta s_\theta}{2a}\right)\left(\frac{s_\theta}{2a}\right) = \left(\frac{\Delta s_\theta}{s_\theta}\right)\left(\frac{s_\theta}{2a}\right)^2$$

Appendix 6.5: Program Listings

- Berni 1 -

```
      PROGRAM BERNI1(INPUT,OUTPUT,TAPE5=INPUT,TAPE6=OUTPUT)
      COMMON/HOLIN/M,E,MUI,B,N,U(21),H(21),K(21),S(21),T(21),G0,G1,R1
      COMMON/BH/BETAH,DELTH,AH(3,3)
      COMMON/ANIS/STR,DIP,DIPDIR,A(3,3)
      DIMENSION KN(3),KS(3),SJ(3),HF(6,6),H1(6,6),H2(6,6),TS(6,6),
     1TTS(6,6),TH(6,6),TE(6,6),TTE(6,6),AA(6,6),BETA(6,6),
     2T1(6,6),T2(6,6),T3(6,6),AX(61,61),CX(61,6),SFO(6,1)
      DIMENSION AAA(7),ROOT(6),MU(3),L2(3),L3(3),L4(3),LANDA(3),
     1PST(3),QST(3),RST(3),FU(6),FV(6),FW(5),DX(67,61),FX(6,6),
     1GX(6,6),FC(61,1),FCT(67,1),WKAREA(153),EX(6,67),
     2T4(6,1),T5(6,1),SIG(6),T6(6,1),T7(6,1),BD(3),Z(3,3),WK(10),
     3HX(61,6),HHX(67,6),EXU(3,67),FXU(3,5),BD1(3,1),BD2(3,1),
     4ROS(6,6),YROS(18,1),XROS(18,6),SFOG(6,1)
      REAL KN,KS,MU21,MU31,MU32,M,MUI,K
      INTEGER DIPDIR
      COMPLEX ROOT,MU,L2,L3,L4,LANDA,DELTA,PST,QST,RST,GAMA,
     1GAMB,GAMC,GAMPA,GAMPB,GAMPC,GAMPPA,GAMPPB,GAMPPC,IM
C
C     ***************************************************************
C
      IUNFC=61
      IUNFCP=67
      READ(5,10) M,E,MUI,B,N
   10 FORMAT(4F10.4,I5)
      WRITE(6,19)
      IF(B.NE.0.) WRITE(6,247)
  247 FORMAT(/10X,≠ FULL INCLUSION ≠,/)
      WRITE(6,20) M,E,MUI,N
   19 FORMAT(10X,≠** INCLUSION PROPERTIES **≠//)
   20 FORMAT(10X,≠ RATIO OUTER/INNER RADII ≠,5X,F10.4/
     110X,≠ YOUNG MODULUS ≠,10X,F10.4/
     210X,≠ POISSON S RATIO ≠,8X,F10.4/
     310X,≠ FOURIER COEFFICIENTS N ≠,12X,I3///)
C
      CALL HOLLOW
C
      NN=N+1
      WRITE(6,21)G0,G1,R1
      WRITE(6,203)
      WRITE(6,22)(I,U(I),H(I),K(I),S(I),T(I),I=1,NN)
   21 FORMAT(10X,≠ G0 ≠,E12.4,≠ G1 ≠,E12.4,≠ R1 ≠,E12.4//)
   22 FORMAT(10X,I3,2X,E12.4,2X,E12.4,2X,E12.4,2X,E12.4,2X,E12.4)
  203 FORMAT(/10X,≠  I≠,7X,≠ U ≠,11X,≠ H ≠,11X,≠ K ≠,11X,≠ S ≠,11X,
     1≠ T ≠,/)
C
C     ***************************************************************
C
      READ(5,11) STR,DIP,DIPDIR,BETAH,DELTH,INDEX
      READ(5,12) E1,E2,E3,G12,G13,G23,MU21,MU31,MU32
      READ(5,13)(KN(I),I=1,3),(KS(I),I=1,3),(SJ(I),I=1,3)
      IF(INDEX.GT.0) GOTO 40
      READ(5,28)(SFO(I,1),I=1,6),NRAINC,TETINC
      READ(5,239) RAMAX,NRAROC,TETROC
   28 FORMAT(6F10.4,I5,F10.4)
  239 FORMAT(F10.4,I5,F10.4)
      GOTO 31
   40 CONTINUE
```

```
      READ(5,32) NROS,TSTU
      READ(5,33)((ROS(I,J),J=1,6),I=1,NROS)
   32 FORMAT(I5,F10.4)
   33 FORMAT(3E10.4,3F10.4)
   31 CONTINUE
   11 FORMAT(2F10.4,I5,2F10.4,I5)
   12 FORMAT(9F8.2)
   13 FORMAT(6E10.4,3F6.2)
C
C     ******************************************************************
C
      WRITE(6,23) STR,DIP,DIPDIR,BETAH,DELTH
      WRITE(6,26)
      WRITE(6,25) E1,E2,E3,G12,G13,G23,MU21,MU31,MU32
      WRITE(6,24)(I,KN(I),KS(I),SJ(I),I=1,3)
      IF(INDEX.GT.0) GO TO 41
      WRITE(6,29)(SFO(I,1),I=1,6)
   29 FORMAT(//10X,*** INITIAL STRESS FIELD ***//
     1          10X,*     SIGXO = *,E12.4/
     1          10X,*     SIGYO = *,E12.4/
     1          10X,*     SIGZO = *,E12.4/
     1          10X,*    TAUYZO = *,E12.4/
     1          10X,*    TAUXZO = *,E12.4/
     1          10X,*    TAUXYO = *,E12.4////)
      WRITE(6,241) NRAINC,NRAROC,TETINC,TETROC,RAMAX
  241 FORMAT(//10X,*** SCALING PARAMETERS ***,//
     1          10X,* NRAINC(INCLUSION) *,I5/
     1          10X,* NRAINC(ROCK) *,I5/
     1          10X,* TETINC(INCLUSION) *,F10.4/
     1          10X,* TETINC (ROCK) *,F10.4/
     1          10X,* MAX RA (ROCK )*,F10.4//)
      GOTO 34
   41 CONTINUE
      WRITE(6,35) NROS,TSTU
      WRITE(6,37)
      WRITE(6,36)(I,(ROS(I,J),J=1,6),I=1,NROS)
   35 FORMAT(//10X,*** ROSETTE PROPERTIES ***//
     1        10X,* NUMBER OF ROSETTES *,I5/
     1        10X,* T DISTRIBUTION *,E12.4/)
   36 FORMAT(10X,I5,5X,3E10.4,3F10.4)
   37 FORMAT(12X,*I*,9X,*EPSIA*,5X,*EPSIB*,5X,*EPSIC*,6X,*PSI*,8X,
     1*RA*,7X,*TETA*/)
   34 CONTINUE
   26 FORMAT(//10X,*** ROCK PROPERTIES ***//)
   23 FORMAT(//10X,*** ORIENTATION PARAMETERS ***//
     1        10X,* ANISOTROPY ** STRIKE *,F10.4/
     1        10X,*               DIP    *,F10.4/
     2        10X,*               DIPDIR *,I5//
     3        10X,* BOREHOLE **   BETAH  *,F10.4/
     4        10X,*               DELTH  *,F10.4//)
   24 FORMAT(10X,* JOINT SET *,I5,* KN *,E12.4,* KS *,E12.4,* S *,
     1F10.4//)
   25 FORMAT(10X,* YOUNG MODULUS IN N DIRECTION *,F10.4/
     1        10X,* YOUNG MODULUS IN S DIRECTION *,F10.4/
     1        10X,* YOUNG MODULUS IN T DIRECTION *,F10.4/
     1        10X,* SHEAR MODULUS IN NS PLANE *,F10.4/
     1        10X,* SHEAR MODULUS IN NT PLANE *,F10.4/
     1        10X,* SHEAR MODULUS IN ST PLANE *,F10.4/
     1        10X,* POISSON RATIO MU21 *,F10.4/
     1        10X,* POISSON RATIO MU31 *,F10.4/
     1        10X,* POISSON RATIO MU32 *,F10.4///)
C
```

```
C     ******************************************************************
C
C     CALCULATE THE COMPLIANCE MATRIX HP AS THE SUM OF TWO MATRICES
C     H1 AND H2
C
C     INITIALIZATION OF MATRICES H1 AND H2
C
      DO 3 I=1,6
      DO 3 J=1,6
      H1(I,J)=0.
    3 H2(I,J)=0.
C
C     CALCULATE THE MATRIX H1 WITH INTACT MATERIAL CONSTANTS
C
      H1(1,1)=1./E1
      H1(2,2)=1./E2
      H1(3,3)=1./E3
      H1(4,4)=1./G23
      H1(5,5)=1./G13
      H1(6,6)=1./G12
      H1(1,2)=-MU21/E2
      H1(1,3)=-MU31/E3
      H1(2,3)=-MU32/E3
      H1(2,1)=H1(1,2)
      H1(3,1)=H1(1,3)
      H1(3,2)=H1(2,3)
C
      WRITE(6,104)
      WRITE(6,105)((H1(I,J),J=1,6),I=1,6)
  104 FORMAT(///30X,≠ MATRIX H1 ≠,//)
  105 FORMAT(6E12.4)
C
C     CALCULATE THE MATRIX H2 WITH JOINT PROPERTIES
C
      DO 16 I=1,3
      H2(I,I)=1./(KN(I)*SJ(I))
   16 CONTINUE
      H2(4,4)=1./(KS(2)*SJ(2))+1./(KS(3)*SJ(3))
      H2(5,5)=1./(KS(1)*SJ(1))+1./(KS(3)*SJ(3))
      H2(6,6)=1./(KS(1)*SJ(1))+1./(KS(2)*SJ(2))
      WRITE(6,112)
  112 FORMAT(///30X,≠ MATRIX H2 ≠//)
      WRITE(6,113)((H2(I,J),J=1,6),I=1,6)
  113 FORMAT(6E12.4)
C
C     THEN, CALCULATE THE COMPLIANCE MATRIX HP BY ADDING MATRICES
C     H1 AND H2
C
      DO 14 I=1,6
      DO 15 J=1,6
      HP(I,J)=H1(I,J)+H2(I,J)
   15 CONTINUE
   14 CONTINUE
C
C     FOR THE ANISOTROPY CALCULATE THE TRANSFORMATION MATRICES
C     TS AND TTS
C
      CALL ORIEN2
      CALL TRANSF(A,TS)
      CALL TRANS(TS,TTS,6,6,6,6)
      WRITE(6,100)
  100 FORMAT(///30X,≠  TRANSFORMATION MATRIX TS≠,//)
```

```
      WRITE(6,1C1)((TS(I,J),J=1,6),I=1,6)
  101 FORMAT(6E12.4)
      WRITE(6,1C2)
  102 FORMAT(///30X,*  TRANSFORMATION MATRIX TTS *,//)
      WRITE(6,1C3)((TTS(I,J),J=1,6),I=1,6)
  103 FORMAT(6E12.4)
C
C     FOR THE HOLE CALCULATE THE TRANSFORMATION MATRIX TH
C
      CALL CRIENH
      CALL TRANSF(AH,TH)
      WRITE(6,1C6)
  106 FORMAT(///30X,*  TRANSFORMATICN MATRIX TH *,//)
      WRITE(6,1C7)((TH(I,J),J=1,6),I=1,6)
  107 FORMAT(6E12.4)
C
C     FOR THE HOLE CALCULATE THE TRANSFORMATICN MATRICES TE AND TTE
C
      DO 4 I=1,6
      DO 5 J=1,6
      TE(I,J)=TH(I,J)
      IF((I.LE.3).AND.(J.GT.3)) TE(I,J)=0.5*TH(I,J)
      IF((I.GT.3).AND.(J.LE.3)) TE(I,J)=2.*TH(I,J)
    5 CONTINUE
    4 CONTINUE
      CALL TRANS(TE,TTE,6,6,6,6)
      WRITE(6,108)
  108 FORMAT(///30X,*  TRANSFORMATION MATRIX TE *,//)
      WRITE(6,109)((TE(I,J),J=1,6),I=1,6)
  109 FORMAT(6E12.4)
      WRITE(6,110)
  110 FORMAT(///30X,*  TRANSFORMATION MATRIX TTE *,//)
      WRITE(6,111)((TTE(I,J),J=1,6),I=1,6)
  111 FORMAT(6E12.4)
C
C      CALCULATE THE STRAIN-STRESS MATRIX AA IN THE COORDINATE SYSTEM
C     ATTACHED TO THE HOLE
C
      CALL PRCMAT(TE,TTS,T1,6,6,6,6,6,6)
      CALL PRCMAT(T1,HP,T2,6,6,6,6,6,6)
      CALL PRCMAT(T2,TS,T3,6,6,6,6,6,6)
      CALL PRCMAT(T3,TTE,AA,6,6,6,6,6,6)
      WRITE(6,114)
  114 FORMAT(//30X,* STRESS STRAIN MATRIX AA *,//)
      WRITE(6,115)((AA(I,J),J=1,6),I=1,6)
  115 FORMAT(6F15.9)
C
C     CALCULATE THE COEFFICIENTS OF MATRIX BETA
C
      DO 1 I=1,6
      DO 2 J=1,6
      BETA(I,J)=AA(I,J)-AA(I,3)*AA(J,3)/AA(3,3)
    2 CONTINUE
    1 CONTINUE
      WRITE(6,121)
  121 FORMAT(/30X,* MATRIX BETA *//)
      WRITE(6,120)((BETA(I,J),J=1,6),I=1,6)
  120 FORMAT(6F15.9)
C
C     CALCULATE THE COEFFICIENTS AAA(I) OF THE 6TH DEGREE POLYNOMIAL
C
      AAA(1)=BETA(1,1)*BETA(5,5)-BETA(1,5)*BETA(1,5)
```

```
      AAA(2)=2.*BETA(1,5)*(BETA(1,4)+BETA(5,6))
      AAA(2)=AAA(2)-2.*(BETA(1,()*BETA(5,5)+BETA(1,1)*BETA(4,5))
      AAA(3)=BETA(5,5)*(2.*BETA(1,2)+BETA(6,6))+4.*BETA(1,6)*BETA(4,5)+
     1BETA(1,1)*BETA(4,4)
      AAA(3)=AAA(3)-(BETA(1,4)+BETA(5,6))**2-2.*BETA(1,5)*(BETA(2,5)+
     1BETA(4,6))
      AAA(4)=-2.*BETA(2,6)*BETA(5,5)-2.*BETA(4,5)*(2.*BETA(1,2)+
     1BETA(6,6))-2.*BETA(1,6)*BETA(4,4)
      AAA(4)=AAA(4)+2.*BETA(1,5)*BETA(2,4)+2.*(BETA(1,4)+BETA(5,6))*
     1(BETA(2,5)+BETA(4,6))
      AAA(5)=BETA(2,2)*BETA(5,5)+4.*BETA(2,6)*BETA(4,5)+BETA(4,4)*
     1(2.*BETA(1,2)+BETA(6,6))
      AAA(5)=AAA(5)-2.*BETA(2,4)*(BETA(1,4)+BETA(5,6))-(BETA(2,5)+
     1BETA(4,6))**2
      AAA(6)=-2.*BETA(2,2)*BETA(4,5)-2.*BETA(2,6)*BETA(4,4)+2.*BETA(2,4)
     1*(BETA(2,5)+BETA(4,6))
      AAA(7)=BETA(2,2)*BETA(4,4)-BETA(2,4)*BETA(2,4)
C
C     CALCULATE THE ROOTS OF THE 6TH DEGREE CHARACTERISTIC EQUATION
C     USING SUBROUTINE ZPCLR
C
      CALL ZPCLR(AAA,6,ROOT,IER)
      WRITE(6,201)IER,(ROOT(I),I=1,6)
  201 FORMAT(///,I5,6(F10.4,F10.4))
C
C     FIND THE ROOTS MU1,MU2,MU3
C
      MU(1)=ROOT(5)
      MU(2)=ROOT(1)
      MU(3)=ROOT(3)
      WRITE(6,116)(MU(I),I=1,3)
  116 FORMAT(//10X,≠ MU1 ≠,(F15.9,F15.9)/
     1          10X,≠ MU2 ≠,(F15.9,F15.9)/
     1          10X,≠ MU3 ≠,(F15.9,F15.9)////)
C
C     CALCULATE THE COEFFICIENTS LANDA1,LANDA2,LANDA3 AND DELTA
C
      DO 17 I=1,3
      L3(I)=BETA(1,5)*(MU(I)**3)-(BETA(1,4)+BETA(5,6))*(MU(I)**2)+(
     1BETA(2,5)+BETA(4,6))*MU(I)-BETA(2,4)
      L2(I)=BETA(5,5)*(MU(I)**2)-2.*BETA(4,5)*MU(I)+BETA(4,4)
      L4(I)=BETA(1,1)*(MU(I)**4)-2.*BETA(1,6)*(MU(I)**3)+(2.*BETA(1,2)+
     1BETA(6,6))*(MU(I)**2)-2.*BETA(2,6)*MU(I)+BETA(2,2)
      LANDA(I)=-L3(I)/L2(I)
      IF(I.EQ.3) LANDA(I)=-L3(I)/L4(I)
      WRITE(6,117) I,L2(I),L3(I),L4(I)
  117 FORMAT(10X,≠ MU ≠,I3,≠ L2 ≠,(E12.4,E12.4),≠ L3 ≠,(E12.4,E12.4),
     1≠ L4 ≠,(E12.4,E12.4),//)
   17 CONTINUE
      DELTA=MU(2)-MU(1)+LANDA(2)*LANDA(3)*(MU(1)-MU(3))+LANDA(1)*
     1LANDA(3)*(MU(3)-MU(2))
C
C     CALCUALTE THE COEFFICIENTS PST(I),QST(I),RST(I)
C
      DO 18 I=1,2
      PST(I)=BETA(1,1)*MU(I)*MU(I)+BETA(1,2)-BETA(1,6)*MU(I)+LANDA(I)*
     1(BETA(1,5)*MU(I)-BETA(1,4))
      QST(I)=BETA(1,2)*MU(I)+LANDA(I)*(BETA(2,5)-BETA(2,4)/MU(I))-
     1BETA(2,6)+BETA(2,2)/MU(I)
      RST(I)=BETA(1,4)*MU(I)+BETA(2,4)/MU(I)-BETA(4,6)+LANDA(I)*(BETA(4,
     15)-BETA(4,4)/MU(I))
   18 CONTINUE
```

```
      PST(3)=LANDA(3)*(BETA(1,1)*(MU(3)**2)+BETA(1,2)-BETA(1,6)*MU(3))+
     1BETA(1,5)*MU(3)-BETA(1,4)
      QST(3)=LANDA(3)*(BETA(1,2)*MU(3)-BETA(2,6)+BETA(2,2)/MU(3))+BETA(
     12,5)-BETA(2,4)/MU(3)
      RST(3)=LANDA(3)*(BETA(1,4)*MU(3)-BETA(4,6)+BETA(2,4)/MU(3))+BETA(
     14,5)-BETA(4,4)/MU(3)
C
C     CALCULATE THE COEFFICIENTS FU,FV,FW
C
      IM=(0.,1.)
      GAMA   =PST(1)*(MU(2)-MU(3)*LANDA(2)*LANDA(3))+
     1          PST(2)*(LANDA(1)*LANDA(3)*MU(3)-MU(1))+
     2          PST(3)*(MU(1)*LANDA(2)-MU(2)*LANDA(1))
      GAMB   =PST(1)*(LANDA(2)*LANDA(3)-1.)+
     2          PST(2)*(1.-LANDA(1)*LANDA(3))+
     2          PST(3)*(LANDA(1)-LANDA(2))
      GAMC   =PST(1)*LANDA(3)*(MU(3)-MU(2))+
     1          PST(2)*LANDA(3)*(MU(1)-MU(3))+
     2          PST(3)*(MU(2)-MU(1))
      GAMPA =QST(1)*(MU(2)-MU(3)*LANDA(2)*LANDA(3))+
     1          QST(2)*(LANDA(1)*LANDA(3)*MU(3)-MU(1))+
     2          QST(3)*(MU(1)*LANDA(2)-MU(2)*LANDA(1))
      GAMPB =QST(1)*(LANDA(2)*LANDA(3)-1.)+
     2          QST(2)*(1.-LANDA(1)*LANDA(3))+
     2          QST(3)*(LANDA(1)-LANDA(2))
      GAMPC =QST(1)*LANDA(3)*(MU(3)-MU(2))+
     1          QST(2)*LANDA(3)*(MU(1)-MU(3))+
     2          QST(3)*(MU(2)-MU(1))
      GAMPPA=RST(1)*(MU(2)-MU(3)*LANDA(2)*LANDA(3))+
     1          RST(2)*(LANDA(1)*LANDA(3)*MU(3)-MU(1))+
     2          RST(3)*(MU(1)*LANDA(2)-MU(2)*LANDA(1))
      GAMPPB=RST(1)*(LANDA(2)*LANDA(3)-1.)+
     2          RST(2)*(1.-LANDA(1)*LANDA(3))+
     2          RST(3)*(LANDA(1)-LANDA(2))
      GAMPPC=RST(1)*LANDA(3)*(MU(3)-MU(2))+
     1          RST(2)*LANDA(3)*(MU(1)-MU(3))+
     2          RST(3)*(MU(2)-MU(1))
C
      FU(1)=-REAL(GAMA/DELTA)
      FU(2)=-REAL(IM*GAMA/DELTA)
      FU(3)=-REAL(GAMB/DELTA)
      FU(4)=-REAL(IM*GAMB/DELTA)
      FU(5)=-REAL(GAMC/DELTA)
      FU(6)=-REAL(IM*GAMC/DELTA)
C
      FV(1)=-REAL(GAMPA/DELTA)
      FV(2)=-REAL(IM*GAMPA/DELTA)
      FV(3)=-REAL(GAMPB/DELTA)
      FV(4)=-REAL(IM*GAMPB/DELTA)
      FV(5)=-REAL(GAMPC/DELTA)
      FV(6)=-REAL(IM*GAMPC/DELTA)
C
      FW(1)=-REAL(GAMPPA/DELTA)
      FW(2)=-REAL(IM*GAMPPA/DELTA)
      FW(3)=-REAL(GAMPPB/DELTA)
      FW(4)=-REAL(IM*GAMPPB/DELTA)
      FW(5)=-REAL(GAMPPC/DELTA)
      FW(6)=-REAL(IM*GAMPPC/DELTA)
      WRITE(6,202)(FU(I),FV(I),FW(I),I=1,6)
  202 FORMAT(3E12.4)
C
C     CALCULATE THE COEFFICIENTS S,T,SP,TP,SPP,TPP
```

```
C
      S11  =-FU(4)
      S22  =-FU(1)
      S12  =-(FU(2)+FU(3))
      S23  =-FU(5)
      S13  =-FU(6)
      T11  =FU(3)
      T22  =-FU(2)
      T12  =FU(1)-FU(4)
      T23  =-FU(6)
      T13  =FU(5)
C
      SP11=-FV(4)
      SP22=-FV(1)
      SP12=-(FV(2)+FV(3))
      SP23=-FV(5)
      SP13=-FV(6)
      TP11=FV(3)
      TP22=-FV(2)
      TP12=FV(1)-FV(4)
      TP23=-FV(6)
      TP13=FV(5)
C
      SPP11=-FW(4)
      SPP22=-FW(1)
      SPP12=-(FW(2)+FW(3))
      SPP23=-FW(5)
      SPP13=-FW(6)
      TPP11=FW(3)
      TPP22=-FW(2)
      TPP12=FW(1)-FW(4)
      TPP23=-FW(6)
      TPP13=FW(5)
C
C     CALCULATE THE COEFFICIENTS OF MATRIX AX(6N+1,6N+1)
C
      NFC=6*N+1
      DO 6 I=1,NFC
      DO 7 J=1,NFC
      AX(I,J)=0.
    7 CONTINUE
    6 CONTINUE
C
      AX(3,4)=1.
      AX(3,3)=-1.
      AX(2,2)=GO-(FU(4)+FV(3))
      AX(2,3)=FU(3)-FV(4)
      AX(2,4)=FU(2)+FV(1)
      AX(2,5)=GO+(FV(2)-FU(1))
      AX(2,6)=-(FU(6)+FV(5))
      AX(2,7)=FU(5)-FV(6)
      AX(1,1)=2.
      AX(1,2)=FU(3)-FV(4)
      AX(1,3)=FV(3)+FU(4)
      AX(1,4)=FV(2)-FU(1)
      AX(1,5)=-(FV(1)+FU(2))
      AX(1,6)=FU(5)-FV(6)
      AX(1,7)=FV(5)+FU(6)
      AX(6,2)=-FW(4)
      AX(6,3)=FW(3)
      AX(6,4)=FW(2)
      AX(6,5)=-FW(1)
```

```
      AX(6,6)=-FW(6)+U(1)
      AX(6,7)=FW(5)
      AX(7,2)=FW(3)
      AX(7,3)=FW(4)
      AX(7,4)=-FW(1)
      AX(7,5)=-FW(2)
      AX(7,6)=FW(5)
      AX(7,7)=FW(6)-U(1)
      AX(4,8)=FU(3)-FV(4)
      AX(4,9)=FV(3)+FU(4)-G1+R1
      AX(4,10)=FV(2)-FU(1)+G1-R1
      AX(4,11)=-(FV(1)+FU(2))
      AX(4,12)=FU(5)-FV(6)
      AX(4,13)=FV(5)+FU(6)
      AX(5,8)=-(FU(4)+FV(3))+G1-R1
      AX(5,9)=FU(3)-FV(4)
      AX(5,10)=FU(2)+FV(1)
      AX(5,11)=FV(2)-FU(1)+G1-R1
      AX(5,12)=-(FU(6)+FV(5))
      AX(5,13)=FU(5)-FV(6)
C
      NNN=N-1
      DO 8 I=2,NNN
      RI=I
      AX(6*I-4,6*I-10)=(FV(3)-FU(4))/(RI-1.)+(H(I)+K(I))
      AX(6*I-4,6*I-9)=(FU(3)+FV(4))/(RI-1.)
      AX(6*I-4,6*I-8)=(FU(2)-FV(1))/(RI-1.)
      AX(6*I-4,6*I-7)=-(FU(1)+FV(2))/(RI-1.)-(H(I)+K(I))
      AX(6*I-4,6*I-6)=(FV(5)-FU(6))/(RI-1.)
      AX(6*I-4,6*I-5)=(FU(5)+FV(6))/(RI-1.)
      AX(6*I-4,6*I+2)=-(FU(4)+FV(3))/(RI+1.)+(H(I)-K(I))
      AX(6*I-4,6*I+3)=(FU(3)-FV(4))/(RI+1.)
      AX(6*I-4,6*I+4)=(FU(2)+FV(1))/(RI+1.)
      AX(6*I-4,6*I+5)=(FV(2)-FU(1))/(RI+1.)+(H(I)-K(I))
      AX(6*I-4,6*I+6)=-(FU(6)+FV(5))/(RI+1.)
      AX(6*I-4,6*I+7)=(FU(5)-FV(6))/(RI+1.)
      AX(6*I-3,6*I-10)=-(FV(4)+FU(3))/(RI-1.)
      AX(6*I-3,6*I-9)=(FV(3)-FU(4))/(RI-1.)+(S(I)+T(I))
      AX(6*I-3,6*I-8)=(FV(2)+FU(1))/(RI-1.)+(S(I)+T(I))
      AX(6*I-3,6*I-7)=(FU(2)-FV(1))/(RI-1.)
      AX(6*I-3,6*I-6)=-(FV(6)+FU(5))/(RI-1.)
      AX(6*I-3,6*I-5)=(FV(5)-FU(6))/(RI-1.)
      AX(6*I-3,6*I+2)=(FU(3)-FV(4))/(RI+1.)
      AX(6*I-3,6*I+3)=(FV(3)+FU(4))/(RI+1.)+(S(I)-T(I))
      AX(6*I-3,6*I+4)=(FV(2)-FU(1))/(RI+1.)+(T(I)-S(I))
      AX(6*I-3,6*I+5)=-(FV(1)+FU(2))/(RI+1.)
      AX(6*I-3,6*I+6)=(FU(5)-FV(6))/(RI+1.)
      AX(6*I-3,6*I+7)=(FV(5)+FU(6))/(RI+1.)
      AX(6*I-2,6*I-10)=-(FV(4)+FU(3))/(RI-1.)
      AX(6*I-2,6*I-9)=(FV(3)-FU(4))/(RI-1.)+(H(I)+K(I))
      AX(6*I-2,6*I-8)=(FV(2)+FU(1))/(RI-1.)+(H(I)+K(I))
      AX(6*I-2,6*I-7)=(FU(2)-FV(1))/(RI-1.)
      AX(6*I-2,6*I-6)=-(FV(6)+FU(5))/(RI-1.)
      AX(6*I-2,6*I-5)=(FV(5)-FU(6))/(RI-1.)
      AX(6*I-2,6*I+2)=-(FU(3)-FV(4))/(RI+1.)
      AX(6*I-2,6*I+3)=-(FV(3)+FU(4))/(RI+1.)+(H(I)-K(I))
      AX(6*I-2,6*I+4)=-(FV(2)-FU(1))/(RI+1.)+(K(I)-H(I))
      AX(6*I-2,6*I+5)=(FV(1)+FU(2))/(RI+1.)
      AX(6*I-2,6*I+6)=-(FU(5)-FV(6))/(RI+1.)
      AX(6*I-1,6*I-10)=-(FV(3)-FU(4))/(RI-1.)-(S(I)+T(I))
      AX(6*I-1,6*I-9)=-(FU(3)+FV(4))/(RI-1.)
      AX(6*I-2,6*I+7)=-(FV(5)+FU(6))/(RI+1.)
```

```
      AX(6*I-1,6*I-8)=-(FU(2)-FV(1))/(RI-1.)
      AX(6*I-1,6*I-7)=(FU(1)+FV(2))/(RI-1.)+(S(I)+T(I))
      AX(6*I-1,6*I-6)=-(FV(5)-FU(6))/(RI-1.)
      AX(6*I-1,6*I-5)=-(FU(5)+FV(6))/(RI-1.)
      AX(6*I-1,6*I+2)=-(FU(4)+FV(3))/(RI+1.)+(T(I)-S(I))
      AX(6*I-1,6*I+3)=(FU(3)-FV(4))/(RI+1.)
      AX(6*I-1,6*I+4)=(FU(2)+FV(1))/(RI+1.)
      AX(6*I-1,6*I+5)=(FV(2)-FU(1))/(RI+1.)+(T(I)-S(I))
      AX(6*I-1,6*I+6)=-(FU(6)+FV(5))/(RI+1.)
      AX(6*I,6*I-4)=-FW(4)/RI
      AX(6*I-1,6*I+7)= (FU(5)-FV(6))/(RI+1.)
      AX(6*I,6*I-3)= FW(3)/RI
      AX(6*I,6*I-2)= FW(2)/RI
      AX(6*I,6*I-1)=-FW(1)/RI
      AX(6*I,6*I)=-FW(6)/RI+U(I)
      AX(6*I,6*I+1)=FW(5)/RI
      AX(6*I+1,6*I-4)= FW(3)/RI
      AX(6*I+1,6*I-3)= FW(4)/RI
      AX(6*I+1,6*I-2)=-FW(1)/RI
      AX(6*I+1,6*I-1)=-FW(2)/RI
      AX(6*I+1,6*I)= FW(5)/RI
      AX(6*I+1,6*I+1)= FW(6)/RI-U(I)
    8 CONTINUE
C
      AX(6*N-4,6*N-4)=-FW(4)/FLOAT(N)
      AX(6*N-4,6*N-3)= FW(3)/FLOAT(N)
      AX(6*N-4,6*N-2)= FW(2)/FLOAT(N)
      AX(6*N-4,6*N-1)=-FW(1)/FLOAT(N)
      AX(6*N-4,6*N)=-FW(6)/FLOAT(N)+U(N)
      AX(6*N-4,6*N+1)= FW(5)/FLOAT(N)
      AX(6*N-3,6*N-4)= FW(3)/FLOAT(N)
      AX(6*N-3,6*N-3)= FW(4)/FLOAT(N)
      AX(6*N-3,6*N-2)=-FW(1)/FLOAT(N)
      AX(6*N-3,6*N-1)=-FW(2)/FLOAT(N)
      AX(6*N-3,6*N)=FW(5)/FLOAT(N)
      AX(6*N-3,6*N+1)=FW(6)/FLOAT(N)-U(N)
      AX(6*N-2,6*N-10)=(FV(3)-FU(4))/FLOAT(NNN)+(H(N)+K(N))
      AX(6*N-2,6*N-9)=(FU(3)+FV(4))/FLOAT(NNN)
      AX(6*N-2,6*N-8)=(FU(2)-FV(1))/FLOAT(NNN)
      AX(6*N-2,6*N-7)=-(FU(1)+FV(2))/FLOAT(NNN)-(H(N)+K(N))
      AX(6*N-2,6*N-6)=(FV(5)-FU(6))/FLOAT(NNN)
      AX(6*N-2,6*N-5)=(FU(5)+FV(6))/FLOAT(NNN)
      AX(6*N-1,6*N-10)=-(FV(4)+FU(3))/FLOAT(NNN)
      AX(6*N-1,6*N-9)=(FV(3)-FU(4))/FLOAT(NNN)+(H(N)+K(N))
      AX(6*N-1,6*N-8)=(FV(2)+FU(1))/FLOAT(NNN)+(H(N)+K(N))
      AX(6*N-1,6*N-7)=(FU(2)-FV(1))/FLOAT(NNN)
      AX(6*N-1,6*N-6)=-(FV(6)+FU(5))/FLOAT(NNN)
      AX(6*N-1,6*N-5)=(FV(5)-FU(6))/FLOAT(NNN)
      AX(6*N,6*N-4)=(FV(3)-FU(4))/FLOAT(N)+(H(N+1)+K(N+1))
      AX(6*N,6*N-3)=(FU(3)+FV(4))/FLOAT(N)
      AX(6*N,6*N-2)=(FU(2)-FV(1))/FLOAT(N)
      AX(6*N,6*N-1)=-(FU(1)+FV(2))/FLOAT(N)-(H(N+1)+K(N+1))
      AX(6*N,6*N)=(FV(5)-FU(6))/FLOAT(N)
      AX(6*N,6*N+1)=(FU(5)+FV(6))/FLOAT(N)
      AX(6*N+1,6*N-4)=-(FV(4)+FU(3))/FLOAT(N)
      AX(6*N+1,6*N-3)=(FV(3)-FU(4))/FLOAT(N)+(H(N+1)+K(N+1))
      AX(6*N+1,6*N-2)=(FV(2)+FU(1))/FLOAT(N)+(H(N+1)+K(N+1))
      AX(6*N+1,6*N-1)=(FU(2)-FV(1))/FLOAT(N)
      AX(6*N+1,6*N)=-(FV(6)+FU(5))/FLOAT(N)
      AX(6*N+1,6*N+1)=(FV(5)-FU(6))/FLOAT(N)
C
C     CALCULATE THE COEFFICIENTS OF MATRIX CX(6N+1,6)
```

```
C
      DO 50 I=1,NFC
      DO 51 J=1,6
      CX(I,J)=0.
   51 CONTINUE
   50 CONTINUE
C
      CX(2,1)=2.*MUI*AA(3,1)+AA(1,1)+AA(2,1)-S11+TP11
      CX(2,2)=2.*MUI*AA(3,2)+AA(1,2)+AA(2,2)-S22+TP22
      CX(2,3)=2.*MUI*AA(3,3)+AA(1,3)+AA(2,3)
      CX(2,4)=2.*MUI*AA(3,4)+AA(1,4)+AA(2,4)+S23-TP23
      CX(2,5)=2.*MUI*AA(3,5)+AA(1,5)+AA(2,5)-S13+TP13
      CX(2,6)=2.*MUI*AA(3,6)+AA(1,6)+AA(2,6)+S12-TP12
      CX(1,1)=-(SP11+T11)
      CX(1,2)=-(SP22+T22)
      CX(1,3)=0.
      CX(1,4)=SP23+T23
      CX(1,5)=-(SP13+T13)
      CX(1,6)=SP12+T12
      CX(6,1)=AA(5,1)-SPP11
      CX(6,2)=AA(5,2)-SPP22
      CX(6,3)=AA(5,3)
      CX(6,4)=AA(5,4)+SPP23
      CX(6,5)=AA(5,5)-SPP13
      CX(6,6)=AA(5,6)+SPP12
      CX(7,1)=-(AA(4,1)+TPP11)
      CX(7,2)=-(AA(4,2)+TPP22)
      CX(7,3)=-AA(4,3)
      CX(7,4)=-(AA(4,4)-TPP23)
      CX(7,5)=-(AA(4,5)+TPP13)
      CX(7,6)=-(AA(4,6)-TPP12)
      CX(8,1)=AA(1,1)-AA(2,1)-S11-TP11
      CX(8,2)=AA(1,2)-AA(2,2)-S22-TP22
      CX(8,3)=AA(1,3)-AA(2,3)
      CX(8,4)=AA(1,4)-AA(2,4)+S23+TP23
      CX(8,5)=AA(1,5)-AA(2,5)-S13-TP13
      CX(8,6)=AA(1,6)-AA(2,6)+S12+TP12
      CX(9,1)=AA(6,1)-SP11+T11
      CX(9,2)=AA(6,2)-SP22+T22
      CX(9,3)=AA(6,3)
      CX(9,4)=AA(6,4)+SP23-T23
      CX(9,5)=AA(6,5)-SP13+T13
      CX(9,6)=AA(6,6)+SP12-T12
      CX(10,1)=AA(6,1)+T11-SP11
      CX(10,2)=AA(6,2)+T22-SP22
      CX(10,3)=AA(6,3)
      CX(10,4)=AA(6,4)-T23+SP23
      CX(10,5)=AA(6,5)+T13-SP13
      CX(10,6)=AA(6,6)-T12+SP12
      CX(11,1)=AA(2,1)-AA(1,1)+TP11+S11
      CX(11,2)=AA(2,2)-AA(1,2)+TP22+S22
      CX(11,3)=AA(2,3)-AA(1,3)
      CX(11,4)=AA(2,4)-AA(1,4)-TP23-S23
      CX(11,5)=AA(2,5)-AA(1,5)+TP13+S13
      CX(11,6)=AA(2,6)-AA(1,6)-TP12-S12
C
C     CALCULATE THE COEFFICIENTS OF MATRIX DX(6N+5,6N+1)
C
      NFCP=NFC+6
      DO 44 I=1,NFCP
      DO 45 J=1,NFC
      DX(I,J)=0.
```

```
 45 CONTINUE
 44 CONTINUE
    DX(1,2)=-0.5
    DX(1,5)=-0.5
    DX(2,8)=-0.5
    DX(2,11)=-0.5
    DX(3,9)=-0.5
    DX(3,10)=0.5
    DX(4,9)=0.5
    DX(4,10)=-0.5
    DX(5,8)=-0.5
    DX(5,11)=-0.5
    DX(6,6)=-1.
    DX(7,7)=-1.
    DO 46 I=2,NNN
    DX(6*I-4,6*I-10)=-0.5
    DX(6*I-4,6*I-7)=0.5
    DX(6*I-4,6*I+2)=-0.5
    DX(6*I-4,6*I+5)=-0.5
    DX(6*I-3,6*I-9)=-0.5
    DX(6*I-3,6*I-8)=-0.5
    DX(6*I-3,6*I+3)=-0.5

    DX(6*I-2,6*I-9)=-0.5
    DX(6*I-2,6*I-8)=-0.5
    DX(6*I-2,6*I+3)=0.5
    DX(6*I-2,6*I+4)=-0.5
    DX(6*I-1,6*I-10)=0.5
    DX(6*I-1,6*I-7)=-0.5
    DX(6*I-1,6*I+2)=-0.5
    DX(6*I-1,6*I+5)=-0.5
    DX(6*I,6*I)=-1.
    DX(6*I+1,6*I+1)=-1.
 46 CONTINUE
    DX(6*N-4,6*N-10)=-0.5
    DX(6*N-4,6*N-7)=0.5
    DX(6*N-3,6*N-8)=-0.5
    DX(6*N-3,6*N-9)=-0.5
    DX(6*N-2,6*N-9)=-0.5
    DX(6*N-2,6*N-8)=-0.5
    DX(6*N-1,6*N-10)=0.5
    DX(6*N-1,6*N-7)=-0.5
    DX(6*N,6*N)=-1.
    DX(6*N+1,6*N+1)=-1.
    DX(6*N+2,6*N-4)=-0.5
    DX(6*N+2,6*N-1)=0.5
    DX(6*N+3,6*N-3)=-0.5
    DX(6*N+3,6*N-2)=-0.5
    DX(6*N+4,6*N-2)=-0.5
    DX(6*N+4,6*N-3)=-0.5
    DX(6*N+5,6*N-4)=0.5
    DX(6*N+5,6*N-1)=-0.5
    DX(6*N+6,3)=0.5
    DX(6*N+6,4)=-0.5
    DX(6*N+7,1)=1.
C
C
    DO 54 I=1,6
    DO 55 J=1,6
    FX(I,J)=0.
    GX(I,J)=0.
 55 CONTINUE
```

```
   54 CONTINUE
      DO 56 I=1,6
      FX(3,I)=E*AA(3,I)
   56 CONTINUE
      DO 57 I=1,6
      GX(I,I)=1./E
      IF(I.GE.4) GX(I,I)=2.*(1.+MUI)/E
   57 CONTINUE
      GX(1,2)=-MUI/E
      GX(1,3)=-MUI/E
      GX(2,3)=-MUI/E
      GX(2,1)=GX(1,2)
      GX(3,1)=GX(1,3)
      GX(3,2)=GX(2,3)
C
      DO 60 I=1,NFC
      DO 61 J=1,NFC
      AX(I,J)=AX(I,J)*1.E+06
   61 CONTINUE
   60 CONTINUE
      DC 52 I=1,NFC
      DO 53 J=1,6
      CX(I,J)=CX(I,J)*1.E+06
   53 CONTINUE
   52 CONTINUE
C
      IF(INDEX.GT.0) GOTO 58
C
C     ******************************************************************
C
C     FOR THE GIVEN STRESS FIELC CALCULATE THE RIGID BOCY RCTATION
C     AND THE FCURIER COEFFICIENTS FC(6N+1,1)
C
      CALL PRCMAT(CX,SFC,FC,NFC,6,1,IUNFC,5,1)
      D1=1.
      IAA=IUNFC
      CALL LINV3F(AX,FC,2,NFC,IAA,D1,D2,WKAREA,IER)
      WRITE(6,205) D1,D2
  205 FCRMAT(//10X,≠ D1 ≠,E12.4,≠  D2 ≠,E12.4//)
      WRITE(6,118) FC(1,1)
  118 FCRMAT(//10X,≠ RIGIC BODY ROTATION ≠,E12.4//)
      WRITE(6,213)
  213 FCRMAT(/19X,≠AX≠,14X,≠BX≠,14X,≠AY≠,14X,≠BY≠,14X,≠AZ≠,14X,≠BZ≠/)
      WRITE(6,119)(FC(I,1),I=2,NFC)
  119 FORMAT(10X,≠     ≠,E12.4,≠     ≠,E12.4,≠     ≠,E12.4,≠     ≠,
     1E12.4,≠     ≠,E12.4,≠     ≠,E12.4)
C
C     FCR THE GIVEN STRESS FIELC CALCULATE THE COEFFICIENTS
C     A,B,C,D,E,F AND STORE THEM INTO MATRIX FCT(6N+7,1)
C
      CALL PRCMAT(DX,FC,FCT,NFCP,NFC,1,IUNFCP,IUNFC,1)
      WRITE(6,206) FCT(1,1),FCT(NFCP-1,1),FCT(NFCP,1)
      WRITE(6,208)FCT(6*N+2,1),FCT(6*N+3,1),FCT(6*N+4,1),FCT(6*N+5,1)
      WRITE(6,214)
  214 FCRMAT(/19X,≠A≠,11X,≠B≠,11X,≠C≠,11X,≠D≠,11X,≠E≠,11X,≠F≠,/)
      WRITE(6,207)(FCT(I,1),I=2,NFC)
  206 FCRMAT(//10X,≠ AO ≠,E12.4,≠  CO  ≠,E12.4,≠ W-WP ≠,E12.4/)
  207 FCRMAT(/10X,≠ ≠,E12.4,≠ ≠,E12.4,≠ ≠,E12.4,≠ ≠,E12.4,≠ ≠,E12.4,
     2≠ ≠,E12.4)
  208 FCRMAT(/10X,≠ A(N+1)≠,E12.4,≠B(N+1)≠,E12.4,≠C(N+1)≠,E12.4,
     4≠D(N+1)≠,E12.4//)
C
```

```
C      CALCULATE THE STRESSES INDUCED BY THE INCLUSION WITHIN THE ROCK
C
       WRITE(6,242)
  242  FORMAT(//10X,≠ ** ROCK ** ≠,//)
       RAINC=(RAMAX-1.)/FLOAT(NRAROC)
       NRAROC=NRAROC+1
       NTET=IFIX(180./TETROC)+1
       TET=0.
       DO 243 I=1,NTET
       RA=1.
       DO 244 J=1,NRAROC
       CALL RESTR(MU,LANDA,DELTA,FC,TET,RA,IUNFC,N,AA,T6)
       RA=RA+RAINC
  244  CONTINUE
       TET=TET+TETROC
  243  CONTINUE
       WRITE(6,245)
  245  FORMAT(//10X,≠ INCLUSION ≠,//)
C
C      CALCULATE STRESS,STRAIN AND DISPLACEMENT COMPONENTS AT ANY POINT
C      WITHIN THE INCLUSION. RA AND TET ARE CALCULATED FROM THE
C      INPUT COEFFICIENTS NRAINC AND TETINC STARTING FROM THE INNER
C      CONTOUR OF THE INCLUSION.
       CALL PRCMAT(FX,SFC,T4,6,6,1,6,6,1)
C
       RAINC=(1.-1./M)/FLOAT(NRAINC)
       NRAINC=NRAINC+1
       NTET=IFIX(90./TETINC)+1
       RA=1./M
       DO 62 I=1,NRAINC
       TET=0.
       DO 63 J=1,NTET
       WRITE(6,211) RA,TET
  211  FORMAT(///10X,≠** RATIO R/A **≠,F10.4,≠ ** ANGLE TETA **≠,F10.4//)
C
       CALL EXMAT(TET,RA,M,MUI,B,N,IUNFCP,EX)
       CALL PRCMAT(EX,FCT,T5,6,NFCP,1,6,IUNFCP,1)
       DO 64 II=1,6
       T6(II,1)=T5(II,1)+T4(II,1)
   64  CONTINUE
       CALL PRCMAT(GX,T6,T7,6,6,1,6,6,1)
C
       WRITE(6,212) (T6(I1,1),I1=1,6)
  212  FORMAT(10X,≠ SIGR ≠,E12.4/
     1        10X,≠ SIGT ≠,E12.4/
     1        10X,≠ SIGZ ≠,E12.4/
     2        10X,≠TAUTZ ≠,E12.4/
     2        10X,≠TAURZ ≠,E12.4/
     2        10X,≠TAURT ≠,E12.4/)
       WRITE(6,235)(T7(I3,1),I3=1,6)
  235  FORMAT(10X,≠ EPSR ≠,E12.4/
     1        10X,≠ EPST ≠,E12.4/
     1        10X,≠ EPSZ ≠,E12.4/
     2        10X,≠GAMTZ ≠,E12.4/
     2        10X,≠GAMRZ ≠,E12.4/
     2        10X,≠GAMRT ≠,E12.4/)
C
       ZA=0.
       CALL EUMAT(TET,RA,ZA,M,E,MUI,B,N,AA,IUNFCP,EXU,FXU)
       CALL PRCMAT(FXU,SFO,BD1,3,6,1,3,6,1)
       CALL PRCMAT(EXU,FCT,BD2,3,NFCP,1,3,IUNFCP,1)
       DO 71 II=1,3
```

```
      BD1(II,1)=BD1(II,1)+BD2(II,1)
   71 CONTINUE
      WRITE(6,237)(BD1(I,1),I=1,3)
  237 FORMAT(10X,≠ UR/A ≠,E12.4/
     1       10X,≠ VT/A ≠,E12.4/
     2       10X,≠  W/A ≠,E12.4/)
C
C     CALCULATE THE PRINCIPAL STRESSES AND THEIR ORIENTATION
C     WITH RESPECT TO THE LOCAL COORDINATE SYSTEM X,Y,Z ATTACHED
C     TO THE HOLE.
C
      TETA=TET*ATAN(1.)*4./180.
      TETA2=2.*TETA
      SIG(1)=T6(1,1)*COS(TETA)*COS(TETA)+T6(2,1)*SIN(TETA)*SIN(TETA)-
     1T6(6,1)*SIN(TETA2)
      SIG(3)=T6(1,1)*SIN(TETA)*SIN(TETA)+T6(2,1)*COS(TETA)*COS(TETA)+
     1T6(6,1)*SIN(TETA2)
      SIG(6)=T6(3,1)
      SIG(2)=SIN(TETA)*COS(TETA)*(T6(1,1)-T6(2,1))+
     1COS(TETA2)*T6(6,1)
      SIG(5)=COS(TETA)*T6(4,1)+SIN(TETA)*T6(5,1)
      SIG(4)=COS(TETA)*T6(5,1)-SIN(TETA)*T6(4,1)
      WRITE(6,238) SIG(1),SIG(3),SIG(6),SIG(5),SIG(4),SIG(2)
  238 FORMAT(/10X,6E12.4,/)
C
      CALL EIGRS(SIG,3,2,BD,Z,3,WK,IER)
C
      SIG1=AMAX1(BD(1),BD(2),BD(3))
      SIG3=AMIN1(BD(1),BD(2),BD(3))
      DO 65 II=1,3
      IF((BD(II).NE.SIG1).AND.(BD(II).NE.SIG3)) SIG2=BD(II)
   65 CONTINUE
      DO 66 II=1,3
      IF(BD(II).EQ.SIG1) WRITE(6,231) BD(II)
      IF(BD(II).EQ.SIG2) WRITE(6,232) BD(II)
      IF(BD(II).EQ.SIG3) WRITE(6,233) BD(II)
  231 FORMAT(/10X,≠ SIG1 ≠,E12.4/)
  232 FORMAT(/10X,≠ SIG2 ≠,E12.4/)
  233 FORMAT(/10X,≠ SIG3 ≠,E12.4/)
      WRITE(6,234) (Z(JJ,II),JJ=1,3)
  234 FORMAT(10X,≠ L ≠,E12.4,≠ M ≠,E12.4,≠ N ≠,E12.4)
   66 CONTINUE
C
      TET=TET+TETINC
   63 CONTINUE
      RA=RA+RAINC
   62 CONTINUE
      GOTO 59
   58 CONTINUE
C
C     ***************************************************************
C
C
C     CALCULATE THE COEFFICIENTS OF THE MATRIX INVERSE OF AX USING
C     SUBROUTINE LINV3F. AX IS REPLACED BY ITS INVERSE.
C
      D1=1.
      IAA=IUNFC
      CALL LINV3F(AX,BD,1,NFC,IAA,D1,D2,WKAREA,IER)
      WRITE(6,236) D1,D2
  236 FORMAT(//10X,≠ D1 ≠,E12.4,≠ D2 ≠,E12.4//)
C
```

```
C      MULTIPLY AX BY CX AND STORE THE RESJLT INTO HX(6N+1,6)
C      MULTIPLY DX BY HX AND STORE THE RESJLT INTO HHX(6N+7,6)
C
       CALL PRCMAT(AX,CX,HX,NFC,NFC,6,IUNFC,IUNFC,6)
       CALL PRCMAT(DX,HX,HHX,NFCF,NFC,6,IUNFCP,IUNFC,6)
C
C      FOR EACH ROSETTE AT A TIME CALCULATE MATRICES EX AND T3(Q)
C
       DO 72 I=1,NROS
       RA=ROS(I,5)
       TET=ROS(I,6)
       CALL EXMAT(TET,RA,M,MUI,E,N,IUNFCP,EX)
C
C      MULTIPLY EX BY HHX AND STORE THE RESJLT INTO T1(6,6)
C
       CALL PRCMAT(EX,HHX,T1,6,NFCP,6,6,IUNFCP,6)
       DO 262 II=1,6
       DO 263 JJ=1,6
       T2(II,JJ)=T1(II,JJ)+FX(II,JJ)
  263  CONTINUE
  262  CONTINUE
       CALL PRCMAT(GX,T2,T3,6,6,6,6,6,6)
       WRITE(6,264)((T3(I2,J),J=1,6),I2=1,6)
  264  FORMAT(10X,6E12.4)
C
       PSIRO=ROS(I,4)*ATAN(1.)*4./180.
       PSIRO2=2.*PSIRO
       YROS(3*I-2,1)=ROS(I,1)
       YROS(3*I-1,1)=ROS(I,2)
       YROS(3*I,1)=2.*(ROS(I,3)-ROS(I,1)*COS(PSIRO)*COS(PSIRO)-
      1ROS(I,2)*SIN(PSIRO)*SIN(PSIRO))/SIN(PSIRO2)
C
       DO 73 J=1,6
       XROS(3*I-2,J)=T3(2,J)
       XROS(3*I-1,J)=T3(3,J)
       XROS(3*I,J)=T3(4,J)
   73  CONTINUE
   72  CONTINUE
C
       N1=3*NROS
       WRITE(6,74)
   74  FORMAT(///10X,≠ MATRICES  Y AND X ≠,//)
       DO 81 I=1,N1
       YROS(I,1)=-YROS(I,1)
       WRITE(6,75) YROS(I,1),(XROS(I,J),J=1,6)
   75  FORMAT(10X,E12.4,10X,6E12.4)
   81  CONTINUE
       DO 95 I=1,N1
       YROS(I,1)=YROS(I,1)*1.E+06
       DO 96 J=1,6
       XROS(I,J)=XROS(I,J)*1.E+06
   96  CONTINUE
   95  CONTINUE
C
       CALL STAT(YROS,XROS,N1,6,TSTU,SFO,S1ROS,R2ROS,T1,T2)
       S1ROS=S1ROS*1.E-06
       DO 97 I=1,N1
       YROS(I,1)=YROS(I,1)*1.E-06
   97  CONTINUE
       WRITE(6,76) S1ROS,R2ROS,(YROS(I,1),I=1,N1)
       WRITE(6,77) (SFO(I,1),T1(I,1) ,T1(I,2),T1(I,3),I=1,6)
       WRITE(6,79)
```

```
      WRITE(6,78)((T2(I,J),J=1,6),I=1,6)
   76 FORMAT(/10X,* STANDARD DEVIATION ABOUT REGRESSION *,E12.4/
     1         10X,* MULTIPLE CORRELATION COEFFICIENT *,E12.4/
     2         10X,* RESIDUALS */
     3         10X,9E10.4////)
   77 FORMAT(10X,* SIGXO*,E12.4,* ESE*,E12.4,* UL*,E12.4,* LL*,E12.4/
     1        10X,* SIGYO*,E12.4,* ESE*,E12.4,* UL*,E12.4,* LL*,E12.4/
     1        10X,* SIGZO*,E12.4,* ESE*,E12.4,* UL*,E12.4,* LL*,E12.4/
     2        10X,*TAUYZO*,E12.4,* ESE*,E12.4,* UL*,E12.4,* LL*,E12.4/
     2        10X,*TAUXZO*,E12.4,* ESE*,E12.4,* UL*,E12.4,* LL*,E12.4/
     2        10X,*TAUXYO*,E12.4,* ESE*,E12.4,* UL*,E12.4,* LL*,E12.4)
   78 FORMAT(10X,6E12.4)
   79 FORMAT(//10X,* VARIANCE COVARIANCE MATRIX *//)
C
C     CALCULATE THE STRESS FIELD IN THE GLOBAL COORDINATE SYSTEM
C
      CALL PRCMAT(TTE,SFO,SFOG,6,6,1,6,6,1)
      WRITE(6,82)
      WRITE(6,83)(SFOG(I,1),I=1,6)
   82 FORMAT(//10X,* STRESS FIELD IN GLOBAL COORDINATE SYSTEM *,//)
   83 FORMAT(10X,* SIGXO *,E12.4/
     1        10X,* SIGYO *,E12.4/
     1        10X,* SIGZO *,E12.4/
     1        10X,*TAUYZO *,E12.4/
     1        10X,*TAUXZO *,E12.4/
     1        10X,*TAUXYO *,E12.4/)
C
C     CALCULATE THE PRINCIPAL STRESSES AND THEIR ORIENTATION IN
C     THE GLOBAL COORDINATE SYSTEM
C
      SIG(1)=SFOG(1,1)
      SIG(2)=SFOG(6,1)
      SIG(3)=SFOG(2,1)
      SIG(4)=SFOG(5,1)
      SIG(5)=SFOG(4,1)
      SIG(6)=SFOG(3,1)
C
      CALL EIGRS(SIG,3,2,BD,Z,3,WK,IER)
C
      SIG1=AMAX1(BD(1),BD(2),BD(3))
      SIG3=AMIN1(BD(1),BD(2),BD(3))
      DO 84 II=1,3
      IF((BD(II).NE.SIG1).AND.(BD(II).NE.SIG3)) SIG2=BD(II)
   84 CONTINUE
      DO 85 II=1,3
      IF(BD(II).EQ.SIG1) WRITE(6,91) BD(II)
      IF(BD(II).EQ.SIG2) WRITE(6,92) BD(II)
      IF(BD(II).EQ.SIG3) WRITE(6,93) BD(II)
   91 FORMAT(/10X,* SIG1 *,E12.4/)
   92 FORMAT(/10X,* SIG2 *,E12.4/)
   93 FORMAT(/10X,* SIG3 *,E12.4/)
      WRITE(6,94)(Z(JJ,II),JJ=1,3)
   94 FORMAT(10X,* L *,E12.4,* M *,E12.4,* N *,E12.4)
   85 CONTINUE
   59 CONTINUE
      STOP
      END
C
C
      SUBROUTINE HOLLOW
      COMMON/HOLIN/M,E,MU,B,N,U(21),H(21),C(21),S(21),T(21),GO,G1,R1
      COMMON/SUF/TI(21),TI11(21),TI12(21),TI21(21),TI22(21),
```

```
     1TI31(21),TI32(21),TI41(21),TI42(21)
      REAL M,MU,K
C
C     M      RATIO OUTER/INNER RADII OF THE INCLUSION
C     E      YOUNG MODULUS OF THE INCLUSION
C     MU      POISSON#S RATIO OF THE INCLUSION
C     B      0. FOR AN HOLLOW INCLUSION
C            1. FOR A FULL INCLUSION
C     N      NUMBER OF FOURIER COEFFICIENTS
C
C
C     INITIALIZATION OF MATRICES U,H,K,S,T
C
      NN=N+1
      DO 1 I=1,NN
      U(I)=0.
      H(I)=0.
      K(I)=0.
      S(I)=0.
      T(I)=0.
      TI(I)=0.
      TI11(I)=0.
      TI12(I)=0.
      TI21(I)=0.
      TI22(I)=0.
      TI31(I)=0.
      TI32(I)=0.
      TI41(I)=0.
      TI42(I)=0.
    1 CONTINUE
      DEN3=(1.+MU)/E
      IF(B.NE.0.) GOTO 5
C
C     CALCULATE G0,G1,R1
C
      X1=1./(1.-M*M)
      X2=-0.5*M*M*X1
      X3=0.5/(M**4-1.)
      X4=X3*(M**4)
      G0=-DEN3*(-X1+2.*X2*(1.-2.*MU))
      G1=-DEN3*(X3+X4*(1.-4.*MU))
      R1=DEN3*(X3+X4*(5.-4.*MU))
C
C     CALCULATE THE COMPONENTS OF MATRIX J FOR I=1,N
C
      DO 2 I=1,N
      RI=I
      MP=2*I
      X5=M**MP
      U(I)=2.*DEN3*(1.+X5)/(RI*(1.-X5))
    2 CONTINUE
C
C     CALCULATE THE COMPONENTS OF MATRICES H,K,S,T
C
      DO 3 J=2,NN
      RJ=J
      M1=2*J-2
      M2=-(2*J+2)
      TT=4.*(M**M1)-4.*RJ*RJ/(M**4)-4.*RJ*RJ
     1+4.*(M**M2)+(8.*RJ*RJ-8.)/(M**2)
C
      M3=-(J+2)
```

```
      M4=J-2
      TT11=-2.*RJ*(M**J)+2.*RJ*(1.-RJ)/(M**J)
     1+(2.*RJ*RJ)*(M**M3)
      TT12=-2.*(RJ-2.)*(M**J)-2.*(RJ+2)*(1.-RJ)/(M**J)
     1-2.*RJ*RJ*(M**M3)
      TT21=-2.*(M**M4)+2.*(RJ+1.)*(M**M3)-2.*RJ/(M**J)
      TT22=-2.*(M**M4)-2.*(RJ+1.)*(M**M3)+2.*(RJ+2.)/(M**J)
      TT31=2.*RJ*RJ*(M**M4)-2.*(RJ+1.)*RJ*(M**J)+2.*RJ/(M**J)
      TT32=2.*RJ*RJ*(M**M4)-2.*(RJ-2.)*(RJ+1.)*(M**J)-2.*(RJ+2.)
     1/(M**J)
      TT41=-2.*(1.-RJ)*(M**M4)-2.*RJ*(M**J)+2.*(M**M3)
      TT42=-2.*(1.-RJ)*(M**M4)-2.*(RJ-2.)*(M**J)-2.*(M**M3)
      TI(J)=TT
      TI11(J)=TT11
      TI12(J)=TT12
      TI21(J)=TT21
      TI22(J)=TT22
      TI31(J)=TT31
      TI32(J)=TT32
      TI41(J)=TT41
      TI42(J)=TT42
C
      S11=TT11*(M**M4)/(RJ*(RJ-1.)*TT)
      S21=-TT21*(M**J)/((RJ+1.)*TT)
      S31=TT31*(M**M3)/(RJ*(RJ+1.)*TT)
      S41=-TT41/((RJ-1.)*(M**J)*TT)
      S12=-TT12*(M**M4)/(RJ*(RJ-1.)*TT)
      S22=TT22*(M**J)/((RJ+1.)*TT)
      S32=-TT32*(M**M3)/(RJ*(RJ+1.)*TT)
      S42=TT42/((RJ-1.)*(M**J)*TT)
C
      H(J)=DEN3*(RJ*S11+(RJ+4.*MU-2.)*S21-RJ*S31
     1-(RJ-4.*MU+2.)*S41)
      K(J)=DEN3*(RJ*S12+(RJ+4.*MU-2.)*S22-RJ*S32
     1-(RJ-4.*MU+2.)*S42)
      S(J)=DEN3*(RJ*S11+(RJ+4.-4.*MU)*S21+RJ*S31
     1+(RJ-4.+4.*MU)*S41)
      T(J)=DEN3*(RJ*S12+(RJ+4.-4.*MU)*S22+RJ*S32
     1+(RJ-4.+4.*MU)*S42)
    3 CONTINUE
      GOTO 6
C
    5 CONTINUE
      GC=-DEN3*(1.-2.*MU)
      G1=-0.5*DEN3*(1.-4.*MU)
      R1=0.5*DEN3*(5.-4.*MU)
      DO 4 J=2,NN
      RJ=J
      H(J)=0.5*DEN3*((RJ+4.*MU-2.)/(RJ+1.)-RJ/(RJ-1.))
      K(J)=0.5*DEN3*((RJ-2.)/(RJ-1.)-(RJ+4.*MU-2.)/(RJ+1.))
      S(J)=0.5*DEN3*((RJ+4.-4.*MU)/(RJ+1.)-RJ/(RJ-1.))
      T(J)=0.5*DEN3*((RJ-2.)/(RJ-1.)-(RJ+4.-4.*MU)/(RJ+1.))
C
      KK=J-1
      RK=KK
      U(KK)=-2.*DEN3/RK
    4 CONTINUE
    6 CONTINUE
      DO 7 J=N,NN
      S1=H(J)+K(J)
      S2=S(J)+T(J)
      S3=S1-S2
```

```
      WRITE(6,30) J,S1,S2,S3
   30 FORMAT(/10X,≠J≠,I5,≠ W+K ≠,E15.8,≠ T+S ≠,E15.8,≠ S3 ≠,E15.8//)
    7 CONTINUE
      RETURN
      END
C
      SUBROUTINE PROMAT(U,V,W,L,N,M,IU,IV,IW)
      DIMENSION U(IU,IV),V(IV,IW),W(IU,IW)
      DO 14 I=1,L
      DO 15 J=1,M
      S=0.
      DO 16 K=1,N
      S=S+U(I,K)*V(K,J)
      W(I,J)=S
   16 CONTINUE
   15 CONTINUE
   14 CONTINUE
      RETURN
      END
C
      SUBROUTINE TRANS(U,V,N,M,IU,IV)
      DIMENSION U(IU,IV),V(IV,IU)
      DO 5 I=1,N
      DO 6 J=1,M
      V(J,I)=U(I,J)
    6 CONTINUE
    5 CONTINUE
      RETURN
      END
C
      SUBROUTINE TRANSF(A1,T)
      DIMENSION A1(3,3),T(6,6)
      REAL LX,MX,NX,LY,MY,NY,LZ,MZ,NZ
      LX=A1(1,1)
      MX=A1(1,2)
      NX=A1(1,3)
      LY=A1(2,1)
      MY=A1(2,2)
      NY=A1(2,3)
      LZ=A1(3,1)
      MZ=A1(3,2)
      NZ=A1(3,3)
C
      T(1,1)=LX*LX
      T(1,2)=MX*MX
      T(1,3)=NX*NX
      T(1,4)=2.*MX*NX
      T(1,5)=2.*NX*LX
      T(1,6)=2.*LX*MX
      T(2,1)=LY*LY
      T(2,2)=MY*MY
      T(2,3)=NY*NY
      T(2,4)=2.*MY*NY
      T(2,5)=2.*NY*LY
      T(2,6)=2.*LY*MY
      T(3,1)=LZ*LZ
      T(3,2)=MZ*MZ
      T(3,3)=NZ*NZ
      T(3,4)=2.*MZ*NZ
      T(3,5)=2.*NZ*LZ
      T(3,6)=2.*LZ*MZ
      T(4,1)=LY*LZ
```

```
      T(4,2)=MY*MZ
      T(4,3)=NY*NZ
      T(4,4)=MY*NZ+MZ*NY
      T(4,5)=NY*LZ+NZ*LY
      T(4,6)=LY*MZ+LZ*MY
      T(5,1)=LZ*LX
      T(5,2)=MZ*MX
      T(5,3)=NZ*NX
      T(5,4)=MX*NZ+MZ*NX
      T(5,5)=NX*LZ+NZ*LX
      T(5,6)=LX*MZ+LZ*MX
      T(6,1)=LX*LY
      T(6,2)=MX*MY
      T(6,3)=NX*NY
      T(6,4)=MX*NY+MY*NX
      T(6,5)=NX*LY+NY*LX
      T(6,6)=LX*MY+LY*MX
      RETURN
      END
      SUBROUTINE ORIENZ
      COMMON/ANIS/STR,DIP,DIPDIR,A(3,3)
      INTEGER DIPDIR
C
C     SUBROUTINE TO CALCULATE THE DIRECTION COSINES OF THE UNIT
C     VECTORS OF A LOCAL COORDINATE SYSTEM ATTACHED TO A PLANE
C     WITH RESPECT TO A GLOBAL COORDINATE SYSTEM X,Y,Z. THE ORIENTATION
C     OF THE PLANE IS DEFINED BY ITS STRIKE,DIP AND DIPDIRECTION
C     AS FOLLOWS
C     STR      STRIKE OF THE PLANE DEFINED FROM NORTH OR -Z AXIS IN
C              THE CLOCKWISE DIRECTION AND IN THE EAST QUADRANT
C     DIP      ANGLE OF DIP
C     DIPDIR   DIRECTION OF THE NORMAL OF THE PLANE
C                     1 IN N-NE OR-Z,X(-Z) DIRECTION
C                     2 IN E-SE OR X,XZ DIRECTION
C                     3 IN S-SW OR Z,Z(-X) DIRECTION
C                     4 IN W-NW OR (-X),(-X)(-Z) DIRECTION
C                     5 HORIZONTAL PLANE
C
      PI=ATAN(1.)*4.
      PI=PI/180.
      ST=STR
      TETA=DIP
C
C     CALCULATE MATRIX A
C
      GOTO (10,20,20,10,50) DIPDIR
C
   10 BETAX=ST+180.
      BETAY=ST
      BETAZ=ST+270.
      GOTO 1
   20 BETAX=ST
      BETAY=ST+180.
      BETAZ=ST+90.
    1 DELTAX=90.-TETA
      DELTAY=TETA
      DELTAZ=0.
C
      A(1,1)=COS(PI*DELTAX)*COS(PI*BETAX)
      A(1,2)=SIN(DELTAX*PI)
      A(1,3)=COS(PI*DELTAX)*SIN(PI*BETAX)
      A(2,1)=COS(PI*DELTAY)*COS(PI*BETAY)
```

```
      A(2,2)=SIN(DELTAY*PI)
      A(2,3)=COS(PI*DELTAY)*SIN(PI*BETAY)
      A(3,1)=COS(PI*DELTAZ)*COS(PI*BETAZ)
      A(3,2)=SIN(DELTAZ*PI)
      A(3,3)=COS(PI*DELTAZ)*SIN(PI*BETAZ)
      GOTO 2
   50 A(1,1)=0.
      A(1,2)=1.
      A(1,3)=0.
      A(2,1)=-1.
      A(2,2)=0.
      A(2,3)=0.
      A(3,1)=0.
      A(3,2)=0.
      A(3,3)=1.
C
    2 CONTINUE
      PRINT 90
   90 FORMAT(///10X,≠ MATRIX A≠///)
      PRINT 100,((A(I,J),J=1,3),I=1,3)
  100 FORMAT(10X,3F10.4)
      RETURN
      END
      SUBROUTINE ORIENH
      COMMON/BH/BETAH,DELTH,AH(3,3)
C
C     SUBROUTINE TO CALCULATE THE DIRECTIONS COSINES OF THE UNIT
C     VECTORS OF THE COORDINATE SYSTEM ATTACHED TO THE HOLE WITH
C     RESPECT TO A GLOBAL COORDINATE SYSTEM X,Y,Z
C     BETAH      ANGLE BETWEEN THE X AXIS AND THE PROJECTION OF THE
C                BOREHOLE ONTO THE XZ PLANE(POSITIVE IN THE CLOCKWISE
C                DIRECTION)
C     DELTH       PLUNGE OF THE AXIS OF THE HOLE(POSITIVE UPWARDS)
C
      IF(DELTH.EQ.90.) GOTO 1
      IF(DELTH.EQ.-90.) GOTO 1
      PI=ATAN(1.)*4./180.
      BETAX=(BETAH+270.)*PI
      BETAY=(BETAH+180.)*PI
      BETAZ=BETAH*PI
      DELTAX=0.
      DELTAY=(90.-DELTH)*PI
      DELTAZ=DELTH*PI
      AH(1,1)=COS(DELTAX)*COS(BETAX)
      AH(1,2)=SIN(DELTAX)
      AH(1,3)=COS(DELTAX)*SIN(BETAX)
      AH(2,1)=COS(DELTAY)*COS(BETAY)
      AH(2,2)=SIN(DELTAY)
      AH(2,3)=COS(DELTAY)*SIN(BETAY)
      AH(3,1)=COS(DELTAZ)*COS(BETAZ)
      AH(3,2)=SIN(DELTAZ)
      AH(3,3)=COS(DELTAZ)*SIN(BETAZ)
      GOTO 2
    1 CONTINUE
      AH(1,1)=0.
      AH(1,2)=0.
      AH(1,3)=-1.
      AH(2,1)=-1.
      AH(2,2)=0.
      AH(2,3)=0.
      AH(3,1)=0.
      AH(3,2)=1.
```

```
      AH(3,3)=0.
    2 CONTINUE
      WRITE(6,10)
   10 FORMAT(///10X,* MATRIX AH *///)
      WRITE(6,20)((AH(I,J),J=1,3),I=1,3)
   20 FORMAT(10X,3F10.4)
      RETURN
      END
      SUBROUTINE EXMAT(TET,RA,M,MUI,B,N,IUCP,EX)
C
C     SUBROUTINE TO CALCULATE THE COEFFICIENTS OF MATRIX (E)
C     AT ANY POINT (R/A,TET) WITHIN THE INCLUSION. MATRIX (E)
C     IS USED IN THE CALCULATION OF THE STATE OF STRESS AT THAT
C     POINT.
C     TET      ANGLE IN DEGREES FROM X AXIS
C     RA       RATIO R/A WHERE A IS THE OUTER RADIUS OF THE INCLUSION
C     M        RATIO A/B WHERE B IS THE INNER RADIUS OF THE INCLUSION
C     MUI      POISSON*S RATIO OF THE INCLUSION
C     B        =0 FOR AN HOLLOW INCLUSION
C              =1 FOR A FULL INCLUSION
C     N        NUMBER OF FOURIER COEFFICIENTS
C     IUCP     COLUMN DIMENSION OF MATRIX EX AS SPECIFIED IN THE
C              DIMENSION STATEMENT IN THE MAIN PROGRAM
      COMMON/SUF/TI(21),TI11(21),TI12(21),TI21(21),TI22(21),
     1TI31(21),TI32(21),TI41(21),TI42(21)
      DIMENSION EX(6,IUCP)
      REAL M,MUI
C
C     INITIALIZATION OF MATRIX EX
      NN=N+1
      NFCP=6*N+7
      DO 1 I=1,6
      DO 2 J=1,NFCP
      EX(I,J)=0.
    2 CONTINUE
    1 CONTINUE
C
      TETA=TET*ATAN(1.)*4./180.
      IF(B.NE.0.) GOTO 3
C
C     HOLLOW INCLUSION
C
      EX(1,1)=((1./RA)**2.-M**2.)/(1.-M**2.)
      X1=(RA*(M**4.)-(1./RA)**3.)/(M**4.-1.)
      EX(1,2)=X1*COS(TETA)
      EX(1,3)=X1*SIN(TETA)
      EX(2,1)=-((1./RA)**2.+M**2.)/(1.-M**2.)
      X2=((1./RA)**3.+3.*RA*(M**4.))/(M**4.-1.)
      EX(2,2)=X2*COS(TETA)
      EX(2,3)=X2*SIN(TETA)
C
      DO 4 I=1,3
      EX(3,I)=MUI*(EX(1,I)+EX(2,I))
    4 CONTINUE
      EX(6,2)=EX(1,3)
      EX(6,3)=-EX(1,2)
C
      DO 5 I=2,NN
      RI=I
      M1=I-2
      M2=-(I+2)
      Y1=(RA*M)**M1
```

```
      Y2=(RA*M)**I
      Y3=(RA*M)**M2
      Y4=1./Y2
      Z1=(Y1*TI11(I)-(RI-2.)*Y2*TI21(I)+Y3*TI31(I)
     1-(RI+2.)*Y4*TI41(I))/TI(I)
      Z2=(-Y1*TI12(I)+(RI-2.)*Y2*TI22(I)-Y3*TI32(I)
     1+(RI+2.)*Y4*TI42(I))/TI(I)
      EX(1,6*I-4)=-COS(RI*TETA)*Z1
      EX(1,6*I-3)=-SIN(RI*TETA)*Z1
      EX(1,6*I-2)=-SIN(RI*TETA)*Z2
      EX(1,6*I-1)=COS(RI*TETA)*Z2
C
      Z3=(Y1*TI11(I)-(RI+2.)*Y2*TI21(I)+Y3*TI31(I)
     1-(RI-2.)*Y4*TI41(I))/TI(I)
      Z4=(-Y1*TI12(I)+(RI+2.)*Y2*TI22(I)-Y3*TI32(I)
     1+(RI-2.)*Y4*TI42(I))/TI(I)
      EX(2,6*I-4)=COS(RI*TETA)*Z3
      EX(2,6*I-3)=SIN(RI*TETA)*Z3
      EX(2,6*I-2)=SIN(RI*TETA)*Z4
      EX(2,6*I-1)=-COS(RI*TETA)*Z4
C
      EX(3,6*I-4)=MUI*(EX(1,6*I-4)+EX(2,6*I-4))
      EX(3,6*I-3)=MUI*(EX(1,6*I-3)+EX(2,6*I-3))
      EX(3,6*I-2)=MUI*(EX(1,6*I-2)+EX(2,6*I-2))
      EX(3,6*I-1)=MUI*(EX(1,6*I-1)+EX(2,6*I-1))
      Z5=(Y1*TI11(I)-RI*Y2*TI21(I)-Y3*TI31(I)+
     1RI*Y4*TI41(I))/TI(I)
      Z6=(-Y1*TI12(I)+RI*TI22(I)*Y2+Y3*TI32(I)-
     1RI*Y4*TI42(I))/TI(I)
      EX(6,6*I-4)=SIN(RI*TETA)*Z5
      EX(6,6*I-3)=-COS(RI*TETA)*Z5
      EX(6,6*I-2)=-COS(RI*TETA)*Z6
      EX(6,6*I-1)=-SIN(RI*TETA)*Z6
    5 CONTINUE
      DO 6 I=1,N
      RI=I
      M1=I-1
      M2=I+1
      M3=2*I
      Z7=((RA**M1)*(M**M3)+(1./RA)**M2)/(M**M3-1.)
      Z8=((RA**M1)*(M**M3)-(1./RA)**M2)/(M**M3-1.)
      EX(4,6*I)=-Z7*SIN(RI*TETA)
      EX(4,6*I+1)=Z7*COS(RI*TETA)
      EX(5,6*I)=Z8*COS(RI*TETA)
      EX(5,6*I+1)=Z8*SIN(RI*TETA)
    6 CONTINUE
      GOTO 10
C
    3 CONTINUE
C
C     FULL INCLUSION
      EX(1,1)=1.
      EX(1,2)=RA*COS(TETA)
      EX(1,3)=RA*SIN(TETA)
      EX(2,1)=1.
      EX(2,2)=3.*EX(1,2)
      EX(2,3)=3.*EX(1,3)
      EX(6,2)=EX(1,3)
      EX(6,3)=-EX(1,2)
      DO 7 I=1,3
      EX(3,I)=MUI*(EX(1,I)+EX(2,I))
    7 CONTINUE
```

```
      DO 8 I=2,NN
      RI =I
      M1=I-2
      Z1=((RI-2.)*(RA**I)-(RA**M1)*RI)/2.
      Z2=(-(RI-2.)*(RA**I)+(RA**M1)*(RI-2.))/2.
      EX(1,6*I-4)=-Z1*COS(RI*TETA)
      EX(1,6*I-3)=-Z1*SIN(RI*TETA)
      EX(1,6*I-2)=-Z2*SIN(RI*TETA)
      EX(1,6*I-1)=Z2*COS(RI*TETA)
C
      Z3=((RI+2.)*(RA**I)-RI*(RA**M1))/2.
      Z4=(-(RI+2.)*(RA**I)+(RI-2.)*(RA**M1))/2.
      EX(2,6*I-4)=Z3*COS(RI*TETA)
      EX(2,6*I-3)=Z3*SIN(RI*TETA)
      EX(2,6*I-2)=Z4*SIN(RI*TETA)
      EX(2,6*I-1)=-Z4*COS(RI*TETA)
      EX(3,6*I-4)=MUI*(EX(1,6*I-4)+EX(2,6*I-4))
      EX(3,6*I-3)=MUI*(EX(1,6*I-3)+EX(2,6*I-3))
      EX(3,6*I-2)=MUI*(EX(1,6*I-2)+EX(2,6*I-2))
      EX(3,6*I-1)=MUI*(EX(1,6*I-1)+EX(2,6*I-1))
      Z5=(RI*(RA**I)-RI*(RA**M1))/2.
      Z6=(-RI*(RA**I)+(RI-2.)*(RA**M1))/2.
      EX(6,6*I-4)=Z5*SIN(RI*TETA)
      EX(6,6*I-3)=-Z5*COS(RI*TETA)
      EX(6,6*I-2)=-Z6*COS(RI*TETA)
      EX(6,6*I-1)=-Z6*SIN(RI*TETA)
    8 CONTINUE
      DO 9 I=1,N
      RI=I
      M1=I-1
      Z7=RA**M1
      EX(4,6*I)=-Z7*SIN(RI*TETA)
      EX(4,6*I+1)=Z7*COS(RI*TETA)
      EX(5,6*I)=EX(4,6*I+1)
      EX(5,6*I+1)=-EX(4,6*I)
    9 CONTINUE
   10 CONTINUE
      RETURN
      END
      SUBROUTINE EUMAT(TET,RA,ZA,M,E,MUI,B,N,AA,IUCP,EU,FU)
C
C     SUBROUTINE TO CALCULATE THE COEFFICIENTS OF MATRICES (EU)
C     AND (FU) AT ANY POINT (R/A,TET,Z/A) WITHIN THE INCLUSION .
C     MATRICES (EU) AND (FU) ARE USED IN THE CALCULATION OF THE
C     DISPLACEMENTS COMPONENTS AT THAT POINT.
C     TET,RA,M,MUI,B,N,IUCP SEE EXMAT
C     ZA      RATIO Z/A WHEREZ IS THE COORDINATE ALONG THE Z AXIS
C     E       YOUNG MODULUS OF THE INCLUSION
      COMMON/SUF/TI(21),TI11(21),TI12(21),TI21(21),TI22(21),
     1TI31(21),TI32(21),TI41(21),TI42(21)
      DIMENSION EU(3,IUCP),FU(3,6),AA(6,6)
      REAL M,MUI
C
C     INITIALIZATION OF MATRICES EU AND FU
C
      NN=N+1
      NFCP=6*N+7
      DO 1 I=1,3
      DO 2 J=1,NFCP
      EU(I,J)=0.
    2 CONTINUE
      DO 3 K=1,6
```

```
      FU(I,K)=0.
    3 CONTINUE
    1 CONTINUE
C
      DEN=(1.+MUI)/E
      TETA=TET*ATAN(1.)*4./180.
      IF(B.NE.0.) GOTO 4
C
C     HOLLOW INCLUSION
C
      EU(1,1)=DEN*(1./RA+RA*M*M*(1.-2.*MUI))/(1.-M*M)
      X1=-DEN*0.5*((1./RA)**2.+RA*RA*(1.-4.*MUI)*(M**4.))/(M**4.-1.)
      EU(1,2)=X1*COS(TETA)
      EU(1,3)=X1*SIN(TETA)
      X2=-DEN*0.5*((1./RA)**2.+RA*RA*(5.-4.*MUI)*(M**4.))/(M**4.-1.)
      EU(2,2)=X2*SIN(TETA)
      EU(2,3)=-X2*COS(TETA)
      EU(2,6*N+7)=-RA
C
      DO 5 I=2,NN
      RI=I
      M1=I-1
      M2=I-2
      M3=I+1
      M4=I+2
      Y1=(RA**M1)*(M**M2)/(RI-1.)
      Y2=(RA**M3)*(M**I)/(RI+1.)
      Y3=((1./RA)**M3)*((1./M)**M4)/(RI+1.)
      Y4=((1./RA)**M1)*((1./M)**I)/(RI-1.)
C
      Z1=(Y1*TI11(I)-(RI+4.*MUI-2.)*Y2*TI21(I)-Y3*TI31(I)+
     1(RI-4.*MUI+2.)*Y4*TI41(I))*DEN/TI(I)
      Z2=(-Y1*TI12(I)+(RI+4.*MUI-2.)*Y2*TI22(I)+Y3*TI32(I)-
     1(RI-4.*MUI+2.)*Y4*TI42(I))*DEN/TI(I)
      EU(1,6*I-4)=Z1*COS(RI*TETA)
      EU(1,6*I-3)=Z1*SIN(RI*TETA)
      EU(1,6*I-2)=Z2*SIN(RI*TETA)
      EU(1,6*I-1)=-Z2*COS(RI*TETA)
      Z3=(Y1*TI11(I)-(RI+4.-4.*MUI)*Y2*TI21(I)+Y3*TI31(I)-
     1(RI-4.+4.*MUI)*Y4*TI41(I))*DEN/TI(I)
      Z4=(-Y1*TI12(I)+(RI+4.-4.*MUI)*Y2*TI22(I)-Y3*TI32(I)+
     1(RI-4.+4.*MUI)*Y4*TI42(I))*DEN/TI(I)
      EU(2,6*I-4)=-Z3*SIN(RI*TETA)
      EU(2,6*I-3)= Z3*COS(RI*TETA)
      EU(2,6*I-2)= Z4*COS(RI*TETA)
      EU(2,6*I-1)= Z4*SIN(RI*TETA)
    5 CONTINUE
      DO 6 I=1,N
      RI=I
      M1=2*I
      Z5=((RA**I)*(M**M1)+(1./RA)**I)/(RI*(M**M1-1.))
      Z5=2.*DEN*Z5
      EU(3,6*I)=-Z5*COS(RI*TETA)
      EU(3,6*I+1)=-Z5*SIN(RI*TETA)
    6 CONTINUE
      GOTO 10
    4 CONTINUE
C
C     FULL INCLUSION
C
      EU(1,1)=-DEN*RA*(1.-2.*MUI)
      EU(1,2)=-0.5*DEN*RA*RA*(1.-4.*MUI)*COS(TETA)
```

```
      EU(1,3)=-0.5*DEN*FA*RA*(1.-4.*MUI)*SIN(TETA)
      EU(2,2)=-0.5*DEN*FA*RA*(5.-4.*MUI)*SIN(TETA)
      EU(2,3)= 0.5*DEN*FA*RA*(5.-4.*MUI)*COS(TETA)
      EU(2,6*N+7)=-RA
      DO 7 I=2,NN
      RI=I
      M1=I-1
      M2=I+1
       Z1=((RI+4.*MUI-2.)*(RA**M2)/(RI+1.)-RI*(RA**M1)/(RI-1.))*
     10.5*DEN
       Z2=(-(RI+4.*MUI-2.)*(RA**M2)/(RI+1.)+(RI-2.)*(RA**M1)/(RI-1.))
     1*0.5*DEN
       Z3=(-RI*(RA**M1)/(RI-1.)+(RI+4.-4.*MUI)*(RA**M2)/(RI+1.))
     1*0.5*DEN
       Z4=((RI-2.)*(RA**M1)/(RI-1.)-(RI+4.-4.*MUI)*(RA**M2)/(RI+1.))*
     10.5*DEN
      EU(1,6*I-4)=Z1*COS(RI*TETA)
      EU(1,6*I-3)=Z1*SIN(RI*TETA)
      EU(1,6*I-2)=Z2*SIN(RI*TETA)
      EU(1,6*I-1)=-Z2*COS(RI*TETA)
      EU(2,6*I-4)=-Z3*SIN(RI*TETA)
      EU(2,6*I-3)=Z3*COS(RI*TETA)
      EU(2,6*I-2)=Z4*COS(RI*TETA)
      EU(2,6*I-1)=Z4*SIN(RI*TETA)
    7 CONTINUE
      DO 8 I=1,N
      RI=I
      EU(3,6*I)=-2.*DEN*(RA**I)*COS(RI*TETA)/RI
      EU(3,6*I+1)= -2.*DEN*(RA**I)*SIN(RI*TETA)/RI
    8 CONTINUE
   10 CONTINUE
C
      DO 11 I=1,6
      FU(1,I)=MUI*RA*AA(3,I)
      FU(3,I)=-ZA*AA(3,I)
   11 CONTINUE
      RETURN
      END
      SUBROUTINE RESTR(MU,LANDA,DELTA,FC,TET,RA,IUNFC,N,AA,SIG)
      DIMENSION MU(3),LANDA(3),FC(IUNFC,1),KSI(3),SQ(3),SIG(6,1)
      DIMENSION AA(6,6)
      COMPLEX MU,LANDA,DELTA,KSI,IM,SQ,PHIP1,PHIP2,PHIP3,Z1,Z2,Z3,
     1Z11,Z22,Z33,Z4
      WRITE(6,20) RA,TET
   20 FORMAT(///10X,≠** RATIO R/A **≠,F10.4,≠ ** ANGLE TETA **≠,F10.4//)
      TETA=TET*ATAN(1.)*4./180.
      IF(TET.EQ.90.) TETA=89.999*ATAN(1.)*4./180.
      IF(TET.EQ.270.) TETA=270.001*ATAN(1.)*4./180.
      IM=(0.,1.)
      DO 1 I=1,3
      Z4=COS(TETA)+MU(I)*SIN(TETA)
      SQ(I)=CSQRT(RA*RA*Z4*Z4-1.-MU(I)*MU(I))
      IF((TET.GT.90.).AND.(TET.LT.270.)) SQ(I)=-SQ(I)
      KSI(I)=(SQ(I)+RA*(COS(TETA)+MU(I)*SIN(TETA)))/(1.-IM*MU(I))
    1 CONTINUE
C
      PHIP1=(0.,0.)
      PHIP2=(0.,0.)
      PHIP3=(0.,0.)
      DO 2 M=1,N
      Z11=FC(6*M-3,1)-IM*FC(6*M-4,1)
      Z22=FC(6*M-1,1)-IM*FC(6*M-2,1)
```

```
      Z33=FC(6*M+1,1)-IM*FC(6*M,1)
      Z1=Z11*(LANDA(2)*LANDA(3)-1.)-
     1Z22*(MU(2)-MU(3)*LANDA(2)*LANDA(3))+
     2Z33*LANDA(3)*(MU(3)-MU(2))
      Z2=Z11*(1.-LANDA(1)*LANDA(3))-
     1Z22*(LANDA(1)*LANDA(3)*MU(3)-MU(1))
     2+Z33*(MU(1)-MU(3))*LANDA(3)
      Z3=Z11*(LANDA(1)-LANDA(2))-
     1Z22*(MU(1)*LANDA(2)-MU(2)*LANDA(1))+
     2Z33*(MU(2)-MU(1))
      PHIP1=PHIF1+Z1/(KSI(1)**M)
      PHIP2=PHIF2+Z2/(KSI(2)**M)
      PHIP3=PHIF3+Z3/(KSI(3)**M)
    2 CONTINUE
C
      PHIP1=-PHIP1/(DELTA*SQ(1))
      PHIP2=-PHIP2/(DELTA*SQ(2))
      PHIP3=-PHIP3/(DELTA*SQ(3))
C
      SIGX=REAL(MU(1)*MU(1)*PHIF1+MU(2)*MU(2)*PHIP2+LANDA(3)*
     1MU(3)*MU(3)*PHIP3)
      SIGY=REAL(PHIP1+PHIP2+LANDA(3)*PHIP3)
      TAUYZ=-REAL(LANDA(1)*PHIP1+LANDA(2)*PHIP2+PHIP3)
      TAUXZ=REAL(LANDA(1)*MU(1)*FHIF1+LANDA(2)*MU(2)*PHIP2+MU(3)*
     1PHIP3)
      TAUXY=-REAL(MU(1)*PHIP1+MU(2)*PHIP2+LANDA(3)*MU(3)*PHIP3)
      SIGZ=AA(3,1)*SIGX+AA(3,2)*SIGY+AA(3,4)*TAUYZ+AA(3,5)*TAUXZ+
     1AA(3,6)*TAUXY
      SIGZ=-SIGZ/AA(3,3)
      TETA2=2.*TETA
      SIG(1,1)=SIGX*COS(TETA)*COS(TETA)+SIGY*SIN(TETA)*SIN(TETA)+
     1TAUXY*SIN(TETA2)
      SIG(2,1)=SIGX*SIN(TETA)*SIN(TETA)+SIGY*COS(TETA)*COS(TETA)-
     1TAUXY*SIN(TETA2)
      SIG(3,1)=SIGZ
      SIG(4,1)=TAUYZ*COS(TETA)-TAUXZ*SIN(TETA)
      SIG(5,1)=TAUYZ*SIN(TETA)+TAUXZ*COS(TETA)
      SIG(6,1)=SIN(TETA)*COS(TETA)*(SIGY-SIGX)+TAUXY*COS(TETA2)
      WRITE(6,10) (SIG(I,1),I=1,6)
   10 FORMAT(10X,≠ SIGR ≠,E12.4/
     1        10X,≠ SIGT ≠,E12.4/
     1        10X,≠ SIGZ ≠,E12.4/
     2        10X,≠TAUTZ ≠,E12.4/
     2        10X,≠TAURZ ≠,E12.4/
     2        10X,≠TAURT ≠,E12.4/)
      RETURN
      END
C
      SUBROUTINE STAT(Y,X,N,M,T,B,S1,R2,S,X1)
      DIMENSION Y(18,1),X(18,6),B(6,1),XT(6,18),X1(6,6),
     1YHAT(18,1),S(6,6),WKAREA(20)
C
C     SUBROUTINE TO CALCULATE THE LEAST SQUARE SOLUTION OF THE
C     SYSTEM Y=X*BETA+EPSI. B IS THE LEAST SQUARE ESTIMATE OF BETA
C
C     Y      VECTOR OBSERVATION
C     X      MATRIX OF KNOWN FORM
C     N      NUMBER OF OBSERVATIONS AND NUMBER OF ROWS IN MATRICES
C            Y AND X
C     M      NUMBER OF ROWS IN MATRIX B AND NUMBER OF COLUMNS IN
C            MATRIX X
C     T      VALUE OF T(N-M,1-0.5*ALPHA) I.E (1.-0.5*ALPHA) PERCENTAGE
```

```
C           POINT OF A T DISTRIBUTION WITH N-M DEGREES OF FREEDOM
C           WITH 100(1-ALPHA) PERCENTS CONFIDENCE LIMITS.
C
C     CALCUALTE MATRIX B
C
      CALL TRANS(X,XT,N,M,18,6)
      CALL PRCMAT(XT,X,X1,M,N,M,6,18,6)
      CALL PRCMAT(XT,Y,B,M,N,1,6,18,1)
      D1=1.
      CALL LINV3F(X1,B,3,M,6,D1,D2,WKAREA,IER)
      WRITE(6,10) D1,D2
   10 FORMAT(//10X,≠ D1 ≠,E12.4,≠ D2 ≠,E12.4//)
C
C     CALCULATE THE VARIANCE , THE STANDARD DEVIATION ABOUT THE
C     REGRESSION (S2,S1) AND THE MULTIPLE CORRELATION COEFFICIENT R2
C
      CALL PRCMAT(X,B,YHAT,N,M,1,18,6,1)
      YBAR=0.
      DO 1 I=1,N
      YBAR=YBAR+Y(I,1)
    1 CONTINUE
      YBAR=YBAR/FLOAT(N)
C
      SSAR=0.
      SSDR=0.
      SSAM=0.
      DO 2 I=1,N
      SSAR=SSAR+(Y(I,1)-YHAT(I,1))*(Y(I,1)-YHAT(I,1))
      SSDR=SSDR+(YHAT(I,1)-YBAR)*(YHAT(I,1)-YBAR)
      SSAM=SSAM+(Y(I,1)-YBAR)*(Y(I,1)-YBAR)
    2 CONTINUE
      S2=SSAR/FLOAT(N-M)
      R2=SSDR/SSAM
      S1=SQRT(S2)
C
C     CALCULATE THE VECTOR OF RESIDUALS AND STORE IT INTO Y
C
      DO 6 I=1,N
      Y(I,1)=Y(I,1)-YHAT(I,1)
    6 CONTINUE
C
C     CALCULATE THE VARIANCE COVARIANCE MATRIX
C
      DO 3 I=1,M
      DO 4 J=1,M
      X1(I,J)=S2*X1(I,J)
    4 CONTINUE
    3 CONTINUE
C
C     FOR EACH COEFFICIENT OF MATRIX B CALCULATE ITS ESTIMATED
C     STANDARD ERROR AND THE CONFIDENCE INTERVAL USING T. STORE THE
C     RESULTS INTO MATRIX S
C
      DO 5 I=1,M
      S(I,1)=SQRT(X1(I,I))
      S(I,2)=B(I,1)+T*S(I,1)
      S(I,3)=B(I,1)-T*S(I,1)
    5 CONTINUE
      RETURN
      END
```

- Berni 2 -

```
      PROGRAM BERNI2(INFUT,OUTFUT,TAPE5=INPUT,TAPE6=OUTFUT)
      COMMON/BH/BETAH,DELTH,AH(3,3)
      COMMON/ANIS/STR,DIP,DIPDIF,A(3,3)
      DIMENSION KN(3),KS(3),SJ(3),HF(6,6),H1(6,6),H2(6,6),TS(6,6),
     1TTS(6,6),TH(6,6),TE(6,6),TTE(6,6),A1(6,6),BETA(6,6),
     2T1(6,6),T2(6,6),T3(6,6),SFO(6,1),FX(6,6),FQX(6,1),FXUH(3,6),
     3FXUD(3,6),FQXU(3,1)
      DIMENSION AAA(7),ROOT(6),MU(3),L2(3),L3(3),L4(3),LANDA(3),
     1PST(3),QST(3),RST(3),T5(6,1),SIG(6),T6(6,1),T7(6,1),T4(6,1),
     2BD(3),Z(3,3),WK(10),ROS(6,6),YROS(18,1),XROS(18,6),SFCG(6,1),
     3BDH(3,1)
      REAL KN,KS,MU21,MU31,MU32
      INTEGER DIPDIR
      COMPLEX ROOT,MU,L2,L3,L4,LANDA,DELTA,PST,QST,RST
C
  C     8****************************M****************************************MM
C
      READ(5,11) STR,DIP,DIPDIF,BETAH,DELTH,INDEX
      READ(5,12) E1,E2,E3,G12,G13,G23,MU21,MU31,MU32
      READ(5,13)(KN(I),I=1,3),(KS(I),I=1,3),(SJ(I),I=1,3)
      IF(INDEX.GT.0) GOTO 40
      READ(5,28)(SFO(I,1),I=1,6),QINT
      READ(5,239) RAMAX,NRAROC,TETROC
   28 FORMAT(7F10.4)
  239 FORMAT(F10.4,I5,F10.4)
      GOTO 31
   40 CONTINUE
      READ(5,32) NROS,TSTU
      READ(5,33)((ROS(I,J),J=1,6),I=1,NROS)
   32 FORMAT(I5,F10.4)
   33 FORMAT(3E10.4,3F10.4)
   31 CONTINUE
   11 FORMAT(2F10.4,I5,2F10.4,I5)
   12 FORMAT(9F8.2)
   13 FORMAT(6E10.4,3F8.2)
C
C     ******************************************************************
C
      WRITE(6,23) STR,DIP,DIPDIF,BETAH,DELTH
      WRITE(6,26)
      WRITE(6,25) E1,E2,E3,G12,G13,G23,MU21,MU31,MU32
      WRITE(6,24)(I,KN(I),KS(I),SJ(I),I=1,3)
      IF(INDEX.GT.0) GOTO 41
      WRITE(6,29)(SFO(I,1),I=1,6),QINT
   29 FORMAT(//10X,*** INITIAL STRESS FIELD ***//
     1          10X,*     SIGXO = *,E12.4/
     1          10X,*     SIGYO = *,E12.4/
     1          10X,*     SIGZO = *,E12.4/
     1          10X,*    TAUYZO = *,E12.4/
     1          10X,*    TAUXZO = *,E12.4/
     1          10X,*    TAUXYO = *,E12.4//
     1          10X,*   INTERNAL PRESSURE *,E12.4///)
      WRITE(6,241) NRAROC,TETROC,RAMAX
  241 FORMAT(//10X,*** SCALING PARAMETERS ***,//
     1          10X,* NRAINC(ROCK) *,I5/
     1          10X,* TETINC (ROCK) *,F10.4/
     1          10X,* MAX RA (ROCK )*,F10.4//)
```

```
      GOTO 34
   41 CONTINUE
      WRITE(6,35) NROS,TSTU
      WRITE(6,37)
      WRITE(6,36)(I,(ROS(I,J),J=1,6),I=1,NROS)
   35 FCRMAT(//10X,≠** FOSETTE FROPERTIES **≠//
     1       10X,≠ NUMBER OF ROSETTES ≠,I5/
     1       10X,≠ T DISTRIBUTICN ≠,E12.4/)
   36 FCRMAT(10X,I5,5X,3E10.4,3F10.4)
   37 FCRMAT(12X,≠I≠,9X,≠EPSIA≠,5X,≠EPSIB≠,5X,≠EPSIC≠,6X,≠PSI≠,8X,
     1≠RA≠,7X,≠TETA≠/)
   34 CONTINUE
   26 FCRMAT(//10X,≠** ROCK PROPERTIES **≠//)
   23 FCRMAT(//10X,≠** CRIENTATICN PARAMETERS **≠//
     1       10X,≠ ANISCTROPY ** STRIKE ≠,F10.4/
     1       10X,≠               DIF    ≠,F10.4/
     2       10X,≠               DIFDIR ≠,I5//
     3       10X,≠ BOREHOLE **   EETAH  ≠,F10.4/
     4       10X,≠               DELTH  ≠,F10.4//)
   24 FCRMAT(10X,≠ JOINT SET ≠,I5,≠ KN ≠,E12.4,≠ KS ≠,E12.4,≠ S ≠,
     1F10.4//)
   25 FCRMAT(10X,≠ YOUNG MODULUS IN N DIRECTION ≠,F10.4/
     1       10X,≠ YOUNG MODULUS IN S DIRECTION ≠,F10.4/
     1       10X,≠ YOUNG MODULUS IN T DIRECTION ≠,F10.4/
     1       10X,≠ SHEAR MODULUS IN NS PLANE ≠,F10.4/
     1       10X,≠ SHEAR MODULUS IN NT PLANE ≠,F10.4/
     1       10X,≠ SHEAR MODULUS IN ST PLANE ≠,F10.4/
     1       10X,≠ PCISSCN RATIC MU21 ≠,F10.4/
     1       10X,≠ POISSCN RATIC MU31 ≠,F10.4/
     1       10X,≠ PCISSCN RATIC MU32 ≠,F10.4///)
C
C     ******************************************************************
C
C     CALCULATE THE COMPLIANCE MATRIX HP AS THE SUM OF TWO MATRICES
C     H1 AND H2
C
C     INITIALIZATION CF MATRICES H1 AND H2
C
      DC 3 I=1,6
      DC 3 J=1,6
      H1(I,J)=0.
    3 H2(I,J)=0.
C
C     CALCULATE THE MATRIX H1 WITH INTACT MATERIAL CONSTANTS
C
      H1(1,1)=1./E1
      H1(2,2)=1./E2
      H1(3,3)=1./E3
      H1(4,4)=1./G23
      H1(5,5)=1./G13
      H1(6,6)=1./G12
      H1(1,2)=-MU21/E2
      H1(1,3)=-MU31/E3
      H1(2,3)=-MU32/E3
      H1(2,1)=H1(1,2)
      H1(3,1)=H1(1,3)
      H1(3,2)=H1(2,3)
C
      WRITE(6,104)
      WRITE(6,105)((H1(I,J),J=1,6),I=1,6)
  104 FCRMAT(///30X,≠ MATRIX H1 ≠,//)
  105 FCRMAT(6E12.4)
```

```
C
C     CALCULATE THE MATRIX H2 WITH JOINT PROPERTIES
C
      DO 16 I=1,3
      H2(I,I)=1./(KN(I)*SJ(I))
   16 CONTINUE
      H2(4,4)=1./(KS(2)*SJ(2))+1./(KS(3)*SJ(3))
      H2(5,5)=1./(KS(1)*SJ(1))+1./(KS(3)*SJ(3))
      H2(6,6)=1./(KS(1)*SJ(1))+1./(KS(2)*SJ(2))
      WRITE(6,112)
  112 FORMAT(///30X,≠ MATRIX H2 ≠//)
      WRITE(6,113)((H2(I,J),J=1,6),I=1,6)
  113 FORMAT(6E12.4)
C
C     THEN, CALCULATE THE COMPLIANCE MATRIX HF BY ADDING MATRICES
C     H1 AND H2
C
      DO 14 I=1,6
      DO 15 J=1,6
      HF(I,J)=H1(I,J)+H2(I,J)
   15 CONTINUE
   14 CONTINUE
C
C     FOR THE ANISOTROPY CALCULATE THE TRANSFORMATION MATRICES
C     TS AND TTS
C
      CALL ORIEN2
      CALL TRANSF(A,TS)
      CALL TRANS(TS,TTS,6,6,6,6)
      WRITE(6,100)
  100 FORMAT(///30X,≠  TRANSFORMATION MATRIX TS≠,//)
      WRITE(6,101)((TS(I,J),J=1,6),I=1,6)
  101 FORMAT(6E12.4)
      WRITE(6,102)
  102 FORMAT(///30X,≠  TRANSFORMATION MATRIX TTS ≠,//)
      WRITE(6,103)((TTS(I,J),J=1,6),I=1,6)
  103 FORMAT(6E12.4)
C
C     FOR THE HOLE CALCULATE THE TRANSFORMATION MATRIX TH
C
      CALL ORIENH
      CALL TRANSF(AH,TH)
      WRITE(6,106)
  106 FORMAT(///30X,≠  TRANSFORMATION MATRIX TH ≠,//)
      WRITE(6,107)((TH(I,J),J=1,6),I=1,6)
  107 FORMAT(6E12.4)
C
C     FOR THE HOLE CALCULATE THE TRANSFORMATION MATRICES TE AND TTE
C
      DO 4 I=1,6
      DO 5 J=1,6
      TE(I,J)=TH(I,J)
      IF((I.LE.3).AND.(J.GT.3)) TE(I,J)=0.5*TH(I,J)
      IF((I.GT.3).AND.(J.LE.3)) TE(I,J)=2.*TH(I,J)
    5 CONTINUE
    4 CONTINUE
      CALL TRANS(TE,TTE,6,6,6,6)
      WRITE(6,108)
  108 FORMAT(///30X,≠  TRANSFORMATION MATRIX TE ≠,//)
      WRITE(6,109)((TE(I,J),J=1,6),I=1,6)
  109 FORMAT(6E12.4)
      WRITE(6,110)
```

```
  110 FORMAT(///30X,*  TRANSFORMATION MATRIX TTE *,//)
      WRITE(6,111)((TTE(I,J),J=1,6),I=1,6)
  111 FORMAT(6E12.4)
C
C      CALCULATE THE STRAIN-STRESS MATRIX AA IN THE COORDINATE SYSTEM
C     ATTACHED TO THE HOLE
C
      CALL PRCMAT(TE,TTS,T1,6,6,6,6,6,6)
      CALL PRCMAT(T1,HP,T2,6,6,6,6,6,6)
      CALL PRCMAT(T2,TS,T3,6,6,6,6,6,6)
      CALL PRCMAT(T3,TTE,AA,6,6,6,6,6,6)
      WRITE(6,114)
  114 FORMAT(//30X,* STRESS STRAIN MATRIX AA *,//)
      WRITE(6,115)((AA(I,J),J=1,6),I=1,6)
  115 FORMAT(6F15.9)
C
C     CALCULATE THE COEFFICIENTS OF MATRIX BETA
C
      DO 1 I=1,6
      DO 2 J=1,6
      BETA(I,J)=AA(I,J)-AA(I,3)*AA(J,3)/AA(3,3)
    2 CONTINUE
    1 CONTINUE
      WRITE(6,121)
  121 FORMAT(/30X,* MATRIX BETA *//)
      WRITE(6,120)((BETA(I,J),J=1,6),I=1,6)
  120 FORMAT(6F15.9)
C
C     CALCULATE THE COEFFICIENTS AAA(I) OF THE 6TH DEGREE POLYNOMIAL
C
      AAA(1)=BETA(1,1)*BETA(5,5)-BETA(1,5)*BETA(1,5)
      AAA(2)=2.*BETA(1,5)*(BETA(1,4)+BETA(5,6))
      AAA(2)=AAA(2)-2.*(BETA(1,6)*BETA(5,5)+BETA(1,1)*BETA(4,5))
      AAA(3)=BETA(5,5)*(2.*BETA(1,2)+BETA(6,6))+4.*BETA(1,6)*BETA(4,5)+
     1BETA(1,1)*BETA(4,4)
      AAA(3)=AAA(3)-(BETA(1,4)+BETA(5,6))**2-2.*BETA(1,5)*(BETA(2,5)+
     1BETA(4,6))
      AAA(4)=-2.*BETA(2,6)*BETA(5,5)-2.*BETA(4,5)*(2.*BETA(1,2)+
     1BETA(6,6))-2.*BETA(1,6)*BETA(4,4)
      AAA(4)=AAA(4)+2.*BETA(1,5)*BETA(2,4)+2.*(BETA(1,4)+BETA(5,6))*
     1(BETA(2,5)+BETA(4,6))
      AAA(5)=BETA(2,2)*BETA(5,5)+4.*BETA(2,6)*BETA(4,5)+BETA(4,4)*
     1(2.*BETA(1,2)+BETA(6,6))
      AAA(5)=AAA(5)-2.*BETA(2,4)*(BETA(1,4)+BETA(5,6))-(BETA(2,5)+
     1BETA(4,6))**2
      AAA(6)=-2.*BETA(2,2)*BETA(4,5)-2.*BETA(2,6)*BETA(4,4)+2.*BETA(2,4)
     1*(BETA(2,5)+BETA(4,6))
      AAA(7)=BETA(2,2)*BETA(4,4)-BETA(2,4)*BETA(2,4)
C
C     CALCULATE THE ROOTS OF THE 6TH DEGREE CHARACTERISTIC EQUATION
C     USING SUBROUTINE ZPOLR
C
      CALL ZPOLR(AAA,6,ROOT,IER)
      WRITE(6,201)IER,(ROOT(I),I=1,6)
  201 FORMAT(///,I5,6(F10.4,F10.4))
C
C     FIND THE ROOTS MU1,MU2,MU3
C
      MU(1)=ROOT(5)
      MU(2)=ROOT(1)
      MU(3)=ROOT(3)
      WRITE(6,116)(MU(I),I=1,3)
```

```
  116 FORMAT(//10X,≠ MU1 ≠,(F15.9,F15.9)/
     1          10X,≠ MU2 ≠,(F15.9,F15.9)/
     1          10X,≠ MU3 ≠,(F15.9,F15.9)////)
C
C     CALCULATE THE COEFFICIENTS LANDA1,LANDA2,LANDA3 AND DELTA
C
      DO 17 I=1,3
      L3(I)=BETA(1,5)*(MU(I)**3)-(BETA(1,4)+BETA(5,6))*(MU(I)**2)+(
     1BETA(2,5)+BETA(4,6))*MU(I)-BETA(2,4)
      L2(I)=BETA(5,5)*(MU(I)**2)-2.*BETA(4,5)*MU(I)+BETA(4,4)
      L4(I)=BETA(1,1)*(MU(I)**4)-2.*BETA(1,6)*(MU(I)**3)+(2.*BETA(1,2)+
     1BETA(6,6))*(MU(I)**2)-2.*BETA(2,6)*MU(I)+BETA(2,2)
      LANDA(I)=-L3(I)/L2(I)
      IF(I.EQ.3) LANDA(I)=-L3(I)/L4(I)
      WRITE(6,117) I,L2(I),L3(I),L4(I)
  117 FORMAT(10X,≠ MU ≠,I3,≠ L2 ≠,(E12.4,E12.4),≠ L3 ≠,(E12.4,E12.4),
     1≠ L4 ≠,(E12.4,E12.4),//)
   17 CONTINUE
      DELTA=MU(2)-MU(1)+LANDA(2)*LANDA(3)*(MU(1)-MU(3))+LANDA(1)*
     1LANDA(3)*(MU(3)-MU(2))
C
C     CALCUALTE THE COEFFICIENTS PST(I),QST(I),RST(I)
C
      DO 18 I=1,2
      PST(I)=BETA(1,1)*MU(I)*MU(I)+BETA(1,2)-BETA(1,6)*MU(I)+LANDA(I)*
     1(BETA(1,5)*MU(I)-BETA(1,4))
      QST(I)=BETA(1,2)*MU(I)+LANDA(I)*(BETA(2,5)-BETA(2,4)/MU(I))-
     1BETA(2,6)+BETA(2,2)/MU(I)
      RST(I)=BETA(1,4)*MU(I)+BETA(2,4)/MU(I)-BETA(4,6)+LANDA(I)*(BETA(4,
     15)-BETA(4,4)/MU(I))
   18 CONTINUE
      PST(3)=LANDA(3)*(BETA(1,1)*(MU(3)**2)+BETA(1,2)-BETA(1,6)*MU(3))+
     1BETA(1,5)*MU(3)-BETA(1,4)
      QST(3)=LANDA(3)*(BETA(1,2)*MU(3)-BETA(2,6)+BETA(2,2)/MU(3))+BETA(
     12,5)-BETA(2,4)/MU(3)
      RST(3)=LANDA(3)*(BETA(1,4)*MU(3)-BETA(4,6)+BETA(2,4)/MU(3))+BETA(
     14,5)-BETA(4,4)/MU(3)
C
C     ****************************************************************
C
      IF(INDEX.GT.0) GOTO 58
C
C     CALCULATE STRESS STRAIN AND DISPLACEMENT COMPONENTS AT ANY
C     POINT RA,TET WITHIN THE ROCK. RA AND TET ARE CALCULATED FROM
C     THE INPUT COEFFICIENTS NRAROC AND TETROC
C
      RAINC=(RAMAX-1.)/FLOAT(NRAROC)
      NRAROC=NRAROC+1
      NTET=IFIX(90./TETROC)+1
      TET=0.
      DO 243 I=1,NTET
      RA=1.
      DO 244 J=1,NRAROC
C
      WRITE(6,211) RA,TET
  211 FORMAT(///10X,≠** RATIO R/A **≠,F10.4,≠ ** ANGLE TETA **≠,F10.4//)
      CALL FXMAT(INDEX,MU,LANDA,DELTA,TET,RA,AA,FX,FOX)
C
C     CALCULATE THE PRODUCT OF FX AND SFO AND STORE THE RESULT
C     INTO MATRIX T4
C
      CALL PROMAT(FX,SFO,T4,6,6,1,6,6,1)
```

```
C
C     CALCULATE THE TOTAL STRESS COMPONENTS IN THE X,Y,Z COORDINATE
C     SYSTEM T5
      DO 6 II=1,6
      T5(II,1)=T4(II,1)+QINT*FOX(II,1)
    6 CONTINUE
C
C     CALCULATE THE TOTAL STRESS COMPONENTS IN THE R,TET,Z COORDINATE
C     SYSTEM AND STORE THE RESULT INTO MATRIX T4
C
      CALL TETSS(TET,T1,T2)
      CALL PRCMAT(T2,T5,T4,6,6,1,6,6,1)
C
C     CALCULATE THE STRAIN COMPONENTS IN THE X,Y,Z COORDINATE SYSTEM
C
      CALL PRCMAT(AA,T5,T7,6,6,1,6,6,1)
C
C     CALCULATE THE STRAIN COMPONENTS IN THE R,TET,Z COORDINATE
C     SYSTEM
C
      CALL PRCMAT(T1,T7,T6,6,6,1,6,6,1)
C
      WRITE(6,212) (T4(I1,1),I1=1,6)
  212 FORMAT(10X,≠ SIGR ≠,E12.4/
     1       10X,≠ SIGT ≠,E12.4/
     1       10X,≠ SIGZ ≠,E12.4/
     2       10X,≠TAUTZ ≠,E12.4/
     2       10X,≠TAURZ ≠,E12.4/
     2       10X,≠TAURT ≠,E12.4/)
      WRITE(6,235)(T6(I3,1),I3=1,6)
  235 FORMAT(10X,≠ EPSR ≠,E12.4/
     1       10X,≠ EPST ≠,E12.4/
     1       10X,≠ EPSZ ≠,E12.4/
     2       10X,≠GAMTZ ≠,E12.4/
     2       10X,≠GAMRZ ≠,E12.4/
     2       10X,≠GAMRT ≠,E12.4/)
C
C     CALCULATE THE PRINCIPAL STRESSES AND THEIR ORIENTATION
C     WITH RESPECT TO THE LOCAL COORDINATE SYSTEM X,Y,Z ATTACHED
C     TO THE HOLE.
C
      SIG(1)=T5(1,1)
      SIG(2)=T5(6,1)
      SIG(3)=T5(2,1)
      SIG(4)=T5(5,1)
      SIG(5)=T5(4,1)
      SIG(6)=T5(3,1)
C
      CALL EIGRS(SIG,3,2,BD,Z,3,WK,IER)
C
      SIG1=AMAX1(BD(1),BD(2),BD(3))
      SIG3=AMIN1(BD(1),BD(2),BD(3))
      DO 65 II=1,3
      IF((BD(II).NE.SIG1).AND.(BD(II).NE.SIG3)) SIG2=BD(II)
   65 CONTINUE
      DO 66 II=1,3
      IF(BD(II).EQ.SIG1) WRITE(6,231) BD(II)
      IF(BD(II).EQ.SIG2) WRITE(6,232) BD(II)
      IF(BD(II).EQ.SIG3) WRITE(6,233) BD(II)
  231 FORMAT(/10X,≠ SIG1 ≠,E12.4/)
  232 FORMAT(/10X,≠ SIG2 ≠,E12.4/)
  233 FORMAT(/10X,≠ SIG3 ≠,E12.4/)
```

```
      WRITE(6,234) (Z(JJ,II),JJ=1,3)
  234 FCRMAT(10X,≠ L ≠,E12.4,≠ M ≠,E12.4,≠ N ≠,E12.4)
   66 CONTINUE
C
C     CALCULATE THE DISPLACEMENT COMPONENTS INDUCED BY THE HOLE AND
C     THE INTERNAL PRESSURE QINT
C
      ZA=0.
      CALL FUMAT(INDEX,MU,LANDA,DELTA,PST,QST,RST,TET,RA,ZA,AA,
     1FXUH,FXUC,FQXU)
      CALL PRCMAT(FXUH,SFC,BDH,3,6,1,3,6,1)
      DC 7 IJ=1,3
      BDH(IJ,1)=BDH(IJ,1)+QINT*FQXU(IJ,1)
    7 CONTINUE
      WRITE(6,237)(BDH(I3,1),I3=1,3)
  237 FCRMAT(10X,≠ URH/A ≠,E12.4/
     1         10X,≠ VTH/A ≠,E12.4/
     2         10X,≠  WH/A ≠,E12.4/)
C
C     CALCULATE THE DISPLACEMENT COMPONENTS INDUCED BY THE INITIAL
C     STRESS FIELD WITHOUT ANY HOLE
C
      CALL PRCMAT(FXUC,SFC,BDH,3,6,1,3,6,1)
      WRITE(6,236) (BDH(I3,1),I3=1,3)
  236 FCRMAT(10X,≠ URO/A ≠,E12.4/
     1         10X,≠ VTO/A ≠,E12.4/
     2         10X,≠  WO/A ≠,E12.4/)
C
      RA=RA+RAINC
  244 CONTINUE
      TET=TET+TETROC
  243 CONTINUE
      GCTO 59
C
C     *********************************************************************
C
   58 CONTINUE
C
C     FOR EACH ROSETTE AT A TIME CALCULATE MATRIX FX
C
      DC 72 I=1,NROS
      RA=ROS(I,5)
      TET=ROS(I,6)
      CALL FXMAT(INDEX,MU,LANDA,DELTA,TET,RA,AA,FX,FQX)
C
C     MULTIPLY AA BY FX AND STORE THE RESULT INTO T1
C
      CALL PRCMAT(AA,FX,T1,6,6,6,6,6,6)
      CALL TETSS(TET,T2,T3)
C
C     MULTIPLY T2 BY T1 AND STORE THE RESULT INTO T3
C
      CALL PRCMAT(T2,T1,T3,6,6,6,6,6,6)
C
      WRITE(6,264)((T3(I2,J),J=1,6),I2=1,6)
  264 FCRMAT(10X,6E12.4)
C
      PSIRC=ROS(I,4)*ATAN(1.)*4./180.
      PSIRC2=2.*PSIRC
      YROS(3*I-2,1)=ROS(I,1)
      YROS(3*I-1,1)=ROS(I,2)
      YROS(3*I,1)=2.*(ROS(I,3)-ROS(I,1)*COS(PSIRC)*COS(PSIRC)-
```

```
     1ROS(I,2)*SIN(PSIRC)*SIN(PSIRO))/SIN(PSIRC2)
C
      DO 73 J=1,6
      XROS(3*I-2,J)=T3(2,J)
      XROS(3*I-1,J)=T3(3,J)
      XROS(3*I,J)=T3(4,J)
   73 CONTINUE
   72 CONTINUE
C
      N1=3*NROS
      WRITE(6,74)
   74 FORMAT(///10X,* MATRICES Y AND X *,//)
      DO 81 I=1,N1
      YROS(I,1)=-YROS(I,1)
      WRITE(6,75) YROS(I,1),(XROS(I,J),J=1,6)
   75 FORMAT(10X,E12.4,10X,6E12.4)
   81 CONTINUE
      DO 95 I=1,N1
      YROS(I,1)=YROS(I,1)*1.E+06
      DO 96 J=1,6
      XROS(I,J)=XROS(I,J)*1.E+06
   96 CONTINUE
   95 CONTINUE
C
      CALL STAT(YROS,XROS,N1,6,TSTU,SFO,S1ROS,R2ROS,T1,T2)
      S1ROS=S1ROS*1.E-06
      DO 97 I=1,N1
      YROS(I,1)=YROS(I,1)*1.E-06
   97 CONTINUE
      WRITE(6,76) S1ROS,R2ROS,(YROS(I,1),I=1,N1)
      WRITE(6,77) (SFO(I,1),T1(I,1) ,T1(I,2),T1(I,3),I=1,6)
      WRITE(6,79)
      WRITE(6,78)((T2(I,J),J=1,6),I=1,6)
   76 FORMAT(/10X,* STANDARD DEVIATION ABOUT REGRESSION *,E12.4/
     1          10X,* MULTIPLE CORRELATION COEFFICIENT *,E12.4/
     2          10X,* RESIDUALS */
     3          10X,9E10.4////)
   77 FORMAT(10X,* SIGXC*,E12.4,* ESE*,E12.4,* UL*,E12.4,* LL*,E12.4/
     1         10X,* SIGYC*,E12.4,* ESE*,E12.4,* UL*,E12.4,* LL*,E12.4/
     1         10X,* SIGZC*,E12.4,* ESE*,E12.4,* UL*,E12.4,* LL*,E12.4/
     2         10X,*TAUYZC*,E12.4,* ESE*,E12.4,* UL*,E12.4,* LL*,E12.4/
     2         10X,*TAUXZC*,E12.4,* ESE*,E12.4,* UL*,E12.4,* LL*,E12.4/
     2         10X,*TAUXYC*,E12.4,* ESE*,E12.4,* UL*,E12.4,* LL*,E12.4)
   78 FORMAT(10X,6E12.4)
   79 FORMAT(//10X,* VARIANCE COVARIANCE MATRIX *//)
C
C     CALCULATE THE STRESS FIELD IN THE GLOBAL COORDINATE SYSTEM
C
      CALL PROMAT(TTE,SFC,SFOG,6,6,1,6,6,1)
      WRITE(6,82)
      WRITE(6,83)(SFOG(I,1),I=1,6)
   82 FORMAT(//10X,* STRESS FIELD IN GLOBAL COORDINATE SYSTEM *,//)
   83 FORMAT(10X,* SIGXC *,E12.4/
     1         10X,* SIGYC *,E12.4/
     1         10X,* SIGZC *,E12.4/
     1         10X,*TAUYZC *,E12.4/
     1         10X,*TAUXZC *,E12.4/
     1         10X,*TAUXYC *,E12.4/)
C
C     CALCULATE THE PRINCIPAL STRESSES AND THEIR ORIENTATION IN
C     THE GLOBAL COORDINATE SYSTEM
C
```

```
      SIG(1)=SFCG(1,1)
      SIG(2)=SFCG(6,1)
      SIG(3)=SFCG(2,1)
      SIG(4)=SFCG(5,1)
      SIG(5)=SFCG(4,1)
      SIG(6)=SFCG(3,1)
C
      CALL EIGRS(SIG,3,2,BD,Z,3,WK,IER)
C
      SIG1=AMAX1(BD(1),BD(2),BD(3))
      SIG3=AMIN1(BD(1),BD(2),BD(3))
      DO 84 II=1,3
      IF((BD(II).NE.SIG1).AND.(BD(II).NE.SIG3)) SIG2=BD(II)
   84 CONTINUE
      DO 85 II=1,3
      IF(BD(II).EQ.SIG1) WRITE(6,91) BD(II)
      IF(BD(II).EQ.SIG2) WRITE(6,92) BD(II)
      IF(BD(II).EQ.SIG3) WRITE(6,93) BD(II)
   91 FORMAT(/10X,≠ SIG1 ≠,E12.4/)
   92 FORMAT(/10X,≠ SIG2 ≠,E12.4/)
   93 FORMAT(/10X,≠ SIG3 ≠,E12.4/)
      WRITE(6,94)(Z(JJ,II),JJ=1,3)
   94 FORMAT(10X,≠ L ≠,E12.4,≠ M ≠,E12.4,≠ N ≠,E12.4)
   85 CONTINUE
   59 CONTINUE
      STOP
      END
C


C
      SUBROUTINE TETSS(TET,TES,TSS)
      DIMENSION TES(6,6),TSS(6,6)
C
C     SUBROUTINE TO CALCULATE MATRICES TES AND TSS USED RESPECTIVELY
C     IN THE TRANSFORMATION OF STRAIN AND STRESS COMPONENTS FROM
C     A CARTESIAN COORDINATE SYSTEM X,Y,Z TO A POLAR COORDINATE
C     SYSTEM R,TET,Z
C
      TETA=TET*ATAN(1.)*4./180.
      TETA2=2.*TETA
      DO 1 I=1,6
      DO 2 J=1,6
      TES(I,J)=0.
      TSS(I,J)=0.
    2 CONTINUE
    1 CONTINUE
      TSS(1,1)=COS(TETA)*COS(TETA)
      TSS(2,2)=TSS(1,1)
      TSS(1,2)=SIN(TETA)*SIN(TETA)
      TSS(2,1)=TSS(1,2)
      TSS(3,3)=1.
      TSS(1,6)=SIN(TETA2)
      TSS(2,6)=-TSS(1,6)
      TSS(6,1)=-0.5*TSS(1,6)
      TSS(6,2)=-TSS(6,1)
```

```
      TSS(6,6)=COS(TETA2)
      TSS(4,4)=COS(TETA)
      TSS(5,5)=TSS(4,4)
      TSS(4,5)=-SIN(TETA)
      TSS(5,4)=-TSS(4,5)
      DO 3 I=1,6
      DO 4 J=1,6
      TES(I,J)=TSS(I,J)
      IF((I.LE.3).AND.(J.GT.3)) TES(I,J)=0.5*TSS(I,J)
      IF((I.GT.3).AND.(J.LE.3)) TES(I,J)=2.*TSS(I,J)
    4 CONTINUE
    3 CONTINUE
      RETURN
      END
C
      SUBROUTINE FUMAT(INDEX,MU,LANDA,DELTA,PST,QST,RST,TET,RA,ZA,AA,
     1FXUH,FXUO,FOXU)
      DIMENSION MU(3),LANDA(3),PST(3),QST(3),RST(3),FXUH(3,6),
     1FXUO(3,6),AA(6,6),FOXU(3,1),KSI(3)   ,SQ(3),DELT(3),T2(3,3),
     2T1(3,6),T3(3,1)
      COMPLEX MU,LANDA,DELTA,PST,QST,RST,IM,KSI,SQ,Z4,DELT,Z1,Z2,Z3
C
C     SUBROUTINE TO CALCULATE THE COEFFICIENTS OF MATRICES FUH,FUO,FUQ.
C     THOSE MATRICES ARE USED IN THE CALCULATION OF INDUCED AND TOTAL
C     DISPLACEMENTS AT A POINT RA,ZA,TET
C
      TETA=TET*ATAN(1.)*4./180.
      IF(TET.EQ.90.) TETA=89.999*ATAN(1.)*4./180.
      IF(TET.EQ.270.) TETA=270.001*ATAN(1.)*4./180.
      TETA2=2.*TETA
      IM=(0.,1.)
      DO 1 I=1,3
      FOXU(I,1)=0.
      DO 2 J=1,6
      FXUH(I,J)=0.
      FXUO(I,J)=0.
      T1(I,J)=0.
    2 CONTINUE
      DO 3 K=1,3
      T2(I,K)=0.
    3 CONTINUE
    1 CONTINUE
C
      DO 4 I=1,3
      Z4=COS(TETA)+MU(I)*SIN(TETA)
      SQ(I)=CSQRT(RA*RA*Z4*Z4-1.-MU(I)*MU(I))
      IF((TET.GT.90.).AND.(TET.LT.270.)) SQ(I)=-SQ(I)
      KSI(I)=(SQ(I)+RA*(COS(TETA)+MU(I)*SIN(TETA)))/(1.-IM*MU(I))
      DELT(I)=1./(DELTA*KSI(I))
    4 CONTINUE
C
      T1(1,1)=REAL(IM*DELT(1)*PST(1)*(LANDA(2)*LANDA(3)-1.)+
     1             IM*DELT(2)*PST(2)*(1.-LANDA(1)*LANDA(3))+
     1             IM*DELT(3)*PST(3)*(LANDA(1)-LANDA(2)))
      T1(2,1)=REAL(IM*DELT(1)*QST(1)*(LANDA(2)*LANDA(3)-1.)+
     1             IM*DELT(2)*QST(2)*(1.-LANDA(1)*LANDA(3))+
     1             IM*DELT(3)*QST(3)*(LANDA(1)-LANDA(2)))
      T1(3,1)=REAL(IM*DELT(1)*RST(1)*(LANDA(2)*LANDA(3)-1.)+
     1             IM*DELT(2)*RST(2)*(1.-LANDA(1)*LANDA(3))+
     1             IM*DELT(3)*RST(3)*(LANDA(1)-LANDA(2)))
      T1(1,2)=REAL(DELT(1)*PST(1)*(MU(2)-MU(3)*LANDA(2)*LANDA(3))+
     1             DELT(2)*PST(2)*(LANDA(1)*LANDA(3)*MU(3)-MU(1))+
```

```
     1              DELT(3)*PST(3)*(MU(1)*LANDA(2)-MU(2)*LANCA(1)))
      T1(2,2)=REAL(DELT(1)*QST(1)*(MU(2)-MU(3)*LANDA(2)*LANCA(3))+
     1              DELT(2)*QST(2)*(LANDA(1)*LANDA(3)*MU(3)-MU(1))+
     1              DELT(3)*QST(3)*(MU(1)*LANDA(2)-MU(2)*LANCA(1)))
      T1(3,2)=REAL(DELT(1)*RST(1)*(MU(2)-MU(3)*LANDA(2)*LANCA(3))+
     1              DELT(2)*RST(2)*(LANDA(1)*LANDA(3)*MU(3)-MU(1))+
     1              DELT(3)*RST(3)*(MU(1)*LANDA(2)-MU(2)*LANDA(1)))
      T1(1,4)=-REAL(DELT(1)*PST(1)*LANDA(3)*(MU(3)-MU(2))+
     1               DELT(2)*PST(2)*LANDA(3)*(MU(1)-MU(3))+
     1               DELT(3)*PST(3)*(MU(2)-MU(1)))
      T1(2,4)=-REAL(DELT(1)*QST(1)*LANDA(3)*(MU(3)-MU(2))+
     1               DELT(2)*QST(2)*LANDA(3)*(MU(1)-MU(3))+
     1               DELT(3)*QST(3)*(MU(2)-MU(1)))
      T1(3,4)=-REAL(DELT(1)*RST(1)*LANDA(3)*(MU(3)-MU(2))+
     1               DELT(2)*RST(2)*LANDA(3)*(MU(1)-MU(3))+
     1               DELT(3)*RST(3)*(MU(2)-MU(1)))
      T1(1,5)=REAL(IM*CELT(1)*FST(1)*LANDA(3)*(MU(3)-MU(2))+
     1              IM*CELT(2)*FST(2)*LANDA(3)*(MU(1)-MU(3))+
     1              IM*DELT(3)*FST(3)*(MU(2)-MU(1)))
      T1(2,5)=REAL(IM*CELT(1)*QST(1)*LANDA(3)*(MU(3)-MU(2))+
     1              IM*CELT(2)*QST(2)*LANDA(3)*(MU(1)-MU(3))+
     1              IM*CELT(3)*QST(3)*(MU(2)-MU(1)))
      T1(3,5)=REAL(IM*CELT(1)*RST(1)*LANDA(3)*(MU(3)-MU(2))+
     1              IM*DELT(2)*RST(2)*LANDA(3)*(MU(1)-MU(3))+
     1              IM*DELT(3)*RST(3)*(MU(2)-MU(1)))
      T1(1,6)=-REAL(DELT(1)*PST(1)*(LANDA(2)*LANDA(3)-1.+
     1IM*(MU(2)-MU(3)*LANDA(2)*LANCA(3)))+
     2DELT(2)*PST(2)*(1.-LANDA(1)*LANDA(3)+
     3IM*(LANCA(1)*LANCA(3)*MU(3)-MU(1)))+
     4DELT(3)*FST(3)*(LANDA(1)-LANDA(2)+
     5IM*(MU(1)*LANDA(2)-MU(2)*LANCA(1))))
      T1(2,6)=-REAL(DELT(1)*QST(1)*(LANDA(2)*LANDA(3)-1.+
     1IM*(MU(2)-MU(3)*LANDA(2)*LANCA(3)))+
     2DELT(2)*QST(2)*(1.-LANDA(1)*LANDA(3)+
     3IM*(LANCA(1)*LANCA(3)*MU(3)-MU(1)))+
     4DELT(3)*QST(3)*(LANDA(1)-LANCA(2)+
     5IM*(MU(1)*LANDA(2)-MU(2)*LANCA(1))))
      T1(3,6)=-REAL(DELT(1)*RST(1)*(LANDA(2)*LANDA(3)-1.+
     1IM*(MU(2)-MU(3)*LANDA(2)*LANDA(3)))+
     2DELT(2)*RST(2)*(1.-LANDA(1)*LANDA(3)+
     3IM*(LANCA(1)*LANDA(3)*MU(3)-MU(1)))+
     4DELT(3)*RST(3)*(LANDA(1)-LANCA(2)+
     5IM*(MU(1)*LANDA(2)-MU(2)*LANCA(1))))
      T2(1,1)=CCS(TETA)
      T2(2,2)=T2(1,1)
      T2(1,2)=SIN(TETA)
      T2(2,1)=-T2(1,2)
      T2(3,3)=1.
      CALL PRCMAT(T2,T1,FXUH,3,3,6,3,3,6)
C
      DC 5 I=1,6
      FXUO(1,I)=-RA*(0.5*(AA(1,I)+AA(2,I))+0.5*(AA(1,I)-AA(2,I))*
     1COS(TETA2)+0.5*AA(6,I)*SIN(TETA2))
      FXUO(2,I)=-RA*(0.5*(AA(2,I)-AA(1,I))*SIN(TETA2)+0.5*AA(6,I)*
     1CCS(TETA2))
      FXUO(3,I)=-ZA*AA(3,I)-RA*(SIN(TETA)*AA(4,I)+COS(TETA)*AA(5,I))
    5 CONTINUE
      IF (INDEX.GT.0) GCTC 6
C
      Z1=MU(2)-MU(3)*LANDA(2)*LANDA(3)+IM*(LANDA(2)*LANCA(3)-1.)
      Z2=LANDA(1)*LANDA(3)*MU(3)-MU(1)+IM*(1.-LANDA(1)*LANDA(3))
      Z3=MU(1)*LANDA(2)-MU(2)*LANDA(1)+IM*(LANDA(1)-LANCA(2))
```

```
      T3(1,1)=-REAL(DELT(1)*PST(1)*Z1+DELT(2)*PST(2)*Z2+
     1DELT(3)*PST(3)*Z3)
      T3(2,1)=-REAL(DELT(1)*QST(1)*Z1+DELT(2)*QST(2)*Z2+
     1DELT(3)*QST(3)*Z3)
      T3(3,1)=-REAL(DELT(1)*RST(1)*Z1+DELT(2)*RST(2)*Z2+
     1DELT(3)*RST(3)*Z3)
      CALL PRCMAT(T2,T3,FCXU,3,3,1,3,3,1)
    6 CONTINUE
      RETURN
      END
C
      SUBROUTINE FXMAT(INDEX,MU,LANDA,DELTA,TET,RA,AA,FX,FQX)
      DIMENSION MU(3),LANDA(3),AA(6,6),FX(6,6),KSI(3),SQ(3),
     1GAMA(3),FQX(6,1)
      COMPLEX MU,LANDA,DELTA,IM,KSI,SQ,GAMA,Z4,Z1,Z2,Z3
C
C     SUBROUTINE TO CALCULATE THE COEFFICIENTS OF MATRICES F AND
C     F* AT ANY POINT RA,TET,WITHIN THE ROCK. MATRICES F AND F* ARE
C     USED IN THE CALCULATION OF THE STATE OF STRESS AT THAT POINT.
C
      TETA=TET*ATAN(1.)*4./180.
      IF(TET.EQ.90.) TETA=89.999*ATAN(1.)*4./180.
      IF(TET.EQ.270.) TETA=270.001*ATAN(1.)*4./180.
      IM=(0.,1.)
C
C     INITIALIZATION OF MATRICES FX AND FQX
C
      DO 1 I=1,6
      FQX(I,1)=0.
      DO 2 J=1,6
      FX(I,J)=0.
    2 CONTINUE
    1 CONTINUE
C
C     CALCULATE KSI(I) AND GAMA(I) FOR I=1,2,3
C
      DO 3 I=1,3
      Z4=COS(TETA)+MU(I)*SIN(TETA)
      SQ(I)=CSQRT(RA*RA*Z4*Z4-1.-MU(I)*MU(I))
      IF((TET.GT.90.).AND.(TET.LT.270.)) SQ(I)=-SQ(I)
      KSI(I)=(SQ(I)+RA*(COS(TETA)+MU(I)*SIN(TETA)))/(1.-IM*MU(I))
      GAMA(I)=-1./(DELTA*KSI(I)*SQ(I))
    3 CONTINUE
C
      FX(1,1)=-REAL(IM*GAMA(1)*MU(1)*MU(1)*(LANDA(2)*LANDA(3)-1.)+
     1                 IM*GAMA(2)*MU(2)*MU(2)*(1.-LANDA(1)*LANDA(3))+
     1                 IM*GAMA(3)*MU(3)*MU(3)*LANDA(3)*(LANDA(1)-
     2                 LANDA(2)))
      FX(1,2)=-REAL(GAMA(1)*MU(1)*MU(1)*(MU(2)-MU(3)*LANDA(3)*
     1               LANDA(2))+
     1                GAMA(2)*MU(2)*MU(2)*(LANDA(1)*LANDA(3)*MU(3)-
     1                MU(1))+
     1                GAMA(3)*MU(3)*MU(3)*LANDA(3)*(MU(1)*LANDA(2)-
     2                MU(2)*LANDA(1)))
      FX(1,6)=REAL(GAMA(1)*MU(1)*MU(1)*(LANDA(2)*LANDA(3)-1.+
     1               IM*MU(2)-IM*MU(3)*LANDA(2)*LANDA(3))+
     1               GAMA(2)*MU(2)*MU(2)*(1.-LANDA(1)*LANDA(3)+
     2               IM*MU(3)*LANDA(1)*LANDA(3)-IM*MU(1))+
     2               GAMA(3)*MU(3)*MU(3)*LANDA(3)*(LANDA(1)-LANDA(2)+
     2               IM*MU(1)*LANDA(2)-IM*MU(2)*LANDA(1)))
      FX(1,4)=REAL(MU(1)*MU(1)*GAMA(1)*LANDA(3)*(MU(3)-MU(2))+
     1               MU(2)*MU(2)*GAMA(2)*LANDA(3)*(MU(1)-MU(3))+
```

```
1              MU(3)*MU(3)*GAMA(3)*LANDA(3)*(MU(2)-MU(1)))
 FX(1,5)=-REAL(IM*GAMA(1)*MU(1)*MU(1)*LANDA(3)*(MU(3)-MU(2))+
1               IM*GAMA(2)*MU(2)*MU(2)*LANDA(3)*(MU(1)-MU(3))+
1               IM*GAMA(3)*MU(3)*MU(3)*LANDA(3)*(MU(2)-MU(1)))
 FX(2,1)=-REAL(IM*GAMA(1)*(LANDA(2)*LANDA(3)-1.)+
1               IM*GAMA(2)* (1.-LANDA(1)*LANDA(3))+
1              IM*GAMA(3)*LANDA(3)*(LANDA(1)-LANDA(2)))
 FX(2,2)=-REAL(GAMA(1)*(MU(2)-MU(3)*LANDA(3)*LANDA(2))+
1               GAMA(2)*(LANDA(1)*LANDA(3)*MU(3)-MU(1))+
1               GAMA(3)*LANDA(3)*(MU(1)*LANDA(2)-MU(2)*LANDA(1)))
 FX(2,6)=REAL(GAMA(1)*(LANDA(2)*LANDA(3)-1.+IM*MU(2)-IM*MU(3)*
1              LANDA(3)*LANDA(2))+
1              GAMA(2)*(1.-LANDA(1)*LANDA(3)+IM*MU(3)*LANDA(1)*
2              LANDA(3)-IM*MU(1))+
3              GAMA(3)*LANDA(3)*(LANDA(1)-LANDA(2)+IM*MU(1)*
1              LANDA(2)-IM*MU(2)*LANDA(1)))
 FX(2,4)=REAL(GAMA(1)*LANDA(3)*(MU(3)-MU(2))+
1              GAMA(2)*LANDA(3)*(MU(1)-MU(3))+
1              GAMA(3)*LANDA(3)*(MU(2)-MU(1)))
 FX(2,5)=-REAL(IM*GAMA(1)*LANDA(3)*(MU(3)-MU(2))+
1               IM*GAMA(2)*LANDA(3)*(MU(1)-MU(3))+
1               IM*GAMA(3)*LANDA(3)*(MU(2)-MU(1)))
 FX(6,1)=REAL(IM*GAMA(1)*MU(1)*(LANDA(2)*LANDA(3)-1.)+
1              IM*GAMA(2)*MU(2)*(1.-LANDA(1)*LANDA(3))+
1              IM*GAMA(3)*MU(3)*LANDA(3)*(LANDA(1)-LANDA(2)))
 FX(6,2)=REAL(GAMA(1)*MU(1)*(MU(2)-MU(3)*LANDA(2)*LANDA(3))+
1              GAMA(2)*MU(2)*(LANDA(1)*LANDA(3)*MU(3)-MU(1))+
1              GAMA(3)*MU(3)*LANDA(3)*(MU(1)*LANDA(2)-MU(2)*
2              LANDA(1)))
 FX(6,6)=-REAL(GAMA(1)*MU(1)*(LANDA(2)*LANDA(3)-1.+IM*MU(2)-
1               IM*MU(3)*LANDA(2)*LANDA(3))+
1               GAMA(2)*MU(2)*(1.-LANDA(1)*LANDA(3)+IM*
2               LANDA(1)*LANDA(3)*MU(3)-IM*MU(1))+
2               GAMA(3)*MU(3)*LANDA(3)*(LANDA(1)-LANDA(2)+
3               IM*MU(1)*LANDA(2)-IM*MU(2)*LANDA(1)))
 FX(6,4)=-REAL(MU(1)*GAMA(1)*LANDA(3)*(MU(3)-MU(2))+
1               MU(2)*GAMA(2)*LANDA(3)*(MU(1)-MU(3))+
1               MU(3)*GAMA(3)*LANDA(3)*(MU(2)-MU(1)))
 FX(6,5)=REAL(IM*MU(1)*GAMA(1)*LANDA(3)*(MU(3)-MU(2))+
1              IM*MU(2)*GAMA(2)*LANDA(3)*(MU(1)-MU(3))+
1              IM*MU(3)*GAMA(3)*LANDA(3)*(MU(2)-MU(1)))
 FX(5,1)=-REAL(IM*GAMA(1)*LANDA(1)*MU(1)*(LANDA(2)*LANDA(3)-1.)+
1               IM*GAMA(2)*LANDA(2)*MU(2)*(1.-LANDA(1)*LANDA(3))+
1               IM*GAMA(3)*MU(3)*(LANDA(1)-LANDA(2)))
 FX(5,2)=-REAL(GAMA(1)*LANDA(1)*MU(1)*(MU(2)-MU(3)*LANDA(3)*
1               LANDA(2))+
2               GAMA(2)*LANDA(2)*MU(2)*(LANDA(1)*LANDA(3)*MU(3)-
2               MU(1))+
2               GAMA(3)*MU(3)*(MU(1)*LANDA(2)-MU(2)*LANDA(1)))
 FX(5,6)=REAL(GAMA(1)*MU(1)*LANDA(1)*(LANDA(2)*LANDA(3)-1.+
1              IM*MU(2)-IM*MU(3)*LANDA(3)*LANDA(2))+
1              GAMA(2)*MU(2)*LANDA(2)*(1.-LANDA(1)*LANDA(3)+
2              IM*LANDA(1)*LANDA(3)*MU(3)-IM*MU(1))+
2              GAMA(3)*MU(3)*(LANDA(1)-LANDA(2)+IM*MU(1)*LANDA(2)
3              -IM*MU(2)*LANDA(1)))
 FX(5,4)=REAL(GAMA(1)*MU(1)*LANDA(1)*LANDA(3)*(MU(3)-MU(2))+
1              GAMA(2)*MU(2)*LANDA(2)*LANDA(3)*(MU(1)-MU(3))+
2              MU(3)*GAMA(3)*(MU(2)-MU(1)))
 FX(5,5)=-REAL(IM*GAMA(1)*MU(1)*LANDA(1)*LANDA(3)*(MU(3)-MU(2))+
1               IM*GAMA(2)*MU(2)*LANDA(2)*LANDA(3)*(MU(1)-MU(3))+
1               IM*GAMA(3)*MU(3)*(MU(2)-MU(1)))
 FX(4,1)=REAL(IM*GAMA(1)*LANDA(1)*(LANDA(2)*LANDA(3)-1.)+
```

```
     2                 IM*GAMA(2)*LANDA(2)*(1.-LANDA(1)*LANCA(3))+
     2                 IM*GAMA(3)*(LANDA(1)-LANDA(2)))
      FX(4,2)=REAL(GAMA(1)*LANCA(1)*(MU(2)-MU(3)*LANDA(2)*LANDA(3))+
     1                 GAMA(2)*LANDA(2)*(LANDA(1)*LANDA(3)*MU(3)-MU(1))+
     1                 GAMA(3)*(MU(1)*LANDA(2)-MU(2)*LANDA(1)))
      FX(4,6)=-REAL(GAMA(1)*LANTA(1)*(LANDA(2)*LANDA(3)-1.+IM*MU(2)-
     1                  IM*MU(3)*LANCA(3)*LANDA(2))+
     3                  GAMA(2)*LANTA(2)*(1.-LANDA(1)*LANCA(3)+
     1                  IM*LANDA(1)*LANCA(3)*MU(3)-IM*MU(1))+
     1                  GAMA(3)*(LANCA(1)-LANDA(2)+IM*MU(1)*LANCA(2)-
     1                  IM*MU(2)*LANCA(1)))
      FX(4,4)=-REAL(LANCA(1)*GAMA(1)*LANDA(3)*(MU(3)-MU(2))+
     1                  LANCA(2)*GAMA(2)*LANDA(3)*(MU(1)-MU(3))+
     1                  GAMA(3)*(MU(2)-MU(1)))
      FX(4,5)=REAL(IM*GAMA(1)*LANDA(1)*LANDA(3)*(MU(3)-MU(2))+
     1                 IM*GAMA(2)*LANDA(2)*LANDA(3)*(MU(1)-MU(3))+
     1                 IM*GAMA(3)*(MU(2)-MU(1)))
      DC 4 I=1,6
      IF(I.EQ.3) GOTO 4
      FX(3,I)=-(AA(3,1)*FX(1,I)+AA(3,2)*FX(2,I)+AA(3,4)*FX(4,I)+
     1AA(3,5)*FX(5,I)+AA(3,6)*FX(6,I))/AA(3,3)
    4 CCNTINUE
C
      DC 5 I=1,6
      FX(I,I)=1.+FX(I,I)
    5 CCNTINUE
C
      IF(INDEX.GT.0) GCTO 6
C
      Z1=GAMA(1)*(MU(2)-MU(3)*LANDA(2)*LANDA(3)+
     1IM*(LANCA(2)*LANCA(3)-1.))
      Z2=GAMA(2)*(LANDA(1)*LANCA(3)*MU(3)-MU(1)+
     1IM*(1.-LANDA(1)*LANCA(3)))
      Z3=GAMA(3)*(MU(1)*LANDA(2)-MU(2)*LANDA(1)+
     1IM*(LANCA(1)-LANDA(2)))
C
      FCX(1,1)=REAL(MU(1)*MU(1)*Z1+MU(2)*MU(2)*Z2+LANCA(3)*MU(3)*
     1MU(3)*Z3)
      FQX(2,1)=REAL(Z1+Z2+LANDA(3)*Z3)
      FCX(6,1)=-REAL(MU(1)*Z1+MU(2)*Z2+LANDA(3)*MU(3)*Z3)
      FCX(5,1)=REAL(LANCA(1)*MU(1)*Z1+LANDA(2)*MU(2)*Z2+MU(3)*Z3)
      FQX(4,1)=-REAL(LANCA(1)*Z1+LANDA(2)*Z2+Z3)
      FCX(3,1)=-(FQX(1,1)*AA(3,1)+FCX(2,1)*AA(3,2)+
     1FCX(4,1)*AA(3,4)+FOX(5,1)*AA(3,5)+
     2FCX(6,1)*AA(3,6))/AA(3,3)
    6 CCNTINUE
      RETURN
      END
C
```

- Berni 3 -

```
      PROGRAM BERNI3(INPUT,OUTPUT,TAPE5=INPUT,TAPE6=OUTPUT)
      COMMON/HOLIN/M,E,MUI,B,N,U(11),H(11),K(11),S(11),T(11),GO,G1,R1
      COMMON/BH/BETAH,DELTH,AH(3,3)
      COMMON/ANIS/STR,DIF,DIPDIF,A(3,3)
      DIMENSION KN(3),KS(3),SJ(3),HF(6,6),H1(6,6),H2(6,6),TS(6,6),
     1TTS(6,6),TH(6,6),TE(6,6),TTE(6,6),AA(6,6),BETA(6,6),
     2T1(6,6),T2(6,6),T3(6,6),AX(61,61),CX(61,6)
      DIMENSION AAA(7),ROOT(6),MU(3),L2(3),L3(3),L4(3),LANDA(3),
     1PST(3),QST(3),RST(3),FU(6),FV(6),FW(6),DX(67,61),FX(6,6),
     1GX(6,6),WKAREA(150),EX(6,67),SIG(6),BD(3),Z(3,3),WK(10),
     3HX(61,6),HHX(67,6),EXU(3,67),FXU(3,6),SFOG(6,1),
     4BETH(3),DETH(3),NROS(3),ROS(4,6,3),YROS(36,1),XROS(36,6),
     5NGAGE(3),GAGE(6,4,3),T4(3,6),T5(3,6),T6(1,3),T7(1,6)
      REAL KN,KS,MU21,MU31,MU32,M,MUI,K
      INTEGER DIPDIR
      COMPLEX ROOT,MU,L2,L3,L4,LANDA,DELTA,PST,QST,RST,GAMA,
     1GAMB,GAMC,GAMPA,GAMPB,GAMPC,GAMPPA,GAMPPB,GAMPPC,IM
C
C     ******************************************************************
C
      IUNFC=61
      IUNFCP=67
      READ(5,10) M,E,MUI,B,N
   10 FORMAT(4F10.4,I5)
      WRITE(6,19)
      IF(B.NE.0.) WRITE(6,247)
  247 FORMAT(/10X,≠ FULL INCLUSION ≠,/)
      WRITE(6,20) M,E,MUI,N
   19 FORMAT(10X,≠** INCLUSION PROPERTIES **≠//)
   20 FORMAT(10X,≠ RATIO OUTER/INNER RADII ≠,5X,F10.4/
     110X,≠ YOUNG MODULUS ≠,10X,F10.4/
     210X,≠ POISSON S RATIO ≠,8X,F10.4/
     310X,≠ FOURIER COEFFICIENTS N ≠,12X,I3///)
C
      CALL HOLLOW
C
      NN=N+1
      WRITE(6,21)GO,G1,R1
      WRITE(6,203)
      WRITE(6,22)(I,U(I),H(I),K(I),S(I),T(I),I=1,NN)
   21 FORMAT(10X,≠ GO ≠,E12.4,≠ G1 ≠,E12.4,≠ R1 ≠,E12.4//)
   22 FORMAT(10X,I3,2X,E12.4,2X,E12.4,2X,E12.4,2X,E12.4,2X,E12.4)
  203 FORMAT(/10X,≠  I≠,7X,≠ U ≠,11X,≠ H ≠,11X,≠ K ≠,11X,≠ S ≠,11X,
     1≠ T ≠,/)
C
C     ******************************************************************
C
      READ(5,11) STR,DIF,DIPDIR,INDEX
      READ(5,12) E1,E2,E3,G12,G13,G23,MU21,MU31,MU32
      READ(5,13)(KN(I),I=1,3),(KS(I),I=1,3),(SJ(I),I=1,3)
   11 FORMAT(2F10.4,2I5)
   12 FORMAT(9F8.2)
   13 FORMAT(6E10.4,3F6.2)
C
      READ(5,28) NHOLES,TSTU,(BETH(I),DETH(I),I=1,NHOLES)
   28 FORMAT(I5,7F10.4)
      IF(INDEX.GT.0) GOTO 40
```

```
      READ(5,32)(NROS(I),I=1,NHOLES)
   32 FORMAT(3I5)
      DO 64 K1=1,NHOLES
      N2=NROS(K1)
      READ(5,33)((ROS(I,J,K1),J=1,6),I=1,N2)
   33 FORMAT(3E10.4,3F10.4)
   64 CONTINUE
      GOTO 31
   40 CONTINUE
      READ(5,239)(NGAGE(I),I=1,NHOLES)
  239 FORMAT(3I5)
      DO 65 K1=1,NHOLES
      N2=NGAGE(K1)
      READ(5,66)((GAGE(I,J,K1),J=1,4),I=1,N2)
   66 FORMAT(3E10.4,F10.4)
   65 CONTINUE
   31 CONTINUE
C
C     ******************************************************************
C
      WRITE(6,23)STR,DIP,DIPDIR
      WRITE(6,26)
      WRITE(6,25) E1,E2,E3,G12,G13,G23,MU21,MU31,MU32
      WRITE(6,24)(I,KN(I),KS(I),SJ(I),I=1,3)
   26 FORMAT(//10X,*** ROCK PROPERTIES ***//)
   23 FORMAT(//10X,*** ORIENTATION PARAMETERS ***//
     1        10X,* ANISOTROPY ** STRIKE *,F10.4/
     1        10X,*               DIP    *,F10.4/
     2        10X,*               DIPDIR *,I5//)
   24 FORMAT(10X,* JOINT SET *,I5,* KN *,E12.4,* KS *,E12.4,* S *,
     1F10.4//)
   25 FORMAT(10X,* YOUNG MODULUS IN N DIRECTION *,F10.4/
     1        10X,* YOUNG MODULUS IN S DIRECTION *,F10.4/
     1        10X,* YOUNG MODULUS IN T DIRECTION *,F10.4/
     1        10X,* SHEAR MODULUS IN NS PLANE *,F10.4/
     1        10X,* SHEAR MODULUS IN NT PLANE *,F10.4/
     1        10X,* SHEAR MODULUS IN ST PLANE *,F10.4/
     1        10X,* POISSON RATIO MU21 *,F10.4/
     1        10X,* POISSON RATIO MU31 *,F10.4/
     1        10X,* POISSON RATIO MU32 *,F10.4///)
C
      WRITE(6,29) NHOLES,TSTU
      WRITE(6,248)(I,BETH(I),DETH(I),I=1,NHOLES)
   29 FORMAT(////10X,* NUMBER OF BOREHOLES *,I5/
     1             10X,* T DISTRIBUTION *,F13.4//)
  248 FORMAT(10X,* BOREHOLE*,I5,* BETAH*,F10.4,* DELTH*,F10.4)
      IF(INDEX.GT.0) GOTO 41
      WRITE(6,249)
  249 FORMAT(///10X,* STRESS MEASUREMENT USING STRAINS *,////)
      DO 62 K1=1,NHOLES
      N2=NROS(K1)
      WRITE(6,35) K1,N2
      WRITE(6,250)(I,(ROS(I,J,K1),J=1,6),I=1,N2)
   35 FORMAT(10X,* *BOREHOLE * *,I5,* NUMBER OF ROSETTES *,I5)
  250 FORMAT(10X,I5,5X,E10.4,5X,E10.4,5X,E10.4,5X,F10.4,5X,F10.4,
     15X,F10.4)
   62 CONTINUE
      GOTO 34
   41 CONTINUE
      WRITE(6,251)
  251 FORMAT(///10X,* STRESS MEASUREMENT USING DISTANCES *,////)
      DO 63 K1=1,NHOLES
```

```
      N2=NGAGE(K1)
      WRITE(6,37) K1,N2
      WRITE(6,252)(I,(GAGE(I,J,K1),J=1,4),I=1,N2)
   37 FORMAT(10X,≠ * BOREHOLE * ≠,I5,≠ NUMBER OF DISTANCES ≠,I5)
  252 FORMAT(10X,I5,5X,E10.4,5X,E10.4,5X,E10.4,5X,F10.4)
   63 CONTINUE
   34 CONTINUE
C
C     ******************************************************************
C
C     CALCULATE THE COMPLIANCE MATRIX HP AS THE SUM OF TWO MATRICES
C     H1 AND H2
C
C     INITIALIZATION OF MATRICES H1 AND H2
C
      DO 3 I=1,6
      DO 3 J=1,6
      H1(I,J)=0.
    3 H2(I,J)=0.
C
C     CALCULATE THE MATRIX H1 WITH INTACT MATERIAL CONSTANTS
C
      H1(1,1)=1./E1
      H1(2,2)=1./E2
      H1(3,3)=1./E3
      H1(4,4)=1./G23
      H1(5,5)=1./G13
      H1(6,6)=1./G12
      H1(1,2)=-MU21/E2
      H1(1,3)=-MU31/E3
      H1(2,3)=-MU32/E3
      H1(2,1)=H1(1,2)
      H1(3,1)=H1(1,3)
      H1(3,2)=H1(2,3)
C
      WRITE(6,104)
      WRITE(6,105)((H1(I,J),J=1,6),I=1,6)
  104 FORMAT(///30X,≠ MATRIX H1 ≠,//)
  105 FORMAT(6E12.4)
C
C     CALCULATE THE MATRIX H2 WITH JOINT PROPERTIES
C
      DO 16 I=1,3
      H2(I,I)=1./(KN(I)*SJ(I))
   16 CONTINUE
      H2(4,4)=1./(KS(2)*SJ(2))+1./(KS(3)*SJ(3))
      H2(5,5)=1./(KS(1)*SJ(1))+1./(KS(3)*SJ(3))
      H2(6,6)=1./(KS(1)*SJ(1))+1./(KS(2)*SJ(2))
      WRITE(6,112)
  112 FORMAT(///30X,≠ MATRIX H2 ≠//)
      WRITE(6,113)((H2(I,J),J=1,6),I=1,6)
  113 FORMAT(6E12.4)
C
C     THEN, CALCULATE THE COMPLIANCE MATRIX HP BY ADDING MATRICES
C     H1 AND H2
C
      DO 14 I=1,6
      DO 15 J=1,6
      HP(I,J)=H1(I,J)+H2(I,J)
   15 CONTINUE
   14 CONTINUE
C
```

```
C       FOR THE ANISOTROPY CALCULATE THE TRANSFORMATION MATRICES
C       TS AND TTS
C
        CALL ORIEN2
        CALL TRANSF(A,TS)
        CALL TRANS(TS,TTS,6,6,6,6)
        WRITE(6,100)
  100   FORMAT(///30X,≠  TRANSFORMATION MATRIX TS≠,//)
        WRITE(6,101)((TS(I,J),J=1,6),I=1,6)
  101   FORMAT(6E12.4)
        WRITE(6,102)
  102   FORMAT(///30X,≠  TRANSFORMATION MATRIX TTS ≠,//)
        WRITE(6,103)((TTS(I,J),J=1,6),I=1,6)
  103   FORMAT(6E12.4)
C
C       ******************************************************************
        N1=0
        DO 36 K1=1,NHOLES
C       ******************************************************************
C
C       FOR THE HOLE CALCULATE THE TRANSFORMATION MATRIX TH
C
        BETAH=BETH(K1)
        DELTH=DETH(K1)
        CALL ORIENH
        CALL TRANSF(AH,TH)
        WRITE(6,106)
  106   FORMAT(///30X,≠  TRANSFORMATION MATRIX TH ≠,//)
        WRITE(6,107)((TH(I,J),J=1,6),I=1,6)
  107   FORMAT(6E12.4)
C
C       FOR THE HOLE CALCULATE THE TRANSFORMATION MATRICES TE AND TTE
C
        DO 4 I=1,6
        DO 5 J=1,6
        TE(I,J)=TH(I,J)
        IF((I.LE.3).AND.(J.GT.3)) TE(I,J)=0.5*TH(I,J)
        IF((I.GT.3).AND.(J.LE.3)) TE(I,J)=2.*TH(I,J)
    5   CONTINUE
    4   CONTINUE
        CALL TRANS(TE,TTE,6,6,6,6)
        WRITE(6,108)
  108   FORMAT(///30X,≠  TRANSFORMATION MATRIX TE ≠,//)
        WRITE(6,109)((TE(I,J),J=1,6),I=1,6)
  109   FORMAT(6E12.4)
        WRITE(6,110)
  110   FORMAT(///30X,≠  TRANSFORMATION MATRIX TTE ≠,//)
        WRITE(6,111)((TTE(I,J),J=1,6),I=1,6)
  111   FORMAT(6E12.4)
C
C        CALCULATE THE STRAIN-STRESS MATRIX AA IN THE COORDINATE SYSTEM
C       ATTACHED TO THE HOLE
C
        CALL PRCMAT(TE,TTS,T1,6,6,6,6,6,6)
        CALL PRCMAT(T1,MF,T2,6,6,6,6,6,6)
        CALL PRCMAT(T2,TS,T3,6,6,6,6,6,6)
        CALL PRCMAT(T3,TTE,AA,6,6,6,6,6,6)
        WRITE(6,114)
  114   FORMAT(//30X,≠ STRESS STRAIN MATRIX AA ≠,//)
        WRITE(6,115)((AA(I,J),J=1,6),I=1,6)
  115   FORMAT(6F15.9)
C
```

```
C     CALCULATE THE COEFFICIENTS OF MATRIX BETA
C
      DO 1 I=1,6
      DO 2 J=1,6
      BETA(I,J)=AA(I,J)-AA(I,3)*AA(J,3)/AA(3,3)
    2 CONTINUE
    1 CONTINUE
      WRITE(6,121)
  121 FORMAT(/30X,≠ MATRIX BETA ≠//)
      WRITE(6,120)((BETA(I,J),J=1,6),I=1,6)
  120 FORMAT(6F15.9)
C
C     CALCULATE THE COEFFICIENTS AAA(I) OF THE 6TH DEGREE POLYNOMIAL
C
      AAA(1)=BETA(1,1)*BETA(5,5)-BETA(1,5)*BETA(1,5)
      AAA(2)=2.*BETA(1,5)*(BETA(1,4)+BETA(5,6))
      AAA(2)=AAA(2)-2.*(BETA(1,6)*BETA(5,5)+BETA(1,1)*BETA(4,5))
      AAA(3)=BETA(5,5)*(2.*BETA(1,2)+BETA(6,6))+4.*BETA(1,6)*BETA(4,5)+
     1BETA(1,1)*BETA(4,4)
      AAA(3)=AAA(3)-(BETA(1,4)+BETA(5,6))**2-2.*BETA(1,5)*(BETA(2,5)+
     1BETA(4,6))
      AAA(4)=-2.*BETA(2,6)*BETA(5,5)-2.*BETA(4,5)*(2.*BETA(1,2)+
     1BETA(6,6))-2.*BETA(1,6)*BETA(4,4)
      AAA(4)=AAA(4)+2.*BETA(1,5)*BETA(2,4)+2.*(BETA(1,4)+BETA(5,6))*
     1(BETA(2,5)+BETA(4,6))
      AAA(5)=BETA(2,2)*BETA(5,5)+4.*BETA(2,6)*BETA(4,5)+BETA(4,4)*
     1(2.*BETA(1,2)+BETA(6,6))
      AAA(5)=AAA(5)-2.*BETA(2,4)*(BETA(1,4)+BETA(5,6))-(BETA(2,5)+
     1BETA(4,6))**2
      AAA(6)=-2.*BETA(2,2)*BETA(4,5)-2.*BETA(2,6)*BETA(4,4)+2.*BETA(2,4)
     1*(BETA(2,5)+BETA(4,6))
      AAA(7)=BETA(2,2)*BETA(4,4)-BETA(2,4)*BETA(2,4)
C
C     CALCULATE THE ROOTS OF THE 6TH DEGREE CHARACTERISTIC EQUATION
C     USING SUBROUTINE ZPOLR
C
      CALL ZPOLR(AAA,6,ROOT,IER)
      WRITE(6,201)IER,(ROOT(I),I=1,6)
  201 FORMAT(///,I5,6(F10.4,F10.4))
C
C     FIND THE ROOTS MU1,MU2,MU3
C
      MU(1)=ROOT(5)
      MU(2)=ROOT(1)
      MU(3)=ROOT(3)
      WRITE(6,116)(MU(I),I=1,3)
  116 FORMAT(//10X,≠ MU1 ≠,(F15.9,F15.9)/
     1           10X,≠ MU2 ≠,(F15.9,F15.9)/
     1           10X,≠ MU3 ≠,(F15.9,F15.9)////)
C
C     CALCULATE THE COEFFICIENTS LANDA1,LANDA2,LANDA3 AND DELTA
C
      DO 17 I=1,3
      L3(I)=BETA(1,5)*(MU(I)**3)-(BETA(1,4)+BETA(5,6))*(MU(I)**2)+(
     1BETA(2,5)+BETA(4,6))*MU(I)-BETA(2,4)
      L2(I)=BETA(5,5)*(MU(I)**2)-2.*BETA(4,5)*MU(I)+BETA(4,4)
      L4(I)=BETA(1,1)*(MU(I)**4)-2.*BETA(1,6)*(MU(I)**3)+(2.*BETA(1,2)+
     1BETA(6,6))*(MU(I)**2)-2.*BETA(2,6)*MU(I)+BETA(2,2)
      LANDA(I)=-L3(I)/L2(I)
      IF(I.EQ.3) LANDA(I)=-L3(I)/L4(I)
      WRITE(6,117) I,L2(I),L3(I),L4(I)
  117 FORMAT(10X,≠ MU ≠,I3,≠ L2 ≠,(E12.4,E12.4),≠ L3 ≠,(E12.4,E12.4),
```

```
     1* L4 *,(E12.4,E12.4),//)
   17 CONTINUE
      DELTA=MU(2)-MU(1)+LANDA(2)*LANDA(3)*(MU(1)-MU(3))+LANDA(1)*
     1LANDA(3)*(MU(3)-MU(2))
C
C     CALCUALTE THE COEFFICIENTS PST(I),QST(I),RST(I)
C
      DO 18 I=1,2
      PST(I)=BETA(1,1)*MU(I)*MU(I)+BETA(1,2)-BETA(1,6)*MU(I)+LANDA(I)*
     1(BETA(1,5)*MU(I)-BETA(1,4))
      QST(I)=BETA(1,2)*MU(I)+LANDA(I)*(BETA(2,5)-BETA(2,4)/MU(I))-
     1BETA(2,6)+BETA(2,2)/MU(I)
      RST(I)=BETA(1,4)*MU(I)+BETA(2,4)/MU(I)-BETA(4,6)+LANDA(I)*(BETA(4,
     15)-BETA(4,4)/MU(I))
   18 CONTINUE
      PST(3)=LANDA(3)*(BETA(1,1)*(MU(3)**2)+BETA(1,2)-BETA(1,6)*MU(3))+
     1BETA(1,5)*MU(3)-BETA(1,4)
      QST(3)=LANDA(3)*(BETA(1,2)*MU(3)-BETA(2,6)+BETA(2,2)/MU(3))+BETA(
     12,5)-BETA(2,4)/MU(3)
      RST(3)=LANDA(3)*(BETA(1,4)*MU(3)-BETA(4,6)+BETA(2,4)/MU(3))+BETA(
     14,5)-BETA(4,4)/MU(3)
C
C     CALCULATE THE COEFFICIENTS FU,FV,FW
C
      IM=(0.,1.)
      GAMA   =PST(1)*(MU(2)-MU(3)*LANDA(2)*LANDA(3))+
     1        PST(2)*(LANDA(1)*LANDA(3)*MU(3)-MU(1))+
     2        PST(3)*(MU(1)*LANDA(2)-MU(2)*LANDA(1))
      GAMB   =PST(1)*(LANDA(2)*LANDA(3)-1.)+
     2        PST(2)*(1.-LANDA(1)*LANDA(3))+
     2        PST(3)*(LANDA(1)-LANDA(2))
      GAMC   =PST(1)*LANDA(3)*(MU(3)-MU(2))+
     1        PST(2)*LANDA(3)*(MU(1)-MU(3))+
     2        PST(3)*(MU(2)-MU(1))
      GAMPA  =QST(1)*(MU(2)-MU(3)*LANDA(2)*LANDA(3))+
     1        QST(2)*(LANDA(1)*LANDA(3)*MU(3)-MU(1))+
     2        QST(3)*(MU(1)*LANDA(2)-MU(2)*LANDA(1))
      GAMPB  =QST(1)*(LANDA(2)*LANDA(3)-1.)+
     2        QST(2)*(1.-LANDA(1)*LANDA(3))+
     2        QST(3)*(LANDA(1)-LANDA(2))
      GAMPC  =QST(1)*LANDA(3)*(MU(3)-MU(2))+
     1        QST(2)*LANDA(3)*(MU(1)-MU(3))+
     2        QST(3)*(MU(2)-MU(1))
      GAMPPA=RST(1)*(MU(2)-MU(3)*LANDA(2)*LANDA(3))+
     1        RST(2)*(LANDA(1)*LANDA(3)*MU(3)-MU(1))+
     2        RST(3)*(MU(1)*LANDA(2)-MU(2)*LANDA(1))
      GAMPPB=RST(1)*(LANDA(2)*LANDA(3)-1.)+
     2        RST(2)*(1.-LANDA(1)*LANDA(3))+
     2        RST(3)*(LANDA(1)-LANDA(2))
      GAMPPC=RST(1)*LANDA(3)*(MU(3)-MU(2))+
     1        RST(2)*LANDA(3)*(MU(1)-MU(3))+
     2        RST(3)*(MU(2)-MU(1))
C
      FU(1)=-REAL(GAMA/DELTA)
      FU(2)=-REAL(IM*GAMA/DELTA)
      FU(3)=-REAL(GAMB/DELTA)
      FU(4)=-REAL(IM*GAMB/DELTA)
      FU(5)=-REAL(GAMC/DELTA)
      FU(6)=-REAL(IM*GAMC/DELTA)
C
      FV(1)=-REAL(GAMPA/DELTA)
      FV(2)=-REAL(IM*GAMPA/DELTA)
```

```
      FV(3)=-REAL(GAMPB/DELTA)
      FV(4)=-REAL(IM*GAMPB/DELTA)
      FV(5)=-REAL(GAMPC/DELTA)
      FV(6)=-REAL(IM*GAMPC/DELTA)
C
      FW(1)=-REAL(GAMPPA/DELTA)
      FW(2)=-REAL(IM*GAMPPA/DELTA)
      FW(3)=-REAL(GAMPPB/DELTA)
      FW(4)=-REAL(IM*GAMPPB/DELTA)
      FW(5)=-REAL(GAMPPC/DELTA)
      FW(6)=-REAL(IM*GAMPPC/DELTA)
      WRITE(6,202)(FU(I),FV(I),FW(I),I=1,6)
  202 FORMAT(3E12.4)
C
C     CALCULATE THE COEFFICIENTS S,T,SP,TP,SPP,TPP
C
      S11  =-FU(4)
      S22  =-FU(1)
      S12  =-(FU(2)+FU(3))
      S23  =-FU(5)
      S13  =-FU(6)
      T11  =FU(3)
      T22  =-FU(2)
      T12  =FU(1)-FU(4)
      T23  =-FU(6)
      T13  =FU(5)
C
      SP11=-FV(4)
      SP22=-FV(1)
      SP12=-(FV(2)+FV(3))
      SP23=-FV(5)
      SP13=-FV(6)
      TP11=FV(3)
      TP22=-FV(2)
      TP12=FV(1)-FV(4)
      TP23=-FV(6)
      TP13=FV(5)
C
      SPP11=-FW(4)
      SPP22=-FW(1)
      SPP12=-(FW(2)+FW(3))
      SPP23=-FW(5)
      SPP13=-FW(6)
      TPP11=FW(3)
      TPP22=-FW(2)
      TPP12=FW(1)-FW(4)
      TPP23=-FW(6)
      TPP13=FW(5)
C
C     CALCULATE THE COEFFICIENTS OF MATRIX AX(6N+1,6N+1)
C
      NFC=6*N+1
      DO 6 I=1,NFC
      DO 7 J=1,NFC
      AX(I,J)=0.
    7 CONTINUE
    6 CONTINUE
C
      AX(3,4)=1.
      AX(3,3)=-1.
      AX(2,2)=GC-(FU(4)+FV(3))
      AX(2,3)=FU(3)-FV(4)
```

```
      AX(2,4)=FU(2)+FV(1)
      AX(2,5)=GC+(FV(2)-FU(1))
      AX(2,6)=-(FU(6)+FV(5))
      AX(2,7)=FU(5)-FV(6)
      AX(1,1)=2.
      AX(1,2)=FU(3)-FV(4)
      AX(1,3)=FV(3)+FU(4)
      AX(1,4)=FV(2)-FU(1)
      AX(1,5)=-(FV(1)+FU(2))
      AX(1,6)=FU(5)-FV(6)
      AX(1,7)=FV(5)+FU(6)
      AX(6,2)=-FW(4)
      AX(6,3)=FW(3)
      AX(6,4)=FW(2)
      AX(6,5)=-FW(1)
      AX(6,6)=-FW(6)+U(1)
      AX(6,7)=FW(5)
      AX(7,2)=FW(3)
      AX(7,3)=FW(4)
      AX(7,4)=-FW(1)
      AX(7,5)=-FW(2)
      AX(7,6)=FW(5)
      AX(7,7)=FW(6)-U(1)
      AX(4,8)=FU(3)-FV(4)
      AX(4,9)=FV(3)+FU(4)-G1+R1
      AX(4,10)=FV(2)-FU(1)+G1-R1
      AX(4,11)=-(FV(1)+FU(2))
      AX(4,12)=FU(5)-FV(6)
      AX(4,13)=FV(5)+FU(6)
      AX(5,8)=-(FU(4)+FV(3))+G1-R1
      AX(5,9)=FU(3)-FV(4)
      AX(5,10)=FU(2)+FV(1)
      AX(5,11)=FV(2)-FU(1)+G1-R1
      AX(5,12)=-(FU(6)+FV(5))
      AX(5,13)=FU(5)-FV(6)
C
      NNN=N-1
      DO 8 I=2,NNN
      RI=I
      AX(6*I-4,6*I-10)=(FV(3)-FU(4))/(RI-1.)+(H(I)+K(I))
      AX(6*I-4,6*I-9)=(FU(3)+FV(4))/(RI-1.)
      AX(6*I-4,6*I-8)=(FU(2)-FV(1))/(RI-1.)
      AX(6*I-4,6*I-7)=-(FU(1)+FV(2))/(RI-1.)-(H(I)+K(I))
      AX(6*I-4,6*I-6)=(FV(5)-FU(6))/(RI-1.)
      AX(6*I-4,6*I-5)=(FU(5)+FV(6))/(RI-1.)
      AX(6*I-4,6*I+2)=-(FU(4)+FV(3))/(RI+1.)+(H(I)-K(I))
      AX(6*I-4,6*I+3)=(FU(3)-FV(4))/(RI+1.)
      AX(6*I-4,6*I+4)=(FU(2)+FV(1))/(RI+1.)
      AX(6*I-4,6*I+5)=(FV(2)-FU(1))/(RI+1.)+(H(I)-K(I))
      AX(6*I-4,6*I+6)=-(FU(6)+FV(5))/(RI+1.)
      AX(6*I-4,6*I+7)=(FU(5)-FV(6))/(RI+1.)
      AX(6*I-3,6*I-10)=-(FV(4)+FU(3))/(RI-1.)
      AX(6*I-3,6*I-9)=(FV(3)-FU(4))/(RI-1.)+(S(I)+T(I))
      AX(6*I-3,6*I-8)=(FV(2)+FU(1))/(RI-1.)+(S(I)+T(I))
      AX(6*I-3,6*I-7)=(FU(2)-FV(1))/(RI-1.)
      AX(6*I-3,6*I-6)=-(FV(6)+FU(5))/(RI-1.)
      AX(6*I-3,6*I-5)=(FV(5)-FU(6))/(RI-1.)
      AX(6*I-3,6*I+2)=(FU(3)-FV(4))/(RI+1.)
      AX(6*I-3,6*I+3)=(FV(3)+FU(4))/(RI+1.)+(S(I)-T(I))
      AX(6*I-3,6*I+4)=(FV(2)-FU(1))/(RI+1.)+(T(I)-S(I))
      AX(6*I-3,6*I+5)=-(FV(1)+FU(2))/(RI+1.)
      AX(6*I-3,6*I+6)=(FU(5)-FV(6))/(RI+1.)
```

```
      AX(6*I-3,6*I+7)=(FV(5)+FU(6))/(RI+1.)
      AX(6*I-2,6*I-10)=-(FV(4)+FU(3))/(RI-1.)
      AX(6*I-2,6*I-9)=(FV(3)-FU(4))/(RI-1.)+(H(I)+K(I))
      AX(6*I-2,6*I-8)=(FV(2)+FU(1))/(RI-1.)+(H(I)+K(I))
      AX(6*I-2,6*I-7)=(FU(2)-FV(1))/(RI-1.)
      AX(6*I-2,6*I-6)=-(FV(6)+FU(5))/(RI-1.)
      AX(6*I-2,6*I-5)=(FV(5)-FU(6))/(RI-1.)
      AX(6*I-2,6*I+2)=-(FU(3)-FV(4))/(RI+1.)
      AX(6*I-2,6*I+3)=-(FV(3)+FU(4))/(RI+1.)+(H(I)-K(I))
      AX(6*I-2,6*I+4)=-(FV(2)-FU(1))/(RI+1.)+(K(I)-H(I))
      AX(6*I-2,6*I+5)=(FV(1)+FU(2))/(RI+1.)
      AX(6*I-2,6*I+6)=-(FU(5)-FV(6))/(RI+1.)
      AX(6*I-2,6*I+7)=-(FV(5)+FU(6))/(RI+1.)
      AX(6*I-1,6*I-10)=-(FV(3)-FU(4))/(RI-1.)-(S(I)+T(I))
      AX(6*I-1,6*I-9)=-(FU(3)+FV(4))/(RI-1.)
      AX(6*I-1,6*I-8)=-(FU(2)-FV(1))/(RI-1.)
      AX(6*I-1,6*I-7)=(FU(1)+FV(2))/(RI-1.)+(S(I)+T(I))
      AX(6*I-1,6*I-6)=-(FV(5)-FU(6))/(RI-1.)
      AX(6*I-1,6*I-5)=-(FU(5)+FV(6))/(RI-1.)
      AX(6*I-1,6*I+2)=-(FU(4)+FV(3))/(RI+1.)+(T(I)-S(I))
      AX(6*I-1,6*I+3)=(FU(3)-FV(4))/(RI+1.)
      AX(6*I-1,6*I+4)=(FU(2)+FV(1))/(RI+1.)
      AX(6*I-1,6*I+5)=(FV(2)-FU(1))/(RI+1.)+(T(I)-S(I))
      AX(6*I-1,6*I+6)=-(FU(6)+FV(5))/(RI+1.)
      AX(6*I-1,6*I+7)= (FU(5)-FV(6))/(RI+1.)
      AX(6*I,6*I-4)=-FW(4)/RI
      AX(6*I,6*I-3)= FW(3)/RI
      AX(6*I,6*I-2)= FW(2)/RI
      AX(6*I,6*I-1)=-FW(1)/RI
      AX(6*I,6*I)=-FW(6)/RI+U(I)
      AX(6*I,6*I+1)=FW(5)/RI
      AX(6*I+1,6*I-4)= FW(3)/RI
      AX(6*I+1,6*I-3)= FW(4)/RI
      AX(6*I+1,6*I-2)=-FW(1)/RI
      AX(6*I+1,6*I-1)=-FW(2)/RI
      AX(6*I+1,6*I)= FW(5)/RI
      AX(6*I+1,6*I+1)= FW(6)/RI-U(I)
    8 CONTINUE
C
      AX(6*N-4,6*N-4)=-FW(4)/FLOAT(N)
      AX(6*N-4,6*N-3)= FW(3)/FLOAT(N)
      AX(6*N-4,6*N-2)= FW(2)/FLOAT(N)
      AX(6*N-4,6*N-1)=-FW(1)/FLOAT(N)
      AX(6*N-4,6*N)=-FW(6)/FLOAT(N)+U(N)
      AX(6*N-4,6*N+1)= FW(5)/FLOAT(N)
      AX(6*N-3,6*N-4)= FW(3)/FLOAT(N)
      AX(6*N-3,6*N-3)= FW(4)/FLOAT(N)
      AX(6*N-3,6*N-2)=-FW(1)/FLOAT(N)
      AX(6*N-3,6*N-1)=-FW(2)/FLOAT(N)
      AX(6*N-3,6*N)=FW(5)/FLOAT(N)
      AX(6*N-3,6*N+1)=FW(6)/FLOAT(N)-U(N)
      AX(6*N-2,6*N-10)=(FV(3)-FU(4))/FLOAT(NNN)+(H(N)+K(N))
      AX(6*N-2,6*N-9)=(FU(3)+FV(4))/FLOAT(NNN)
      AX(6*N-2,6*N-8)=(FU(2)-FV(1))/FLOAT(NNN)
      AX(6*N-2,6*N-7)=-(FU(1)+FV(2))/FLOAT(NNN)-(H(N)+K(N))
      AX(6*N-2,6*N-6)=(FV(5)-FU(6))/FLOAT(NNN)
      AX(6*N-2,6*N-5)=(FU(5)+FV(6))/FLOAT(NNN)
      AX(6*N-1,6*N-10)=-(FV(4)+FU(3))/FLOAT(NNN)
      AX(6*N-1,6*N-9)=(FV(3)-FU(4))/FLOAT(NNN)+(H(N)+K(N))
      AX(6*N-1,6*N-8)=(FV(2)+FU(1))/FLOAT(NNN)+(H(N)+K(N))
      AX(6*N-1,6*N-7)=(FU(2)-FV(1))/FLOAT(NNN)
      AX(6*N-1,6*N-6)=-(FV(6)+FU(5))/FLOAT(NNN)
```

```
      AX(6*N-1,6*N-5)=(FV(5)-FU(6))/FLOAT(NNN)
      AX(6*N,6*N-4)=(FV(3)-FU(4))/FLOAT(N)+(H(N+1)+K(N+1))
      AX(6*N,6*N-3)=(FU(3)+FV(4))/FLOAT(N)
      AX(6*N,6*N-2)=(FU(2)-FV(1))/FLOAT(N)
      AX(6*N,6*N-1)=-(FU(1)+FV(2))/FLOAT(N)-(H(N+1)+K(N+1))
      AX(6*N,6*N)=(FV(5)-FU(6))/FLOAT(N)
      AX(6*N,6*N+1)=(FU(5)+FV(6))/FLOAT(N)
      AX(6*N+1,6*N-4)=-(FV(4)+FU(3))/FLOAT(N)
      AX(6*N+1,6*N-3)=(FV(3)-FU(4))/FLOAT(N)+(H(N+1)+K(N+1))
      AX(6*N+1,6*N-2)=(FV(2)+FU(1))/FLOAT(N)+(H(N+1)+K(N+1))
      AX(6*N+1,6*N-1)=(FU(2)-FV(1))/FLOAT(N)
      AX(6*N+1,6*N)=-(FV(6)+FU(5))/FLOAT(N)
      AX(6*N+1,6*N+1)=(FV(5)-FU(6))/FLOAT(N)
C
C     CALCULATE THE COEFFICIENTS OF MATRIX CX(6N+1,6)
C
      DO 50 I=1,NFC
      DO 51 J=1,6
      CX(I,J)=0.
   51 CONTINUE
   50 CONTINUE
C
      CX(2,1)=2.*MUI*AA(3,1)+AA(1,1)+AA(2,1)-S11+TP11
      CX(2,2)=2.*MUI*AA(3,2)+AA(1,2)+AA(2,2)-S22+TP22
      CX(2,3)=2.*MUI*AA(3,3)+AA(1,3)+AA(2,3)
      CX(2,4)=2.*MUI*AA(3,4)+AA(1,4)+AA(2,4)+S23-TP23
      CX(2,5)=2.*MUI*AA(3,5)+AA(1,5)+AA(2,5)-S13+TP13
      CX(2,6)=2.*MUI*AA(3,6)+AA(1,6)+AA(2,6)+S12-TP12
      CX(1,1)=-(SP11+T11)
      CX(1,2)=-(SP22+T22)
      CX(1,3)=0.
      CX(1,4)=SP23+T23
      CX(1,5)=-(SP13+T13)
      CX(1,6)=SP12+T12
      CX(6,1)=AA(5,1)-SPP11
      CX(6,2)=AA(5,2)-SPP22
      CX(6,3)=AA(5,3)
      CX(6,4)=AA(5,4)+SPP23
      CX(6,5)=AA(5,5)-SPP13
      CX(6,6)=AA(5,6)+SPP12
      CX(7,1)=-(AA(4,1)+TPP11)
      CX(7,2)=-(AA(4,2)+TPP22)
      CX(7,3)=-AA(4,3)
      CX(7,4)=-(AA(4,4)-TPP23)
      CX(7,5)=-(AA(4,5)+TPP13)
      CX(7,6)=-(AA(4,6)-TPP12)
      CX(8,1)=AA(1,1)-AA(2,1)-S11-TP11
      CX(8,2)=AA(1,2)-AA(2,2)-S22-TP22
      CX(8,3)=AA(1,3)-AA(2,3)
      CX(8,4)=AA(1,4)-AA(2,4)+S23+TP23
      CX(8,5)=AA(1,5)-AA(2,5)-S13-TP13
      CX(8,6)=AA(1,6)-AA(2,6)+S12+TP12
      CX(9,1)=AA(6,1)-SP11+T11
      CX(9,2)=AA(6,2)-SP22+T22
      CX(9,3)=AA(6,3)
      CX(9,4)=AA(6,4)+SP23-T23
      CX(9,5)=AA(6,5)-SP13+T13
      CX(9,6)=AA(6,6)+SP12-T12
      CX(10,1)=AA(6,1)+T11-SP11
      CX(10,2)=AA(6,2)+T22-SP22
      CX(10,3)=AA(6,3)
      CX(10,4)=AA(6,4)-T23+SP23
```

```
      CX(10,5)=AA(6,5)+T13-SP13
      CX(10,6)=AA(6,6)-T12+SP12
      CX(11,1)=AA(2,1)-AA(1,1)+TP11+S11
      CX(11,2)=AA(2,2)-AA(1,2)+TP22+S22
      CX(11,3)=AA(2,3)-AA(1,3)
      CX(11,4)=AA(2,4)-AA(1,4)-TP23-S23
      CX(11,5)=AA(2,5)-AA(1,5)+TP13+S13
      CX(11,6)=AA(2,6)-AA(1,6)-TP12-S12
C
C     CALCULATE THE COEFFICIENTS OF MATRIX DX(6N+6,6N+1)
C
      NFCP=NFC+6
      DO 44 I=1,NFCP
      DO 45 J=1,NFC
      DX(I,J)=0.
   45 CONTINUE
   44 CONTINUE
      DX(1,2)=-0.5
      DX(1,5)=-0.5
      DX(2,8)=-0.5
      DX(2,11)=-0.5
      DX(3,9)=-0.5
      DX(3,10)=0.5
      DX(4,9)=0.5
      DX(4,10)=-0.5
      DX(5,8)=-0.5
      DX(5,11)=-0.5
      DX(6,6)=-1.
      DX(7,7)=-1.
      DO 46 I=2,NNN
      DX(6*I-4,6*I-10)=-0.5
      DX(6*I-4,6*I-7)=0.5
      DX(6*I-4,6*I+2)=-0.5
      DX(6*I-4,6*I+5)=-0.5
      DX(6*I-3,6*I-9)=-0.5
      DX(6*I-3,6*I-8)=-0.5
      DX(6*I-3,6*I+3)=-0.5
      DX(6*I-3,6*I+4)=0.5
      DX(6*I-2,6*I-9)=-0.5
      DX(6*I-2,6*I-8)=-0.5
      DX(6*I-2,6*I+3)=0.5
      DX(6*I-2,6*I+4)=-0.5
      DX(6*I-1,6*I-10)=0.5
      DX(6*I-1,6*I-7)=-0.5
      DX(6*I-1,6*I+2)=-0.5
      DX(6*I-1,6*I+5)=-0.5
      DX(6*I,6*I)=-1.
      DX(6*I+1,6*I+1)=-1.
   46 CONTINUE
      DX(6*N-4,6*N-10)=-0.5
      DX(6*N-4,6*N-7)=0.5
      DX(6*N-3,6*N-8)=-0.5
      DX(6*N-3,6*N-9)=-0.5
      DX(6*N-2,6*N-9)=-0.5
      DX(6*N-2,6*N-8)=-0.5
      DX(6*N-1,6*N-10)=0.5
      DX(6*N-1,6*N-7)=-0.5
      DX(6*N,6*N)=-1.
      DX(6*N+1,6*N+1)=-1.
      DX(6*N+2,6*N-4)=-0.5
      DX(6*N+2,6*N-1)=0.5
      DX(6*N+3,6*N-3)=-0.5
```

```
      DX(6*N+3,6*N-2)=-0.5
      DX(6*N+4,6*N-2)=-0.5
      DX(6*N+4,6*N-3)=-0.5
      DX(6*N+5,6*N-4)=0.5
      DX(6*N+5,6*N-1)=-0.5
      DX(6*N+6,3)=0.5
      DX(6*N+6,4)=-0.5
      DX(6*N+7,1)=1.
C
C
      DO 54 I=1,6
      DO 55 J=1,6
      FX(I,J)=0.
      GX(I,J)=0.
   55 CONTINUE
   54 CONTINUE
      DO 56 I=1,6
      FX(3,I)=E*AA(3,I)
   56 CONTINUE
      DO 57 I=1,6
      GX(I,I)=1./E
      IF(I.GE.4) GX(I,I)=2.*(1.+MUI)/E
   57 CONTINUE
      GX(1,2)=-MUI/E
      GX(1,3)=-MUI/E
      GX(2,3)=-MUI/E
      GX(2,1)=GX(1,2)
      GX(3,1)=GX(1,3)
      GX(3,2)=GX(2,3)
C
C     ******************************************************************
C
      DO 60 I=1,NFC
      DO 61 J=1,NFC
      AX(I,J)=AX(I,J)*1.E+06
   61 CONTINUE
   60 CONTINUE
      DO 52 I=1,NFC
      DO 53 J=1,6
      CX(I,J)=CX(I,J)*1.E+06
   53 CONTINUE
   52 CONTINUE
C
C
C     CALCULATE THE COEFFICIENTS OF THE MATRIX INVERSE OF AX USING
C     SUBROUTINE LINV3F. AX IS REPLACED BY ITS INVERSE.
C
      D1=1.
      IAA=IUNFC
      CALL LINV3F(AX,BC,1,NFC,IAA,D1,D2,WKAREA,IER)
      WRITE(6,236) D1,D2
  236 FORMAT(//10X,* D1 *,E12.4,* D2 *,E12.4//)
C
C     MULTIPLY AX BY CX AND STORE THE RESULT INTO HX(6N+1,6)
C     MULTIPLY DX BY HX AND STORE THE RESULT INTO HHX(6N+7,6)
C
      CALL PRCMAT(AX,CX,HX,NFC,NFC,6,IUNFC,IUNFC,6)
      CALL PRCMAT(DX,HX,HHX,NFCP,NFC,6,IUNFCP,IUNFC,6)
      IF(INDEX.GT.0) GOTO 58
C
C     ******************************************************************
C
```

```
C     FOR EACH ROSETTE AT A TIME CALCULATE MATRICES EX AND T3(Q)
C
      NK1=NROS(K1)
      DO 72 I=1,NK1
      RA=ROS(I,5,K1)
      TET=ROS(I,6,K1)
      CALL EXMAT(TET,RA,M,MUI,E,A,IUNFCP,EX)
C
C     MULTIPLY EX BY HHX AND STORE THE RESULT INTO T1(6,6)
C
      CALL PRCMAT(EX,HHX,T1,6,NFCP,6,6,IUNFCP,6)
      DO 262 II=1,6
      DO 263 JJ=1,6
      T2(II,JJ)=T1(II,JJ)+FX(II,JJ)
  263 CONTINUE
  262 CONTINUE
C
      CALL PRCMAT(GX,T2,T1,6,6,6,6,6,6)
      CALL PRCMAT(T1,TH,T3,6,6,6,6,6,6)
C
      PSIRO=ROS(I,4,K1)*ATAN(1.)*4./180.
      PSIRO2=2.*PSIRO
      YROS(N1+3*I-2,1)=ROS(I,1,K1)
      YROS(N1+3*I-1,1)=ROS(I,2,K1)
      YROS(N1+3*I,1)=2.*(ROS(I,3,K1)-ROS(I,1,K1)*COS(PSIRO)*
     1COS(PSIRO)-ROS(I,2,K1)*SIN(PSIRO)*SIN(PSIRO))/SIN(PSIRO2)
      DO 73 J=1,6
      XROS(N1+3*I-2,J)=T3(2,J)
      XROS(N1+3*I-1,J)=T3(3,J)
      XROS(N1+3*I,J)=T3(4,J)
   73 CONTINUE
   72 CONTINUE
      N1=N1+3*NK1
      GOTO 36
C
C     ****************************************************************
C
   58 CONTINUE
C
C     FOR EACH GAGE AT A TIME CALCULATE MATRICES EXU,FXU,T5(QU)
C
      NK1=NGAGE(K1)
      DO 231 I=1,NK1
      RA=1./M
      TET=GAGE(I,4,K1)
      ZA=GAGE(I,3,K1)
      CALL EUMAT(TET,RA,ZA,M,E,MUI,E,N,AA,IUNFCP,EXU,FXU)
C
C     MULTIPLY EXU BY HHX AND STORE THE RESULT  INTO T4
C
      CALL PRCMAT(EXU,HHX,T4,3,NFCP,6,3,IUNFCP,6)
      DO 232 II=1,3
      DO 233 JJ=1,6
      T5(II,JJ)=T4(II,JJ)+FXU(II,JJ)
  233 CONTINUE
  232 CONTINUE
C
C     MULTIPLY T5 BY TH AND STORE THE RESULT INTO T4
C
      CALL PRCMAT(T5,TH,T4,3,6,6,3,6,6)
C
      T6(1,1)=1./M
```

```
      T6(1,2)=0.
      T6(1,3)=GAGE(I,3,K1)
      CALL PRCMAT(T6,T4,T7,1,3,6,1,3,6)
      YROS(N1+I,1)=GAGE(I,1,K1)*GAGE(I,2,K1)*GAGE(I,2,K1)
      DO 234 J=1,6
      XROS(N1+I,J)=T7(1,J)
  234 CONTINUE
  231 CONTINUE
      N1=N1+NK1
C
C     ******************************************************************
C
   36 CONTINUE
      WRITE(6,74)
   74 FORMAT(///10X,* MATRICES Y AND X *,//)
      DO 81 I=1,N1
      YROS(I,1)=-YROS(I,1)
      WRITE(6,75) YROS(I,1),(XROS(I,J),J=1,6)
   75 FORMAT(10X,E12.4,10X,6E12.4)
   81 CONTINUE
      DO 95 I=1,N1
      YROS(I,1)=YROS(I,1)*1.E+06
      DO 96 J=1,6
      XROS(I,J)=XROS(I,J)*1.E+06
   96 CONTINUE
   95 CONTINUE
C
      IF(N1.NE.6) GOTO 254
      D1=1.
      CALL LINV3F(XROS,YROS,2,6,36,D1,D2,WKAREA,IER)
      WRITE(6,255) D1,D2
  255 FORMAT(//10X,* D1 *,E12.4,* D2 *,E12.4)
      DO 256 I=1,6
      SFOG(I,1)=YROS(I,1)
  256 CONTINUE
      GOTO 257
  254 CONTINUE
      CALL STAT(YROS,XROS,N1,6,TSTU,SFOG,S1ROS,R2ROS,T1,T2)
      S1ROS=S1ROS*1.E-06
      DO 97 I=1,N1
      YROS(I,1)=YROS(I,1)*1.E-06
   97 CONTINUE
      WRITE(6,76)S1ROS,R2ROS
   76 FORMAT(/10X,* STANDARD DEVIATION ABOUT REGRESSION *,E12.4/
     1        10X,* MULTIPLE CORRELATION COEFFICIENT *,E12.4/
     2        10X,* RESIDUALS *)
      WRITE(6,253)(YROS(I,1),I=1,N1)
  253 FORMAT(10X,9E10.4)
      WRITE(6,77)(SFOG(I,1),T1(I,1) ,T1(I,2),T1(I,3),I=1,6)
      WRITE(6,79)
      WRITE(6,78)((T2(I,J),J=1,6),I=1,6)
   77 FORMAT(10X,* SIGXC*,E12.4,* ESE*,E12.4,* UL*,E12.4,* LL*,E12.4/
     1        10X,* SIGYC*,E12.4,* ESE*,E12.4,* UL*,E12.4,* LL*,E12.4/
     1        10X,* SIGZC*,E12.4,* ESE*,E12.4,* UL*,E12.4,* LL*,E12.4/
     2        10X,*TAUYZC*,E12.4,* ESE*,E12.4,* UL*,E12.4,* LL*,E12.4/
     2        10X,*TAUXZC*,E12.4,* ESE*,E12.4,* UL*,E12.4,* LL*,E12.4/
     2        10X,*TAUXYC*,E12.4,* ESE*,E12.4,* UL*,E12.4,* LL*,E12.4)
   78 FORMAT(10X,6E12.4)
   79 FORMAT(//10X,* VARIANCE COVARIANCE MATRIX *//)
C
  257 CONTINUE
      WRITE(6,82)
```

```
      WRITE(6,83)(SFOG(I,1),I=1,6)
   82 FCRMAT(//10X,≠ STRESS FIELC IN GLOBAL CCORDINATE SYSTEM ≠,//)
   83 FCRMAT(10X,≠ SIGXC ≠,E12.4/
     1        10X,≠ SIGYC ≠,E12.4/
     1        10X,≠ SIGZC ≠,E12.4/
     1        10X,≠TAUYZC ≠,E12.4/
     1        10X,≠TAUXZC ≠,E12.4/
     1        10X,≠TAUXYC ≠,E12.4/)
C
C     ***********************************************************************
C
C     CALCULATE THE PRINCIPAL STRESSES AND THEIR ORIENTATION IN
C     THE GLCBAL COORDINATE SYSTEM
C
      SIG(1)=SFCG(1,1)
      SIG(2)=SFCG(6,1)
      SIG(3)=SFOG(2,1)
      SIG(4)=SFCG(5,1)
      SIG(5)=SFCG(4,1)
      SIG(6)=SFCG(3,1)
C
      CALL EIGRS(SIG,3,2,BD,Z,3,WK,IER)
C
      SIG1=AMAX1(BD(1),BD(2),BC(3))
      SIG3=AMIN1(ED(1),ED(2),BC(3))
      DC 84 II=1,3
      IF((BD(II).NE.SIG1).AND.(BD(II).NE.SIG3)) SIG2=BC(II)
   84 CCNTINUE
      DO 85 II=1,3
      IF(BC(II).EQ.SIG1) WRITE(6,91) BD(II)
      IF(BC(II).EQ.SIG2) WRITE(6,92) BD(II)
      IF(BC(II).EQ.SIG3) WRITE(6,93) BD(II)
   91 FCRMAT(/10X,≠ SIG1 ≠,E12.4/)
   92 FCRMAT(/10X,≠ SIG2 ≠,E12.4/)
   93 FCRMAT(/10X,≠ SIG3 ≠,E12.4/)
      WRITE(6,94)(Z(JJ,II),JJ=1,3)
   94 FCRMAT(10X,≠ L ≠,E12.4,≠ M ≠,E12.4,≠ N ≠,E12.4)
   85 CCNTINUE
      STOP
      END
C
```

AUTHOR INDEX

Agarwal, R., 205
Aggson, J.R., 193
Aires-Barros, L., 27-49
Akai, K., 50, 52
Allirot, D., 50, 69
Amadei, B., 19, 41, 45
Archambault, G., 54, 68
Arioka, M., 50, 52
Atkinson, R.H., 37
Attewell, P.B., 50
Barla, G., 3, 55, 204
Barron, K., 55, 68, 78, 196
Becker, R.M., 39, 215
Berry, D.S., 133, 134, 141, 144, 213, 314, 216
Bielenstein, H.U., 196
Bickel, D.L., 193
Bieniawski, Z.T., 41
Blackwood, R.L., 194, 212
Boehler, J.P., 50, 69
Bonnechere, F.J., 193, 205, 210
Brace, W.F., 67
Brady, B.H.G., 106, 113
Bray, J.W., 59, 61, 106, 113
Brekke, T.L., 19
Brown, E.T., 62
Brown, J.W., 50, 55
Buchholt, L., 29, 38, 50
Byerlee, J.D., 20
Carabelli, E., 42
Carrillo, N., 64
Carroll, M.M., 20
Casagrande, A., 64
Cook, H.E., 29
Cornet, F.M., 193, 210
Crouch, S.L., 19, 193, 205
Dayre, M., 29, 52, 57
Deklotz, E.T., 50, 55
De la Cruz, R.V., 41, 42, 57, 126, 20
Delgado, J.S., 194, 213
Donath, F.A., 50
Douglass, P.M., 27, 49
Draper, N.R., 208
Drozd, K., 57
Duncan, J.M., 19
Duncan-Fama, M.E., 144, 173, 174, 200
Duvall, W.I., 38, 189, 193, 194
Einstein, H.M., 39, 54
Eissa, E.S.A., 19, 29, 134
Eshelby, J.D., 143, 247
Fairhurst, C., 130, 133, 134, 141, 189, 193, 196, 213
Fayed, L.A., 50
Filon, L.N.G., 101, 119

Frenk, B.W., 189
Gay, N.C., 189
Gerrard, C.M., 16, 18
Gerstle, K.H., 29, 37
Goffi, L., 42, 55
Goodier, J.N., 103, 109, 120, 153
Goodman, R.E., 19, 41, 42, 45, 46, 57, 49, 88, 126, 189
Gray, K.E., 50, 55, 65, 67, 78
Gray, W.M., 204, 208, 209
Green, A.E., 119
Grossman, N., 47
Handin, J., 69
Hast, N., 198
Hata, S., 42
Hayes, D.J., 193, 210
Heuze, F.E., 41, 42, 57, 126
Hill, R., 69
Hiltscher, R., 193, 198
Hiramatsu, Y., 130, 210
Harashima, K., 134, 144, 215, 216
Hirschfeld, R.C., 39, 54
Hobbs, D.W., 52, 55, 78
Hocking, G., 102, 205
Hoek, E., 50, 62, 68
Hooker, V.E., 39, 193, 215
ISRM, 29
Jaeger, J.C., 52, 58, 68
Johnson, C.F., 215
Johnson, J.N., 29, 38, 50
Kawamoto, T., 42, 43, 215
Kimura, K., 42
Ko, H.-Y., 29, 37
Kobayashi, S., 61, 134, 215
Koga, A., 215
Kreyszig, E., 368
Kulhawy, F.H., 42
Ladanyi, B., 54, 62
Leeman, E.R., 189, 193, 209, 210
Lekhnitskii, S.G., 26, 81, 86, 91, 93, 95, 101, 106, 119, 123, 134, 143, 215
Lerau, J., 29
Liakhovitski, F.M., 38
Little, R.W., 153
Love, H.E.H., 103, 109, 119, 120
Mahmoud, A., 52
Martino, D., 38
Martna, J., 193, 198
Masure, P., 29, 52
McGarr, A., 189
McLamore, R., 50, 55, 65, 67, 78
McClintock, F.A., 67
Merrill, R.H., 193
Michell, J.H., 152
Milne-Thomson, L.M., 101, 106, 133, 245
Morgan, T.A., 193
Morland, L.W., 19
Muskhelishvili, N.I., 83
Nelson, R.A., 39
Nevski, M.V., 38

Nichols, T.C., 198
Niwa, Y., 134, 144, 215, 216
Nova, R., 68
Nur, A., 20
Obert, L., 189, 193, 194
Oberti, G., 42
Oka, Y., 130, 210
Okubo, H., 39, 78
Oliveira, E., 47
Panek, L.A., 208, 209
Pariseau, W.G., 70, 72, 74
Patrick, W.C., 41, 42
Paulman, H.G., 55
Pedro, J.O., 194, 213
Pender, M.J., 144, 173, 174, 200, 213
Peres-Rodrigues, F., 27, 28, 49
Peterson, J.R., 193
Pickering, D.J., 14
Pinto, J.L., 16, 29, 31, 39
Pomeroy, C.D., 52
Raphael, J.M., 42
Reik, G., 39, 54
Ribacchi, R., 38, 204
Rocha, M., 47, 194, 212, 213
Rossi, P.P., 42
Royea, M.J., 193
De Saint-Venant, B., 38
Saint-Leu, C., 29
Salamon, M.D.G., 16, 17
Sandford, M.R., 50
Savage, W.Z., 198
Sawczuck, A., 69
Schmidt, C.M., 189
Serafim, J.L., 57
Shidomoto, Y., 47
Silveira, A., 47
Silverio, A., 194, 212, 213
Simonson, E.R., 29, 38, 50
Singh, B., 19
Sirieys, P., 29
Skempton, A.W., 20
Smith, H., 208
Sokolnikoff, I.S., 103, 109
Starfield, A.M., 19
Stemler, O.A., 50, 55
Strindell, L., 193, 198
Takano, M., 47
Tanimoto, Ch., 42
Taylor, R.L., 19
Teodorescu, P.D., 102
Timoshenko, S.P., 103, 109, 120, 153
Titchmarsch, E.C., 357
Toews, N.A., 204, 208, 209
Turner, J.T., 198
Van, T.K., 41, 57, 126
Van Heerden, W.L., 205
Voight, B., 27, 49, 194, 198
Voss, C.F., 41, 42
von Mises, R., 69

Walsh, J.B., 67
Walton, R.J., 194, 213
Wane, M.T., 204
Wardle, L.J., 16, 18
Weiss, L.E., 198
Worotnicki, G., 194, 213
Yamamoto, K., 50, 52
Zacas, M., 39, 54
Zerna, W., 119

SUBJECT INDEX

Anisotropy
Definition 2
Classification 3
Antiplane Problems 101, 151, 156

Basic Equations for the Elastic Equilibrium of Anisotropic Media 83-88, 148-151
Beltrami-Michell Equations of Compatibility 90, 150
Bonding of an Inclusion with Anisotropic Media 143, 159, 174, 202, 216, 249
Bray's Diagram 59, 79

Continuous Models for the Strength of Anisotropic Rocks 62-78

Dilatometer Tests 42-47

Effective Stress Laws 20
Elastic Constants
Definition 8-10, 17
Bounds for 14, 17
Determination 33-37
Transformation 21-26
Elasticity
Definition 2
Elastic Symmetries in Anisotropic Media 10-13, 98-100, 129
Equivalent Continuum Concept
For Stratified Media 16
For Regularly Jointed Rocks 19, 272-276

Forces
Body Forces 84, 143
Surface Forces 83, 146
Fourier Series Boundary Conditions 121-122, 124-126, 146, 151

Generalized Hooke's Law 6, 86

Hole Eccentricity 205
Homogeneity/Heterogeneity 2

Inclusion Theory 143-188
Influence of Rock Anisotropy on the Determination of the In-Situ Stress Field 233-239

Joints
    Equivalent Anisotropic Medium 19
    Behavior in Shear 270-271
    Behavior in Normal Compression 269-270

Measurement of Joint Stiffness <u>in-situ</u> 45-47
Modulus of Deformation 41
Mohr-Coulomb Failure Criterion 62, 65, 68, 73

Number of Boreholes
    Isotropic Solution 209-213
    Anisotropic Solution 222-233
Number of Independent Measurements in a Single Borehole
    Isotropic Solution 209-213
    Anisotropic Solution 221-222
Numerical Methods 134-135

Overcoring Diameter; Finite Character 199-201, 216, 249
Overcoring Techniques
    General Procedure 190-193
    General Formulas for 199-205

Physical Models for Regularly Jointed Rocks 39, 54
Plane of Weakness Theory
    For One Joint Set 58-61
    For Several Joint Sets 61-62
Plane Problems of the Theory of Elasticity 101-102, 245
Plane Strain Formulations
    Classical 103-105
    Generalized 105-107, 127, 131, 339
    Complete 106, 339-340
Plane Stress Formulations
    Classical 107-111
    Generalized 115-119, 132, 133, 206
    Complete 111-115

Scale Effect 41, 57, 242
Sign Convention for Stress, Strain and Displacement 88
St- Venant Relations 38
Strain Energy 13
Stratified Media 16
Strength of Anisotropic Rocks
    Compressive 49-55
    Tensile 55, 78, 244
    Indirect Tensile 55, 244
    Qualitative Tests 55-56
Stress
    Stress Functions 89
    Components 6, 83, 96
    Stress Field at Infinity 113, 125, 144, 201, 206
    Stress Field Induced by an Inclusion in an
      Infinite Medium 171 , 178, 186, 200, 247-248
    In Situ Stress Field; Nomenclature 194-198
    Changes in Stress Field 207, 216
    Least Square Estimate of In-Situ Stress Components 208

Testing of Anisotropic Rocks
In-Situ 41-47, 57
Laboratory
for Deformability 26-41
for Strength 48-57

Undercoring Technique
General Procedure 194
General Formulas for 205-207